ESSAYS ON

THE HISTORY OF

ORGANIC CHEMISTRY

IN THE

UNITED STATES, 1875-1955

ESSAYS ON

THE HISTORY OF

ORGANIC CHEMISTRY

IN THE

UNITED STATES, 1875-1955

By

DEAN STANLEY TARBELL
ANN TRACY TARBELL

FOLIO PUBLISHERS
NASHVILLE, TENNESSEE

Portions of the essay "The Origin of 'Organic Syntheses' " are quoted from *Roger Adams: Scientist and Statesman* (1981) by D. S. Tarbell and Ann T. Tarbell by permission of the American Chemical Society. The table of undergraduate origins of Johns Hopkins Ph.D.s in chemistry is quoted by permission of *Isis*.

Address inquiries and orders to
D. Stanley Tarbell or Ann T. Tarbell
Department of Chemistry
1520 Station B
Vanderbilt University
Nashville, TN 37235

Library of Congress Cataloging-in-Publication Data

Tarbell, Dean Stanley, 1913-
    Essays on the history of organic chemistry in the United States, 1875-1955.

    Bibliography: passim.
    Includes index.
    1. Chemistry, Organic—United States—History.
I. Tarbell, Ann Tracy, 1916-   . II. Title.
QD248.5.U6T37 1986   547'.009   86-12062
ISBN 0-939454-03-3

Manufactured in the United States of America
Printed by the K & S PRESS, Nashville, TN

Table of Contents

Illustrations and Credits

Abbreviations of Frequently Quoted Journals

ACJ - *American Chemical Journal*

Ann. - *Liebigs Annalen der Chemie*

Ber. - *Berichte der Deutschen Chemischen Gesellschaft*

JACS - *Journal of the American Chemical Society*

JBC - *Journal of Biological Chemistry*

JCE - *Journal of Chemical Education*

JCS - *Journal of the Chemical Society (London)*

JOC - *Journal of Organic Chemistry*

JPr - *Journal für Praktische Chemie*

Acknowledgments

We have utilized a number of libraries, particularly the Stevenson Science Library and the Jean and Alexander Heard Library of Vanderbilt University. We are grateful to the Linda Hall Library of Science in Kansas City for the opportunity to work for several months in its rich collections. We have also made profitable use of the E. F. Smith Collection at the University of Pennsylvania. While working in the Illinois Library at Urbana on the Roger Adams papers, we found valuable books and catalogs in that library's very large general collections on science and education.

We are indebted to many friends and colleagues both here and elsewhere for encouragement and for reading parts of the manuscript. We are particularly grateful to Dr. R. M. Joyce for reading the manuscript critically in its various versions and for valuable comments. Our thanks go to him in addition for his work in drawing all of the structural formulas. Others to whom we are indebted for reading parts of the manuscript include Drs. J. E. Bunnett, R. C. Elderfield, A. H. Ihde, J. A. Kampmeier, Erwin Klingsberg, E. S. Lewis, J. C. Martin, T. W. Martin, C. S. Marvel, L. J. Schaad, J. L. Sturchio, J. D. Roberts, E. H. Volwiler, and H. E. Zaugg. We are also grateful to the officers and members of the Division of the History of Chemistry of the American Chemical Society for their interest in our attempts to develop a general view of organic chemistry in the United States. Among others, we should mention Drs. Ernest Eliel, Natalie Foster, Leon Gortler, Ned Heindel, George Kauffman, Jeffrey Sturchio, Arnold Thackray, Bertrand Ramsey, and John Wotiz. We are, of course, responsible for any errors.

We have used biographical information from *American Men of Science; Who's Who,* and *Who Was Who* (both American and British); W. D. Miles, Ed., *American Chemists and Chemical Engineers; The Dictionary of Scientific Biography; The Dictionary of National Biography; The Dictionary of American Biography; Chemical Abstracts;* and the *Biographical Memoirs of the National Academy of Sciences.* We have normally quoted the latter in our bibliography when biographies are available because they usually contain a complete list of the subject's scientific publications. The *National Union Catalog* has been a valuable source for bibliographic information. We have also used relevant research articles and books for biographical and other kinds of information. Unfortunately, the extensive compilation of statistical information, *Chemistry in America, 1876-1976,* by Arnold Thackray, J. L. Sturchio, P. T. Carroll, and Robert Bud, Dordrecht, 1985, appeared too late for our work.

We have identified women authors of papers and we have not usually listed more than two names as authors of papers, unless there seemed a special reason to do so.

Grants from the Centennial Fund of Vanderbilt University and from the Division of Organic Chemistry of the American Chemical Society have defrayed some of the expense of this project. The secretarial staff of the chemistry department has typed the various drafts of these essays over the years, and we appreciate the diligence of Ruby Moore, computer technician, who has set the text using TEX, on Vanderbilt's Digital 1099 computer. We acknowledge Vanderbilt's contribution of computer resources.

We are grateful to many friends and colleagues, both here and elsewhere, for their encouragement and advice.

D. Stanley Tarbell
Ann Tracy Tarbell
Department of Chemistry
Vanderbilt University
Nashville, TN 37235

Introduction

Our object in these essays is to describe the development of organic chemistry in the United States over the period approximately limited by 1875 and 1955. Our emphasis will be placed on the growth of basic ideas of the science, on the interrelation of organic chemistry to other fields, on the contributions of key personalities, and to some extent on the institutional aspects of the development. We hope to avoid the compilation of a long and tedious catalog of facts and names and to make the account intelligible to others besides specialists in organic chemistry. We shall consider interaction of social and economic factors on a developing science.

The limitation of a study of the growth of a science to a single country may seem contradictory to the fact that science is, or tries to be, international in its outlook and scope. We accept the internationalism of science; however, the limitation of this enquiry to the United States allows us to study in detail the manner in which a discipline can take root, grow, and mature in an undeveloped country. The United States certainly was undeveloped scientifically in 1875. Advances in the science in other countries will, to some extent, be a necessary part of the story, as well as the interaction of overseas ideas and personalities with the evolution of an independent and indigenous American science.

Organic chemistry is the study of compounds containing carbon; these range in complexity from carbon monoxide, CO, to synthetic or naturally occurring materials of very large molecular size, such as dacron, nylon, cellulose, proteins, deoxyribonucleic acid (DNA, carrier of the genetic code in living systems, the "double helix"), viruses, and a host of other materials of intermediate molecular size. The field offers a fascinating intellectual and experimental challenge in itself, but there has been remarkably little writing about it for the non-expert. There is no comprehensive account of the development of organic chemistry which is not many decades out of date. The importance of the field for other branches of science, particularly the life sciences, and its applications in fields such as plastics, dyes, synthetic fibers, rubber, antibiotics, drugs, such as anti-cancer agents and compounds useful in treating many other pathological conditions, are all part of the story. Although these aspects of the application of organic chemistry are of pressing importance, they represent only a small part of the content and impact of the science.

There are available many discussions of relativity, quantum theory, nuclear physics, astrophysics, plate tectonics, biochemistry, genetics, and other life sciences, all directed to the non-specialist, but there is very little comparable for organic chemistry. This lack of accounts concerning the nature and structure of the field is both surprising and unfortunate. Scientists in closely related disciplines are usually not very conversant with the contributions of organic chemistry to their own work. Students and teachers of organic chemistry frequently have little appreciation of the long and varied history of the subject, or of careers and personalities of scholars responsible for our present knowledge. Students in particular are apt to feel that organic chemistry emerged fully grown in the enormous text books which they study. Some sense of the slow development over many decades of the ideas and experimental methods of the science might materially enliven their studies. Furthermore, the fact that a science, or any other field of learning, is dynamic, not static, and that important new experimental observations or theories are constantly altering our views is a stimulating and salutary idea that should be emphasized for students.

The fact that advances in knowledge in the life sciences depend on increased knowledge of organic chemistry and on the application of new techniques developed by work on organic compounds is almost never made clear. The intellectual beauty and coherence of organic chemistry as a discipline is rarely discussed.

Organic chemistry has many excellent text books, and the present essays are not intended to form another one. In spite of the enormous scope and bulk of the subject matter, and in spite of the intricate mathematical and experimental techniques which are presently available for use in research on current problems in organic chemistry, the basic ideas of the field are simple, and special terms requiring definition are limited. The systematic nomenclature of organic compounds is complex and troublesome, even for experts, but discussion of nomenclature is not necessary for our presentation.

Our interest in writing these essays arose after several decades of active research and teaching in organic chemistry. The centennial of the founding of the American Chemical Society in 1976 directed increased attention to the hundred years of achievement in the field, and we contributed two summaries of the hundred years of organic chemistry. One appeared in the volume *A Century of Chemistry*, and the other in the Centennial number of *Chemical and Engineering News*. These were based on our studies to that point and were necessarily highly selective. Our approach to the project had as its first step an examination of American research publications in organic chemistry in American or foreign scientific journals. This gave unique insight into the state of the science: who was doing what, and where; the quality and originality of the American work compared to that in other countries; the paths of influence from one worker and institution to another; the spread of advanced training and research in academic, government, and industrial laboratories; and a sense of the number of people engaged in research and advanced study.

In following this plan, we have examined practically every paper on organic chemistry published by an American in native or foreign publications from the years before 1875 to 1939 as far as they can be identified. Significant papers we read carefully and noted their contents and authors for discussion. For the period from 1940 to 1955 we relied on our own reading and review articles or monographs to identify key papers.

Some of our reading suggested more detailed researches on special topics, which we published as journal articles or gave as papers, hitherto unpublished, before various groups. We had completed rough drafts of many of the following essays when we interrupted our work on this project for several years to write a book-length biography of Roger Adams. We necessarily learned much more about American organic chemistry during this pleasant interlude because of Adams' broad activities in the field for several decades.

We have called this book a collection of essays, rather than a history, because we have not approached the ideal history of a science as closely as we would have wished. It is impossible to divide a history into tight chronological units; some essays therefore extend beyond the periods stated. Occasional repetition may add to the clarity of some essays.

The history of a science, in addition to recounting the genesis and development of important ideas and experiments, must take account of the social milieu in which the science is done. The social and economic conditions, the climate of intellectual opinion, the rise of technology, and institutional developments all are part of the story.

MOSES GOMBERG

The Classical Structural Theory of
Organic Chemistry, 1875

Observations accumulated over many centuries led to the belief that substances derived from the plant and animal kingdoms, "organic" substances, were fundamentally different from "mineral" compounds, thought to play no part in living nature. Organic substances were believed to result from a Vital Force. The necessity for this Vital Force was disproved early in the 19th century, particularly by Friedrich Wöhler, who in 1824 and 1828 synthesized oxalic acid and urea in his laboratory, ushering in the recognition of organic chemistry as a true laboratory science. As more organic compounds were isolated, investigators realized by mid-century that the element carbon was characteristic of all organic substances. By 1875 European chemists had worked out a *structural theory* which explained how carbon and other elements could combine to form the increasing numbers and astonishing diversity of organic compounds. Although the great European masters based the theory on limited experimental material, their remarkable insight provided a satisfactory and enduring foundation for the development of organic chemistry.[1,2]

This theory made it possible to assign three-dimensional arrangements of the atoms in molecules ranging in size from methane 1, CH_4, to proteins and nucleic acids containing many thousands or millions of atoms of the elements carbon, C; hydrogen, H; nitrogen, N; oxygen, O; phosphorus, P; sulfur, S; and a few others. The great American chemist Gilbert Newton Lewis (1875-1946) regarded the structural theory as one of the most successful qualitative generalizations[3] in all of science, a judgment which is fully justified.[4] Its principles have not only correlated an extraordinary mass of observations but have required few fundamental additions during the last hundred years. The classical structural theory of organic chemistry is of the same importance in the history of science as the development of the two laws of thermodynamics around 1850, the quantum theory and the theory of general relativity after 1900, and the explanation of the molecular basis of genetics after 1950.

In simultaneous publications in 1858 the German Friedrich August Kekulé (1829-1896) and the Scotsman Archibald Couper (1831-1892) proposed the simple structural theory. Illness prevented Couper from further scientific work, and the later development was due to Kekulé and to the Russian chemist Alexander M. Butlerov (1827-1886).[5]

The two fundamental postulates of the simple structural theory were based on the novel combining power of carbon: (1) carbon was *tetravalent* and hence could combine with four monovalent atoms or groups such as hydrogen, chlorine, or hydroxyl; (2) carbon could combine with itself to form stable chains. Thus structural formulas could be written showing the way the atoms were joined in such *saturated* compounds as methane 1, CH_4; methyl chloride 2, CH_3Cl; methyl alcohol 3, CH_3OH; and ethane 4, C_2H_6. It is this ability of carbon to form stable chemical bonds between two or more carbon atoms that makes possible the existence of vast numbers of carbon compounds, of which several million have been prepared and characterized.

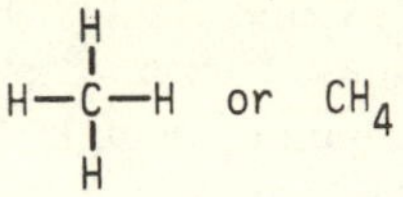

$$
\begin{array}{c}
\text{H} \\
| \\
\text{H} - \text{C} - \text{H} \quad \text{or} \quad CH_4 \\
| \\
\text{H}
\end{array}
\qquad
\begin{array}{c}
\text{H} \\
| \\
\text{H} - \text{C} - \text{Cl} \quad \text{or} \quad CH_3Cl \\
| \\
\text{H}
\end{array}
$$

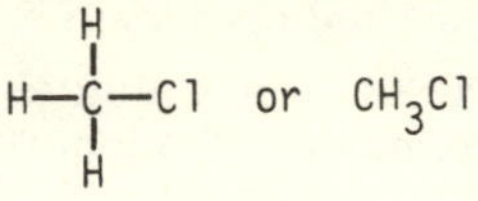

$$\underline{1} \qquad\qquad\qquad \underline{2}$$

$$
\begin{array}{c}
\text{H} \\
| \\
\text{H} - \text{C} - \text{OH} \quad \text{or} \quad CH_3OH \\
| \\
\text{H}
\end{array}
\qquad
\begin{array}{c}
\text{H} \quad \text{H} \\
| \quad | \\
\text{H} - \text{C} - \text{C} - \text{H} \quad \text{or} \quad CH_3CH_3 \quad \text{or} \quad C_2H_6 \\
| \quad | \\
\text{H} \quad \text{H}
\end{array}
$$

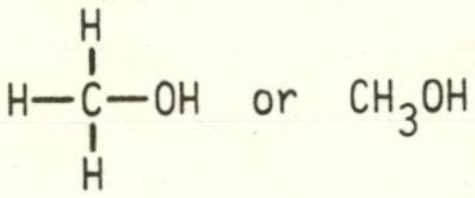

$$\underline{3} \qquad\qquad\qquad \underline{4}$$

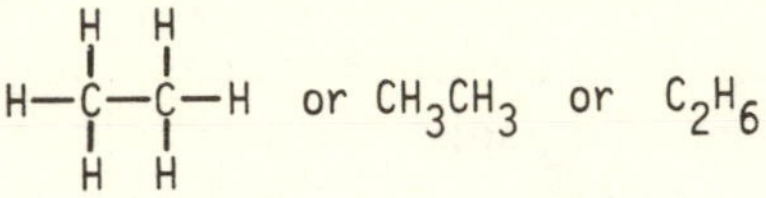

The determination of molecular formulas, for example C_2H_6O for ethyl alcohol, was made possible by the application of the postulate of the Italian Amadeo Avogadro (1776-1856) that equal volumes of gases contained equal numbers of molecules; another Italian, Stanislao Cannizzaro (1826-1912), revived the idea in publications from 1857 to 1858. He publicized the idea further at the famous Congress of Karlsruhe organized by Kekulé in 1860, a most timely development for the confirmation and extension of the structural theory.

The structural theory accounted for the existence of *position isomers*, compounds having the same molecular formula but different structural formulas. The simplest example among the saturated hydrocarbons, C_nH_{2n+2}, occurs when n = 4; there are two different compounds known with formulas C_4H_{10}, *n*-butane 5 and *iso*butane 6, and these correspond to the only two possible arrangements of the atoms in which the valence rules are followed.

$$
CH_3CH_2CH_2CH_3
\qquad\qquad
\begin{array}{c}
\text{H} \\
| \\
CH_3 - \text{C} - CH_3 \\
| \\
CH_3
\end{array}
$$

$$\underline{5} \qquad\qquad\qquad \underline{6}$$

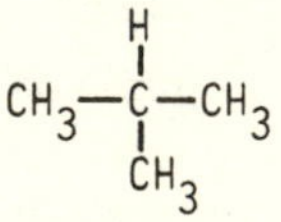

Thus there are three possible arrangements for the pentanes, C_5H_{12}; all three isomers, and no more, are known. The possible hexanes, C_6H_{14}, number five; again five, and no more, are known as distinct compounds. The number of possible isomers (or structural arrangements) increases very rapidly with increasing number of carbon atoms, and in each case the number of isomers predicted by the structural theory is the total number of different compounds which can be synthesized. With still larger numbers of carbons the number of possible isomers becomes astronomical, and not all of them have been obtained beyond the C_9H_{20} group, but a *greater* number of isomers than that predicted by the simple structural theory has never been found. Furthermore, by using a combination of chemical and physical properties, each pure isomer which has been obtained can be assigned unambiguously one of the possible structures written down following the valence rules. This one-to-one correspondence of structural formulas and chemical compounds was particularly emphasized by Butlerov.

The common elements were assigned the following valences: hydrogen and the halogens (fluorine, F, chlorine, Cl, bromine, Br, and and iodine, I), one; oxygen and sulfur, two (usually); nitrogen, three (usually); phosphorus, three, four, and five. Substitution of a hydrogen by a monovalent group in the saturated hydrocarbons leads to a larger number of possible isomers; thus there are four isomeric butyl alcohols, C_4H_9OH, derived from the two isomeric butanes, C_4H_{10}. This statement is subject to qualification when account is taken of the arrangements of atoms in three dimensions.

Compounds of the general molecular formula C_nH_{2n} were known, such as ethylene, C_2H_4, propylene, C_3H_6, and higher homologs of the *unsaturated* or *ethylenic* hydrocarbons. Structural formulas for these compounds could be written which preserved the postulate of the tetravalence of carbon only by writing a carbon-carbon double bond, as in ethylene **7** and propylene **8**. This postulate of *carbon-carbon unsaturation* was justified by the reactions which these compounds undergo and, many years later, by several types of spectroscopic measurements.

$$\underline{7} \quad \overset{H}{\underset{H}{>}}C=C\overset{H}{\underset{H}{<}} \quad \text{or} \quad CH_2=CH_2 \qquad CH_3CH=CH_2 \quad \underline{8}$$

Unsaturation involving carbon and other elements was also postulated and found. Thus, acetaldehyde 9, the product of oxidation of ethyl alcohol, and its trichloro derivative, chloral **10**, were early recognized before satisfactory structures could be written for them, as were other *carbonyl* $(C=O)$ compounds, such as acetone **11**. Multiple linkages between carbon and other elements like nitrogen also occur.[6]

$$CH_3-\overset{\overset{O}{\|}}{C}-H \qquad\qquad CCl_3-\overset{\overset{O}{\|}}{C}-H \qquad\qquad CH_3-\overset{\overset{O}{\|}}{C}-CH_3$$

$$\underline{9} \qquad\qquad\qquad \underline{10} \qquad\qquad\qquad \underline{11}$$

The structural theory outlined so far allowed assignment of rational structures to compounds such as acetic acid **12** (from vinegar), and citric acid **13** (from lemons and oranges), both containing the *carboxyl group*, COOH, characteristic of organic acids.

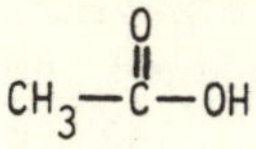

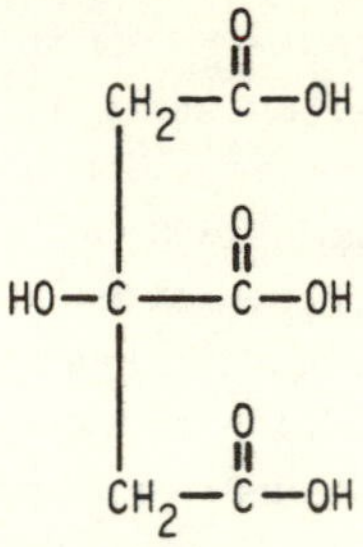

The success of the simple structural theory in predicting the number of possible isomers and in providing rational structural formulas for many known compounds was soon followed by its extension to three dimensions to explain the existence of *stereoisomers* (isomers due to differing three-dimensional arrangements).

The occasion for this extension was the isolation of three lactic acids, all having the two-dimensional structural formula **14a**. One of these, *l*-lactic acid, isolated from sour milk, had the property of rotating plane-polarized light to the left (levorotatory, hence *l*); another, *d*-lactic acid, isolated from muscle, rotated plane-polarized light to the right (dextrorotatory, hence *d*). The third lactic acid, synthesized in the laboratory, had no effect on plane-polarized light and was, therefore, *dl* (*racemic* or *optically inactive*). All of the physical and chemical properties of the *d*- and *l*-lactic acids were identical except their optical activity.

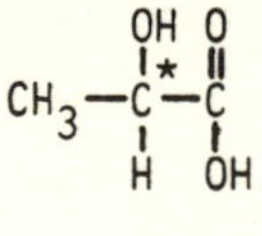

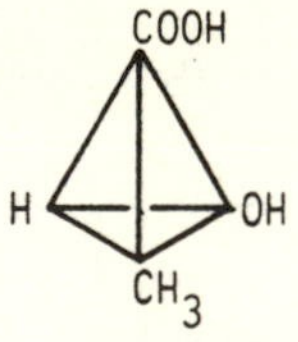

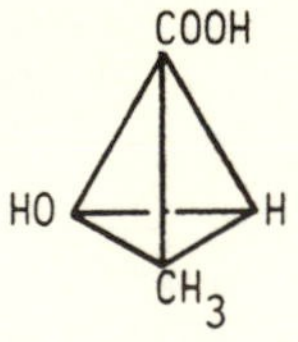

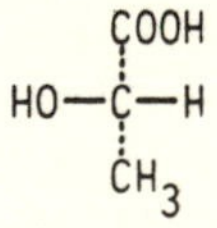

A Dutch chemist, J. H. van't Hoff (1852-1911), and a Frenchman, J. A. Le Bel (1847-1930), suggested the correct explanation independently in 1874. Valuable preliminary insights were due to Butlerov, Kekulé, Louis Pasteur (1822-1895), and Johannes Wislicenus (1835-1902). The centennial of this fundamental advance in the structural theory was observed with numerous papers and lectures.[7] Van't Hoff's statement in the rare pamphlet *Le Chimie dans l'Espace*[8] gives a striking sense of his originality and clarity of thought. It is not surprising that he became one of the founders of modern physical chemistry.

Van't Hoff and Le Bel suggested that the four valences of carbon were directed to the corners of a regular tetrahedron, as in **1c**, rather than lying

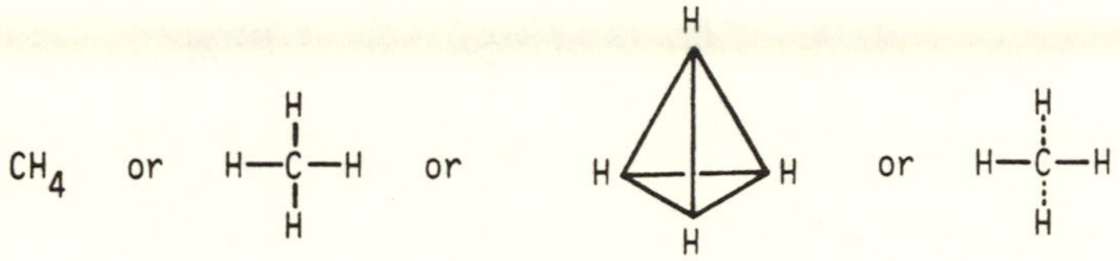

| CH_4 | or | H—C—H | or | (tetrahedron) | or | H—C—H |
| **1a** | | **1b** | | **1c** | | **1d** |

in one plane, as shown in **1b**. They pointed out that it was a fact of geometry that a tetrahedron with four different groups attached to the corners could exist in two, and only two, different arrangements which have a mirror image relationship to each other. Therefore a molecule containing a carbon carrying four different groups, such as the starred carbon in **14a** (an *asymmetric carbon*), could exist in three dimensions in two forms which would be mirror images and which might be expected to affect plane-polarized light because of the asymmetric nature of the molecule. These three-dimensional relationships are represented on paper by various conventions shown for methane in **1**. The tetrahedral carbon atom may be drawn out, as for methane in **1c**, where its fourth corner is back of the plane of the paper. More often as in **1d** and **14d**, the plane of the paper is considered to pass through the tetrahedral carbon atom. The groups in front of the plane of the paper are indicated by solid lines, those back of the plane by dotted lines, and all four groups attached to the tetrahedral carbon are considered to be attached to the apexes of a regular tetrahedron. The two mirror image arrangements for lactic acid are **14b** and **14c**.

It is apparent that as the number of asymmetric carbons in a molecule increases, the number of possible three-dimensional arrangements increases very rapidly; the formula for the general case is $x = 2^n$, where x is the number of possible optically active forms, and n is the number of dissimilar asymmetric carbons. There will be $2^n/2$ *dl* or racemic pairs of mirror images possible, which are optically inactive because they consist of equal numbers of the two mirror images.

A complicated molecule like cholesterol **15**, which occurs very widely distributed in mammalian cells, has eight asymmetric carbons (starred) and hence could exist in 2^8 or 256 optically active forms. The one which occurs naturally (in our arteries, among other places) is one of these 256 possible three-dimensional arrangements. The relative configurations (three-dimensional arrangements) of each of the eight asymmetric carbons and the absolute configuration (which mirror image it is) are known, and the correct stereoisomer has been synthesized in the laboratory. The single lines in this structure represent bonds between saturated carbon atoms, each carbon with two hydrogens attached unless otherwise indicated. The left-hand ring, if written in full, would be **15a**.

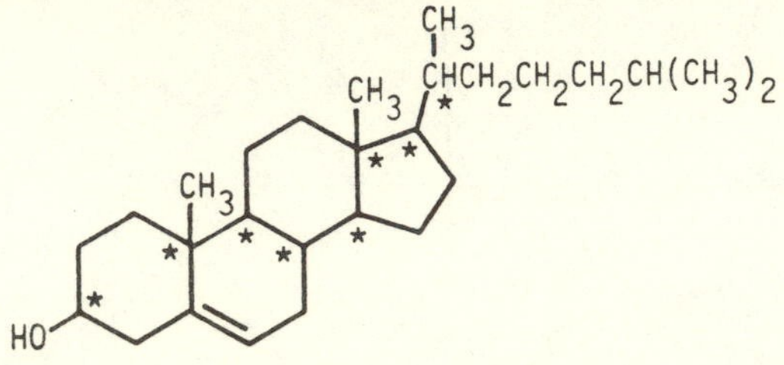

15 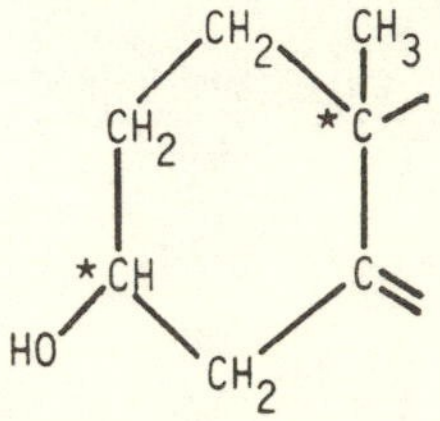   15a

Just as in the case of the position isomers, like butane and isobutane, the three-dimensional structural theory (*stereoisomerism* or *stereochemical theory*) leads to the prediction of the correct number of stereoisomers, and the very extensive studies of *stereochemistry* in many series of compounds have found no facts which are in conflict with it.

A situation that we shall meet later is the case where asymmetric carbons are present but are substituted in the same way so that there is a plane of symmetry, i.e., a plane in which each half of the molecule is the mirror image of the other. Such *meso* forms are optically inactive; this can be regarded as a cancellation of the rotation of one half of the molecule by an equal and opposite rotation of the other half. The first recognized and classical case is optically inactive *meso*-tartaric acid **16**. *Meso* forms cannot be separated into optically active forms, and their occurrence is of fundamental importance in assigning configurations.

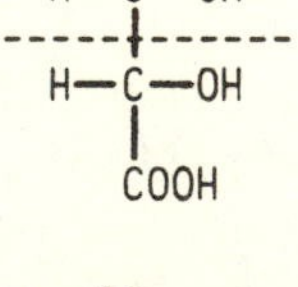

```
        COOH
         |
   H—C—OH            meso-tartaric acid
  -----|------
   H—C—OH            -----designates the plane
         |                 of symmetry, or mirror plane
        COOH
```

16

It is apparent from the structure **17a** that one of the butyl alcohols, C_4H_9OH, contains an asymmetric carbon and hence can be obtained in optically active forms **17b** and **17c**, as well as in the inactive (*dl*, or racemic)

form, consisting of equal amounts of the two active forms. The tetrahedral nature of the saturated carbon atom has been amply confirmed by many types of physical measurements unknown in 1874, and now the easiest way of establishing the structure of the three-dimensional arrangement of all the atoms in a complex molecule is by X-ray diffraction with computer analysis of the results.

$$CH_3\overset{*}{C}HCH_2CH_3 \\ \quad | \\ \quad OH$$

$$\begin{array}{c} CH_3 \\ | \\ CH_2 \\ | \\ H-C-OH \\ | \\ CH_3 \end{array}$$

$$\begin{array}{c} CH_3 \\ | \\ CH_2 \\ | \\ HO-C-H \\ | \\ CH_3 \end{array}$$

$$\underline{17a} \qquad\qquad \underline{17b} \qquad\qquad \underline{17c}$$

The examples of optically active compounds cited above–lactic acid, cholesterol, and others–illustrate an essential fact. This is that the organic compounds produced by action of the enzymes associated with living systems are optically active; all of the asymmetric carbons produced in such processes have only one of the two possible spatial arrangements of the groups. All processes in living systems are thus asymmetric, and optical activity is a characteristic of life. Although optically active materials can be produced in the laboratory by separating mixtures of mirror images, such separations (*resolution*, first done by Pasteur) almost always require the use of optically active reagents, which are obtained, directly or indirectly, from materials produced in optically active form by living organisms. The origin of optical activity is thus inseparably connected with the origin of life itself; the origin of optical activity in living systems is not known, but it may have been caused by reactions at the surface of an asymmetric inorganic crystal.

The carbon atoms involved in a double bond are not tetrahedral, but their valences all lie in one plane, and hence they do not have optical activity. There can be stereoisomerism associated with a disubstituted double bond, as van't Hoff suggested in his first paper of 1874; the illustration first used was the *cis* dicarboxylic acid, maleic acid **18**, and the *trans* acid, fumaric acid **19**. Because it requires considerable energy to cause rotation around a carbon-carbon double bond, in contrast to a carbon-carbon single bond where the energy of rotation about a single bond is normally very small, *cis-trans* isomers such as **18** and **19** usually can be isolated as stable compounds, although they may be interconverted by high temperature, light, or various reagents. The Latin prefix *cis* means "on this side" (as in Caesar's "cisalpine Gaul") and *trans* means "across". The theory of *cis-trans* stereoisomerism was not generally accepted until some years after 1875, and, as will appear later, early American organic chemists viewed it with attitudes varying from reserve to extreme suspicion.

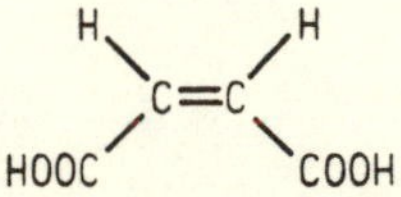

18, cis 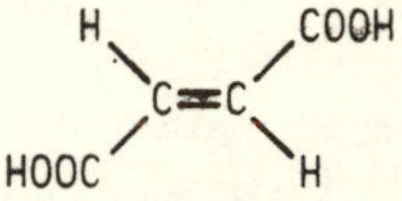   19, trans

Just as in the case of the tetrahedral carbon atom, modern experimental and theoretical methods confirm the existence both of *cis-trans* isomerism and the planar disposition of the carbon valences in the double bond. It is usually possible to establish configurations of *cis-trans* isomers (show what is *cis* or *trans* to what) by modern spectroscopic or X-ray methods.

Acetylene **20**, C_2H_2, and members of the related *homologous series*, C_nH_{2n-2}, such as **21**, were assigned structures containing a carbon-carbon triple bond; the triple bond was based on analogy to the ethylene series and on the additional reactions of the acetylene. Again, modern physical methods which show that the four atoms involved in the triple bond are linear (in **20**, H-C≡C-H) confirm this structure.

$$H-C≡C-H \qquad\qquad H-C≡C-CH_3$$

20   21

Kekulé's great contributions to the structural theory culminated in his proposal of the structure of benzene, C_6H_6. This hydrocarbon was isolated and characterized in 1825 by Michael Faraday (1791-1867). Its molecular formula showed it to be highly unsaturated, but its chemical reactions did not correspond to those of compounds containing carbon-carbon double or triple bonds. Benzene was found to be the parent substance of a huge family of analogs and derivatives, many of them of great chemical, biochemical, and technical importance. The whole family is frequently called the *aromatic series*, after Kekulé, because of the sweet smell of some naturally occurring derivatives of benzene. The pure compound, C_6H_6, was first called *benzin(e)*, but the name was later changed to *benzol* in German, and *benzene* in English. The term *benzin* has since been used in English to denote a mixture of saturated hydrocarbons obtained from petroleum (sometimes called *naphtha*) and chemically inhomogeneous and ill-defined.[9]

Kekulé suggested in 1865[10] that benzene contained a ring of six carbons each carrying a hydrogen and joined by alternate single and double bonds; he subsequently wrote his formula as **22**. In response to various alternative structural proposals for benzene and to chemical evidence that the benzene molecule was symmetrical, all the hydrogens being equivalent, Kekulé proposed in 1872 that the double bonds oscillated rapidly between the two "Kekulé structures" **23** and **24**, thus in effect making all positions and all carbon-carbon bonds equivalent. Although later theoretical concepts have modified Kekulé's original ideas, physical measurements have shown that the benzene ring is symmetrical, and thus Kekulé's insight has been confirmed. The Kekulé benzene structure opened up a vast domain of fundamental and applied research in aromatic compounds, in which American chemists were to play an important role.

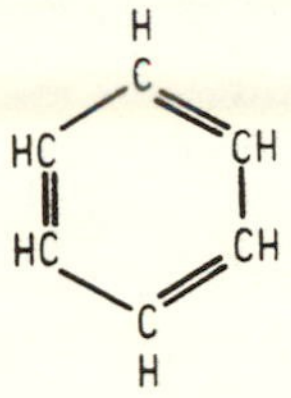

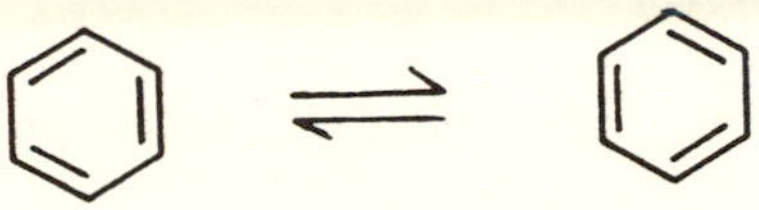

22   23   24

Kekulé's flash of intuitive genius that resulted in his benzene structure was described by him many years later in words that have become traditional examples of the mystery of the creative imagination.[11] He was dozing and watching the fire in an interval of relaxation from work on his textbook. In his reverie he saw forms of snakes, one of which formed a ring by taking its tail in its mouth. Kekulé realized that the ring thus suggested offered a solution to the problem of the benzene structure about which he had been puzzling. It is possible that Kekulé's account was a later embellishment of the actual discovery, but its metaphoric truth makes it worth preserving.

With the success of the Kekulé structure for benzene, the classical structural theory of organic chemistry may be considered complete. Before undertaking a general critique of the theory as it stood in 1875, some exceptions to the rules of valence, on which the theory was based, may be noted.

The most striking exceptions to the postulate of the tetravalence of carbon are the stable compounds carbon monoxide, CO, and the isocyanides, such as phenyl isocyanide, $C_6H_5N{=}C$. (The *phenyl* group, C_6H_5, represents a benzene ring with one hydrogen missing and acts like a monovalent group.) One could write both CO and the isocyanides with tetravalent carbon, and this was sometimes done, but there are convincing reasons not known in 1875 for preferring the bivalent carbon structure in these compounds. Around 1900 an American chemist, John Ulrich Nef, of Chicago, was to develop a whole system of reaction mechanisms based on very reactive bivalent carbon intermediates. Some of Nef's ideas are now much more attractive then they were during his lifetime.

The first stable trivalent carbon compound, the *free radical triphenyl-methyl*, was reported by Moses Gomberg of Ann Arbor in 1900. Much more reactive and shorter-lived free radicals were to form the basis for a whole realm of organic chemistry, including important industrial processes starting about 1925, and American chemists were to be in the forefront of this field.

Numerous free radicals containing nitrogen of abnormal valence, some rather stable and useful as *labels* in biochemical systems, have become known since 1875, as have free radicals derived from other elements. Free radicals, which contain an odd number of electrons, were called *odd molecules* by G. N. Lewis and were predicted correctly to have characteristic magnetic properties.[12] This "oddness" and their unusual properties merely emphasize how limited are the exceptions to the valence postulates of the simple structural theory.

As shown above, the structural theory has been completely successful in predicting the number of possible isomers, both position isomers and stereoisomers. It also predicted correctly the number of possible isomers derived from benzene, its substitution products, and its higher analogs (*benzologs*). Few additional postulates are required to accommodate later experimental observations.

The concept of isomerism has been refined somewhat by the idea of hydrogen bonding. The occurrence of stereoisomers caused by hindered rotation around bonds that would be considered "single" by the classical structural theory has been shown by the occurrence of optical activity in some cases, and more recently by nuclear magnetic resonance and other modern techniques. These additions to the structural theory of 1875 are really second-order corrections which emphasize the superlative usefulness and generality of the classical theory.

The theory does not predict anything about the stability of any structures; it cannot say which structures may be easy to synthesize in the laboratory or virtually impossible to synthesize. It predicts very little, qualitatively or quantitatively, about the reactions which a given compound may undergo, or about its physical properties. It predicts nothing about the *mechanism*, or detailed course of transformations of organic compounds, and tells nothing of the *rate* (*kinetics*), or *equilibrium* (*thermodynamics*) of such transformations; all information of this kind must be gained by experiment. Once a background of knowledge about the chemical and physical properties of a given series of compounds is available, the behavior of structurally analogous compounds can be predicted with considerable qualitative, and sometimes quantitative, success. All such predictions, however, must have an empirical basis.

There is a logical gap between the structural formulas written on paper or represented by three-dimensional models and the compounds themselves as prepared and studied in the laboratory. Most chemists do not worry much about such philosophical distinctions, but in some cases they have become the subject of heated discussion.[13] The one-to-one correspondence, which holds between the structures which can be written following the structural theory and the compounds which can be prepared or isolated, is valid for such a vast body of chemical observations that the structural theory is accepted as one of the most useful of scientific generalizations.

The structural theory of 1875 has other obvious shortcomings in addition to its lack of predictive value about reactivity and stability. It gives no hint of the basis of the valence forces holding atoms together in molecules. Why is carbon tetravalent, with its tetrahedral arrangement in saturated compounds? Why do the carbon-carbon multiple linkages show their characteristic geometrical and chemical properties? Why is the carboxyl group acidic, and the amine group in some ammonia derivatives basic? Reasonable answers to these and hundreds of similar questions could only come with the development of theories of atomic and molecular structure, and with the quantum mechanical theory of bonding in the twentieth century.

The classical structural theory was particularly unsatisfactory in *conjugated systems*, molecules containing alternate single and multiple linkages, such as acrolein, $CH_2=CHCH(=O)$, or butadiene, $CH_2=CHCH=CH_2$. In the archetypal conjugated system of benzene and the aromatic compounds in general, the Kekulé structures were misleading about the reactions of the systems compared to ethylenic carbon-carbon double bonds, for example. The Kekulé structures for aromatic compounds did predict the correct number of isomers, assuming the rapid oscillations of bonds indicated by $23 \rightleftharpoons 24$, but reasonable explanations of the structures of conjugated and aromatic systems were possible only with the arrival of quantum mechanical ideas of bonding in the 1920-1930 period.

The question is frequently raised, in discussing the history of ideas, whether great advances are due to the individual genius of original thinkers, or whether the advances are due rather to the climate of opinion which gifted individuals can crystallize and concentrate. In the present case, it seems clear that the development of chemical knowledge from 1800 to 1850 had created enough facts so that great unifying ideas occurred to several gifted workers simultaneously. In the development of organic chemistry, simultaneity of discovery seems to be more common than individual isolated leaps.

In the development of ideas in any field, numerous ideas appear at a time when there is no way of testing them critically; a pertinent example is the early suggestion of atomic theories by Democritus, Lucretius, and others in classical times.[14] Such ideas are frequently later rediscovered independently when there are ways of proving or disproving them.

LITERATURE CITED

1. Brief modern reviews: D. S. Tarbell and A. T. Tarbell in H. Skolnik and K. M. Reese, Eds., *A Century of Chemistry*, Washington, 1976, p. 339; D. S. Tarbell, Chem. Eng. News, Centennial Issue, Apr. 6, 1976, p. 110.

2. A. J. Ihde, *The Development of Modern Chemistry*, New York, 1964, gives an excellent general account; hereafter quoted as Ihde. An older, admirable treatment of the development of organic chemistry from 1770 to 1880 is Carl Graebe, *Geschichte der organischen chemie*, Berlin, 1920; reprinted by Springer, 1972, hereafter quoted as Graebe. Various aspects of the development of the structural theory of organic chemistry are discussed in detail in *The Kekulé Centennial*, Washington, 1966, O. Theodore Benfey, Ed., a sym-

posium with ten papers. Those by G. E. Hein, H. M. Leicester, A. Sementsov, A. J. Ihde, and Virginia M. Schelar are particularly pertinent.

3. G. N. Lewis, *Valence and the Structure of Atoms and Molecules*, New York, 1923; reprinted by Dover, 1966, with preface by K. S. Pitzer, pp. 20-21; hereafter quoted as Lewis.

4. Cf. *inter alia*, G. E. Hein in *Kekulé Centennial*, pp. 1 ff.

5. Graebe, pp. 192 ff; Ihde, pp. 222-226; H. M. Leicester, in *Kekulé Centennial*, pp. 13-24; O. T. Benfey, Ed., *Classics in the Theory of Chemistry Combination*, New York, 1963, for commentary and translation of key papers; O. T. Benfey, *From Vital Force to Structural Formulas*, Washington, 1975.

6. Graebe, pp. 262-273 for details.

7. J. Weyer, Angew. Chem. Int. Ed., **13**, 591 (1974); O. B. Ramsay, Ed., *van't Hoff-Le Bel Centennial*, Washington, 1975.

8. Rotterdam, 1875.

9. *Oxford English Dictionary*, and *Supplement*, under *benzene* and *benzin*; M. P. Crosland, *Historical Studies in the Language of Chemistry*, London, 1962, Dover, 1978, pp. 301, 351.

10. A. Kekulé, Bull. Soc. Chim. [2] **3**, 98 (1865); Ann. **137**, 129 (1866).

11. A. Kekulé, Ber., **23**, 1306 (1890).

12. Lewis, pp. 80 ff.

13. For accounts of such discussions which became far from "academic", see L. R. Graham, *Science and Philosophy in the Soviet Union*, New York, 1972, especially pp. 69-111, 297-324.

14. G. Sarton, *Introduction to the History of Science*, Baltimore, 1927, Vol. 1, pp. 88, 205; G. Sarton, *A History of Science*, Cambridge, 1959, Vol. 2, pp. 266, 271.

2
The American Background to 1875

In 1875 the United States was a "developing" country economically and scientifically. The Golden Spike, driven at Promontory, Utah, in 1869, celebrated the first transcontinental railroad. Sioux Indians destroyed part of Custer's command in the Custer Massacre in Montana Territory in 1876, and wild buffalo herds still roamed the open plains in 1880, until ruthless slaughter almost wiped them out a few years later. The last military occupying forces in the South after the Civil War withdrew from North Carolina in 1877.

In 1861 Yale granted the first Ph.D. in any subject in the United States, in English. Josiah Willard Gibbs (1839-1903) received the first American doctorate in science from Yale in 1863 with a thesis in engineering; he was to become the greatest American scientist since Benjamin Franklin.[1] Only a handful of Ph.D.s was awarded in the whole country between 1861 and the opening of the Johns Hopkins University in 1876. Yale granted two Ph.D.s in chemistry from 1865 to 1869, and seven from 1870 to 1879; for the latter period, Johns Hopkins, Harvard, and Michigan gave one each.[2] Chemical research in industry and in government was exceedingly limited in 1875.[3]

The choice of the year 1875 as a starting date for a discussion of organic chemistry in the United States rests upon several timely occurrences that greatly stimulated the emergence of chemistry as a profession. The Johns Hopkins University, America's first major center of research in organic chemistry, opened in 1876, and the American Chemical Society was founded that same year. Two journals for the publication of American chemistry were established in 1879: the *American Chemical Journal* (ACJ) started by Ira Remsen at Johns Hopkins, and the *Journal of the American Chemical Society* (JACS) in New York City.

By 1875 Germany had become the acknowledged leader in organic chemistry; it was the Mecca for aspiring students of chemistry, and its university and industrial laboratories were doing the pioneering work in the field. German hegemony in chemistry was, however, a relatively recent development due primarily to the emergence of Justus Liebig (1803-1873) and Friedrich Wöhler (1800-1882). They gave German chemistry, and particularly organic chemistry, the leading position which it held unchallenged until after World War I. Previously France and the United Kingdom had been the pacesetters. Thus Dr. Benjamin Rush (1746-1813) of Philadelphia, Revolutionary patriot, signer of the Declaration of Independence, physician, and educational statesman, had studied chemistry with Joseph Black at Edinburgh (as well as medical subjects with others) from 1766 to 1768. He returned to become professor of chemistry at the College of Philadelphia in 1769, the first formal professor of chemistry in America, and wrote the first chemical textbook in this country.[4] Later, Benjamin Silliman, Sr., (1779-1864), professor at Yale (1802-1853), founder of the *American Journal of Science and Arts*, and a leading figure in American science for many decades, studied chemistry at Edinburgh around 1800 in preparation for his professorship at Yale. He was unable to read German.[5,6] His son, Benjamin Silliman, Jr., (1816-1885), writing to a student in 1854, advised him to study in Germany for the professional and financial advantages which would result.[7]

The active chemical culture in this country between 1840 and 1875 thus derived its main inspiration from Americans who studied in Germany, rather than in France, England, or Scotland, as at the beginning of the century. Estimates place the number of Americans who studied chemistry in Germany from 1850 to 1920 at about 1000, assuming that ten percent of American students there studied chemistry[7]; another writer [8] gives a total of 10,000 American students in all fields in Germany for 1850-1914. A more recent study[9] infers about 9000 for the period 1820-1920. A list of 101 Americans who studied with Robert Bunsen (1811-1899) at Heidelberg shows that 33 took Ph.D.s with him, and 62 of the total either took doctorates elsewhere in Germany or reached professorial chairs in this country. The best known organic chemists in this list are C. L. Jackson and Arthur Michael (1853-1942), neither of whom took a doctorate.[10] From 1841 to 1852, 16 Americans studied with Liebig at Giessen and 13 more worked in chemistry with him or his colleagues at Munich from 1852 to 1876. Those having important careers in American chemistry included J. Lawrence Smith (1801-1883), E. N. Horsford (1818-1893), Wolcott Gibbs (1822-1908), G. J. Brush (1831-1912), S. W. Johnson (1830-1909), H. A. Weber (1845-1912), E. Renouf (1848-1934), and Ira Remsen (1846-1927).[10] Of these only Remsen became outstanding in organic chemistry.

James F. Magee (1834-1903) exemplifies an American student of chemistry in Germany. After instruction in James Booth's laboratory in Philadelphia for eighteen months, Magee spent the year 1855-1856 with Wöhler at Göttingen and with Bunsen at Heidelberg. Magee described his German studies and European travels in vivid letters to his family.[11] Two chemist brothers, Edward and Joseph Eastwick, accompanied him from Philadelphia to Göttingen, joining nine other Americans at the university.

Wöhler's laboratory building stood in the garden surrounding his home at Göttingen and he managed to find room for Magee and the Eastwicks, not wishing to turn away Americans who had travelled 4000 miles to study with him. Nine Americans elected chemistry and five worked together in one room. The laboratory had no gas and no blowpipe table; spirit lamps, mouth blowpipes, and small charcoal furnaces supplied heat for reactions. Conditions were exceedingly crowded, more so than in Booth's Philadelphia laboratory, but the German master's kindness and attentiveness to his American students were marked. His assistant, Dr. Karl Goessman, also very popular with the young men, was overwhelmed by their gift of a fine balance at Christmas in 1855. Goessman (1827-1910) emigrated to America in 1857 at the invitation of the Eastwick brothers, served for forty years as a well-known professor at the (then) Massachusetts Agricultural College in Amherst, and became president of the American Chemical Society in 1887.

Magee was surprised to find that Göttingen students were subject only to the laws and regulations of the university itself and not to the legal code of the town, a relic of the medieval university system. At Göttingen he concentrated on chemistry and learning German, although his language study suffered from the clannishness of the American group and the lack of congeniality with the hard-driving, hard-drinking German students. Magee had no publications with Wöhler.

At Heidelberg in 1856, Magee with ten Americans and six English students worked in Bunsen's admirably equipped laboratories; there he augmented practical and technical chemistry with botany and horsemanship. Magee's name is included in the list of Bunsen's American students, but without his first name and his later occupation.[10] After extended travels Magee returned without taking a Ph.D. and ran a successful chemical business in Philadelphia.

Prior to 1860 chemistry was frequently taught in conjunction with geology and sometimes with medical or agricultural studies because of its applications to these fields. S. M. Guralnick has shown that there was a notable advance in the teaching of science (including chemistry) in American colleges[12] in the 1820s. Liebig's writings opened up the study of application of chemistry to agriculture, and M. W. Rossiter[10] details the growth of agricultural science in this country based on the work of Liebig and of his American students.

Amos Eaton (1776-1842), a gifted popular lecturer, author, and teacher of geology, botany, and chemistry, established the first student program for actual laboratory instruction in chemistry in 1824 at the school which developed into Rensselaer Polytechnic Institute.[13] He required each student to work up experiments and to lecture with these as demonstrations. This scheme undoubtedly arose from Eaton's experience with popular lectures, in which he followed his mentor, Benjamin Silliman, Sr., an extremely successful and popular lecturer on chemistry and geology.

In 1847 Silliman and his son, Benjamin Silliman, Jr., inaugurated a more systematic program at Yale to teach chemistry, including analytical procedures. This project, supported at first by Silliman personally, eventually grew into the important scientific center, the Sheffield Scientific School.[14] Daniel C. Gilman, the first president of the Johns Hopkins University, was a product of Silliman's Yale and was associated with Sheffield for several years.

The establishment of the Lawrence Scientific School at Harvard in 1847-1848 paralleled the Yale venture. Eben N. Horsford,[15] a student of Eaton at Rensselaer and of Liebig at Giessen from 1844 to 1846, where he became interested in agricultural and industrial chemistry as well as organic chemistry, won the appointment as Rumford Professor in the Lawrence School against strong competition. Horsford soon developed a useful phosphate baking powder and founded the successful Rumford Chemical Co. for its manufacture, but he lost interest in academic chemistry and finally resigned his professorship in 1863. His successor as Rumford Professor was (Oliver) Wolcott Gibbs, trained at Columbia, at Pennsylvania, and in Germany with Liebig and others. Gibbs, a Unitarian, received the appointment at Harvard which accepted his "Boston Faith", although orthodox board members of Columbia in 1854 had refused to vote him a professorship because of his religious views.[16] Scion of a wealthy, scientifically inclined father, Gibbs made significant contributions to inorganic and analytical chemistry and did excellent research at Harvard from 1863 to 1887, in spite of the virtual demise of the Lawrence Scientific School.[14] Sheffield, the Lawrence School, and similar ventures at other institutions had difficult times at first because of lack of university support and recognition, but they furnished trained scientists to work in chemistry and geology, as these sciences became more professional in the 1860s and 1870s.

Private consulting chemical laboratories, which grew up between 1835 and 1860, supplied analytical services to businesses and, to some extent, chemical instruction to students. One of the first of these was started in Boston in 1836 by the versatile and eccentric Charles T. Jackson (1805-1880) to do water and mineral analyses. Jackson, an M.D., was primarily a geologist, and his career represents chemistry in close relation to both medicine and geology.[17]

James C. Booth (1810-1888), a student of Eaton at Rensselaer and of Wöhler in Germany, began his laboratory in Philadelphia and trained fifty students (among them Magee) in analytical and industrial chemistry and in chemical research. Primarily his firm performed analyses for manufacturers and assisted them in developing or improving their chemical processes. Booth had many talents; said to be the first American to study analytical chemistry in Germany, he was also a geologist, assayer of the United States Mint in Philadelphia, editor of the *Encyclopedia of Chemistry, Practical and Theoretical*, and president of the American Chemical Society from 1883 to 1885.[18] His analytical apparatus and reagents are in existence; his blowpipe is of the type used by Berzelius and in the German laboratories of the 1830s and 1840s.[19]

In the decades before the Civil War, numerous attempts failed to establish true universities with graduate schools following the German university model in Washington, Albany, Ann Arbor, New York, and other places. Nevertheless, they left a continuing interest in graduate study and research that favored the success of later ventures.[20]

It is estimated[21] that about 200 chemistry books were published in the United States from 1820 to 1870, although few of them were written here. J. P. Cooke's *Chemical Problems and Reactions*[22] and his *Elements of Chemical Physics*[23] were important American texts, the latter being a pioneer book in the development of physical chemistry. E. L. Youman's *A Class-Book of Chemistry*[24] sold 144,000 copies. Examination of the 1856 edition shows that it was an elementary but reasonably up-to-date presentation, containing discussions of naturally occurring organic compounds, such as sugars, fats, oils, proteins, and plant products, along with some description of "animal chemistry" (nutrition and biochemistry).

The principal American outlets for chemical publications prior to 1879 were the *Proceedings of the American Academy of Arts and Sciences* (of Boston) and the *American Journal of Science and Arts*, founded in 1816 by Benjamin Silliman, Sr., and known familiarly as *"Silliman's Journal"*. Both published articles in all scientific fields. A scrutiny of the latter journal for the years 1850-1885 shows a total of about sixty publications in organic chemistry, most of them of negligible scientific value. After 1870, however, a number of papers by Ira Remsen at Williams College, C. L. Jackson at Harvard, and a few others indicate that chemists were beginning to carry out serious and substantial research in this country. The *Proceedings of the American Academy* show little of import before 1875, but after that a series of respectable papers appeared, principally from Harvard. Although the American Philosophical Society in Philadelphia published little chemistry at any time, its publications, too, indicate traces of the increasing research in organic chemistry about 1875.

American chemists might also publish in foreign journals; those who worked in German laboratories usually saw their results printed in German journals, and some of them who carried out research in this country later continued to publish in German journals. Examinations of the files of *Justus Liebigs Annalen der Chemie* (Ann.) from 1860, of the *Berichte der Deutschen Chemischen Gesellschaft* (Ber.) from its beginning in 1869, and of the *Journal für Praktische Chemie* (JPr) in detail from 1870 and less thoroughly from 1850, show a similar pattern of a sudden burst of activity in organic chemistry in the United States around 1875.

American organic chemists, even after the American journals were established, continued to submit papers to German journals, at times publishing virtually identical manuscripts in an American and a German journal, and some cases of triple publication exist. A German journal, particularly the *Berichte*, frequently provided rapid preliminary publication, with the more complete paper appearing later in an American journal.

As expected, the organic chemists trained in Germany tended to publish more in German journals than those who trained in this country. Indeed, more than a trace of colonial mentality was apparent in American chemists.

An American paper published in the *Berichte,* for example, seemed to acquire additional validity in the author's eyes, and probably, he hoped, in the eyes of his compatriots. This emphasis on double publication and foreign publication corrects and supplements the picture of American chemical activity resulting from an examination of only the American journals.

The establishment of a viable scientific culture in an undeveloped country requires three things: (1) a foreign source for adequate training and inspiration, (2) growing domestic programs to furnish leadership and trained personnel, and (3) an economic base able to provide jobs for graduates and support for educational institutions. In the United States, Germany provided the first, and the Johns Hopkins University furnished the second for several decades after 1876. The economic base evolved from the growth of chemical industry; from the development of laboratories by federal, state, and local governments; and from programs in private colleges and universities as well as those in the land-grant institutions and agricultural experiment stations resulting from the Morrill Act of 1862 and later similar measures. Given all the requirements for the growth of the science, organic chemistry was ready to spurt ahead.

LITERATURE CITED

1. L. P. Wheeler, *Josiah Willard Gibbs,* New Haven, 1951.
2. Figures from P. T. Carroll, "Academic Chemistry in America; Diversification, Growth and Change", Ph.D. Thesis, University of Pennsylvania, 1982, p. 148.
3. For research in industry, see Williams Haynes, *American Chemical Industry: A History,* New York, 1954, Vol. 1, pp. 386-400; K. A. Birr, in D. D. Van Tassel and M. G. Hall, Eds., *Science and Society in the United States,* Homewood, Ill., 1966, pp. 52-62. For research in government laboratories, A. H. Dupree, *Science in the Federal Government,* Cambridge, 1957.
4. N. G. Goodman, *Benjamin Rush, Physician and Citizen,* Philadelphia, 1934, p. 14.
5. G. P. Fischer, *Life of Benjamin Silliman,* Philadelphia, 1866, Vol. 2, p. 36.
6. L. G. Wilson, Ed., *Benjamin Silliman and His Circle,* New York, 1979; J. F. Fulton and Elizabeth H. Thomson, *Benjamin Silliman,* New York, 1947.
7. E. H. Beardsley, *The Rise of the American Chemical Profession, 1850-1900,* Gainesville, 1964 (Universiy of Florida Monographs, Social Sciences, No. 23, Summer 1964), pp. 14,15.
8. C. F. Thwing, *The American and the German University,* New York, 1928, pp. 40-41.
9. Jurgen Herbst, *The German Historical School in American Scholarship,* Ithaca, 1965, pp. 1-3.
10. Bunsen's students: JACS, Proceedings, **22**, 102 (1900); Liebig's students, Margaret W. Rossiter, *The Emergence of Agricultural Science: Justus Liebig and the Americans, 1840-1880,* New Haven, 1975, pp. 184-195. American chemists of the 19th century are discussed by C. A. Browne, JCE, 9, 696 (1932); B. Silliman, Jr., American Chemist, 5, 70, 195, 372, (1874-1875), the first attempt at a historical view of 19th century American chemistry; Fulton and Thomson, pp. 146-158.
11. G. W. Magee, Jr., Ed., *An American Student Abroad,* from the *Letters of James F. Magee,* Philadelphia, 1932, passim. Although the editing of the letters does not meet scholarly standards, the details of Magee's account are convincing and fascinating. Unfortunately he gives no discussion of the chemistry he worked on, because he was writing for his non-chemical family. The Isis bibliographies do not mention the Magee book,

and it seems not to have been used by historians of chemistry.

12. S. M. Guralnick, *Science and the Ante-bellum American College*, Philadelphia, 1975, passim, particularly pp. 94-118; also Frederick Rudolph, *The American College and University*, New York, 1962, pp. 110-130.

13. Ethel M. McAllister, *Amos Eaton*, Philadelphia, 1941, especially pp. 384 ff.

14. For Sheffield, see biographies of Silliman, Sr., above; L. I. Kuslan, in *Benjamin Silliman and His Circle*, pp. 129-157; R. H. Chittenden, *History of the Sheffield Scientific School*, New Haven, 1928, 2 vol. For the Lawrence School, S. E. Morison, *Three Centuries of Harvard*, Cambridge, 1937, pp. 279-280, 324, 334, 371-372. When C. W. Eliot became president of Harvard in 1869, Wolcott Gibbs was relegated to a laboratory in the physics department, and J. P. Cooke (1822-1894) was named Rumford Professor in Harvard College; he reorganized instruction and research in the field. On Cooke, C. L. Jackson, *Biog. Mem. Nat. Acad. Scis.*, 4, 175 (1902); C. L. Jackson in S. E. Morison, Ed., *The Development of Harvard University 1869-1929*, Cambridge, 1930, pp. 253-262; on Wolcott Gibbs, F. W. Clarke, *Biog. Mem. Nat. Acad. Scis.* 7, 3 (1910).

15. H. S. Van Klooster, JCE, **33**, 493 (1956); S. Rezneck, Technology and Culture, **11**, 366 (1970), describes Horsford's career and prints significant extracts from his letters from Germany. Rossiter, pp. 49-90, gives a detailed account of Horsford's education, particularly with Liebig, and of his academic career at Harvard.

16. *The Diary of George Templeton Strong*, Allan Nevins and M. H. Thomas, Eds., New York, 1952, gives a detailed account of this celebrated case, Vol. 2, pp. 145-174. Strong, a devoted Columbia graduate and trustee, was furious at his bigoted colleagues.

17. J. B. Woodworth, Am. Geologist, **20**, 69, especially p. 84 (1897).

18. E. F. Smith, JCE, **20**, 315 (1943); W. F. Miles, Chymia, II, 139 (1966), prints excerpts from Booth's letters from Germany as a student. Booth was a leading candidate for the professorship at Harvard which Horsford obtained. Booth's encyclopedia, published in Philadelphia in 1850, went through five editions, the last in 1883.

19. Described in detail with photographs by S. M. Edelstein, JCE, **26**, 126 (1949). Edelstein obtained the equipment from Booth's aged daughter.

20. R. J. Storr, *The Beginnings of Graduate Education in America*, Chicago, 1953.

21. C. A. Browne, JCE, **9**, 724 (1932).

22. Cambridge, 1857, and five later eds.

23. Boston, 1860, and five later eds.

24. New York, copyright 1851.

IRA REMSEN

3

The Johns Hopkins University, Ira Remsen and Organic Chemistry, 1876-1913

The opening of the Johns Hopkins University in Baltimore in 1876 was a landmark in the advancement of scholarly training and original research in the United States.[1] Its revolutionary effect on chemistry mirrored its challenging impact on other fields of learning. Thus the Hopkins chemistry department, headed by Ira Remsen, played a unique role in the development of American organic chemistry.

Johns Hopkins (1795-1873), wealthy Baltimore merchant and banker, left the bulk of his seven million dollar estate in two bequests, one to found a university and the other to establish a hospital. His ideas may have come from the local benefactions of his contemporary, American-born George Peabody (1795-1869), London banker and philanthropist.

Hopkins appointed civic leaders of the Baltimore region to the boards of trustees for the two institutions. The university trustees received essentially unlimited freedom in implementing his gift, although his will did stipulate a "judicious" number of scholarships for students from Maryland, Virginia, and North Carolina for an education "free of charge". The trustees of the new university proceeded with laudable responsibility and energy, informing themselves about the current state of higher education in the country. They conducted searching interviews with Presidents C. W. Eliot of Harvard, J. B. Angell of Michigan, and A. D. White of the new Cornell University. White recommended dedication to advanced study and research with scholarships for full-time graduate students. This policy was to have far-reaching results. Innovative in American universities at the time, supported enthusiastically by the university's trustees, skillfully implemented by the first president, it was to become Hopkins' matchless contribution to higher education in the nation.

The newly chosen president, Daniel Coit Gilman (1831-1908), whom Eliot, Angell, and White had unanimously recommended, proved an admirable choice. A Yale graduate (1852) with brief graduate attendance at Yale and at Harvard, he indulged his developing interest in geography by travel and study in Europe as well as by diplomatic service in St. Petersburg.

Returning to Yale, where he spent the next seventeen years, Gilman drew up a plan based on his knowledge of European institutions for the organization of what was to become Yale's Sheffield Scientific School. Positions as Yale librarian, professor of geography, and fund-raiser for the new scientific school furnished experience for the rising educational statesman. Although he effected only marginal improvements in the Yale library, one of his appeals for funds contained the seminal statement so significant for his Johns Hopkins achievements, "It is scholars who make a college; not bricks and mortar. It is endowments which secure the time and services of scholars." He realized the significance of the Morrill Act of 1862 and promoted Yale's early use of the act's educational land-grant funds. He also gained extensive administrative experience with public education in the New Haven schools.

In 1872 Gilman became the forceful president of the newly established University of California. Among other achievements he successfully defended the aim of the university to become a broad-based institution of higher education against the state legislature's push to limit it to technical and agricultural programs. Finding the prospect of continued fierce, political controversy uninviting, he welcomed the approach of the Hopkins trustees, who promised enlightened support and freedom from political and sectarian control. At Johns Hopkins in 1875 Gilman undauntedly tackled the difficult task of constructing a university national in scope and endowed with "intellectual freedom in the pursuit of truth." He proved to be a remarkable judge of talent in choosing his faculty, and he effectively fostered an atmosphere of intellectual adventure and excitement, not only within the university but also to a significant extent in the Baltimore community and the country at large.

Gilman utilized his contacts in this country and abroad to recruit his faculty. In mathematics and literature he was able to attract distinguished scholars as professors, J. J. Sylvester from England and B. L. Gildersleeve from the University of Virginia. He picked an able young man in physics, Henry A. Rowland from Rensselaer Polytechnic Institute, who became an internationally recognized experimentalist. Gilman offered the professorship in chemistry to Wolcott Gibbs of Harvard who declined to move to Hopkins. However, Gibbs recommended Ira Remsen of Williams College as an unusually promising young chemist.

Ira Remsen (1846-1927), the son of a prosperous New York family of Dutch and Huguenot forbears, took an M. D. from Columbia in 1867 but soon decided to study chemistry in Germany. At Munich he found that Justus Liebig, with whom he had planned to study, was no longer taking students; he therefore worked with J. Volhard at Munich and later with Rudolf Fittig at Göttingen. His Ph.D. thesis, completed in 1870 when Remsen was twenty-four years old, dealt with aromatic acids, and he continued as research assistant to Fittig, moving with him to Tübingen. In 1872 he became professor of chemistry and physics at Williams College; his zeal for research survived a statement by the Williams president that he should remember that Williams was a college, not a technical school, and that the college's object was culture, not practical knowledge. It was Remsen's research productivity at Williams that rapidly attracted the attention of American chemists. Gilman's appointment of Remsen to the Johns Hopkins faculty in 1876 allowed Remsen to demonstrate his teaching, research, editorial, and administrative abilities.

With Gilman's support he was able to make one of his greatest contributions to the advancement of American chemistry, the founding and editing of the *American Chemical Journal* (ACJ), which furnished a medium for publication of chemical research in this country. In Remsen's later years he served as President of Johns Hopkins from 1901 to 1913.[2]

Before the new university opened in 1876 with a fanfare of publicity by the press, Remsen may have had misgivings. "We had no laboratory. We had less than a handful of students. What was to become of it?"[3] But as he developed his department and furnished his laboratories with the books and apparatus he wanted,[4] the students came, three or four the first year, then more, "drawn to Johns Hopkins by the fame of its faculty and perhaps even more by the belief that it offered unusual opportunities for original scholarship," eager to study in a "research institution like the great European universities,"[5] and recruited by graduate fellowships, that innovation among American universities. Remsen and his newly hired associate, H. N. Morse (Ph.D. 1875, Göttingen), were the two faculty members in chemistry, and three graduate fellows in chemistry had been appointed, E. Hart, L. B. Hall, and M. W. Iles, as Johns Hopkins opened its doors.

The Hopkins students found the laboratories adequate, "located in a busy, noisy, dusty part of old Baltimore" where the clamor and clash of "the city's traffic penetrated the walls," and "the equipment was excellent."[5] The first chemical laboratory occupied one of the old city residences remodeled for the use of the new university. In a few years substantial buildings replaced these, including separate ones for each of the sciences: physics, biology, geology, and chemistry. Dalton Hall, the three-story chemistry building, opened in 1883; it contained laboratories designed for inorganic, analytical, and organic chemistry, offices, and the chemical library which included complete files of the standard journals.[6] Glassware, much equipment, and many chemicals, especially organic compounds, were ordered from Germany each year.

Adding to the luster of the new university in 1880 were the brilliant scholars in other disciplines gathered by President Gilman, as well as the galaxy of distinguished visiting scholars from here and abroad, whose lectures were open to all students and the public alike. The intellectual excitement in nonscientific fields at Hopkins shines out in the experiences of Woodrow Wilson and Walter Hines Page as students.[7]

Gilman and the faculty drew up formal procedures for the Ph.D., setting forth the requirements for independent research and its publication, examination, language, distribution of courses, and residence, not very different from those in effect now.[8] Registration meant signing up for major and minor work in chemistry and for one or more minors in other fields. V. J. Chambers, for example, chose botany,[9] and Walter Jones, geology and mineralogy,[10] a good thing, as the new graduates found, when the available jobs covered a multitude of subjects. The students were expected to be familiar with all the lectures, undergraduate and advanced, in all chemistry offered. Preparations and small research problems done with several faculty members often provided publications in different fields. Remsen modeled his teaching on his German experience with lectures, daily conferences, and laboratory visits, so that the contact between instructor and student was close and personal. Working in the laboratories with the doctoral candidates were zealous undergraduates and postdoctoral fellows, and other graduate students studied for one or two years then left to go to other universities for advanced degrees here or abroad, to medical schools, or to jobs. Although course offerings were limited, the busy and spirited atmosphere of research and the leadership were stimulating.

A chemistry graduate student in need of money might aspire to a fellowship of $500 or less, or find analytical, patent, or other work to do. W. A. Noyes, besides finishing his Ph.D. in one and one-half years, performed water analyses on the side and during the summer to pay his expenses.[11] Others were appointed as assistants in the courses, taught part-time at their home colleges or in nearby institutions, or had worked for a while, usually teaching, before entering graduate school.[12]

In general the Ph.D. degree was awarded after three years of work (formal requirements stipulated two, then three years) and the results were usually promptly published. In some instances the period was shorter, especially if the student possessed a master's degree, or it sometimes stretched to four or five years. Most of the students did not continue their studies in Europe. Although many faculty members and postdoctoral fellows had trained abroad, the necessity for foreign study declined, and, after the turn of the century, only two graduates traveled overseas for study as compared to fourteen before.[12]

Remsen's personality is revealed by the quality of his research and his devotion to teaching, by his writing in research articles, in reviews in the ACJ, and in his textbooks, as well as by accounts of his career. William A. Noyes of Illinois and James F. Norris of MIT, both Remsen students who were organic chemists of outstanding accomplishment and sound judgment, wrote:[2]

> Ira Remsen was for many years the outstanding figure in American chemistry. When the history of the development of the science in this country is written, the fact will be evident that through his influence the serious study of chemistry and the output of new knowledge very rapidly increased. Much had been accomplished by a few gifted men in America before Remsen's day, but he opened up a life work in chemistry as a career to many, and developed a spirit of research that spread over the country. He made it possible for a young man to be adequately and broadly trained at home, whereas Remsen himself and others who sought, at that time, to prepare themselves for work in chemistry had been forced to go to Europe.

And what is equally important, he transplanted the 'atmosphere' of the laboratories of the great masters–the spirit of hard work, the desire to learn and a love of chemistry.

Although Remsen's chemical originality was not comparable to that of some of his American contemporaries like Arthur Michael, Julius Stieglitz, S. F. Acree, and some of his own students, he was well qualified for his role as leader of the development of organic chemistry in this country. His research, carefully and intelligently done, was clearly described in his publications. His students received excellent training in research methods and thinking, as well as in the current developments in the field, particularly through his use of a journal seminar that was part of the German graduate educational system. He obviously possessed one of the indispensable attributes of a great teacher: the ability to inspire his students with an enduring love for the subject and a desire to contribute to it themselves. This is clearly shown by the number of his students who developed significant research programs of their own in both academic and industrial laboratories and by those who became inspiring undergraduate teachers.

Johns Hopkins had an undergraduate school from the beginning,[13] and Remsen's activities included lecturing in what would now be called general and elementary organic chemistry. He wrote a number of very widely used textbooks which went through several editions and were translated into several languages. They are marked by the clarity of presentation and expression that was characteristic of his research papers. The 1909 edition of his organic text has an extremely limited discussion of stereochemistry,[14] although this was thirty-five years after the publication of the van't Hoff-Le Bel theory and many years after Emil Fischer's epoch-making work on the configuration of the sugars. The book as a whole has almost no discussion of theory or of significant new developments. His teaching experience gave Remsen positive views on the most desirable methods of presentation and content of textbooks, which appear most clearly in his book reviews in ACJ. He was very conservative in his approach to new theoretical ideas and to innovations in textbook content.

The comments of a Hopkins Ph.D., E. T. Allen, are in general agreement with the impression one gains from reading Remsen's papers and reviews:[15]

> Remsen succeeded remarkably with beginners, but I think his success with more advanced students was not quite so great. I remember a certain feeling of disappointment which I felt in listening to a course of his lectures in organic chemistry for the older students which was offered only once in two or three years. It all seemed like the older courses somewhat revised and enlarged... He was always quick to head off a student who indulged in reckless speculation... But, irreproachable as this attitude may be, I think his students got little encouragement even in legitimate speculation. A man not

naturally much inclined to speculation himself, I think he rather discouraged theorizing as an exercise dangerous perhaps to all but the greatest minds.

The eminent biochemist, W. Mansfield Clark, a graduate student at Hopkins in 1908-1910, found that the lectures by then President Remsen were not inspiring and that the atmosphere in the department was not congenial.[16]

There is undoubtedly a large element of self-portraiture in Remsen's obituary of Fittig; he says in part:[17]

> There [Tübingen] he [Fittig] found conditions somewhat unfavorable...The laboratory was a sorry affair. Fittig had the most fastidious tastes as far as his laboratory surroundings were concerned. He was extremely neat and orderly... Fittig's work was characterized by great thoroughness and accuracy and he justly prided himself on the fact that no one had ever convicted him of an experimental error!... Fittig did not contribute to any extent to the theories of chemistry, though he furnished the material basis for profitable speculation.... It is doubtful whether any text book of chemistry has ever been written that had less reference to underlying theories than his "Grundriss"... and his edition of Wöhler's "Grundriss".... These books are made up of compact statements of fact and are extremely valuable for the purpose for which they were prepared. He was a clear lecturer and an excellent laboratory teacher. His insistence on thoroughness was characteristic of everything he did, and those who had the privilege of working under him have every reason to be thankful for the influence exerted by him as a teacher. The writer of this sketch was closely associated with Fittig for four years, two years as a student at Göttingen and two years as an assistant at Tübingen, and he always recalls these years with gratitude, conscious of the fact that his association played a large part in such intellectual regeneration as he has been permitted to experience.

Remsen's own voluminous researches, resulting in some 200 papers and in the training of over 100 research collaborators, do not show great originality, being mainly an outgrowth of his work in Germany. His work was principally concerned with aromatic sulfonic carboxylic acids and related compounds, such as 1. The discovery of the sweetening properties of saccharin 2, which was a result of some of his early researches,[18] brought Remsen no financial reward, but it did give his laboratory and chemical research in general much

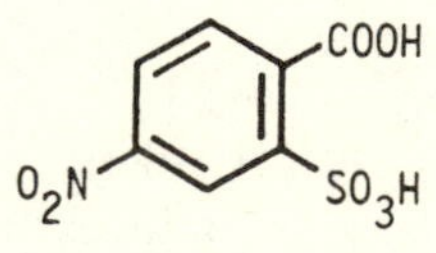

1 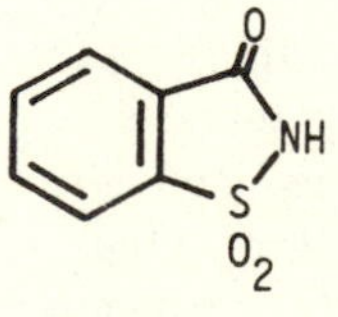   2

favorable popular publicity. Remsen's unpleasant experience with his German assistant, Dr. C. Fahlberg, who patented their joint discovery of saccharin under his own name and started a profitable commercial production without consulting Remsen, may have been one cause for Remsen's reluctance to have connections, consulting or other, with chemical industry that has been noted by several writers.[2] Remsen's stated views were that in the early days at Hopkins his students were not regarded as "practical" enough, and there were few places available in industrial research. Later, although the character of the graduate training had not changed, the demand by industry increased. Remsen did do some consulting for state and municipal governments and helped to design the chemical laboratories for the University of Chicago.

Remsen also worked on problems in inorganic and analytical chemistry, particularly on inorganic double salts,[19] but his research on aromatic compounds related to sulfonates accounted for about eighty percent of all his papers in the ACJ in which most of his work appeared. In Remsen's day the line of demarcation between organic chemistry and other branches of chemistry had not become as distinct as it did later. Remsen, Noyes, J. U. Nef, Moses Gomberg, and their leading contemporaries regarded themselves as complete chemists, quite capable of doing synthetic organic work, physical measurements, or precise analytical determinations. However, the specialization was clear enough to be useful for classification.

For a few years Remsen and Morse published research papers together and with fellows and instructors, including Americans and Germans with degrees from German universities. The meticulous Morse was slow in developing a research program in analytical chemistry, and it was not until the turn of the century that he and his students solved the exacting problems of the measurement of osmotic pressure which was to bring him late recognition, including election to the National Academy of Sciences shortly before his retirement in 1916.[3]

Other faculty members, added later and well-known for their research programs, were H. C. Jones (1895-1916), professor of physical chemistry; S. F. Acree (1906-1914); J. C. W. Frazer (1911-1935), Baker Professor of Analytical Chemistry and chairman of the department following Remsen in 1914; and E. E. Reid (1908-1937, with a three-year interval in industry), Johnson Fellow and later professor of organic chemistry. Three of these men held Hopkins doctorates: Jones and Reid with Remsen, and Frazer with Morse. These later appointments resulted in the flood of publications in physical chemistry after 1900 and the greatly increased number of Ph.D.s granted outside of organic chemistry. Acree took his Ph.D. with Nef at Chicago (1902) and had studied at Berlin. He directed many of the organic students during Remsen's last years as president. J. B. Tingle (Ph.D. 1889, Munich), a much traveled chemist, was assistant in charge of organic chemistry at Hopkins and sub-editor of ACJ from 1904 to 1908.

Both E. Renouf (1848-1934, Ph.D. 1880, Freiburg), who advanced to professor in analytical chemistry from 1885 to 1911, and J. E. Gilpin, at Hopkins from 1892 to 1924, collegiate professor of chemistry, one of Remsen's men who later published on crude petroleum, assisted with the students and with the ACJ. Other Hopkins men were taken on as assistants, associates, or "Fellows by Courtesy" (without stipend) from time to time.[12] Remsen, Acree, Hall, Jones, Frazer, Reid, and Renouf were all "starred chemists" in *American Men of Science*. (The "starred" list, compiled by ten leaders in each field, recognized good research workers. The first edition in 1906 of *American Men of Science* listed 175 starred chemists.)

Remsen apparently did not utilize the full power and prestige of Johns Hopkins to build up his faculty. He failed to make any serious search for talent or to call back any of the exceptionally able young men who passed through. Tingle, Acree, and Reid did not maintain Hopkins' position of leadership in organic chemistry against the competition of other vigorous departments. By this failure to develop his replacements and by retaining the chairmanship while president of the university, Remsen retarded the further development of organic chemistry at Hopkins. Nevertheless, the influence of Remsen on organic chemistry in this country was very great through the students he trained; during the years under review (1876-1913) Hopkins was by far the largest producer of Ph.D.s in organic chemistry in the country.

During the years 1879 to 1913, Johns Hopkins University awarded 202 Ph.D. degrees to students in chemistry.[12] Of these, 130 (64 percent) contributed papers in organic chemistry published in the ACJ and are considered to have received their doctorates in this field. Ira Remsen's name is clearly associated with 107 of these recipients, over 50 percent of the Ph.D.s granted in his department. Practically all of these students worked in organic chemistry, except for a very few who did research on double salts and who are included in the group of Remsen's men for the purposes of this review. Acree published papers with 16 Ph.D. candidates, including 2 women students, and Tingle published with 7 Ph.D. candidates. These three groups make up the sample of Ph.D.s surveyed below.

Remsen's most productive years were from 1889 to 1900, paralleling those of the entire Hopkins faculty. Hopkins granted 4 Ph.D.s in 1879 and awarded a total of 496 by 1902,[20] of which 101 went to Remsen's men. His high point of 11 recipients, almost a quarter of all the Ph.D.s granted that year in the university, occurred in 1895. Remsen, even while president, continued to take some graduate students and publish their results until 1912.

These 202 future Ph.D.s in chemistry came bearing credentials from 87 institutions across the nation and from 2 foreign countries, Canada and Japan.[12,21] Small liberal arts colleges supplied most of the candidates, with a sprinkling from the Ivy League schools, technical institutes, and state universities. Those from the Northeast (New England and the middle Atlantic states) and those from the South (below the Mason-Dixon line) entered in almost equal numbers. A smaller list from the Midwest and West and a substantial number of Johns Hopkins alumni rounded out a geographically diverse and cosmopolitan student body (Table 1).[12,22] These figures represent the total group of doctoral students in chemistry (and probably resemble those in all fields within the university), as well as the smaller group of organic chemists. Of particular interest is the number of southern students, reflecting the geographical location of Baltimore, the wishes of the founder Johns Hopkins and of President Gilman, and the slow recovery of the impoverished institutions of the South after the Civil War.

This broadly national character of the Johns Hopkins graduate school contrasted sharply with that of several other university groups whose research workers were publishing papers in organic chemistry from 1881 to 1915.[23,24] Of 40 Yale Ph.D.s, 70 percent had entered from Yale's Sheffield School, and 62 percent of 26 Harvard Ph.D.s held baccalaureate degrees from Harvard College. Columbia University drew a more diverse student body, yet very few graduate students at any of these older universities were from the South. More like Johns Hopkins was the new University of Chicago, its research goals similar, its auspicious beginning in 1892 financed by Rockefeller money, and its student body cosmopolitan.[24,25] A study of 40 Chicago Ph.D.s in organic chemistry from 1894 to 1913 shows that they entered from 26 colleges with more entrants from the West than at Hopkins, but few from the South.[24]

Table 1. Undergraduate origins of Johns Hopkins Ph.D.s in chemistry, 1879–1913

Northeast:	
63 students	Gettysburg (8); Amherst (5); Lafayette, Lehigh, Westminster
28 colleges	(4 each); Dickinson, Franklin and Marshall, Haverford, Williams (3 each); Bowdoin, Brown, Colby, MIT, Rochester, Swarthmore, Yale (2 each); 12 with 1 each.
Midwest and West:	
33 students	Beloit (4); Oberlin (3); California, Illinois Wesleyan, Kansas,
23 colleges	Knox, Western Reserve (2) each; 16 with 1 each.
South:	
95 students (including	Davidson (6), Richmond (5); Emory and Henry, University
35 Hopkins graduates)	of Georgia, Hampden-Sydney, Kentucky State, Randolph-
31 colleges	Macon, St. Johns, Wake Forest (3 each); Alabama, Charleston, George Washington, Mercer, Newberry, North Carolina College of Agriculture, Wofford (2 each); 14 with 1 each.
Foreign:	9 students; 5 colleges
Total:	200 students (2 unknown); 87 colleges

Upon leaving Hopkins, by far the largest number of newly fledged Ph.D.s in organic chemistry, 97 (75 percent), went directly into college teaching, and almost 80 percent of these (60 percent of the total) made this their life work. Inspired by Remsen's example and fortified by his farewell instructions in the art of lecturing,[5] they fanned out into colleges and universities across the country.[26] In 48 private and 17 state institutions, Hopkins chemists rose to senior appointments of long tenure, many as full professors and department chairmen, and a number entered administration, including 9 deans and 3 presidents. Only 12 returned to long careers at their undergraduate colleges, including the Hopkins chemistry faculty mentioned above, and Walter Jones and E. K. Marshall at the Hopkins Medical School. Secondary school teaching claimed 7.

Striking, but not surprising for the times when jobs were being created and facilities were often poor or lacking, was their mobility. Institutions who had Hopkins men for short periods, mostly in early appointments or in mid-career, totalled 60 private and 34 state colleges and universities. In some cases these were postdoctoral fellowships, others were positions ranging from assistant in several disciplines to president of the college. Most appointments were in small liberal arts colleges with more than one-third in institutions in the South; again Johns Hopkins differed from Harvard, Yale, and Columbia.

The new teacher often faced stiff challenges of limited staff and resources, not infrequently being the whole department or even more. Walter Jones, a "vigorous teacher," in his first job at Wittenberg College in 1891, "offered courses in chemistry, mineralogy, zoology and botany. It is possible that he may have offered a course in crystallography."[27] Chemists who started in organized and productive laboratories or who were given the support needed for good teaching and for developing a department of chemistry were fortunate.

The pressure of increased enrollments in advanced studies[28] (the number of United States undergraduates in all fields rose from 231,800 in 1900 to 346,100 in 1910, and of graduate students from 5,800 to 9,200), the rise of the land-grant institutions, a growing chemical industry, enlarging governmental departments, increased mobility and communication, and the rapid increase in knowledge in the physical sciences led to increasing facilities and demand for training scientists. W. A. Noyes wrote in 1908 from Illinois:[29] "Thirty years ago, very few educational institutions could have been found which had more than three or four chemists on their staff. In the institution with which I am connected, the staff includes more than thirty chemists who are engaged in teaching or research, and I do not think that the institution is unusual in this regard."

Hopkins graduates who developed independent research groups, trained doctoral candidates, and were among the twenty-three starred scientists in *American Men of Science*[24,30] from Remsen's laboratories (with date of Ph.D. from Hopkins in parentheses), included W. A. Noyes (1882), Rose Polytech and Illinois; W. R. Orndorff (1887), Cornell; J. H. Kastle (1888), Kentucky State University; E. P. Kohler (1892), Bryn Mawr and Harvard; E. C. Franklin (1894), Kansas and Stanford; J. F. Norris (1895), MIT; and E. E. Reid (1898), Johns Hopkins; all in organic chemistry. Starred in other fields were Walter Jones (1891), physiological chemistry, and E. K. Marshall (1911), pharmacology, both at the prestigious, new Hopkins Medical School; H. C. Jones (1892), physical chemistry at Hopkins; and Henry Fay (1895), metallography at MIT.[31] All of these were Remsen's men except Marshall, who had worked with Acree, and six were elected to the National Academy of Sciences: Noyes, H. C. Jones, Kohler, Norris, Walter Jones, and Marshall.

Hopkins men, marked by their effective undergraduate teaching and contributions to the development of smaller departments and cited for producing future Ph.D.s,[32] included W. M. Blanchard (1899), DePauw; H. W. Doughty (1904), Amherst; A. M. Patterson (1900), Antioch; J. A. Lyman (1892) and C. J. Robinson (1906), Pomona; and H. N. Holmes (1907), Oberlin. Remembered and honored by their students and contemporaries as influential teachers were E. Hart (1879), Lafayette (Remsen's first Ph.D.);[33] H. M. Ullmann (1892), Lehigh;[34] G. Alleman (1897), Swarthmore;[32] C. E. Caspari (1900), St. Louis College of Pharmacy;[35] C. E. Coates (1891), dean of the Audubon Sugar School of Louisiana State University;[36] and J. N. Swan (1893), Monmouth and the University of Mississippi, who designed many laboratories in various sections of the country.[37] These men also maintained research interests in a variety of fields.

Only two women received Ph.D.s from Hopkins in organic chemistry in this period: Julia P. Harrison (1912) and Bessie Brown (1913). Brown returned to Simmons, her college, where she later worked with J. F. Norris who was planning the scientific departments there, and Harrison taught at Wilson College.[12]

A relatively small number of Hopkins Ph.D.s, 30 (23 percent), ended up in industry or business. The great market for college faculty and the emphasis on teaching careers at Hopkins coupled with Remsen's lack of interest in industrial contacts appeared to outweigh opportunities in the applied fields. Of the 30, about half had academic experience also; only 10 are known to have entered industry directly upon leaving Hopkins, and more than half of these entered after 1900.

Among the industrialists who entered their firms directly were W. M. Burton (1889), who became president of Standard Oil of Indiana in 1918; A. R. L. Dohme (1889), president of Sharp and Dohme, a pharmaceutical house; and H. Bradshaw (1905), who advanced to asssistant director of the Chemical Department at Du Pont in 1925. C. H. Herty (1890), in a lifetime devoted actively to teaching, editing, research, and the promotion of developing chemical industry in the United States particularly after World War I, made important contributions to the utilization of the southern pine forests for naval stores and pulpwood.[38]

Other industrial positions were scattered in consulting, minerals and metallurgy, the expanding food industry, ceramics, retail drugs, and explosives, and five were in non-chemical business. The connections with Standard Oil of Indiana provided a number of jobs for Hopkins men. Only these few, plus H. W. Hillyer (1885) and W. J. Karslake (1895), who transferred into National Aniline and Chemical Company after World War I, hint at the stream of future Ph.D.s into the vast synthetic organic chemical industry about to materialize in the nation.

Government laboratories offered job opportunities to 19 (15 percent) of the graduates, a few entering directly and some after teaching or industrial work. These were the formative years of the Departments of Agriculture and the Interior and for the development of their scientific laboratories; they also coincided with increased federal, state, and municipal involvement in public health.[39] By 1916 the total number of chemists employed by the United States Government was 716, of which 397 were in the USDA and 94 in Interior, the others were mainly in Commerce and Navy.[40]

Within the USDA the Washington Bureaus, as well as the federally supported state experiment stations associated with land-grant institutions since 1875, provided opportunities for careers in research, teaching, and publication. In particular many young graduates looked upon the stations as valuable training grounds for postdoctoral experience. G. S. Fraps (1899), associated with the Texas station for forty years, published extensively in agricultural science;[41] W. W. Garner (1900), starred in plant physiology in *American Men of Science*, rose to high rank in the Bureau of Plant Industry in tobacco chemistry; F. K. Cameron (1894) was chief of the Bureau of Soil Chemistry in Washington for sixteen years,[42] and F. C. Blanck (1907) did research in soil chemistry and with regulations for pure food laws.

The powerful U. S. Geological Survey established within the Department of the Interior in 1879 drew four Hopkins chemists. C. W. Hayes (1887) made the survey his career, being chief from 1907 to 1911; his publications dealt with the geology of the southern Allegheny Mountains and coal measures. D. T. Day (1884), with the Survey and the Bureau of Mines from 1886 to 1920, contributed to the rapid development of the mineral industry, especially in the establishment and study of western naval petroleum and oil-shale reserves.[43] C. S. Palmer (1882) worked as chief chemist with Hayes, and H. N. Stokes (1884) worked in the Survey and in the Bureau of Standards for many years along with C. E. Waters (1889) and with W. A. Noyes, who was there for a brief period.

Information about Yale, Harvard, Columbia, and Chicago Ph.D.s in organic chemistry[23,24,25] shows definite differences from the Hopkins pattern for this period. Of 40 Yale Ph.D.s, 13 taught almost exclusively (many at Yale), 15 went into industry, and 8 worked mainly in government laboratories. The remainder worked in two or more of these categories. Of 26 Harvard Ph.D.s 14 taught permanently, 7 worked mainly in industry, none went into government laboratories, and the balance had mixed careers. Of 20 Columbia doctorates from 1892 to 1912, 9 taught, 7 went into industry, and the few others were in government laboratories, or in the Rockefeller and Mellon Institutes. Of 40 Chicago Ph.D.s, 30 started in teaching and 25 of these remained, mainly in the Midwest; 9 went into industry, none into government research, and careers of the remaining graduates included scientific writing (E. E. Slosson) and research in the Mellon and Carnegie Institutes. Very few graduates of these four schools were employed in the South.

The myriad useful activities that make up the life of a professional chemist also claimed the energies of the Hopkins men. Among them were editors of scientific journals, Hart and Noyes of the JACS; Noyes and Patterson of *Chemical Abstracts*; Marshall of the *Journal of Pharmacology and Experimental Therapeutics*; and Herty of the *Journal of Industrial and Engineering Chemistry*. Many wrote textbooks and monographs and served as editorial specialists in scientific publications, for example, Noyes for the *American Chemical Society Monographs* and *Chemical Reviews*.

Active in numerous scientific societies (five took office as president of the American Chemical Society), vigorous in local sections, recipients of awards and honors in their fields, aiding the government in its war efforts in two World Wars, knowledgeable in many domains of consulting, the early Hopkins chemists extended their personal influence throughout the country for over seventy years.

As Ira Remsen said on receiving the Willard Gibbs medal in Chicago in 1914 (referring to the years at Hopkins):[4]

> Whatever the influences may have been that led to increased activity in chemical research in this country, it is certain that the increase was very marked soon after the time of which I am now speaking. This was due, I think, largely to the fact that such excellent opportunities were given the little band of workers at Johns Hopkins. This led to similar opportunities being given to the workers in other institutions, and many of these profited greatly in consequence. It was not long before there were a number of centers of activity in America, and the number devoting themselves to research work has increased astonishingly...
> Chemical research is in a healthy condition in our country and the signs of the future growth are most promising.

LITERATURE CITED

1. This account is based on four references: Hugh Hawkins, *Pioneer: A History of the Johns Hopkins University, 1874-1889*, Ithaca, 1960; D. C. Gilman, *The Launching of a University*, New York, 1906; Fabian Franklin, *The Life of Daniel Coit Gilman*, New York, 1910; J. C. French, *A History of the University Founded by Johns Hopkins*, Baltimore, 1946.
2. On Remsen, W. A. Noyes and J. F. Norris, *Biog. Mem. Nat. Acad. Scis.*, **14**, 205

(1932); Hawkins, passim; O. Hannaway, Ambix, **23**, 146 (1976); F. A. Getman, *The Life of Ira Remsen*, Easton, 1940; A. H. Corwin, *Welch Foundation Conferences, XX. American Chemistry - Bicentennial*, Houston, 1976, pp. 46 ff.; E. Emmett Reid (one of Remsen's students), *My First Hundred Years*, New York, 1972; D. S. Tarbell and A. T. Tarbell, "Remsen Revisited," paper presented at the Chicago meeting of the American Chemical Society, September 1975.

3. I. Remsen, *Nat. Acad. Scis., XXI, Eleventh Memoir,* on H. N. Morse (1923).
4. I. Remsen, JACS, **37**, 1 (1915).
5. Noyes and Norris, p. 242.
6. Reid, pp. 62-65; a description of the new Dalton Hall, almost certainly written by Remsen, is in Science, 5, 25 (1885).
7. H. W. Bragdon, *Woodrow Wilson: The Academic Years*, Cambridge, 1967, pp. 101-123; B. J. Hendrick, *The Training of an American: The Earlier Life and Letters of Walter H. Page*, Boston, 1928, pp. 68-90.
8. F. Cordasco, *Daniel Coit Gilman and the Protean Ph.D.* , Leiden, 1960, pp. 87-90.
9. R. W. Helmkamp and W. R. Line, Chem. Eng. News, **22**, 1576 (1944).
10. W. M. Clark, *Biog. Mem. Nat. Acad. Scis.*, **20**, 87 (1939).
11. On W. A. Noyes, Roger Adams, *Biog. Mem. Nat. Acad. Scis.*, **27**, 179 (1952); J. W. Mallett, ACJ, **4**, 355 (1883) reports many water analyses run by Noyes as a graduate student.
12. *Johns Hopkins Half Century Directory*, compiled by W. Norman Brown, Baltimore, 1926.
13. Hawkins, pp. 27-28.
14. I. Remsen, *Organic Chemistry*, Boston, 1909, Fifth revision; 1st ed., 1885.
15. Noyes and Norris, p. 244.
16. H. B. Vickery, *Biog. Mem. Nat. Acad. Scis.*, **39**, 4 (1967).
17. I. Remsen, JACS, **33**, Proceedings, p. 46 (1911).
18. I. Remsen and C. Fahlberg, ACJ, **1**, 426 (1879); D. S. Tarbell and A. T. Tarbell, JCE, **55**, 161 (1978).
19. I. Remsen and others, ACJ, **14**, 81-185 (1892).
20. Cordasco, p. 111.
21. Biographical data for the Hopkins men through 1926 are from Ref. 12, supplemented by additional material from *American Men of Science*, 1st ed. (1906) through 10th ed. (1960). A single date after the chemist's name indicates the year in which he received the Ph.D. from Hopkins.
22. Hawkins, pp. 68 and 270.
23. Bulletin of Yale University, Obituary Record of Yale Graduates, 1919-1920, p. 1587; 1921-1922 p. 523; 1939-1940, p. 185.
24. *American Men of Science*, eds. one (1906) through ten (1960).
25. Frederick Rudolph, *The American College and University*, New York, 1962, pp. 336-338; 349-351; R. J. Storr, *Harper's University: the Beginnings: A History of the University of Chicago*, Chicago, 1966.
26. D. S. Tarbell, A. T. Tarbell, and R. M. Joyce, Isis, **71**, 620 (1980), give a detailed comparison of the academic origins and career patterns of the Ph.D.s of Remsen and Roger Adams. Table 1 is used with the permission of *Isis*.
27. W. M. Clark, *Biog. Mem. Nat. Acad. Scis.*, **20**, 88 (1939).
28. W. T. Furniss, Ed., *American Universities and Colleges*, American Council on Education, Washington, 11th ed., 1973, p. 19.
29. W. A. Noyes, Science, **27**, 877 (1908).
30. These figures were arrived at independently from those of S. S. Visher, *Scientists Starred 1903-1943, in "American Men of Science,"* Baltimore, 1947.
31. W. T. Hall, Ing. Eng. Chem. News Ed., **13**, 95 (1935).
32. R. H. Knapp and H. B. Goodrich, *Origins of American Scientists*, Chicago, 1952, passim.

33. Anon., Ind. Eng. Chem. News Ed., **9**, 196(1931).
34. A. E. Buchanan, Jr., ibid., **14**, 287 (1936).
35. F. Sultan, ibid., **18**, 623 (1940).
36. W. L. Owen, J. Ind. Eng. Chem., **23**, 339 (1931).
37. A. Hume, Ind. Eng. Chem. News Ed., **13**, 49 (1935).
38. F. K. Cameron, JACS., **61**, 1619 (1939).
39. A. H. Dupree, *Science in the Federal Government*, Cambridge, 1957, especially chapters 8, 10, and 15.
40. F. E. Breithut, J. Ind. Eng. Chem., **9**, 66 (1917); a very comprehensive article.
41. J. F. Fudge, J. Assoc. Offic. Agri. Chemists, **39**, No. 1, v (1956).
42. W. D. Bancroft, Ind. Eng. Chem. News Ed., **13**, 425 (1935).
43. M. R. Campbell, Trans. Am. Inst. Mining Met. Eng., **71**, 1371 (1925).

ARTHUR MICHAEL

4

Structural Organic Chemistry, 1875-1913

This period of thirty-eight years, dominated at the beginning by the Hopkins school, showed a general growth in organic research in educational, governmental, and industrial laboratories. The range of problems studied was broad, including structural and synthetic work and a surprising amount of what would be later termed "physical organic chemistry". The first kinetic study of an organic reaction in this country was published in 1881,[1] and the general field of reaction mechanisms became a lively one, especially after 1900. Among the biochemically oriented chemists, research and speculation on the mechanism of enzyme-catalyzed reactions stirred. Sterochemistry of organic reactions and the relation between configuration and optical rotation were actively pursued. Both undergraduate and graduate instruction in chemistry improved and spread with better textbooks, better trained teachers, and more adequate laboratories.

The publication of American research found suitable outlets in the newly established American journals. The ACJ contained most of the reports of the Hopkins laboratory and most of those from Remsen's students who had gone on to develop research programs of their own. The following principal centers of research published organic papers in the ACJ during its thirty-four years of existence (1879-1913): Hopkins (Remsen, S. F. Acree); Harvard (C. L. Jackson, H. B. Hill); Yale (W. G. Mixter, H. L. Wheeler, T. B. Johnson); Chicago (J. U. Nef, J. Stieglitz); Tufts (Arthur Michael); Cornell (W. R. Orndorff); Michigan (A. B. Prescott); MIT (J. F. Norris); Bryn Mawr (E. P. Kohler); Purdue (W. E. Stone, R. A. Worstall); Case (C. F. Mabery); and Kentucky (J. H. Kastle). Several dozen laboratories, including various government laboratories, were represented by smaller numbers of publications.

A few American chemists, particularly Michael, Nef, Stieglitz, and M. Gomberg, reported some of their results in German journals. After 1900 many published mainly in the resurgent JACS and some in the JBC. The latter group included the Yale chemists H. L. Wheeler and T. B. Johnson, P. A. Levene of the Rockefeller Institute, and H. D. Dakin of the Herter Laboratory.

The principal institutions represented in the JACS from 1879 to 1913 were Columbia (M. T. Bogert); Michigan (A. B. Prescott, M. Gomberg, W. J. Hale); Illinois (W. A. Noyes, R. S. Curtiss, C. G. Derick); Harvard (H. E. Torrey, Latham Clarke); McMaster in Canada, (J. B. Tingle); Chicago (J. Stieglitz); Stevens Institute (A. R. Leeds, Henry Norton); Cincinnati (H. S. Frye, T. H. Norton); Ohio State (W. L. Evans, W. McPherson); College of the City of New York (L. H. Friedburg); James Stebbins, Jr., (consultant in New York City); and a large number of laboratories with less than ten publications or with no address listed.

The best of this work compared favorably with the good work being done in Germany, but the number of American chemists of this high caliber was small. Most of the research, however, was competent and reasonably professional in ideas and execution and on a par with that from a provincial Germany university. Let us focus on American findings in structural chemistry.

Without question, the single most important piece of work from the United States in this period and the one which aroused the most interest abroad was the discovery by Moses Gomberg (1866-1947) of structures which contained trivalent carbon. Gomberg was an example of the best kind of American success story. His fascinating reminiscences given at the Organic Symposium in Ann Arbor in December 1941 unfortunately were not preserved.[2] Coming to this country from Russia as a child with his penniless father, he eventually worked his way through the University of Michigan at Ann Arbor and received a doctorate in chemistry there in 1894.

During a year of study with Victor Meyer at Heidelberg in 1897, he synthesized tetraphenylmethane[3] 1, and after his return to Michigan, he studied the preparation of the next analog, hexaphenylethane 2, by the action of various metal salts and metals on triphenylmethyl chloride (trityl chloride) 3.

$$Ar_4C \qquad Ar_3CCAr_3 \xleftarrow{\quad\not\quad Ag\quad} Ar_3CCl \qquad Ar_3C\text{-}O\text{-}O\text{-}CAr_3$$

$$\underline{1} \qquad\qquad \underline{2} \qquad\qquad\qquad \underline{3} \qquad\qquad \underline{4}$$

$$Ar = C_6H_5$$

To his surprise, he obtained, instead of the expected hydrocarbon 2, a compound whose composition and chemical properties[4] corresponded to the peroxide 4. He found that the hydrocarbon first formed was exceedingly reactive, taking up oxygen on exposure to air within seconds and reacting very rapidly with iodine to give $(C_6H_5)_3CI$. He suggested in his first paper that the initial, highly reactive substance contained trivalent carbon. He first wrote its structure as the *carbonium ion* 5 rather than the *free radical* 6; at that time the electronic theory of valence had not developed enough to allow a clear distinction between 5 and 6. We would now write 5 with six electrons on the central carbon, and 6 with seven, one of them unpaired (dot in 6), hence it is an "odd molecule" or free radical, the triphenylmethyl radical.

$$Ar_3C^+ \qquad\qquad Ar_3C\bullet$$

$$\underline{5} \qquad\qquad\qquad \underline{6}$$

Gomberg's interpretation of his results ran counter to classical structural theory, which held that carbon was always tetravalent except for a few special instances of bivalence noted before, carbon monoxide and the isocyanides. Naturally this revolutionary claim of departure from classical theory was attacked by many conventionally minded chemists in both America and Europe. Although Gomberg was by nature modest and unassuming, he had confidence in his extraordinarily elegant experimental technique, and he defended stoutly and eventually successfully his discovery of a trivalent carbon compound. Using an oxygen-free system,[5] he isolated the first product from the reaction of silver on trityl chloride and showed by analysis that it had the composition corresponding to hexaphenylethane **2**; this material rapidly took up the calculated amount of oxygen to form the peroxide **4**.

Gomberg's discovery and his successful defense of his interpretation freed the theory of organic chemistry from the dogma of tetravalent carbon. His work opened the door to the concept, developed in detail in the 1920s, that many important organic reactions proceed by way of free-radical, or trivalent carbon, intermediates. Gomberg's work was thus the forerunner of the fundamental studies of free radicals by H. Staudinger in Germany and by M. S. Kharasch at Chicago. It is pleasant to record that while Gomberg was involved in controversy over his trivalent carbon, the grand old man of German organic chemistry, Adolph Baeyer, referred[6] to "der wundervollen Arbeit Gomberg's" ("Gomberg's wonderful work").

Treatment of the supposed hexaphenylethane with acid gave a stable hydrocarbon, which did not yield triphenylmethyl or its peroxide. The structure of this was shown[7] to be **7**, isomeric with **2**; however, the structure of the initial product was not demonstrated by Gomberg or his contemporaries to be hexaphenylethane. Only many years later, application of nuclear magnetic resonance spectroscopy showed[8] that the supposed hexaphenylethane actually had structure **8**, which is the source of the triphenylmethyl radicals. So far, hexaphenylethane has not been synthesized. This in no way detracts from the significance of Gomberg's discovery of the free radical.

Two sidelights on Gomberg's work should be mentioned. During his studies he showed that trityl chloride was a conductor of electricity in liquid sulfur dioxide and benzonitrile solutions, undoubtedly owing to formation of the carbonium ion **5** and the chloride ion. The secondary literature usually credits the German-Baltic chemist Paul Walden[9] with the discovery of electrical conductivity of trityl chloride solutions, but the original papers show that Gomberg was at least an independent codiscoverer.

In 1908, E. P. Kohler at Bryn Mawr prepared a bromotriphenylindene which was converted by metals to a material **9** that behaved like Gomberg's triphenylmethyl and was undoubtedly also a free radical. Kohler did not pursue this finding, apparently out of deference to Gomberg's priority in this field.[10]

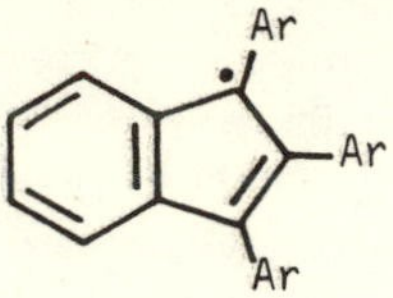

9

Both here and overseas the field of triarylmethyl radicals continued active, aided by newer physical methods: first magnetic susceptibility and later, much more conclusively, electron spin resonance measurements for qualitative and quantitative determination. Gomberg's elegant and imaginative work has not lost its significance in current chemical thought.

Other examples of free radicals turned up after Gomberg's discovery, even though the investigators did not always recognize their structures in that early period. For example, H. D. Dakin in 1907 observed some reactions which undoubtedly involved free hydroxyl radical intermediates: the hydroxylation of benzoic acid by hydrogen peroxide to give all three isomers, and the degradation of aliphatic acids to carbonyl compounds.[11,12] Dakin's work was done to provide model *in vitro* reactions for biochemical oxidation processes and was a forerunner of extensive free-radical work in later years, particularly free-radical attacks on aromatic rings.

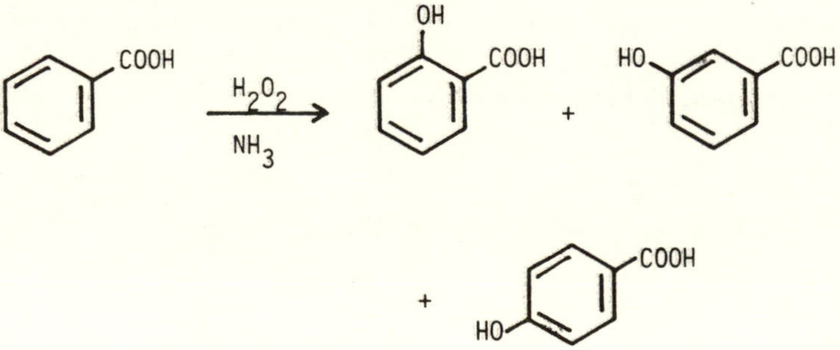

$$RCHCOOH \xrightarrow[Fe^{++}]{H_2O_2} RC=O + CO_2 + NH_3$$

with NH_2 on the left reactant and H on the product.

$$C_6H_5CH_2CH_2COOH \xrightarrow{H_2O_2} C_6H_5CCH_3 + C_6H_5CHO + HOC_6H_4CH_2CH_2COOH$$

Although Arthur Michael (1853-1942) did not do any single piece of work as fundamental as Gomberg's free-radical studies, he was for many years the most versatile and productive of American organic chemists, with unrivaled scope and a highly individual view of the theoretical basis of organic chemistry.[13] The son of a prosperous family of Buffalo, New York, he was educated in Germany, Russia, and France and had no college or university degrees except for honorary doctorates. His chemical training abroad resulted in over sixteen papers from Germany starting[14] in 1876, some with S. Gabriel of the University of Berlin faculty and some with other students. His scientific publications numbered 225 (nearly 300 counting papers in both English and German), appeared over a period of nearly seven decades, and were published from Tufts College, Harvard, or private laboratories which he maintained on the Isle of Wight in the 1890s and at Newton Center, Massachusetts, after 1912.

Michael generalized very greatly the reaction known by his name (actually first discovered by Claisen[15]), the addition to a conjugated system of a negative ion or some other structure seeking a positive center (a nucleophile, in present terminology). Michael's first example appeared in the ACJ in 1887[16]; it involved the addition of a sodium derivative of malonic ester **10** to the conjugated system in ethyl cinnamate **11**. The initial product **12**, a sodium salt with a delocalized negative charge, results from 1,4-addition to the conjugated system in **11**. After acidification the product is **13**, a saturated triester. In addition to its generality and synthetic utility, the Michael reaction contributed to knowledge of the behavior of conjugated systems.

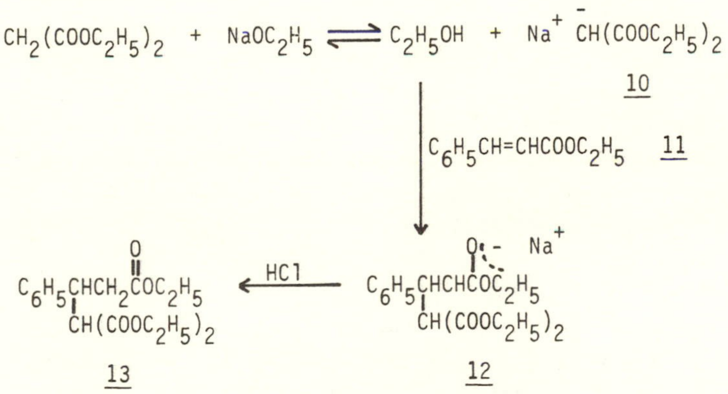

$$CH_2(COOC_2H_5)_2 \; + \; NaOC_2H_5 \rightleftharpoons C_2H_5OH \; + \; Na^+ \; {}^-CH(COOC_2H_5)_2$$

10

$$C_6H_5CH=CHCOOC_2H_5 \quad \underline{11}$$

$$\underset{\underset{\underline{13}}{CH(COOC_2H_5)_2}}{C_6H_5\overset{\overset{O}{\|}}{C}HCH_2COC_2H_5} \quad \xleftarrow{\text{HCl}} \quad \underset{\underset{\underline{12}}{CH(COOC_2H_5)_2}}{C_6H_5CHCHCOC_2H_5} \; O^{\cdot -} \; Na^+$$

Another example of his prolific and useful synthetic work was his synthesis of salicin, a bitter compound of some medicinal interest isolated from willow. It is a member of an important class called glycosides which contain an organic group attached to a sugar, glucose in this case. He developed a general method for preparing phenyl glycosides, and salicin was the first total synthesis of a naturally occurring glycoside.[17]

However, most of Michael's experimental work was directed to the support of his theoretical ideas. He developed a comprehensive system for classifying and rationalizing organic reactions, particularly addition and elimination, categories that Michael thought embraced all or nearly all organic reactions. His theoretical scheme started with the second law of thermodynamics, developed by 19th century physics, which states that the *entropy* of the university increases toward a maximum.[18,19] The notion of entropy, that the changes in the universe were irreversible and all energy would eventually reach an unusable form, was one with far-reaching effects and implications for all thinkers, scientists and non-scientists alike.[20]

Michael stated that reactions occurred in such a manner as to maximize the entropy change, or synonymously, to lead to the maximum neutralization of chemical free energy, or of chemical potential, as he later expressed it, doubtless influenced by Gibbs' nomenclature. In some places,[21] Michael equated the amount of heat evolved during a reaction with the entropy increase involved; at other times he appeared to think there was no direct relationship.[18] Michael's terms "chemical free energy" and "entropy" do not correspond to the later definitions of these terms in systematic thermodynamics.[22] His theory can be criticized with reference to his own period as being a tautology: reactions occur which have a tendency to occur.

Michael's "law of entropy" is further subject to a fundamental criticism, in that many reactions do *not* lead initially to the most stable products (based on Michael's criteria or later more exact thermodynamic criteria). Instead, less stable products may be obtained initially because of a higher rate of formation (reaction under *kinetic control*). The thermodynamically more stable products may form only much more slowly. This dichotomy between kinetics and thermodynamics, between rate and equilibrium, runs through the whole study of organic reactions, and the recognition of the role of rate and equilibrium in any reaction requires a more extensive knowledge of the course (*mechanism*) of the reaction than was available to Michael.[23] Indeed, from a rigorous thermodynamic view, the *most stable* products from most organic compounds reacting in the presence of oxygen are carbon dioxide and water. This simply means that complete combustion has occurred, an undesirable fate for an organic reaction.

Michael's chemical intuition was strong enough to escape occasionally from the bounds of his formalistic system. In some cases, he seemed to recognize that the organic products formed initially are *not* always the most stable ones, but due to "intramolecular resistance of some sort" require additional energy to form the more stable products.[24] From the beginning Michael utilized Kekulé's suggestion[25] that reactions occurred via a combination of reactants called a "polymolecule" by Michael, which broke up to give products. The polymolecule may be considered the precursor of the *transition state* of later years.

Michael's "partition theory" explained the formation of mixtures[26] as due to the ratios of the respective affinities of the participants in the reaction to form the respective products. This amounted to saying that proportions of products in competing reactions were controlled by the relative equilibrium constants, or rate constants, or a combination of the two.[27]

A suggestion of van't Hoff led Michael to the important concept of the chemical plasticity of carbon,[28] by which he meant that the position of carbon in the center of the periodic system allowed it to assume either a more positive or negative character, depending on its environment. As Michael knew, the hydrogens on a carbon adjacent to one or more unsaturated groups, such as the CH in $CH_3CH(COOCH_3)_2$, are acidic. With base such compounds form salts in which the negative charge resides on the carbon structure. However, hydrogens on a carbon farther removed from the unsaturated group, such as the CH_3 in $CH_3CH(COOCH_3)_2$, show no acidic properties. The ability of carbon to become relatively positive by repelling negative charge was deduced by Michael from the greater dissociation constant of formic acid, $HCOOH$, compared to its higher homologues. He recognized that formic acid is a stronger acid than acetic acid, CH_3COOH, because the CH_3 compared to H repels negative charge to a greater extent. Hence the acidic hydrogen in formic acid, being less tightly held, is more easily lost. (Gomberg's salts, $(C_6H_5)_3C^+Cl^-$, discovered shortly after Michael's paper[28] appeared, are the best example of positive carbon.) It is notable that Michael's voluminous writings rarely contain specific indication of charged ions or separation of charges in non-ionic compounds.

With his recognition of the ambivalent nature of carbon, Michael was thus able to give an explanation of the mode of addition of H^+X^- to unsymmetrically substituted olefins (the Markovnikov Rule).[29]

$$\overset{1}{C}H_2=\overset{2}{C}(CH_3)_2 \quad + \quad H^+X^- \quad \longrightarrow \quad \underset{H \quad X}{CH_2C(CH_3)_2}$$

In modern terms, the CH_3 groups tend to repel electrons from C-2 to C-1, and hence C-1 is negative relative to C-2, and the H^+ adds to C-1. The fact that a knowledge of reaction mechanism is a necessary prerequisite to theoretical speculation is emphasized by the discovery by Morris Kharasch in 1930 that addition reactions going via a free-radical intermediate give the other isomer from the one predicted by Michael. This finding of Kharasch was considered by Michael and his research assistants in his last papers, but he was unable to make any adjustments of his ideas.[30]

Michael developed a numerical scheme for predicting the influence of atoms on each other as the number of intervening atoms increased, to which he seemed to attribute *a priori* character.[31] He attempted to apply it to such subtle problems as rationalizing the distribution of isomers obtained by addition of water to 2-pentene, and the chlorination of saturated hydrocarbons.[32] Problems of this type were not solvable with any quantitative accuracy, either theoretically or experimentally, in Michael's time. Later work showed that his numerical order of interatomic influence was not generally valid.

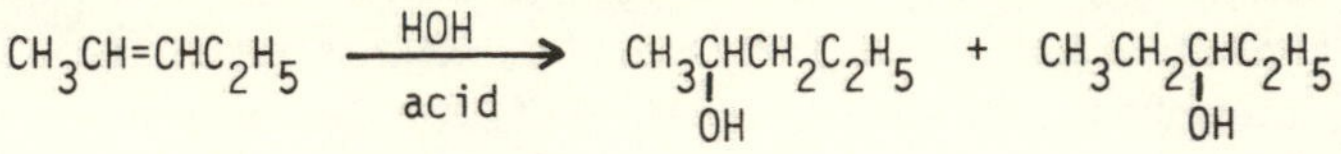

$$CH_3CH=CHC_2H_5 \xrightarrow[\text{acid}]{\text{HOH}} CH_3\underset{OH}{CH}CH_2C_2H_5 \;+\; CH_3CH_2\underset{OH}{CH}C_2H_5$$

Michael, like many chemists of his time, was fascinated by the problem of *tautomerism*: the reversible interconversion of two or more compounds differing only in the position of a hydrogen atom and of a double bond. The classical example of tautomerism is the structure of ethyl acetoacetate, first prepared in 1863; this substance gives some reactions characteristic of structure **14**, the *keto* form (from the keto group, C=O), and some of the *enol* form, **15**. A similar case is that of the amides and isomides **16** and **17**; this is of particular importance because the amide or peptide linkage, -C(=O)NH-, is characteristic of polypeptides, proteins, and enzymes. These structures **14-17** obey the valence rules and represent definite compounds which can be isolated pure in some cases. The same is true of other tautomeric compounds; many examples include cyclic compounds, such as the purine and pyrimidine bases which occur in coenzymes and nucleic acids.

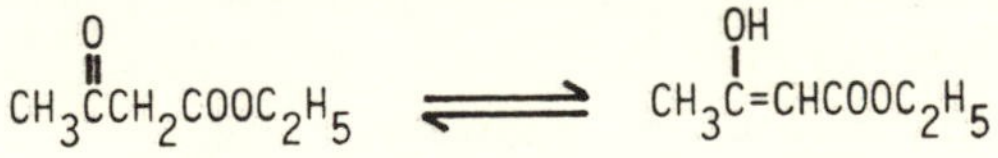

$$CH_3\overset{O}{\overset{\|}{C}}CH_2COOC_2H_5 \;\rightleftharpoons\; CH_3\overset{OH}{\overset{|}{C}}=CHCOOC_2H_5$$

$$\underline{14} \qquad\qquad\qquad \underline{15}$$

$$CH_3\overset{O}{\overset{\|}{C}}NHCH_3 \;\rightleftharpoons\; CH_3\overset{OH}{\overset{|}{C}}=NCH_3$$

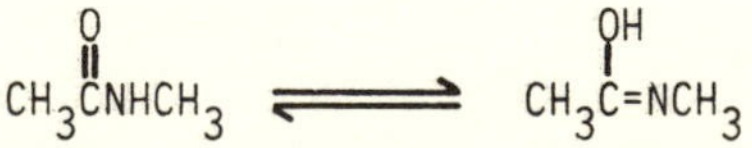

$$\underline{16} \qquad\qquad\qquad \underline{17}$$

To many workers of Michael's time in this country and abroad, however, compounds capable of tautomerism represented an exception to the postulate of the structural theory that each compound corresponded to a single definite written structure. Tautomeric compounds were regarded, therefore, with distinct uneasiness, and their structures were the subject of protracted polemics. What was not recognized was that the interconversion of tautomers (**14** to **15** and vice-versa) was usually very rapid, more rapid than reactions which might be applied to the system to attempt to establish the structure. Furthermore, compounds like **14** and **15** readily formed salts, called enolates, and these salts reacted with methyl iodide, CH_3I, to form derivatives methylated on carbon.

A fundamental difficulty underlies attempts to determine by chemical reactions the structure of compounds in rapid equilibrium with each other.

Treatment of **14**, **15**, or (actually) a mixture of the two, with some reagent *A* which can yield compound *X* from **14** or *Y* from **15**, will *only* yield valid structural information about **14** or **15** if the rate of reaction of *A* with **14** to give *X*, for example, is much greater than the rates of interconversion of **14** and **15**. Because the rates of interconversion of **14** and **15**, as well as the rate of reaction with *A* to form *X* are rarely known (and were not known at all at the time we are considering), attempts to determine the structure of tautomers by reaction were not conclusive. Wheeler and Johnson in 1899 seem to have been the first to recognize this, but S. F. Acree gave a more searching analysis of a more complicated case in 1907.[33] The question of determining the structure of tautomers by reactions thus was not meaningless, but rather insoluble without additional information and experimental techniques.

In considering the problem of tautomerism, Michael perceived that two or more electron-attracting groups, such as $C=O$ and $COOC_2H_5$ in **14**, made the hydrogen on the CH_2 group acidic enough to be removed by bases[34] to form the enolate **15a** or **15b**, which he always wrote with the sodium attached to oxygen. This followed his rule of the "maximum degradation of free energy" in reactions; oxygen is clearly the most negative center, as sodium is the most positive center in the structure **15a**. (The structure **15b** with a delocalized negative charge is the modern picture of the enolate.)

$$CH_3\overset{O}{\overset{\|}{C}}CH_2COOC_2H_5 \rightleftharpoons CH_3\overset{OH}{\overset{|}{C}}=CHCOOC_2H_5$$

<u>14</u>   <u>15</u>

$$\Updownarrow NaOC_2H_5$$

$$CH_3\overset{ONa}{\overset{|}{C}}=CHCOOC_2H_5 \quad\text{or}\quad CH_3\overset{O^-\,Na^+}{\overset{|}{C}}=CHCOOC_2H_5$$

<u>15a</u>   <u>15b</u>

$$\downarrow CH_3I$$

$$\left[CH_3\overset{I\ \ CH_3}{\underset{ONa}{\overset{|\ \ |}{C-CHCOOC_2H_5}}} \right] \xrightarrow{-NaI} CH_3\overset{O\ \ CH_3}{\overset{\|\ \ |}{C-CHCOOC_2H_5}}$$

<u>18</u>   <u>19</u>

Michael explained the experimental fact that action of methyl iodide on **15a** gave carbon-methylation instead of oxygen-methylation[34] by assuming an intermediate addition of methyl iodide to the carbon-carbon double bond to give **18**. The product of this reaction, for which no real analogy exists (the ones quoted by Michael were irrelevant), immediately lost the "well-neutralized" sodium iodide and formed the C-methyl compound **19**. A more reasonable view of the mechanism was given by Nef[35] in 1907; this allowed a rapid shift in charge from oxygen to carbon followed by reaction of methyl iodide with the atom (carbon or oxygen) carrying the negative charge. Although Michael's attitude toward the existence of a mobile equilibrium between the tautomers **14** and **15** varied during his long career, his insistence that sodium was bonded to oxygen in enolates never changed. Although he contributed much good experimental work on this problem, his theoretical views were not useful, and the later development of the quantum mechanical theory of bonding around 1930 rendered the question of the "location" of the negative charge in **15a** or **15b** meaningless, because the negative charge is smeared out (delocalized) over the unsaturated system.

Probably Michael's most important contributions were those on the spatial course of addition to carbon-carbon double bonds. As we noted earlier, the van't Hoff-Wislicenus stereochemistry[36,37] postulated enough hindrance to rotation around a carbon-carbon double bond to make *cis-trans* stereoisomers of the maleic **20** (*cis*) and fumaric **21** (*trans*) types stable, although **20** could be converted to **21** by relatively mild treatment.

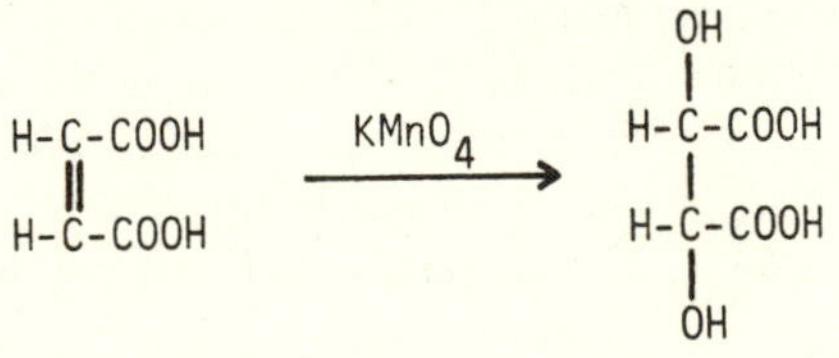

$$
\begin{array}{ccc}
\text{H-C-COOH} & & \text{OH} \\
\| & \xrightarrow{\text{KMnO}_4} & | \\
\text{H-C-COOH} & & \text{H-C-COOH} \\
& & | \\
& & \text{H-C-COOH} \\
& & | \\
& & \text{OH}
\end{array}
$$

$\underline{20 \; cis}$ $\qquad\qquad\qquad\qquad$ $\underline{22}$

$$
\begin{array}{ccc}
\text{HOOC-C-H} & & \text{OH} \\
\| & \xrightarrow{\text{KMnO}_4} & | \\
\text{H-C-COOH} & & \text{HOOC-C-H} \\
& & | \\
& & \text{H-C-COOH} \\
& & | \\
& & \text{OH}
\end{array}
$$

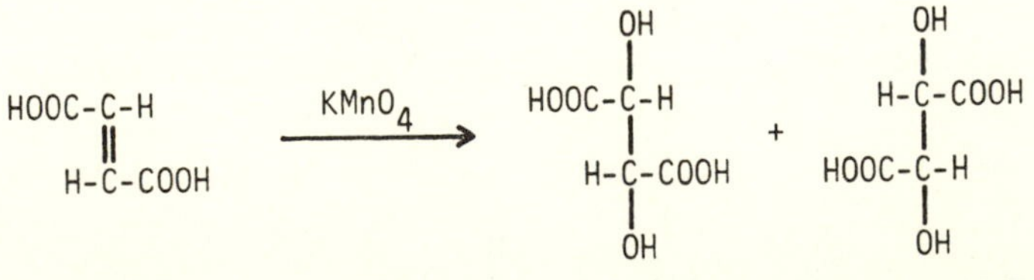

$\underline{21 \; trans}$ $\qquad\qquad\qquad\qquad$ $\underline{23}$ $\qquad$ mirror image

Kekulé showed that hydroxylation of maleic acid gave mesotartaric acid **22**, which cannot be obtained in optically active forms and therefore has a plane of symmetry. The configuration of maleic acid was known to be *cis* because it easily formed a cyclic anhydride; therefore the addition of two hydroxyls to form **22** had taken place by *cis* addition to the double bond. Fumaric acid was therefore *trans*, and *cis* hydroxylation of its double bond gave **23** (and its mirror image), which has no plane of symmetry and therefore can exist in optically active forms.

These configurations rested on conclusive proofs, and hence Wislicenus and van't Hoff assumed that *all* additions to double bonds were *cis*, an extrapolation which seemed intuitively reasonable. If *cis* addition occurs on bromination, maleic acid would yield mesodibromosuccinic acid **24** and fumaric acid would give the *dl*-compound **25**. Wislicenus assumed, therefore, that the dibromo compound from maleic acid was the *meso* isomer **24**, and the one from fumaric was *dl*, corresponding to **25**.

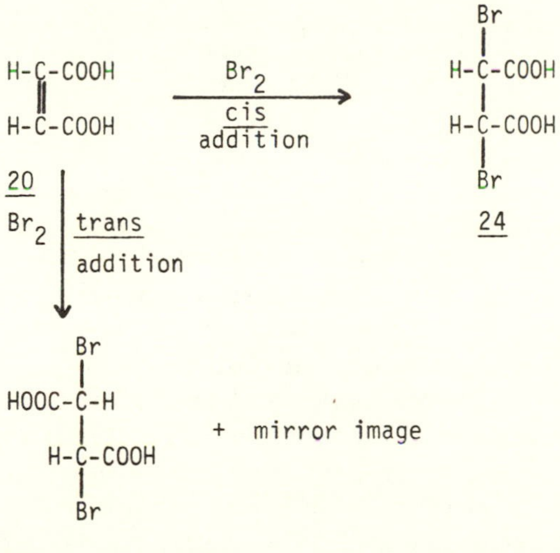

$$
\begin{array}{ccc}
\begin{array}{l} \text{H-C-COOH} \\ \ \ \| \\ \text{H-C-COOH} \end{array}
& \xrightarrow[\substack{\text{cis} \\ \text{addition}}]{Br_2}
& \begin{array}{l} \overset{\displaystyle Br}{\underset{\displaystyle |}{}} \\ \text{H-C-COOH} \\ \ \ | \\ \text{H-C-COOH} \\ \ \ | \\ \ \ Br \end{array} \\[2em]
\underline{20} & & \underline{24}
\end{array}
$$

Br_2 | $\underline{\text{trans}}$ addition

$$
\begin{array}{l}
\ \ \ Br \\
\ \ \ | \\
\text{HOOC-C-H} \\
\ \ \ | \\
\ \ \text{H-C-COOH} \\
\ \ \ | \\
\ \ \ Br
\end{array}
\qquad + \ \text{mirror image}
$$

$\underline{25}$

It was Michael's service to have shown in a long series of vigorous experimental and polemical papers that the assumption of *cis* addition of bromine was not proved.[38] There were several alternative conclusions which Michael might have drawn; (a) the rule of *cis* addition applied only to hydroxylation by permanganate, and other additions were *trans*; (b) the configurations assigned to maleic and fumaric acids were wrong; (c) the whole van't Hoff-Wislicenus scheme of stereochemistry was wrong and should be replaced by a better one.

Much of Michael's work was done with *derivatives* of cinnamic acid **26** or crotonic acid **27** in which configurations had to be assigned on the basis of melting point or other physical properties; these in turn were assigned by analogy to the maleic-fumaric pair. With compounds like **28** and **29** no *meso* form is posssible, and the proof of configurations of starting materials and products rested on physical properties, which were not as conclusive as the proofs of configurations of the tartaric acids.

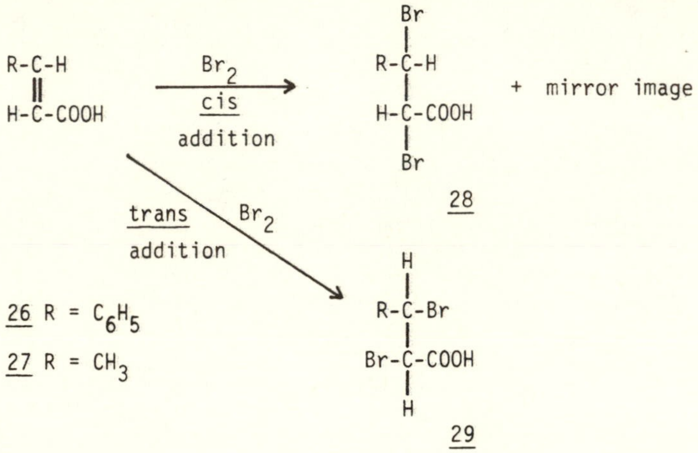

$$
\begin{array}{l}
\text{R-C-H} \\
\quad\| \\
\text{H-C-COOH}
\end{array}
\xrightarrow[\substack{cis \\ \text{addition}}]{\text{Br}_2}
\begin{array}{l}
\quad\text{Br} \\
\quad| \\
\text{R-C-H} \\
\quad| \\
\text{H-C-COOH} \\
\quad| \\
\quad\text{Br}
\end{array}
+ \text{ mirror image}
$$

28

trans Br$_2$
addition

26 R = C$_6$H$_5$

27 R = CH$_3$

$$
\begin{array}{l}
\quad\text{H} \\
\quad| \\
\text{R-C-Br} \\
\quad| \\
\text{Br-C-COOH} \\
\quad| \\
\quad\text{H}
\end{array}
$$

29

Despite some of his later claims, it is very difficult to find in Michael's papers published before 1901 a clear statement that *trans* addition occurred. There are several places where such a conclusion may be read into his discussion. He did show, for example, that addition of HCl to HOOCC≡CCOOH gives chlorofumaric acid, which would mean *trans* addition to the *triple bond*.[39] This, however, is not necessarily relevant to addition to the *double bond*, although Michael is usually credited with having proved *trans* addition to the *double bond*. His views were further complicated by the fact that with cinnamic acid **26** there appeared to be four forms with different melting points, although there should be two, *cis* and *trans*, on the basic of the van't Hoff-Wislicenus ideas. The "extra" isomers were later shown to be different crystalline forms,[40] and there are really only two cinnamic acids, as theory predicted.

Michael's usual conclusion at the time was therefore (b) or (c) above: the original configurations were wrong, or else the whole scheme of stereochemistry was wrong and needed replacement by a better one. Michael never specified what a better one would be. Until the end of his life, long after he had finally accepted *trans* addition as normal, and indeed had claimed credit for suggesting it, he viewed the whole van't Hoff-Wislicenus stereochemistry of unsaturated compounds with deep suspicion.[41]

The question of *trans* addition to double bonds has been discussed in some detail to demonstrate Michael's contributions to this important idea, but more fundamentally, it points up some important aspects of the development of new ideas in science. It is an illustration of A. N. Whitehead's remark that any really new idea (such as *trans* addition in this case) appears somewhat foolish when it is first proposed.

It was obviously difficult, even for an original and gifted man like Michael, who had no respect for authority as such, to accept the idea of *trans* addition, which at that time was troublesome to visualize as a reaction sequence. (Today, *trans* addition is a commonplace; there is no "energy barrier" to accepting it.) An additional point is that it was nearly twenty-five years after the publication of Wislicenus' memoir before anyone did the conclusive experiment, analogous to the hydroxylation of maleic acid, of adding bromine to maleic acid and determining by resolution whether the product was *meso* **24** or *dl* **25**. This experiment, which seems obvious now, was carried out simultaneously by Emil Fischer in Germany and by A. McKenzie in Britain in 1911, and it showed that the dibromo acid **25** formed from maleic acid was resolvable, i.e., had no plane of symmetry and therefore must be the result of *trans* addition.[42] The experimental work in Fischer's laboratory was done by D. D. Van Slyke, who was to be one of America's most distinguished biochemists.

Emil Fischer, in wise and measured words,[42] concluded that all configurations hitherto assigned by assuming *cis* addition to double bonds must now be reconsidered; he shared Michael's objections to Wislicenus' configurations. He added, however, that there was no reason to discard the general scheme of stereochemistry of unsaturated compounds, which, like the stereochemistry of saturated compounds, remained unshaken and was still the best picture available of isomerism in this series.

A further consequence of Michael's work and the establishment of *trans* addition was that elimination reactions, which Michael had also studied in great detail,[38] must take place by a *trans* process, because the original compound, to which the *trans* addition had taken place, was regenerated in the same configuration. Once the idea of *trans* elimination was accepted, perfectly plausible reaction mechanisms could be devised for it; again this illustrates the importance of familiarity in the acceptance of novel scientific ideas.

The only American chemist to develop a comprehensive scheme of organic reactions at all comparable to Michael's was John Ulric Nef (1862-1915), born a Swiss, with undergraduate training at Harvard and a Ph.D. with Baeyer at Munich. He taught at Purdue and at Clark University in Worcester, Massachusetts, and in 1892 became professor of chemistry at the founding of the University of Chicago.[43] An interesting sidelight on the great influence of the German university system on the thinking of American chemists was Nef's remark to President Harper that it should be possible to establish a school of chemistry at Chicago equal in standing to the good German universities.[44]

Nef, basing his ideas on the bivalent carbon in carbon monoxide and the isocyanides, postulated that many, if not all, organic reactions went through a process of "dissociation" to a bivalent carbon compound or methylene intermediate, which then reacted rapidly to complete the process.[45] Nef was a courageous and prolific experimentalist and an excellent teacher, but a doctrinaire and dogmatic thinker. His methylene scheme was applied, for example, to displacement reactions on alkyl halides, as in the synthesis of ethers. He postulated similar methylene intermediates in the Friedel-Crafts reaction of alkyl halides with benzene, in the high temperature decompositions of benzyl ethers and alkoxides, in the base-catalyzed isomerizations and cleavage of monosaccharides, and in numerous other types of reactions.

$$RCH_2Br \longrightarrow R\overset{|}{\underset{|}{C}}H + HBr \xrightarrow{R'ONa} R'OH + NaBr$$

$$R\overset{|}{\underset{|}{C}}H + R'OH \longrightarrow RCH_2OR'$$

Trenchant criticisms of some applications of Nef's ideas were made by Acree, based on kinetic studies of the displacement reaction on alkyl halides, by Michael, by Emil Fischer, and by Burke and Donnan.[46] Michael, who was a caustic and indefatigable controversialist, wrote sarcastically that Nef had "Methyllenbrillen" (methylene spectacles) through which he saw methylene intermediates everywhere.

Nevertheless, Nef's iconoclastic originality, like Michael's, was a liberating influence on American chemical thought. His postulate of a reactive intermediate with two free valences undoubtedly influenced the ideas of his colleague Stieglitz on rearrangements of nitrogen compounds. Work of the past decades has reinstated Nef's reactive intermediates as carbenes and nitrenes in numerous processes, and they are now commonplace in synthetic and mechanistic thinking.

C. Loring Jackson (1847-1935), a Harvard graduate, studied with Bunsen at Heidelberg and then with Hofmann at Berlin in 1873-1875 at the same time as Arthur Michael. He returned to Harvard to teach freshman chemistry, in which he was notably successful, and to carry on an active research program in organic chemistry[47,48] until his retirement in 1912 because of ill health. Jackson carried out significant studies with highly substituted aromatic halogen compounds in which the halogens were rendered readily replaceable by *nucleophilic* agents (agents seeking a relatively positive carbon atom) due to the presence of strongly electron-attracting nitro groups, NO_2.[49] The type of reaction is shown below, the nucleophile being the negative ion (*anion*) derived from malonic ester. The reaction to give **30** involves the displacement of chlorine by the malonate.

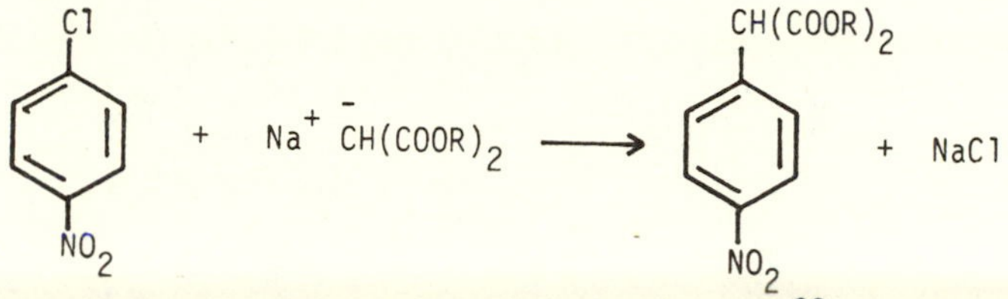

Jackson first recognized the structure of the colored materials **31** formed from bases and polynitro aromatic compounds. These have been called "Meisenheimer complexes" after the German chemist who also suggested the structure, but Jackson deserves equal credit.[50,51]

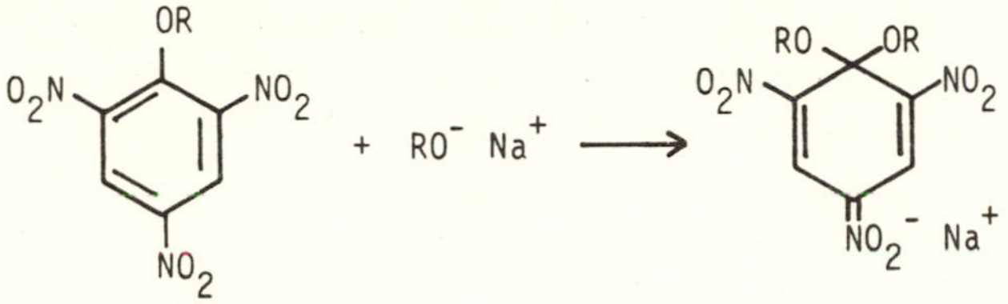

Jackson also pioneered in studying the preparation and complicated reactions of halogenated quinones such as **32** and **33**; the quinones as a class are derived from aromatic compounds, but are not aromatic themselves. Some are important as dyes and as biochemically active compounds, such as the K vitamins, the ubiquinones, and many others. Hydroquinone **34**, representative of a large class, is easily oxidized to a quinone and is used as a photographic developer.

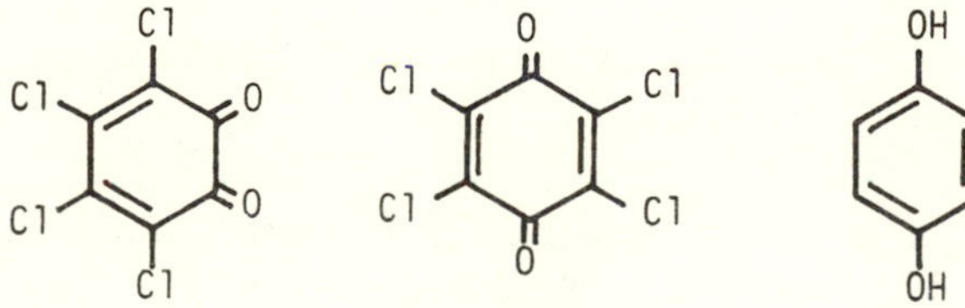

Jackson's colleague, Henry B. Hill[52] (1849-1903), spent a year in Hofmann's Berlin laboratory after his Harvard A.B. in 1869 and taught at Harvard thereafter until his death. A gift of furfural **35** from Dr. E. R. Squibb led Hill to unravel its complicated transformations leading to many open chain compounds, such as nitromalonic dialdehyde **36**, a useful synthetic intermediate.[53]

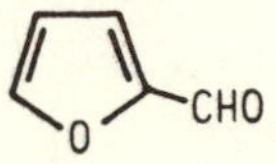

$$H-C=O$$
$$H-C-NO_2$$
$$H-C=O$$

35 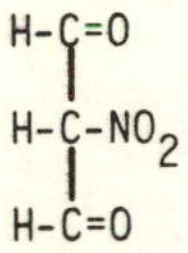   **36**

The addition reactions of conjugated systems, earlier studied by Michael, were extended by E. P. Kohler (1865-1938) at Bryn Mawr College in an impressive series of papers, done mainly with his own hands and in part with a few gifted women students. Kohler (A.B. 1886, Muhlenberg; Ph.D. 1892, Johns Hopkins with Remsen) taught at Bryn Mawr from 1892 to 1912 and at Harvard thereafter. Although he was shy and retiring and never attended chemical meetings or gave public lectures, Kohler had a powerful influence on the development of American organic chemistry through his masterly teaching and research and by the students he trained.[54]

The Kohler reaction[55] involved the 1,4-addition of a *Grignard reagent* (an organic magnesium halide) to a conjugated ketone. He initially wrote the addition product as **37**, with the MgBr bonded to oxygen by a covalent bond, which on hydrolysis and hydrogen shift yielded the saturated ketone **38**.

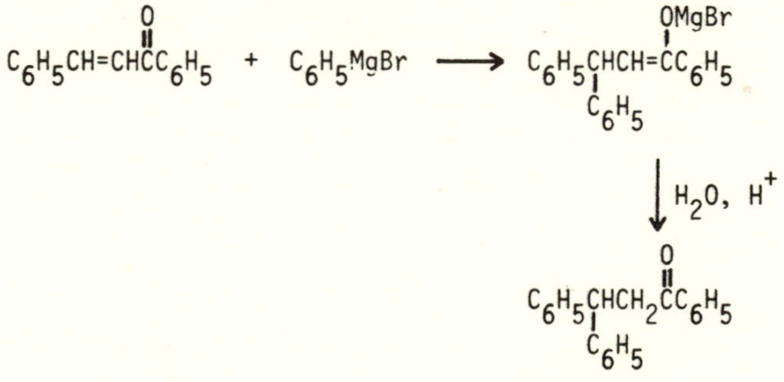

$$C_6H_5CH=CHCC_6H_5 \ + \ C_6H_5MgBr \ \longrightarrow \ C_6H_5CHCH=CC_6H_5 \quad \textbf{37}$$

$$\downarrow H_2O, \ H^+$$

$$C_6H_5CHCH_2CC_6H_5 \quad \textbf{38}$$

The work of Michael and of Kohler suggested that such a conjugated system **39** could be viewed as a delocalized carbonyl group **40** with a relative excess of positive charge on the β-carbon and excess negative charge on oxygen. This picture was not formulated explicitly by them, but their work was significant in judging J. Thiele's attempts to describe conjugated systems by such *partial valence* structures as **41**, which is discussed later.

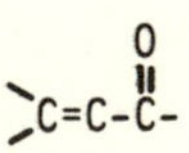

39   or   **40** 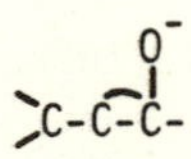   **41**

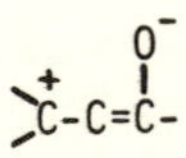

Following German chemists, Remsen at Hopkins and his student W. R. Orndorff at Cornell studied the compounds called phthaleins, formed by condensing phenols and phthalic anhydride. Some of these, such as phenolphthalein **42a** (now used as a laxative), are valuable *indicators*, which change color with changes in acidity in solution.[56] The relationship between chemical structure and color was a problem arising in the early days of organic chemistry. A step toward the answer in this case was given by Remsen's colleague[56] S. F. Acree, who suggested that the red color in basic solution was due to the quinone structure of the salt **42b**, changed from the colorless **42a** in acid solution. Remsen's chief contribution to this problem was to prepare "sulfonephaltheins", with a sulfonic acid group, SO_3H, in place of the COOH.

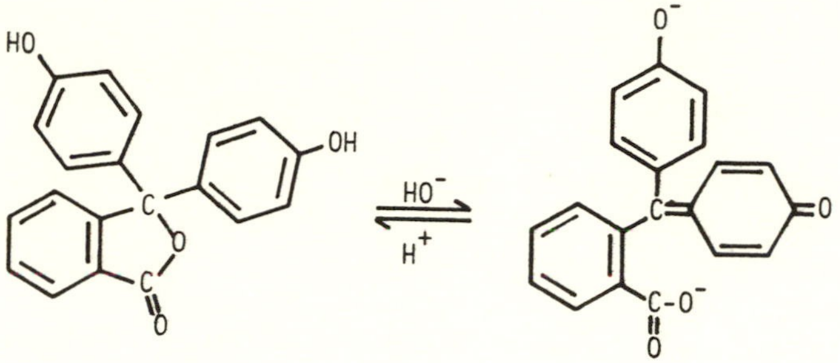

<u>42a</u> phenolphthalein,
colorless form

<u>42b</u> red form

American organic chemists, like their German and English counterparts, studied the structure and synthesis of organic compounds occurring in plant and animal sources (*natural products*); the remaining examples in this essay are of this type.

Samuel C. Hooker isolated colored compounds of the lapachol series from tropical wood;[57] these are α-napthoquinones, containing an aromatic ring and a quinone ring, as in **43**. Hooker became a successful executive in sugar refining, and his long hours in research at work and in his private laboratory at home revealed remarkable transformations in these aromatic compounds. After Hooker's death, L. F. Fieser at Harvard completed and published his results. Still later, these compounds were found to have high antimalarial activity, but unfortunately, in spite of much work, no clinically useful antimalarial compound has resulted.[57]

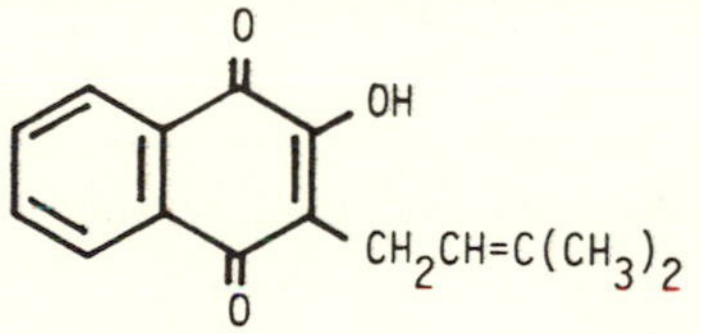

<u>43</u> lapachol

A long series of papers by H. L. Wheeler and T. B. Johnson of Yale dealt with the synthesis and properties of *nitrogen heterocycles* (structures containing carbon and nitrogen in a ring), particularly the pyrimidine series, **44**. Wheeler and Johnson were "second generation" American organic chemists, with their Ph.D. training in this country. Henry L. Wheeler (1867-1914) trained at Yale (Ph.B. 1890, Ph.D. 1893), spent a year each in Munich and Chicago (1893-1895), and returned to teach at Yale until his untimely death. Treat B. Johnson[58] (1875-1947) spent his entire career as student (Ph.B. 1893, Ph.D. 1901 with Wheeler) and teacher at Yale, retiring in 1943.

Although some of the Yale work became rather routine, part was of fundamental importance in the chemistry and biochemistry of the nucleic acids and nucleotides. The Yale group synthesized[59] cytosine **44a**, uracil **44b**, and orotic acid **44c** and studied many reactions of other pyrimidines.

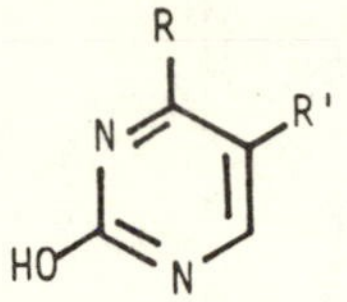

44a R = NH_2, R' = H

44b R = OH, R' = H

44c R = OH, R' = COOH

This work was useful many years later when it became possible to establish the structures of the nucleic acids (DNA and RNA); these are made up of nucleotide units, each containing a heterocyclic base, a sugar (D-ribose or 2-deoxy-D-ribose), and a phosphate group. The pyrmidines, such as **44a**, can exist in tautomeric forms differing only by the position of hydrogen atoms and of double bonds. In general these tautomeric structures must be determined in each case by physical measurement, usually by spectroscopy.

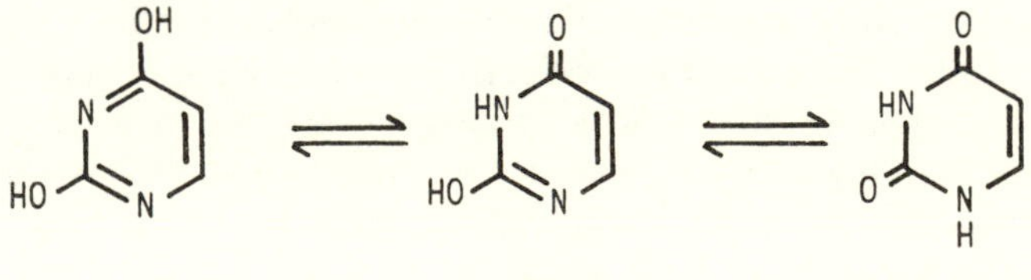

44a

The complicated problem of the structure of the fragrant, natural terpene camphor **45** and its derivatives was attacked energetically by W. A. Noyes[60] at Rose Polytechnic. He published nearly forty papers on camphor derivatives, and some of his work is quoted in reviews. Nevertheless, the difficulty in competing successfully with the large German research groups because of limited facilities and manpower illustrated the common problem in early American organic research. In this instance, only with exceptionally laborious work did even the European workers succeed. The seventy years required from the determination of camphor's empirical formula in 1833 to its structural determination by J. Bredt in Germany in 1893 and its total synthesis in 1903 by the Finnish chemist G. Komppa indicate the puzzling chemistry involved.

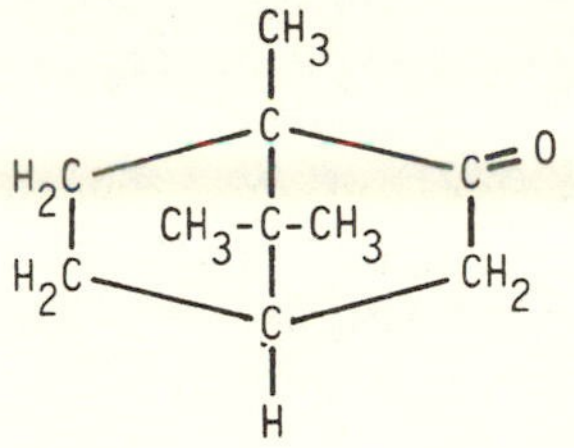

45

H. D. Dakin suggested that the base-catalyzed recemization he observed[61] in 1910 of optically active amino acid derivatives, such as the hydantoin **46**, was due to enolization by the base, or in modern terminology to the formation of a carbanion with loss of asymmetry. Dakin showed that proteins, but not dipeptides, were partially racemized by base at low temperature, and he presciently pointed out that this ease of racemization might mean that a naturally occurring, optically active protein or polypeptide could not be synthesized "by existing methods".[62] Modern methods of polypeptide synthesis must be very carefully devised and tested to avoid racemization.

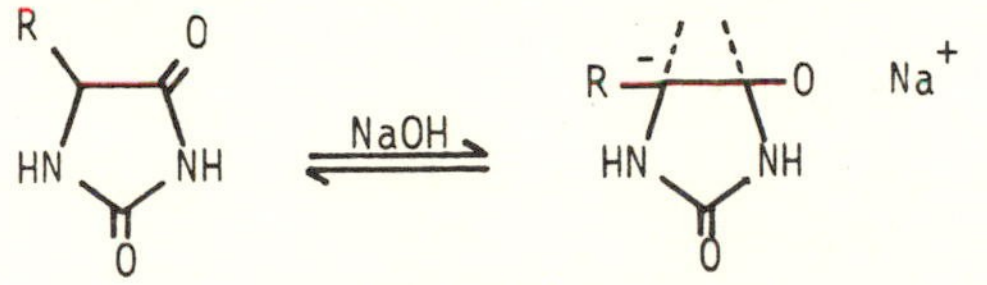

46

The outstanding researches[63] of Claude S. Hudson (1881-1952) in sugar chemistry started around 1900. Hudson,[64] educated at Princeton (A.B. 1901, Ph.D. 1907), planned first to be a preacher, then a physicist, then a physical chemist, but was led by his interest in chemical kinetics and optical activity to a career as a master of sugar chemistry. He spent 1902-1903 in Germany with W. Nernst and with van't Hoff; the influence of the latter is seen in Hudson's lifelong interest in the relation between configuration and the optical rotation of sugars. This led to a number of empirical "Hudson's Rules", showing that the numerical value of the optical rotation of some sugars could be assigned to two contributions, one due to the first carbon in the sugar and the other to the rest of the molecule as a whole. This allowed assignment of configurations to the first (glycosidic) carbon in some series.[65] Hudson's most important structural and synthetic work on sugars was largely done after 1913.

The number and variety of problems tackled in this first period of research and the solid publications forthcoming showed that American organic chemists were moving away from their apprenticeship to European masters and were learning their trade well. Although much of their research was tentative and routine, they made some outstanding contributions to be of importance in later years. The discovery of compounds with trivalent carbon was a fundamental one whose significance increased with time.

Americans boldly undertook problems currently investigated in Germany and elsewhere. These included tautomerism, sterochemistry of addition reactions, nucleophilic aromatic substitution, application of thermodynamic principles, however sketchily, to organic reactions, and behavior of conjugated systems. In purely structural chemistry, American work laid substantial foundations for future discoveries in sugars, purines and pyrimidines, and polypeptides.

Organic chemistry began to attract many very able students in this time of expanding college and university training. It became a lively and stimulating subject, which could lead to good positions for its practioners. Remsen's discovery of saccharin also helped in making the field known.

The striking industrial growth and the vast natural resources available showed that the application of chemistry to technology had great promise for the future. Although American chemists did not yet threaten the German monopoly in dyes and drugs, it was clear that successful competition lay ahead. Chemists overseas accorded increased recognition to American research.

We should emphasize a more general point, illustrated above and throughout our account. This is the inertia that frequently opposes the acceptance of novel scientific ideas. The replacement of familiar concepts by new ones is a slow process, requiring time and overwhelming proof of the validity and usefulness of the new ideas.

LITERATURE CITED

1. D. S. Tarbell and A. T. Tarbell, JCE, 58, 551 (1981).
2. J. C. Bailar, Jr., *Biog. Mem. Nat. Acad. Scis.*, 41, 141 (1970).
3. M. Gomberg, Ber., 30, 2043 (1897); JACS, 20, 773 (1898); an example of double publication.
4. M. Gomberg, ibid., 22, 757 (1900); Ber., 33, 3150 (1900).
5. M. Gomberg, ibid., 34, 2726 (1901).
6. A. Baeyer, ibid., 35, 1196 (1902).
7. A. E. Tschitschibabin, ibid., 37, 4709 (1904).
8. H. Lankamp, W. T. Nauta and C. MacLean, Tet. Letters, 249 (1968); see J. M. McBride, Tetrahedron, 30, 2009 (1974) for an excellent account of "hexaphenylethane". Structure 8 was proposed by P. Jacobsen, Ber. 38, 196 (1905), and was considered as a possibility by Gomberg, Ber., 46, 225 (1913).
9. M. Gomberg, ACJ, 25, 323 (1901); Ber., 35, 2397 (1902); P. Walden, ibid., 35, 2018 (1902); on Walden, D. S. Tarbell, JCE, 51, 7 (1974).
10. E. P. Kohler, ACJ, 40, 217 (1908).
11. H. D. Dakin and Mary D. Herter, JBC, 3, 419 (1907); H. J. H. Fenton and H. O. Jones, JCS, 77, 69 (1900) report oxidation of benzoic acid with Fenton's reagent (H_2O_2 + Fe^{++}) but do not seem to have identified products.
12. H. D. Dakin, JBC, 1, 171 (1905); 4, 418 (1908), and related papers.

13. L. F. Fieser, *Biog. Mem. Nat. Acad. Scis.*, **46**, 331 (1975); W. T. Read, Ind. Eng. Chem., **22**, 1137 (1930).
14. A. Michael and T. H. Norton, Ber., **9**, 1752 (1876).
15. L. Claisen and T. Komenos, Ann., **218**, 158 (1883); Michael admitted Claisen's priority, JPr, **36**, 113 (1887); the editor of the journal remarked that Claisen undoubtedly discovered and recognized the reaction, but Michael established its scope and generality.
16. A. Michael, ACJ, 9, 112 (1887).
17. A. Michael, Ber., **12**, 2260 (1879); **15**, 1922 (1882): ACJ, **1**, 305 (1879); **5**, 171 (1883).
18. Michael invokes the "entropy law" frequently: JACS, **32**, 990 (1910); JPr, **60**, 292 (1899); **68**, 487 (1903).
19. A. B. Costa's conclusions, JCE, **48**, 243 (1971), differ somewhat from ours. Only a superficial account of Michael's ideas is in F. Henrich, *Theories of Organic Chemistry*, transl. by T. B. Johnson and Dorothy A. Hahn, New York, 1922, or in the German editions of the same book.
20. D. S. Tarbell, "Entropy vs. Perfectibility", Science/Technology and the Humanities, **1**, 103 (1978).
21. A. Michael, Ber., **33**, 3731 (1900), for example. For use of term "chemical potential", see A. Michael, Ann., **363**, 21 (1908). Michael's views may owe something to a paper by L. Pfaundler, JPr, **10**, 37 (1874), which, as its title suggests ("The Struggle for Existence among Molecules"), is a mixture of a sort of chemical Darwinism, thermodynamics, and some discussion of entropy increase.
22. For development of a precise definition of "free energy", R. E. Kohler, article on G. N. Lewis, *Dictionary of Scientific Biography*, New York, VIII, pp. 289-294.
23. D. S. Tarbell, Chem. Eng. News, Centennial Ed., April 6, 1976, pp. 111-112.
24. A. Michael and R. F. Brunel, ACJ, **41**, 120 (1909); A. Michael and O. D. E. Bunge, Ber., **41**, 2907 (1908).
25. A. Kekulé, Ann., **106**, 141 (1858).
26. A. Michael, JPr, **60**, 341 (1899).
27. A. Michael and N. Weiner, JOC, **3**, 372 (1938).
28. Ref. 26, pp. 325 ff.
29. Ref. 26, p. 333.
30. A. Michael and N. Weiner, JOC, **5**, 389 (1940); with G. H. Shadinger, **4**, 128 (1939); with H. S. Mason, JACS, **65**, 683 (1943).
31. Ref. 26, pp. 329-333.
32. A. Michael and R. N. Hartman, Ber., **39**, 2149 (1906); with H. J. Turner, p. 1253.
33. H. L. Wheeler and T. B. Johnson, ACJ, **21**, 186 (1899); **28**, 130 (1902); S. F. Acree, ibid., **38**, 1 (1907).
34. A. Michael, JPr, **37**, 487 (1888).
35. J. U. Nef., ACJ, **37**, 296 (1907), footnote 1.
36. J. H. van't Hoff, *La Chimie dans l'Espace*, Rotterdam, 1875; *Die Lagerung der Atome in Raum*, Brunswick, 2nd ed., 1894; J. Wislicenus, "Uber die Räumliche Anordnung der Atome in Organischen Molekulen", Konigl. Sachsischen Gesel. der Wissenschafter, Math. Physische Classe, **14**, 1-79 (1887). We read the first of these rare sources in the Smith Collection at the University of Pennsylvania, and the last two in the Linda Hall Library, Kansas City.
37. A modern review, A. J. Ihde, JCE, **36**, 330 (1959).
38. A representative list of Michael's papers on this problem: JPr, **38**, 6 (1888); with O. Schulthess, **43**, 587 (1891); **46**, 400 (1892); **52**, 344 (1895); with W. R. Whitehorne, Ber., **34**, 3640 (1901); with A. B. Lamb, ACJ, **36**, 552 (1906); with H. D. Smith, **39**, 16 (1908).
39. A. Michael, JPr, **52**, 345, 363, (1895).

40. H. Stobbe, Ber., **44**, 2739 (1911).
41. A. Michael, JACS, **40**, 704 (1918); with J. Ross, **55**, 1638 footnote (1933). In an unpublished note to William A. Noyes (1901), Michael said he planned to write a comprehensive account of stereochemistry (W. A. Noyes Archive, University of Illinois Archives, Urbana, Illinois). This account has not appeared and was probably never completed.
42. E. Fischer, Ann., **386**, 374 (1911); A. McKenzie, Proc. Chem. Soc., 150 (1911); B. Holmberg, JPr, **84**, 145 (1911) had reached similar conclusions on less rigorous grounds.
43. M. L. Wolfrom, *Biog. Mem. Nat. Acad. Scis.*, **34**, 204 (1960).
44. R. J. Storr, *Harper's University*, Chicago, 1966, p. 72.
45. Review, Ann., **298**, 202-374 (1897); JACS, **26**, 1549-1597 (1904). Key publications by his students: A. F. McLeod, **37**, 20 (1907); Miss Willey Denis, **38**, 561 (1907); J. W. E. Glattfeld, **50**, 135 (1913).
46. S. F. Acree and G. H. Shadinger, ibid., **39**, 226 (1908); A. Michael, JPr, **60**, 471 (1899); Ber., **34**, 918 (1901); Ann., **364**, 64 (1908); E. Fischer, **381**, 123 (1911); Katharine A. Burke and F. G. Donnan, JCS, **95**, 555 (1904).
47. G. S. Forbes, *Biog. Mem. Nat. Acad. Scis.*, **37**, 97 (1964).
48. C. L. Jackson in *The Development of Harvard University, 1869-1929*, S. E. Morison Ed., Cambridge, 1930, pp. 260, 263-265.
49. C. L. Jackson and W. S. Robinson, ACJ, **11**, 93 (1889); for a review of aromatic nucleophilic displacement, including much of Jackson's work, J. F. Bunnett and R. E. Zahler, Chem. Rev., **49**, 273 (1951).
50. C. L. Jackson and F. H. Gazzolo, ACJ, **23**, 376 (1900); C. L. Jackson and R. B. Earle, ibid., **29**, 89 (1903).
51. J. Meisenheimer, Ann., **323**, 205 (1902).
52. C. L. Jackson, *Biog. Mem. Nat. Acad. Scis.*, **5**, 255 (1904).
53. H. B. Hill and J. Torrey, Jr., ACJ, **22**, 89 (1899); Hill and C. R. Sanger, Ber., **15**, 1906 (1882).
54. J. B. Conant, *Biog. Mem. Nat. Acad. Scis.*, **27**, 265 (1952).
55. E. P. Kohler, ACJ, **31**, 642 (1904).
56. S. F. Acree, ibid., **39**, 528 (1908), Ira Remsen and C. W. Hayes, ibid., **9**, 372 (1887), W. R. Orndorff and C. E. Brewer, ibid., **23**, 425 (1900); and later papers.
57. S. C. Hooker and W. H. Greene, ibid., **11**, 267, 394 (1889); S. C. Hooker, JACS, **58**, 1163 (1936) and following papers; L. F. Fieser, M. T. Leffler and others, ibid., **70**, 3151 (1948) and following papers on antimalarial properties.
58. H. B. Vickery, *Biog. Mem. Nat. Acad. Scis.*, **27**, 83 (1952). Although Wheeler was also a member of the academy, no biography has appeared.
59. H. L. Wheeler and T. B. Johnson, ACJ, **29**, 492 (1903) (cytosine); H. L. Wheeler and H. F. Merriam, ibid., **29**, 478 (1903) (uracil); H. L. Wheeler, T. B. Johnson and C. O. Johns, ibid., **37**, 392 (1907) (orotic acid).
60. W. A. Noyes, ibid., **18**, 685 (1896), for example. For the chemistry of camphor, J. L. Simonsen, *The Terpenes*, Vol. 2, 2nd ed. revised, Cambridge, 1949, pp. 373-525.
61. H. D. Dakin, ibid., **44**, 48 (1910).
62. H. D. Dakin and H. W. Dudley, JBC, **15**, 263 (1913).
63. *The Collected Papers to C. S. Hudson*, R. M. Hann and N. K. Richtmyer, Eds., New York, 1946; 2 vol.
64. L. F. Small and M. L. Wolfrom, *Biog. Mem. Nat. Acad. Scis.*, **32**, 181 (1958).
65. F. Micheel, *Chemie der Zucker und Polysaccharide*, Leipzig, 1956, p. 221.

5

Studies on Reaction Rates
and Mechanism, 1876-1913

Measurements of *rates* of reaction (*kinetic studies*) and of *mechanism* (path) of organic reactions were significant in number and importance in the 1876-1913 period. *Physical organic chemistry* includes these topics and others, such as the relation between structure and physical properties, and it became a major field of American research after 1913. In the period under review, rate and mechanism studies not only formed a larger fraction of American studies than of German or English researches, but American physical organic chemistry was also better than American synthetic and structural organic chemistry compared with the corresponding researches overseas. Additionally, American development in this lively field has been largely indigenous both before and after 1913; influence from other countries, while important, has not been decisive. By 1890, equations governing rates and equilibria of chemical reactions, fundamental ideas of thermodynamics, views on the dissociation of molecules into charged ions, and theories about catalysis had been developed by many workers.[1] The most important work was done by J. H. van't Hoff (1852-1911), codiscoverer of the arrangements of atoms in space described earlier. The Swedish chemist Svante Arrhenius (1859-1927) proposed that salts, acids, and bases dissociated into charged ions in water. He also gave a satisfactory explanation of the observed fact that a rise in temperature accelerates chemical reactions. In Germany, Wilhelm Ostwald (1853-1932) studied catalysis of reactions by acids and bases. In America, Josiah Willard Gibbs (1839-1903) contributed fundamental studies in thermodynamics. Van't Hoff, Arrhenius, and Gibbs founded physical chemistry.[1,2] Ostwald translated Gibbs' papers into German, and European scientists recognized their significance earlier than Americans.

Satisfactory kinetic measurements on the alkaline hydrolysis of an ester, ethyl acetate, were first carried out by R. B. Warder of Cincinnati in 1881.

$$CH_3COOC_2H_5 \ + \ HO^- \ \xrightarrow{\ H_2O\ } \ CH_3COO^- \ + \ C_2H_5OH$$

$$rate = k[CH_3COOC_2H_5][HO^-]$$

The rate of the reaction was proportional to the concentration of ethyl acetate times the concentration of hydroxyl ion.[3,4] Since water was present in excess, its concentration did not change, and the rate constant k includes the water concentration. Although German chemists later produced more extensive studies on ester hydrolysis, Warder was the pioneer. Warder (1848-1905) was well-grounded in physical chemistry from studies in Germany; he taught at several universities, including Howard, where he spent the years from 1887 until his death.

Around 1900, D. M. Lichty at Michigan (Ann Arbor) measured the enhancing effect of halogen substitution on rates of esterification of carboxylic acids.[5] Thus the first reaction shown here is faster than the second; this is due to the greater acidity (*dissociation constant*) of the chlorinated acid compared to the unchlorinated one, with a consequent higher concentration of hydrogen ion to catalyze the reaction.

$$ClCH_2COOH \ + \ C_2H_5OH \ \underset{}{\overset{H^+}{\rightleftharpoons}} \ ClCH_2COOC_2H_5 \ + \ H_2O$$

$$CH_3COOH \ + \ C_2H_5OH \ \underset{}{\overset{H^+}{\rightleftharpoons}} \ CH_3COOC_2H_5 \ + \ H_2O$$

E. Emmet Reid (1872-1973), a Remsen Ph.D. in 1898 and eventually his successor as the senior organic chemist at Hopkins, published some excellent kinetic measurements on the acid-catalyzed reactions of amides,[6] as well as several equilibrium studies.[7] His most important results are indicated in the equilibria involving sulfur compounds.[8]

$$C_6H_5COSH \ + \ C_2H_5OH \ \rightleftharpoons \ C_6H_5COOC_2H_5 \ + \ H_2S$$

$$C_6H_5COOH \ + \ C_2H_5SH \ \rightleftharpoons \ C_6H_5COSC_2H_5 \ + \ H_2O$$

These led by analogy to the prediction that oxygen esters form by elimination of water; the hydrogen of water comes from the alcohol and the hydroxyl from the carboxyl group. Subsequent tracer and kinetic work has supported this scheme for ester formation.

$$R-\overset{O}{\overset{\|}{C}}\text{-}OH \ + \ H\text{-}OR' \ \rightleftharpoons \ R-\overset{O}{\overset{\|}{C}}\text{-}OR' \ + \ H_2O$$

Reid made many valuable contributions to American chemistry during his very long life; his fascinating autobiography, published when he was 100, is entitled, appropriately, *My First Hundred Years.*[9]

Victor Meyer's demonstration at Heidelberg of the steric effect of adjacent (*ortho*) substituents in benzoic acids in decreasing the rate of HCl-catalyzed esterification[10] led both M. A. Rosanoff and Arthur Michael to study the esterification reaction at high temperature without added HCl.[11] The results showed that effects of ortho substituents on the rates were not important under these conditions. Michael did show[12] that the rates of esterification of trichloracetic acid by various alcohols followed the following

$$CH_3OH > C_2H_5OH > \underline{n}\text{-}C_3H_7OH > (CH_3)_2CHCH_2OH$$

order. Michael found that tertiary alcohols esterified more slowly and the rate constants were poor; there was clearly something different qualitatively and quantitatively about tertiary alcohols as compared to primary and secondary alcohols.

James F. Norris[13] (1871-1940; A.B. 1892, Ph.D. 1895 with Remsen, at Johns Hopkins) taught at Simmons College and MIT, wrote several widely used texts, and published extensively on rate of reaction as affected by change in structure. He showed[14] that *tert*-butyl alcohol reacted with HCl to form the chloride far more rapidly than primary or secondary alcohols. Further, the tertiary chloride was hydrolyzed by water far more rapidly than the corresponding primary or secondary chlorides. *Tert*-butyl chloride and the alcohols thus resembled the triarylmethyl halides and the corresponding alcohols, Ar_3CCl and Ar_3COH (where Ar is an aromatic nucleus), which were known to form relatively stable carbonium ions in ionizing solvents. Norris suggested tentatively that the *tert*-butyl carbonium ion might be an intermediate in the reactions of *tert*-butyl compounds, thus extending the carbonium ion idea proposed first by Julius Stieglitz of Chicago.[15] This was the origin of a concept which was to cause more experimental work and more controversy than other other single one in American chemistry.

$$(CH_3)_3COH \;+\; HCl \;\rightleftharpoons\; [(CH_3)_3C^+ \; Cl^- \;+\; H_2O]$$

$$\Updownarrow$$

$$(CH_3)_3CCl \;+\; H_2O$$

Julius Stieglitz[16] (1867-1937), born in this country but educated in Germany, was at first on the staff at Clark University and then for many years at the University of Chicago. He came from a wealthy and gifted family; Alfred Stieglitz, the famous photographer and art critic, was his brother. A biography of Alfred, written by his (and Julius') grandniece,[16] explains the attitudes and mode of thought of Julius, whose papers are written in a rather lordly manner.[16]

His career commenced brilliantly, but he published little significant research after 1914, although he continued to direct a large group of Ph.D. students and to give excellent lectures to his classes. His first important papers were published with a Chicago colleague, Felix Lengfeld (Ph.D. 1888 with Remsen, Chicago faculty 1892-1901), and dealt with the rearrangements of the N-bromoamide type,[17] which were postulated by Stieglitz to go through a monovalent nitrogen intermediate. This intermediate gave a rearrangement of the R group from carbon to nitrogen, and the resulting isocyanate, $RN=C=O$, reacted with an alcohol to form the urethan, $RNHCOOCH_3$, or with water to give the amine, RNH_2.[18] The monovalent nitrogen intermediate, $RC(=O)N$, was undoubtedly suggested by Nef's bivalent carbon intermediate; Stieglitz generalized his mechanism to include rearrangements of acid azides and hydroxamic acids (Curtius and Lossen reactions).

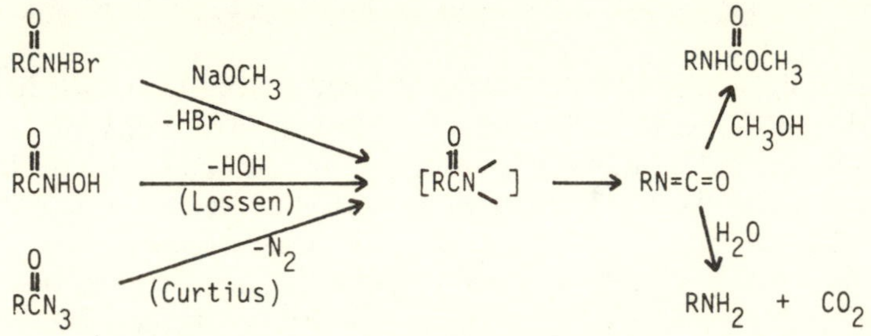

Stieglitz observed the shift of a phenyl group in the triphenylmethyl structure.[19] Much of his later work was in this general field of nitrogen compounds, and it led to his studies on the mechanism of acid- and base-catalyzed reactions.

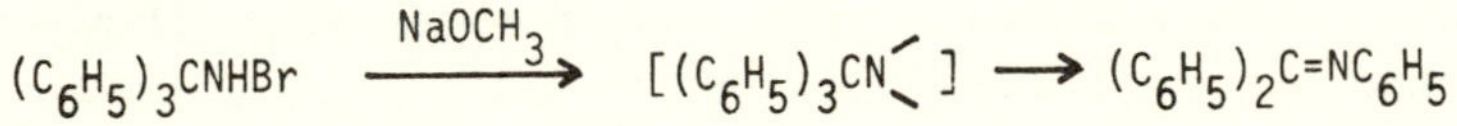

Steiglitz showed by careful kinetic and conductance experiments[20] that the hydrolysis of imido esters was acid-catalyzed and presumably proceeded via the ammonium salt, which was attacked by water. The reason the reaction is acid-catalyzed is that the positively charged ion, $C_6H_5C(=NH_2^+)OCH_3$, is attacked by HOH far more rapidly than the uncharged $C_6H_5C(=NH)OCH_3$ molecule. He also applied his mechanism to the acid-catalyzed hydrolysis of esters, the scheme being in agreement with Reid's observations above.

Stieglitz recognized correctly that the overall rate of these reactions depended on the initial equilibrium with the proton to form the positively charged ion, and *also* on the rate of attack of water on the positive ion. He pointed out that there could be a relationship between the basic strength of the ester, $RC(=O)OR'$, or imido ester, $RC(=NH)OR'$, toward a proton and the dissociation constant of the corresponding acid, RCOOH, and hence there may be an overall parallelism of rate of hydrolysis to dissociation constant of the acid, RCOOH. This is an early suggestion in not very exact terms of a *linear free energy relationship* between rates and equilibria, a relationship to be developed in greater detail by J. N. Brönsted, L. P Hammett, and many others in later years.[21]

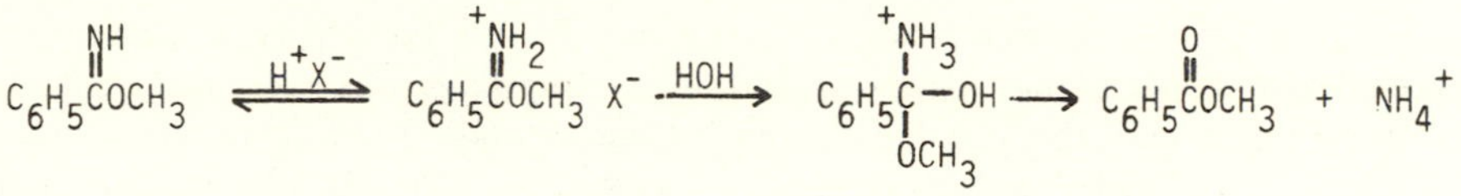

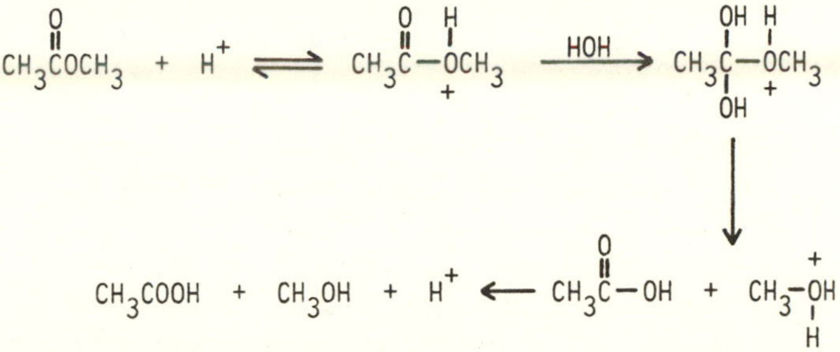

$$C_6H_5\overset{\overset{NH}{\|}}{C}OCH_3 \xrightleftharpoons{H^+X^-} C_6H_5\overset{\overset{+NH_2}{\|}}{C}OCH_3 \ X^- \xrightarrow{HOH} \underset{\underset{OCH_3}{|}}{C_6H_5\overset{+NH_3}{\underset{}{C}}-OH} \longrightarrow C_6H_5\overset{\overset{O}{\|}}{C}OCH_3 + NH_4^+$$

$$CH_3\overset{\overset{O}{\|}}{C}OCH_3 + H^+ \rightleftharpoons CH_3\overset{\overset{O}{\|}}{\underset{+}{C}}\overset{H}{\overset{|}{-O}}CH_3 \xrightarrow{HOH} CH_3\underset{\underset{OH}{|}}{\overset{\overset{OH}{|}}{C}}\overset{H}{\underset{+}{\overset{|}{-O}}}CH_3$$

$$CH_3COOH + CH_3OH + H^+ \longleftarrow CH_3\overset{\overset{O}{\|}}{C}-OH + CH_3\underset{\underset{H}{|}}{\overset{+}{-O}H}$$

Stieglitz was also concerned about the effect of neutral salts on the rate, but in his time, before the use of buffers at constant ionic strength and before the Debye-Hückel treatment of interionic attractions in dilute solutions, little useful information could result from consideration of salt effects, especially in water-organic solvent mixtures.

S. F. Acree (1875-1957) received his Ph.D. at Chicago with Nef in 1902, and from 1904 to 1914 he directed most of the research in organic chemistry at Hopkins. Although he continued to publish after 1914 from several laboratories, his work at Hopkins was his best. He was a brilliant, versatile chemist, the best American in the field of reaction mechanisms before 1920. His work and that of Stieglitz on kinetics and catalysis in organic reactions compare favorably with that being done abroad by Arthur Lapworth at this time.[22]

Acree carried out excellent studies on catalysis of oxime formation from carbonyl compounds and its reversal. He suggested a mechanism which remained satisfactory for many years.[23]

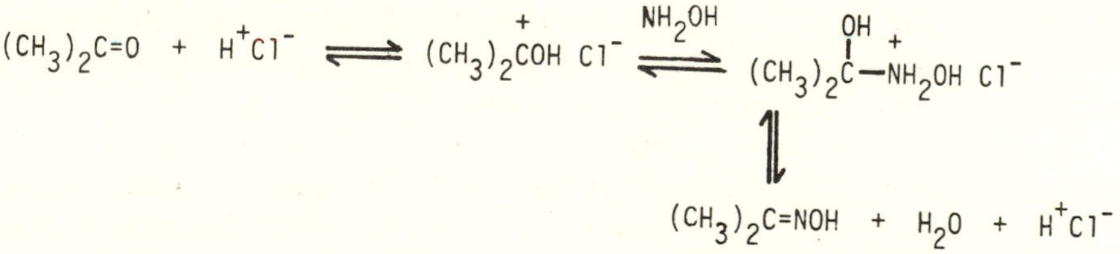

$$(CH_3)_2C=O + H^+Cl^- \rightleftharpoons (CH_3)_2COH \ Cl^- \overset{NH_2OH}{\underset{}{\rightleftharpoons}} (CH_3)_2\underset{}{\overset{\overset{OH}{|}}{C}}-NH_2OH \ Cl^-$$

$$(CH_3)_2C=NOH + H_2O + H^+Cl^-$$

Michael published somewhat later than Acree on the rates of semicarbazone formation as affected by changes in structure;[24] Michael realized that the reaction was acid-catalyzed, but his work is more antiquated in viewpoint than Acree's. The Conant-Bartlett study of the kinetics of semicarbazone formation (1932) is based on modern concepts and techniques and gives a more searching view of the reaction mechanism.

In a detailed kinetic study of a reaction of the following type (later classified as a *displacement reaction*), Acree and Shadinger[25] concluded that the intermediate was the ion RS^-, which then displaced iodide ion, I^-, from C_2H_5I. Nef's postulated methylene intermediate, CH_3CH, shown here was incompatible with Acree's results.

$$RSH + Na^+OH^- + C_2H_5I \longrightarrow RSC_2H_5 + Na^+I^- + H_2O$$

$$C_2H_5I + Na^+OH^- \xrightarrow{\quad//\quad} CH_3\overset{|}{\underset{|}{C}}H + Na^+I^- \xrightarrow{\ RSH\ //\ } C_2H_5SR$$

Additional studies by Acree showed that the reaction rate varied with the halogen being displaced ($RI > RBr > RCl$), and also with the structure of the halide: primary, RCH_2X, faster than secondary, R_2CHX; and this faster than tertiary, R_3CX. These observations agreed with numerous earlier ones.

From detailed kinetic study of several reactions, Acree concluded[26] that reactions involving compounds which could ionize, such as $Na^+OC_2H_5^-$, proceeded partly through the ions $OC_2H_5^-$ and partly through the nonionized or molecular material $NaOC_2H_5$. He expressed this mathematically in what has become known as the Acree Equation, where α is the percentage of ionization, usually determined from electrical conductivity. This Acree Equation, in which k_2 molecular is now considered to be the rate constant for ion pairs (instead of nonionized molecules), has been the subject of active research in recent years[27] and has been confirmed in the cases examined. Acree's ideas have thus had a marked influence on modern physical organic chemistry.

$$\text{Rate} = k_1\alpha + k_2(1-\alpha)$$
$$\quad\ \ \text{ionic} \qquad \text{molecular}$$

The mechanism of enzyme action became a subject of active research and speculation during the years before 1913. The Morrill Act (land-grant college) of 1862 and several later acts of Congress led to the establishment of state university schools of agriculture and of state agricultural research stations. Furthermore, the growth of medical schools, government laboratories, and research institutes, such as the Rockefeller Institute, added to the emphasis on the application of chemistry to the processes of living systems. L. L. Van Slyke (1859-1931) of the New York State Experiment Station at Geneva and T. B. Osborne (1859-1929) of the Connecticut State Experiment Station both published in ACJ and later in biochemical journals. Van Slyke specialized in the chemistry of milk products and Osborne in plant proteins. This thrust toward the chemistry of agriculture and public health problems gave American chemists an incentive to work on the structures of naturally occurring organic compounds and also on the mechanism of enzyme action.

The papers of H. D. Dakin, already mentioned, were part of his efforts to find *in vitro* models for enzymatic processes.

Although pure enzymes were not available before the late 1920s, several workers, particularly J. H. Kastle (1864-1916; A. B. 1884, Kentucky; Ph.D. 1888, Johns Hopkins with Remsen), studied rates of enzyme-catalyzed reactions using enzyme concentrates, particularly the behavior of ester-hydrolyzing enzymes (esterases).[28] Acree evidently read the papers on enzymatic reactions; he suggested that enzymes probably accelerate esterification and hydrolysis of esters by the acidic and basic groups present in the enzymes.[29] He later proposed that an enzymatic reaction goes at a rate determined by the "velocity of formation and change of the intermediate catalyzer-co-catalyzer-substrate double compounds", an idea now basic in analyzing enzyme kinetics.[29]

J. U. Nef, a few months before his death, gave an unpublished lecture at the University of Illinois in May 1915 entitled "The Chemistry of Enzyme Action".[30] Nef had investigated intensively the degradation of glucose by alkali, which he thought proceeded via methylene intermediates, and it may be supposed that his lecture presented his ideas on the enzymatic breakdown of glucose to form ethyl alcohol and carbon dioxide.

Clarence G. Derick (1883-1980; B. S. 1906, Worcester; Ph.D. 1910, Illinois with William A. Noyes), working at Illinois[31] from 1910 to 1916, carried out an ambitious program designed to determine the polarity of atoms and groups from their effects on ionization constants of carboxylic acids containing them. He hoped to be able to explain organic mechanisms and molecular rearrangements. He recognized "negative" groups, such as chlorine, which increased the ionization of acids, the increase falling off rapidly as the chlorine was separated by an increasing number of carbons from the carboxyl group. "Positive" groups, such as methyl, decreased the ionization of carboxylic acids. He applied his procedure to aromatic acids and attempted to explain some reactions of ketones and some rearrangement reactions by his polarity results. These attempts were not entirely successful, but he did correct some of Arthur Michael's quasithermodynamic ideas, and his work provided quantitative information about the polarity or charge associated with halogens and other groups. This was before G. N. Lewis had provided the shared electron bond theory of structure for organic compounds, and hence Derick's polarities were not expressed in electronic terms.[32]

The work by Edward C. Franklin[33] (1862-1937; B.S. 1888, Kansas; Ph.D. 1894 with Remsen) and Charles A. Kraus (1875-1967) on liquid ammonia solutions[34] has been important in extending ideas of acids, bases, and salts to solvents other than water and in broadening the viewpoint of organic reactions. It also supplied a series of important reagents for synthetic organic chemistry. Franklin and Kraus showed that liquid ammonia was similar to water in being a good ionizing solvent, having a high dielectric constant, and forming analogous series of compounds. The properties of solutions of alkali metals in liquid ammonia have remained of great theoretical and practical interest.

To summarize the work from 1876-1913 in physical organic chemistry in this country, it may be said that it was on a higher level of originality and importance than most of the synthetic and structural work, in comparison to what was being done elsewhere. It was less provincial in its approach, more self-reliant, and more significant in its effects on future work in this country. In the years after 1914, the theory of valence was decisively advanced by the work of G. N. Lewis and others. The next generation of researchers after 1914 on mechanisms of reactions, including L. P. Hammett, J. B. Conant, H. J. Lucas, M. S. Kharasch, C. R. Hauser, and others, was to set the standard of performance for world chemistry. Later generations, Saul Winstein, P. D. Bartlett, Cheves Walling, J. D. Roberts, H. C. Brown, D. J. Cram, G. S. Hammond, and many more were to enhance the quality of American work to a still greater degree. This is not to denigrate the achievements of English chemists, particularly Sir Robert Robinson and C. K. Ingold in this field, but American work was just as high in quality and considerably ahead in range and quantity.

LITERATURE CITED

1. J. R. Partington, *A History of Chemistry*, London, Vol. 4, 1964, pp. 569-663, describes the European developments in detail with copious documentation.
2. Arrhenius' ionization theory was proposed in his thesis of 1884 and published in Z. Phys. Chem., **1**, 631 (1887). His "activation energy", explaining the increase of reaction rate with temperature based on equations of van't Hoff, is ibid., **4**, 226 (1889). Ostwald's work on catalysis is described in JPr, **27**, 1, 449 (1883) and later papers.
3. R. B. Warder, Ber, **14**, 1361 (1881); ACJ, **3**, 340 (1882); D. S. Tarbell and A. T. Tarbell, JCE, **58**, 559 (1981).
4. W. Nernst, *Theoretical Chemistry*, revised by H. T. Tizard from the 6th German ed., London, 1911, p. 560, credits Warder, although Partington does not mention him.
5. D. M. Lichty, ACJ, **18**, 590 (1896).
6. I. Remsen and E. E. Reid, ACJ, **19**, 319 (1897); **24**, 397 (1900); **41**, 483 (1909).
7. Ibid., **45**, 38 (1911).
8. Ibid., **43**, 489 (1910) and later papers.
9. E. E. Reid, *My First Hundred Years*, New York, 1972.
10. V. Meyer, Ber, **27**, 510 (1894); V. Meyer and J. J. Sudborough, p. 1580, and later papers by each.
11. M. A. Rosanoff and W. L. Prager, JACS, **30**, 1895, 1908 (1908); A. Michael and K. J. Oechslin, Ber, **42**, 310, 317 (1909).
12. A. Michael and K. Wolgast, ibid., **42**, 3157 (1909).
13. J. D. Roberts, *Biog. Mem. Nat. Acad. Scis.*, **45**, 413 (1974).
14. J. F. Norris, ACJ, **38**, 627 (1907).
15. J. Stieglitz, ibid., **21**, 101 (1899).
16. Biography of Julius Stieglitz, W. A. Noyes, *Biog. Mem. Nat. Acad. Scis.*, **21**, 275 (1942); Sue Davidson Lowe, *Stieglitz, a Memoir/Biography*, New York, 1983, passim, especially p. xx.
17. F. Lengfeld and J. Stieglitz, ACJ, **15**, 215 (1893).

18. J. Stieglitz, ibid., **29**, 49 (1903); **18**, 751 (1896): Stieglitz refers to all these as "Beckmann" rearrangements, a name now reserved for rearrangements of oximes.

19. J. Stieglitz and Isabelle Vosburgh, Ber, **46**, 2151 (1913).

20. Summarizing papers: J. Stieglitz, ACJ, **39**, 29, 166 (1908); JACS, **32**, 221 (1910); **34**, 1687 (1912); experimental papers: with I. H. Derby, ACJ, **39**, 437 (1908); with W. McCracken, p. 586; with H. I. Schlesinger, p. 719; polemical, p. 402.

21. J. N. Brönsted and K. J. Pedersen, Z. Phys. Chem., **108**, 185 (1924); L. P. Hammett, Chem. Rev., **17**, 125 (1935), and many later statements.

22. The outstanding papers of Arthur Lapworth in England on the kinetics of halogenation of acetone, which showed that enolization was the rate-determining step, JCS, **85**, 30(1904) and later papers, and on the formation of cyanhydrins by addition of cyanide ion to carbonyl groups, ibid., **83**, 995 (1903), should be cited, as well as the work of Barrett and Lapworth on oxime formation, referrred to below.

23. S. F. Acree, ACJ, **39**, 300 (1908); S. F. Acree and J. M. Johnson, ibid., **38**, 308 (1907); E. Barrett and A. Lapworth, JCS, **93**, 85 (1908).

24. A. Michael, et al. JACS, **41**, 393 (1919).

25. S. F. Acree and G. H. Shadinger, ibid., **39**, 226 (1908).

26. Summarizing paper, S. F. Acree, H. C. Robertson, E. K. Marshall, Julia P. Harrison, Bessie M. Brown, J. Chandler, S. Nirdlinger, F. M. Rogers, C. N. Myers, and J. H. Schrader, ibid., **48**, 352 (1912).

27. Many papers, such as N. N. Lichtin and K. N. Rao, JACS, **83**, 2417 (1961); P. Beronius and L. Patai, Acta Chem. Scand., **25**, 3705 (1971); C. Germain, Tetrahedron, **32**, 2389 (1976).

28. A. S. Loevenhart and J. H. Kastle, ACJ, **29**, 397 (1903), for example.

29. S. F. Acree, ibid., **39**, 306 (1908); ibid., **41**, 462 (1909).

30. University of Illinois, Circular of Information of the Department of Chemistry, Urbana, 1916, p. 44; Illinois Archives, Series 15/5/0/1, Box 1.

31. Derick left Illinois for industrial research in 1916 and was succeeded by Roger Adams.

32. C. G. Derick, JACS, **32**, 1333 (1910); **33**, 1162 (1911); **34**, 74 (1912) and later papers.

33. Biography by A. Findlay, JCS, 583 (1938); although Franklin was a member of the National Academy of Sciences, no biographical memoir has appeared. Franklin was professor at Kansas and Stanford.

34. E. C. Franklin and C. A. Kraus, ACJ, **20**, 820, 836 (1898), and many later papers, together and individually; E. C. Franklin, *The Nitrogen System of Compounds*, New York, 1935. C. A. Kraus (1875-1967, B.S. 1898, Kansas; Ph.D. 1908, MIT) faculty member at MIT, Clark, and Brown, contributed to physical, inorganic, and organic chemistry over the years 1897-1968. For biography, R. A. Fuoss, *Biog. Mem. Nat. Acad. Scis.*, **42**, 119 (1971).

Solomon Farley Acree

Chemistry

1906 – 1914

SOLOMON F. ACREE

6

Chemical Laboratories, Supplies, Instruction, Journals, Women Students to 1914

We discuss in this essay some institutional aspects of organic chemistry, including the construction of laboratories, chemical supplies and equipment, chemical instruction and textbooks, increase in the number of students, including women students, and the chemical journals available for publishing results of research.

The first American laboratories were adapted houses or other buildings; Amos Eaton's first laboratory at Rensselaer Polytechnic in 1825 was formerly a bank.[1] The Sillimans and J. P. Norton started laboratory work in 1847 in what became the Sheffield Scientific School at Yale in a house built in 1799. This was replaced by a larger more suitable building[2] in 1860. Harvard built Boylston Hall as a chemical laboratory for forty students in 1858, but an additional building was not available until 1913.[3]

At Johns Hopkins in 1876, "a small chemical laboratory was built" for Ira Remsen in an old house to accommodate about forty working students. The success of the Hopkins department soon required more space and the enlarged building, Dalton Hall, dedicated in 1883, housed 100 students. Remsen's description of the building gave other institutions a model for their laboratories.[4] He stated that "the laboratory not only works well on paper, like some of the chemical reactions which students are wont to originate, but, as a matter of fact, it has been found to be extremely convenient and practical...no serious complaint has been made against the working of an essential feature, though a large number of students have been constantly engaged in it during the year." The three-story and basement building contained three large student laboratories: a forty-student undergraduate laboratory, a thirty-student laboratory for incoming graduate students, and Remsen's organic research laboratory for fifteen to twenty students. Later the third floor room, the original location of the chemical and mineralogical museum, was made over to accommodate courses and research in quantitative analysis and physical chemistry. The basement housed storerooms and heating facilities as well as separate rooms for smelting furnaces and assays.

Appropriately sited were three large lecture rooms, two private offices and private laboratories, a fine, well-stocked library, and special rooms for gas analysis, balances, photometric equipment, preparations, and apparatus. For H_2S a special "stink-room" was completely isolated in a "separate, thoroughly ventilated building," to which all work with noxious gases was relegated. In the main laboratory table-tops were of black walnut; light, water, and gas supplies were good; the hoods, furnished by drafts from rows of gas-jets, drew well; and the building "affords facilities for every kind of chemical work." This laboratory served Hopkins for many years, long after the rest of the university was moved to the new Homewood Campus, and introduced hundreds of young men (and a few young women) to the value of laboratory work in chemistry.

During the 1880s and later, many colleges and universities built chemical laboratories.[5] This expansion was stimulated by the establishment of land-grant colleges and agricultural experiment stations by the Morrill Act of 1862 and the Hatch Act of 1887. The Johns Hopkins research school, both through its publications and its widely dispersed graduates, was also important. Another compelling reason was that "governing boards of colleges recognize that after a main building has been erected, the next should be a chemical building. The artfulness of teachers of chemistry, perhaps aided by their fumes, has caused their colleagues to exhibit little regret and display but minor envy in the placing of the chemistry department under a separate roof."[6] One of Remsen's Ph.D.s, J. N. Swan, who taught at Monmouth College and the University of Mississippi, designed many laboratories in various sections of the country.[7] Presumably the Hopkins laboratory had a strong influence on his ideas.

The University of Illinois (then called Illinois Industrial University) built a chemistry laboratory in 1878 for $40,000, which is still standing. It was furnished with city gas and water from a steel tank in the top story. (The previous quarters for chemistry had a kitchen stove as a heat source, and water was brought from the college pump.) The new building was partly destroyed by a lightning-caused fire in 1896; the heavy water tank crashed through two stockrooms to the cellar.[8] The department made the best of it until 1901, when the west portion of what is now the W. A. Noyes Laboratory was completed.

The University of Nebraska opened a new chemical laboratory in 1886, "planned after a careful study of the newest and best arranged laboratories of Europe and this country...Each room has its ventilation hoods, steam and sand baths, drying ovens, filtering pumps etc."[9]

These laboratories are typical of many others that were built in these years. Laboratory construction became common enough so that W. H. Chandler published a monograph in 1893 describing in great detail laboratories at Yale, Lehigh, MIT, Zurich, and Cornell.[10] The Lehigh laboratory built in 1884-1885 cost $200,000, with 24,000 square feet of floor space for chemistry and 5000 for metallurgy and mineralogy. It had steam baths and ventilating hoods activated by convection from steam coils. The building had an assay laboratory, gas analysis room, a spectroscopy room, and three rooms for industrial chemistry. The original plans for the Cornell laboratory called for forced ventilation going through the hoods, but the cost caused replacement of this feature by steam coils in the hood flues.

It is not clear how effective these convection-operated hoods actually were. In the late 1930s the Illinois hoods were almost completely useless,[11] but this may have been due to some structural deterioration in the flues, then nearly forty years old, not to the basic design. Hoods with electrically driven fans are reported[6] from the City College of New York laboratories in 1908. They were probably used earlier elsewhere, as electric power became generally available.

Before the 1850s, laboratory heating was done by alcohol lamps and varied types of coal, coke, or charcoal-fired furnaces. Amos Eaton's first student laboratory was furnished[1] with a "chemical air furnace, nearly on Dr. Black's [Joseph Black, a distinguished professor at Edinburgh] method." Michael Faraday's book of 1831 describes many types of furnace for heating laboratory materials.[12]

Coal gas for heating purposes became generally available around 1850, and Robert Bunsen's invention of the Bunsen burner in 1855 made this an indispensable laboratory tool.[13] It is safe to assume that virtually all laboratories after 1870 had gas available, except for a few in small towns. One ingenious substitute for coal gas is reported from the government laboratories of the Philippine Islands, where gas for laboratory use was generated by dropping coconut oil on red hot iron.[14]

The building of chemical laboratories in the south was of course hindered by the troubled conditions during reconstruction after the Civil War. However, Alabama Polytechnic (later Auburn University) obtained a well-equipped laboratory[15] in 1888. The University of North Carolina at Chapel Hill reported in 1877 that "a large and commodious two-story brick building has recently been fitted up...for an Analytical Laboratory."[16]

A considerable number of laboratories were destroyed or damaged severely by fire, such as the 1877 Chapel Hill laboratory.[16] The pages of *Science* report nine more cases over the years 1891 to 1916; most fires undoubtedly started from laboratory accidents.

Actual laboratory work by students was not always the rule even after 1870. The elder Silliman did not wish to have students work with his scientific equipment, because he used it for lecture demonstrations, and he feared the students would break it.[17] E. Emmet Reid, who graduated from Richmond College (now the University of Richmond) in 1892, taught at a small college for two years, taking his classes through an elementary chemistry textbook with small-scale classroom demonstrations.[18a] When he went to Johns Hopkins for graduate work in 1894, he introduced himself to Remsen, "who did not ask whether I had ever seen a chemistry laboratory, so there was no occasion to tell him I had not." After completing his Ph.D in 1898 and while teaching in his first job at the College of Charleston, South Carolina, he had desks built and started chemical laboratory training. In 1901 he was brought to Baylor University in Texas to plan the chemical laboratories on the second floor of the new science building and to teach the subject in the "best chemical laboratory in the state," next to the University of Texas.

According to Remsen, Williams College in 1872 had no provision for laboratory instruction in chemistry; Remsen soon remedied the situation. Due to lack of space, Harvard did not offer laboratory work in elementary organic chemistry until 1914, when Roger Adams started a one-semester laboratory course.[19] The University of Nebraska in 1888-1889 offered organic chemistry with two lectures and two hours of laboratory practice per week throughout the year.[9]

The rise of the laboratory method of instruction in Germany due to Liebig and Wöhler in organic chemistry greatly stimulated the German development of laboratory equipment and chemicals. The hundreds of students, American and European, trained in these methods provided an increasing market for the familiar German equipment, gradually displacing the earlier preference for English and French material, as Americans studying abroad shifted to German universities. German equipment and chemicals continued to be the major factor in American laboratories until World War I.

Most laboratory equipment[20] used routinely in organic chemistry in America was designed by German chemists or instrument makers, as the following list shows: coal gas burner, aspirator pump (Bunsen), combustion analyses for carbon and hydrogen (Liebig), extraction (Soxhlet), flask (Erlenmeyer), filter funnel (Buchner), burettes (Geissler and Mohr), refractometer (Zeiss and Abbe). The polariscope was developed by Biot in France around 1840, and its design changed little over the years. Analytical balances came first from the German maker Sartorius, although Christopher and Christian Becker from Holland established the manufacture of analytical balances in New York in the 1850s. W. A. Ainsworth, a jeweler, built balances in Denver from 1880. The need for sensitive balances for assay of gold and silver undoubtedly helped the American balance industry.

A few American dealers in imported supplies were established prior to the Civil War and provided materials for colleges, private laboratories, and industry.[17] Eimer and Amend, for example, was founded in New York City in 1851 by Bernard Amend, formerly an assistant to Liebig. Eben Horsford, professor of chemistry at Harvard, invited him to come to this country. Carl Eimer, his old schoolmate, joined him, and by 1874 the firm carried a full line of assay, chemical, and medical laboratory equipment, and chemicals.[21] By 1895 the glass-blowing shops were producing custom apparatus. Other early suppliers included Mallincrodt Chemical Works, 1867; Fisher Scientific Company, 1874, (which took over Eimer and Amend in 1940); Cenco Scientific Company, 1889; Pfaltz and Bauer, Inc., 1900; and J. T. Baker Chemical Company, 1904.

Before 1900 domestic manufacture was limited to simple apparatus turned out by local factories or craftsmen, such as woodenware (test tube racks), hardware (tripods, Bunsen burners, metal baths), ordinary glassware (bottles, tubing), and some scales, weights, balances, and thermometers, but no company devoted its entire output to scientific equipment.

The development of collegiate teaching and research, and the increasing employment of chemists by industry and by the government stimulated domestic production of laboratory equipment. Particular impetus came from the perfection of Babcock's assay for the amount of butterfat in milk in 1890, the Pure Food and Drug Act of 1902, and the analytical requirements of the rising petroleum industry, all of which required specialized laboratory glassware cheaply produced in quantity. Development of scientific glassware followed, and by 1914 the Kimble Glass Company in New Jersey was turning out items of excellent quality. With the propitious development of the borosilicate glass Pryex, the Corning Glass Works (established in 1851) was ready to enter the market in laboratory glassware as the 1914 war shut off shipments of Jena glass from Germany.[21] Pryex glass was so superior to the German product that competition from European articles, even after the war, was eliminated. Modern chemists, who daily take ample supplies of this hard, chemical-resistant glassware for granted, must respect earlier scientific workers who produced such voluminous and significant results with "soft" glass equipment, which required great care in heating and cooling to avoid breakage from thermal shock.

Porcelain ware came from overseas (Wedgwood, Royal Berlin) before World War I. Early American ceramic ware firms were unable to compete with European suppliers, but during the war the U. S. government pushed domestic manufacture of porcelain. An immigrant German brewer, Adolph Coors, and his son worked to provide good American porcelain. The only significant American design in porcelain equipment was the Gooch crucible for analytical chemistry, invented in 1878 by F. C. Gooch of Yale. Fused silica for combustion tubes came from England or Germany until 1912, when domestic material became available. Filter paper was obtained early from England (Whatman) and Germany, but domestic manufacture started around 1890.

Nevertheless, most educational institutions, favored by their right to import educational and instructional materials duty-free, ordered directly from Europe, chiefly from Germany. Usually the large apparatus order for the year was sent in the summer and the shipments received were unpacked and readied for the fall term. At Johns Hopkins basic organic chemicals were ordered from Kahlbaum in the late spring and used as starting materials for other compounds desired for research.[22] Unforeseen items needed later in the year meant long delays, usually of six weeks or more, a frustrating experience for chemists and cited as one of the difficulties of carrying out original research in this country. Arthur Michael, not noted for the paucity of his published research, declared in 1909 that his output of work during the past thirty years had been reduced by fifty percent due to this handicap.[23]

The situation at the University of Kansas in 1884, described by C. A. Kraus,[24] was undoubtedly typical. "Anyone who was not at one of the mid-western universities during the last quarter of the nineteenth century can have no comprehension of the meagerness of the physical facilities...Glass suitable for glass-blowing purposes was unobtainable in this country; chemicals had to be ordered from Germany, and, frequently, had to be made because of the length of time required in obtaining them from abroad."

American dependence on Germany in 1914 for industrial and research organic chemicals is discussed elsewhere and requires no further elaboration here. In 1922 a new tariff act gave protection to the growing American chemical and scientific apparatus industry, and American firms became the major suppliers of American laboratories.

A fundamental fact in any discussion of chemical instruction in organic chemistry in the period 1876 to 1914 is the very great increase in the number of students. This is reflected indirectly in the increase in membership of the American Chemical Society, from 230 in 1876 to 7170 in 1914.[25] According to a survey of 59 educational institutions made in 1901 by William A. Noyes for the 25th Anniversary of the American Chemical Society,[26] only 15 were offering organic chemistry in 1876. They had a total of 294 students taking lectures and 11 taking laboratory work in organic chemistry, these at MIT, Pennsylvania, and Virginia. Only 7 students were doing organic research in 1876, 5 at MIT and 2 at Virginia. In 1901 many hundreds of students were taking lecture and laboratory work in organic chemistry in 59 institutions, and 128 were doing research.[27] The increase in schools and in research students is striking. The largest groups of research students were at Chicago (15), MIT, University of Michigan (9 each), Columbia, Johns Hopkins, University of Wisconsin (8 each), University of Minnesota (5), Yale and several others (3 each). In 1901 to 1904, 9 to 15 Ph.Ds were awarded in organic chemistry each year.

At Chapel Hill in 1875, organic chemistry was listed as the fifth part of the General Chemistry course, in which "instruction is given by textbooks and lectures, and illustrated by numerous experiments." The course met three times a week for ten months. The laboratory course met four hours a week for ten months, although organic chemistry is not mentioned specifically. Apparently blow-pipe and qualitative analysis were the main features of the laboratory work.[16] The textbooks listed were Fownes' *Chemistry*, Fresenius' *Qualitative Analysis*, and for reference, Miller's *Chemistry*. George Fownes (1815-1849), an English chemist with a Ph.D. from Giessen, wrote widely used textbooks, which appeared in American editions from 1845 to 1885. William A. Miller (1817-1870), also English, worked with Liebig in Giessen and was a distinguished spectroscopist. He wrote a three-volume treatise on chemistry, the third volume dealing with organic. This series reached six editions in England, with the last American edition 1879-1885. From the publication record of these books it is clear that they were standard texts in this country for several decades.[28] C. Remigius Fresenius (1818-1897) studied at Bonn and with Liebig at Giessen and was a German university professor. He developed the systematic scheme of qualitative analysis for inorganic ions and founded the standard German journal in analytical chemistry, *Zeitschrift für Analytische Chemie*. His textbook on qualitative analysis went through many editions.

These books by Fownes, Miller, and Fresenius are listed in the 1891 catalogue[29] of the Heil Chemical Co., so they were still in use at that time. This catalogue contains the names of about 175 science texts and reference books, of which fifteen are organic chemistry. These include textbooks by H. E. Armstrong (English), A. Bernthsen (German), M. Berthelot (French), E. Hjelt (Scandinavian), H. E. Roscoe and C. Schorlemmer (English); organic laboratory manuals by Emil Fischer (in English), H. C. Jones (American); and books on qualitative and/or quantitative analysis of organic compounds by Allen (English), F. Beilstein, and A. B. Prescott (American). The catalogue also lists J. P. Cooke's books, several encyclopedias, and Gmelin's eighteen volume handbook of chemistry (translated by Henry Watts).

The first widely used American organic text was Ira Remsen's book[30] of 1885, which contained directions for simple laboratory experiments within the text. In 1893 W. R. Orndorff, an Ira Remsen Ph.D. of 1887, published a laboratory manual to accompany Remsen's text based on his experience in the large teaching laboratory at Cornell. The books went through many editions and foreign language translations, and, compared to more detailed German texts, "Remsen's little book on organic chemistry" was the best for beginning students, in the eyes of one reviewer.[31] The Remsen text, however, is very scant on theory, and even in its later versions is deficient not only in important topics like triphenylmethyl, but in the fundamentals of stereochemistry. A student who had studied Remsen's texts in high school and college called them "good books of their kind, but they were highly descriptive, and, in a factual sense, a 'catalogue of ships.'"[32]

William A. Noyes' textbook of organic chemistry (1903 and several later editions) offered a broader and more searching account with excellent sections on physical properties and measurements and on reaction mechanisms. A. B. Prescott greeted the book as a "refreshing innovation" and a "very valuable contribution" to the primary organic chemistry texts. Remsen was much more reserved, thinking that Noyes' approach was too severe and demanding on the student.[33]

Reference books for organic chemistry included the encyclopedic *Beilstein*, the third edition published in 1892-1899, the fourth from 1918 to 1937.[34] His volumes gave a systematic coverage of the known facts about all organic compounds, and the publication was taken over by the Deutsche Chemische Gesellschaft around 1900. Other German compendia included M. M. Richter's tables and classification of carbon compounds in 1900, and these volumes[35] were revised and continued by R. Stelzner through 1921. Beilstein and these other monumental volumes made possible for the first time a quick and thorough literature survey of any organic compounds and their reactions, advancing organic chemistry as an organized science and promoting its scintillating growth. Not until *Chemical Abstracts*, begun in 1906 with W. A. Noyes as editor and sponsored by the American Chemical Society, had grown in scope and utility could Americans match the contributions of the German chemists to the codification of the chemical literature.

Less extensive than the great compendia, but extremely useful, was the multivolume reference book by Victor Meyer and Paul Jacobsen which appeared (in German) from 1889 on.[36] Other books ranging from reference to textbooks included A. F. Holleman's text,[37] which went through seven English editions to 1930, Bernthsen's text[38] of 1894, and the Perkin and Kipping book[39] of 1903.

Books on practical organic chemistry (laboratory manuals) in German and English multiplied; examples were Emil Fischer's *Exercizes*, 1887; Lassar-Cohn's *Arbeitsmethoden*, first edition 1890; and notably L. Gatterman's *Praktis*, first published in 1894, translated into English by W. B. Schober (a Remsen Ph.D.) in 1896, and by 1927 in its twentieth German edition.[40] Although ability to read chemical German was essential for advanced students, most elementary books were available in English. J. B. Tingle, an associate of Remsen at Hopkins in teaching and editing 1904-1907, translated three: E. Hjelt's text, a laboratory book by Lassar-Cohn, and a manual for analysis of organic compounds by H. Meyer. The experiments in elementary organic laboratories did not change much, in subject matter and equipment, from Emil Fischer's book of 1887 to the middle of the 1930s. The principal difference was in the use of Pyrex glass after World War I in this country. The Instrumental Revolution did not affect undergraduate instruction notably until after 1940.

Monographs on special topics appeared frequently. Heil's catalogue lists A. Pictet's book on alkaloids (in French), for example. Two two-volume treatises by English chemists were widely used for advanced organic courses. Julius B. Cohen's volumes [41] gave reviews of stereochemistry, various classes of natural products, and detailed accounts of reaction mechanisms and physical organic chemistry, especially the work of Arthur Lapworth. Alfred W. Stewart's reviews were similar, but less comprehensive in physical chemistry.[42]

The use of qualitative organic analysis and the identification of unknowns in teaching organic chemistry was largely an American development. The first text was a pamphlet[43] by A. A. Noyes and S. P. Mulliken of MIT of 1896. Noyes soon turned to physical chemistry, but Mulliken devoted a lifetime of patient scholarship to the preparation of four large volumes[44] which appeared from 1905 to 1922. They described the identification of over 10,000 compounds, plus many dyes. Mulliken (1864-1934, B. S. 1887, MIT; Ph.D. 1890, Leipzig) taught at MIT from 1895 on. His son, R. S. Mulliken, of Chicago, received the Nobel Prize in 1966 for his work on molecular structure.

H. T. Clarke (later professor of biochemistry at Columbia) published a frequently reprinted text[45] in qualitative organic analysis. Oliver Kamm at Illinois taught a course and wrote a well-known volume.[46] Like Kamm's the classic text of Shriner and Fuson[47] gave valuable experience in handling organic compounds in the laboratory. The Instrumental Revolution changed "qual organic" into instrumental analysis courses after 1940.

The *American Chemical Journal* (ACJ) was the most important American chemical journal from 1879 to about 1900. Remsen's account of his founding it states:[48]

> Up to the year 1878, the present writer had sent his chemical articles to the "American Journal of Science and Arts", and they had been kindly received. In the year named, however, a number of articles were sent in together and, in the course of a few days, they were returned with a long and extremely friendly letter from Professor Dana, in which he made it clear that he could not publish the articles without entirely changing the character of his journal; and, at the same time, he strongly urged the starting of a chemical journal. The letter led to correspondence which resulted in the foundation of this Journal.

Ira Remsen thus started the ACJ at Hopkins in 1879 and continued to edit it until the journal was combined with the *Journal of the American Chemical Society* (JACS) in 1913. The two chemical journals began in the same year, but until 1899 the papers in the JACS were of slight scientific importance, whereas the ACJ published most of the good work in the United States, particularly in organic chemistry. The high quality of papers maintained in the ACJ until well after 1900 (when the JACS began to excel) had decisive influence on the development of organic chemistry in this country. *The Journal of Biological Chemistry* (JBC), founded in the United States in 1905, also published much good work in organic chemistry from its inception.

Remsen's editorial policies in conducting his journal had a definite didactic cast, in addition to his aim in providing an avenue of publication for original research. Textbooks and monographs, both foreign and domestic, were reviewed; important advances in fundamental or applied chemistry[49] were reported, summarized, or sometimes reprinted. Reports on special topics, such as stereochemistry,[50] Emil Fischer's carbohydrate work,[51] and tautomerism[52] were also part of Remsen's effort to keep American chemists abreast of current advances.

Summer schools at the large universities provided oases for the teachers in small colleges. Reid, teaching at the College of Charleston and at Baylor, lamented his inability to plan research projects; his personal library was inadequate and little was available in the colleges.[18b] He welcomed summer school at Hopkins and Chicago for the purpose of library research. In small colleges it was not uncommon for the professor's own shelf to furnish the only reference books and journals. Rachael Lloyd performed her collegiate work in chemistry at the Harvard summer school, which began in 1874 and offered courses, laboratory instruction, and some summer teaching positions to chemists from smaller institutions. Louis J. Bircher, of Vanderbilt University, attended summer sessions for many years at Chicago to achieve his Ph.D. in chemistry in 1924.

During this period another movement in education was reflected in the pages of the ACJ: the broadening of advanced education for women. Over thirty papers in organic chemistry were published with more than twenty women as co-authors between 1881 and 1913, and half as many again in other journals.[53] In this progress Johns Hopkins was not a leader. Although by 1900 coeducation was widely accepted in the United States,[54] Hopkins excluded women from its graduate school until 1907 and granted only two Ph.D.s in organic chemistry to women, Bessie Brown and Julia Harrison, in this interval.[55]

The first woman to publish organic chemistry in the ACJ was Rachel Lloyd, a co-worker with C. F. Mabery at Harvard Summer School for three years from 1881. Lloyd took her Ph.D. at Zurich in 1886 and accepted an appointment at Nebraska to teach and to help establish the beet-sugar industry of that state; she rose to full professor in 1888.[56] Scattered publications came from women in the Midwest and New England, where the women's colleges were training women in chemistry to teach chemistry or home economics in women's colleges or state institutions. A number of women took advanced degrees, taught, and performed research, for example, Dorothy Hahn (Ph.D. 1916, Yale) at Mount Holyoke in organic chemistry, and Jennie Tilt (M.S. 1910, Purdue) and Marie L. Fossler (A.M. 1898 Nebraska) in physiological chemistry and nutrition. Helen Abbott Michael (the wife of Arthur Michael) was privately educated and widely traveled; between 1883 and 1896 she published fifteen papers on organic constituents of plants and in pure organic chemistry.[57] Her account of visits to various European laboratories and their attitudes toward women students is enlightening. The most important centers in America for women in organic chemistry were Chicago, open to women since its inception in 1892, and Bryn Mawr College.

At Chicago from 1892 to 1913, seven women took doctoral degrees with J. U. Nef and J. Stieglitz, and six of these established reputations for teaching and research: Willey G. Denis, professor and head of biochemistry at Tulane; Emma P. Carr, in physical chemistry at Mount Holyoke and chairman of its well-recognized department; Ethel Terry McCoy and Edith Barnard, organic chemistry at Chicago; and in nutrition, Katharine Blunt at Vassar and Chicago and Nellie Goldthwaite at Illinois and Colorado.

Bryn Mawr, opening for women and offering graduate degrees in 1885, was modeled on Johns Hopkins by the indomitable M. Carey Thomas, dean and later president, who had once been refused full participation in the Hopkins graduate school. A steady procession of Hopkins chemists from 1885 to 1924 emphasized this scholarly connection and published with a few students, but E. P. Kohler's vigorous research in organic chemistry from 1892 to 1912 produced thirty papers, some of them joint papers with undergraduates, and several doctorates. Among these Margaret MacDonald taught and conducted biochemical research at several colleges, experiment stations, and the Johns Hopkins Medical School, while Marie Reimer headed the chemistry department at Barnard College and with her students published original studies in conjugated systems. Encouraging as these beginnings were for women in chemistry, their contributions and advancement remained relatively limited.[53]

In all of the various subjects discussed in this essay, we see the United States initially as a scientific colony, dependent for models and equipment on European sources. Gradually, as universities develop advanced training, chemical industry grows, and government laboratories increase in number, the country becomes more and more an independent scientific power. This process was accelerated in some ways by World War I, but it would have come about anyway.

LITERATURE CITED

1. Ethel M. McAllister, *Amos Eaton*, Philadelphia, 1941, pp. 374ff.
2. R. H. Chittenden, *History of the Sheffield Scientific School*, New Haven, 1928, pp. 47, 73.
3. S. E. Morison, Ed., *The Development of Harvard University*, Cambridge, 1930, pp. 262, 271.
4. Science, 5, 25(1885); although unsigned, clearly indicates Remsen as the author.
5. Documented by reports in Science, passim.
6. C. Baskerville, Science, 28, 665(1908).
7. A. Hume, Ind. Eng. Chem. News Ed., 13, 49(1935).
8. S. W. Parr, *University of Illinois, Circular of Information of the Department of Chemistry*, 1916-1917, pp. 18ff.
9. *University of Nebraska Catalogue and Register*, 1888, p. 86.
10. W. H. Chandler, *The Construction of Chemical Laboratories*, Washington, D. C., 1893. This contains floor plans for the buildings described.
11. DST, personal experience, 1937-1938; according to E. H. Volwiler, the Illinois hoods were ineffective in 1914.
12. Michael Faraday, *Chemical Manipulations*, from the last London ed., Philadelphia, 1831, pp. 92ff.
13. *A History of Technology*, Charles Singer et al., Eds., Oxford, Vol. IV, pp.272 ff.
14. P. C. Freer, Science, 20, 105(1904).
15. Anon., ibid., 11, 126(1888).
16. M. M. Bursey, *Carolina Chemists*, Chapel Hill, 1982, pp. 29 ff. This book contains a large amount of information about the Chapel Hill department, including faculty, students, curricula, and student life from 1816 to 1980.
17. C. A. Browne, JCE, 9, 712(1932).
18. E. Emmet Reid, *My First Hundred Years*, New York, 1972, a) p. 70, b) pp. 85,91.

19. D. S. Tarbell and A. T. Tarbell, *Roger Adams*, Washington, 1981, p. 35.

20. Our account is based largely, unless otherwise indicated, on Ernest Child, *The Tools of the Chemist*, New York, 1940, and on F. Kraissl, Sr., JCE, **10**, 519(1933), an interesting article on the history of the chemical apparatus industry by an ex-producer of apparatus of the late 19th century.

21. W. Haynes, *The American Chemical Industry*, New York, 1949, Vol. VI, pp. 165 ff; for Corning Pyrex glass, p. 89.

22. E. E. Reid, op. cit., p. 65, for a vivid account.

23. M. A. Rosanoff, Science, **30**, 645(1909).

24. C. A. Kraus, ibid., **85**, 232(1937).

25. H. Skolnik and K. M. Reese, Eds., *A Century of Chemistry*, Washington, 1976, p. 456.

26. W. A. Noyes, "Report of Census Committee," supplement to JACS, **24**, 1902, p. 114. Harvard apparently did not reply to the questionnaire; H. B. Hill had started the first organic course there in 1873.

27. The totals of students taking organic chemistry include an unspecified number in courses primarily for agricultural students, ref. 16.

28. Information from the *National Union Catalog* and the *Dictionary of National Biography*.

29. *Catalogue and Price-List of Chemical and Physical Apparatus*, Henry Heil Chemical Co., St. Louis, 1891; in the University of Illinois library.

30. I. Remsen, *An Introduction to the Study of Organic Chemistry*, Boston, 1885; fifth revision, with W. R. Orndorff, 1909.

31. L. L. Van Slyke, JACS, **17**, 500(1895).

32. A. Klock, JCE, **30**, 40(1953). The "catalogue of ships" refers to Homer's tedious listing of the ships and tribes taking part in the siege of Troy (*Iliad*, book 2).

33. Reviews: A. B. Prescott, JACS, **25**, 774(1903); I. Remsen, ACJ, **31**, 85(1904).

34. *75 Jahre Beilsteins Handbuch der Organischen Chemie*, F. Richter, Ed., Berlin, 1957. K. F. Beilstein (1838-1906) was born in Russia of German parents, was educated in Germany, and taught at the Technical Institute of St. Petersburg from 1866 until his death.

35. M. M. Richter, *Lexikon der Kohlenstoff-Verbindungen*, 2 vol., Hamburg and Leipzig, 1900, a new edition of Richter's *Tabellen* 1884; R. Stelzner, *Literatur-Register der Organischen Chemie Nach M. M. Richter's Formelsystem*, Braunschweig, 1910-11.

36. *Lehrbuch der Organischen Chemie*, Vol. I, Leipzig, 1889; publication of a revised version by P. Jacobsen and R. Stelzner started in 1907, and this was reprinted in 1922.

37. Holleman was professor at Amsterdam; the second English edition of his *A Textbook of Organic Chemistry* was reviewed by J. Stieglitz, JACS, **30**, 158(1908) and the seventh by R. C. Fuson JCE, **7**, 3023(1930). By that time it had gone through 52 editions in nine languages.

38. A. Bernthsen, *A Textbook of Organic Chemistry*, 4th ed., transl. by G. McGowan, New York, 1894, and many later eds.

39. W. H. Perkin and F. S. Kipping, *Organic Chemistry*, 1st ed., London, 1894, 2nd ed., Philadelphia, 1903, and many later revisions.

40. E. Fischer, *Exercizes in the Preparation of Organic Compounds*, Würzburg (1887); transl. by A. Kling, Glasgow, 1889; Lassar-Cohn, *Arbeitsmethoden für Organische Chemische Laboratorium*, Berlin and Leipzig, 1890. Many eds. in German; transl. by Alexander Smith, 1895, and by R. E. Oesper, 5th revised ed., Baltimore, 1928; L. Gatterman, *The Practical Methods of Organic Chemistry*, Heidelberg, 1894, transl. by W. B. Schober, New York, 1896.

41. J. B. Cohen, *Organic Chemistry for Advanced Students*, London, 1907, 2 vol. and later eds. Cohen (1859-1935) was professor at Leeds. E. H. Volwiler used Cohen at Illinois in 1914.

42. A. W. Stewart, *Recent Advances in Organic Chemistry*, London, 1908, 2 vol., and later eds. The 1927 version has a good account of G. N. Lewis' electronic theory of valence.

43. A. A. Noyes and S. P. Mulliken, *Laboratory Experiments on the Class Reactions of Organic Substances and their Identification*, Boston, 1896; Easton, Pa., 1897, 1915. The 1897 ed. was favorably reviewed by Remsen in ACJ, **20**, 251(1898).

44. S. P. Mulliken, *A Method for the Identification of Pure Organic Compounds*, New York, Vol.1, 1905; Vol. 4, 1922.

45. H. T. Clarke, *A Handbook of Organic Analysis, Qualitative and Quantitative*, London, 1912, and later eds.

46. Oliver Kamm, *Qualitative Organic Analysis*, New York, 1923, revised ed. 1932.

47. R. L. Shriner and R. C. Fuson, *The Systematic Identification of Organic Compounds*, New York, 1935, and several later eds.

48. I. Remsen, ACJ, **17**, 470(1895).

49. ACJ, **8**, 51 ff. (1886), Victor Meyer's discovery of thiophene and proof of structure; p. 63 ff., the evaluation of commercial phosphate; p. 291 ff., the constitution of levulose and dextrose; 9, 453(1887), a very reserved review of J. Wislicenus' proposal of *cis-trans* isomerism; 11, 124 (1889), recent progress on sulfuric acid manufacture; p. 437, the structure of benzene; these reviews and reports are representative. The statement of Beardsley, *American Chemical Profession*, op cit., that the ACJ seldom noticed material from the field of applied chemistry is thus an exaggeration.

50. A. Eiloart, ACJ, **13**, 573(1891); **14**, 58(1892).

51. E. H. Keiser, ibid., **11**, 277(1889); **12**, 357 (1890).

52. W. R. Orndorff, ibid., **14**, 238(1892).

53. This section was written before the appearance of Margaret W. Rossiter's *Women Scientists in America*, Baltimore, 1982. Our information was obtained from papers published by women in chemical journals and from *American Men of Science*.

54. Rudolph, *American College*, pp. 306 ff.

55. *Johns Hopkins Half-Century Directory*.

56. A. T. and D. S. Tarbell, JCE, **59**, 743(1982).

57. Helen Abbott Michael, *Studies in Plant and Organic Chemistry, and Literary Papers*, Cambridge, 1907; A. T. Tarbell and D. S. Tarbell, JCE, **59**, 548(1982). Neither Rachel Lloyd nor Helen Michael is listed in E. T. James, Ed., *Notable American Women*, Cambridge, 1971, Vols. I and II. For more on Blunt, Carr, Hahn, Lloyd, McCoy, and Michael, see Patricia J. Siegel and K. T. Finley, *Women in the Scientific Search: An American Bio-bibliography, 1724-1979*, Metuchen, N. J., 1985.

MORRIS S. KHARASCH

7

The Development of Ideas of Valence
in Organic Chemistry

In an earlier essay, the development of the basic ideas of the structural theory of organic molecules was discussed. These ideas were simple: the tetravalent nature of carbon; the occurrence of linkages between carbon atoms, single, double, or triple; the tetrahedral nature of the saturated carbon atom, with the consequent possibility of stereoisomerism for compounds containing asymmetric carbons; and the occurrence of *cis-trans* isomerism for suitably substituted carbon-carbon double bonds. The occurrence of single or multiple linkages of carbon with other elements was also postulated, and by 1875 the structural theory was complete. It succeeded brilliantly in predicting the maximum number of isomers or stereoisomers and in allowing the assignment of structural formulas to the many thousands of organic compounds of synthetic or natural origin. Its greatest triumph and corroboration undoubtedly was the determination by Emil Fischer (1852-1919) of the three-dimensional structures for the sixteen aldohexose sugars, which contain four asymmetric carbons. In addition to this spectacular achievement (both for Fischer and for the theory), the theory provided a reasonably satisfactory theoretical foundation at first for the vast structure of organic chemistry which workers in Germany and other countries erected between 1860 and 1900.

The theory nevertheless had serious weaknesses, both aesthetic and practical. It predicted the patterns in which atoms could be combined to form molecules, but it gave no information about the *how* and *why* of bond formation. It indicated bonds or valences by a line, and three-dimensional structures by tetrahedra or planar formulas with suitable conventions to indicate three dimensions, but it gave no clue to the nature of bonds, the reason for the tetrahedral structure of saturated carbon, or the constitution of unsaturated linkages. This complete lack of explanation for the reasons behind the myriad observations about structure and reactions was apparently felt at first, in a fashion not usually articulated, as an aesthetic shortcoming. The more reflective early chemists, like Arthur Michael, responded to this limited character of the structural theory by developing their own systems of additional theoretical concepts.

What had started as a feeling of the intellectual bareness of the structural theory soon developed into a practical matter affecting the development of the science. The very success of the structural theory in facilitating synthesis and in allowing assignments of structure led to such a proliferation of compounds and reactions that the correlation of the information at hand and the prediction of new results urgently required more extensive basic knowledge. The elucidation of reaction mechanisms (the paths by which reactions occur) and the correlations of chemical structure with both chemical reactivity and biological activity required something more fundamental than dashes and tetrahedra.

It is sometimes thought that Emil Fischer was interested only in structure and synthesis, a field in which he was the acknowledged master. He was in fact actively committed to the application of organic chemistry to biological problems,[1] and he also made fundamental contributions to reaction mechanisms, particularly to the explanation of the Walden Inversion and to the proof of *trans* addition of bromine to double bonds.[2] He expressed his concern with the limitations of the classical structural theory.[2]

There was one class of bonding for which the structural theory had never been satisfactory, that in benzene and other aromatic compounds, and in conjugated systems of unsaturated linkages in general. In benzene the valence rules could be satisfied formally by Kekulé structures **1a** or **1b**. These correctly predicted the number of isomeric substituted benzenes, assuming that the double bonds oscillated rapidly enough so that all positions on the benzene ring were equivalent. However, the stability of benzene to oxidation and the prevalence of substitution over addition reactions, **1 → 3**, showed that if double bonds were actually present in benzene, their behavior differed sharply from that of the double bond in ethylene or cyclohexene **2**. The many other formulas proposed for benzene, aside from Ladenburg's prism formula which was disproved by chemical evidence, merely said, in effect, that there was something unusual about benzene and its reactions which did not fit in with the classical structural theory.[3] These formulas, aside from Ladenburg's, did predict correctly the number of isomers in substituted benzenes.

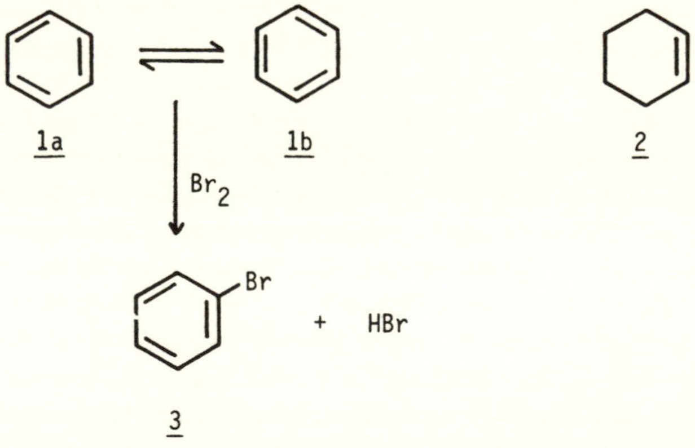

The difficulties with formulating a satisfactory structure in classical terms for benzene were encountered in a less aggravated form with *conjugated* systems like butadiene **4** and other conjugated systems containing a hetero atom. Benzalacetophenone **5** is the prototype for conjugated carbonyl compounds.

In some reactions these conjugated systems behaved "normally", as in the addition of Br_2 to **4** to give the 1,2-dibromo compound **6a**, but with simultaneous 1,4-addition to give **6b**. Similarly, the conjugated ketone **5** could give reactions of the carbonyl group, reactions of the carbon-carbon double bond, or the reaction illustrated, addition of the Grignard reagent, C_6H_5MgBr, to form (after hydrolysis) the product **7**, presumably via 1,4-addition. This reaction, like the formation of **6b**, involves the whole conjugated system; the Michael reaction undoubtedly takes a similar course.

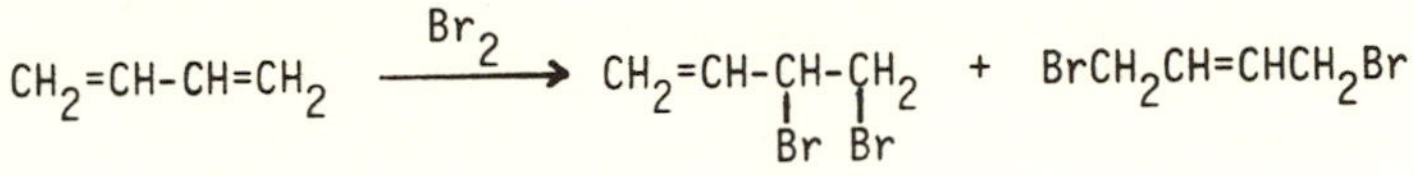

$$CH_2=CH-CH=CH_2 \xrightarrow{\ Br_2\ } \underset{\underset{Br\ Br}{|\ \ |}}{CH_2=CH-CH-CH_2} \ + \ BrCH_2CH=CHCH_2Br$$

4   **6a**   **6b**

$$\overset{\overset{O}{\|}}{C_6H_5CH=CHCC_6H_5} \xrightarrow[\ \ 2.\ H_2O\ \ ]{1.\ C_6H_5MgBr} \underset{\underset{C_6H_5}{|}}{C_6H_5CHCH_2}\overset{\overset{O}{\|}}{CC_6H_5}$$

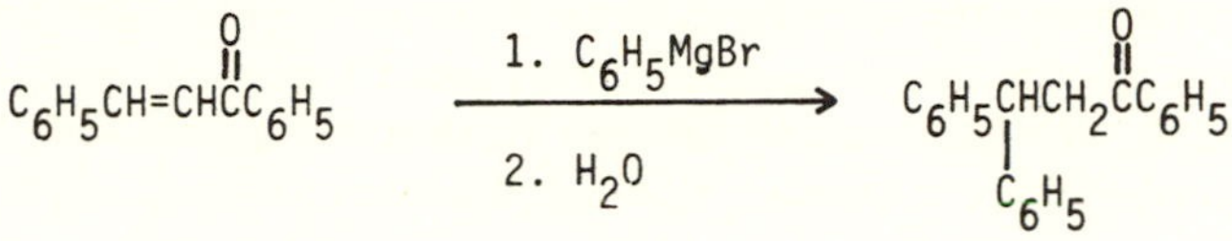

5   **7**

Aromatic compounds like benzene and conjugated systems like **4** and **5** also show characteristic absorption in the ultraviolet spectra entirely different from that of non-aromatic compounds, such as **2** with isolated double bonds. These and many other physical and chemical properties showed the need for more detailed theories.

The most successful treatment of aromatic and conjugated systems, before the advent of quantum mechanics, was Johannes Thiele's partial valence theory.[4] His idea was that double bonds did not use all the combining power of the atoms but left them with an unused *partial valence*, indicated by dotted lines as in **8** and **9**. For conjugated systems, as in **10** and **11** (or **4** and **5**), the partial valences of the two interior atoms would saturate each other, indicated by a curved line, and the two remaining partial valences would be left hanging over the ends of the conjugated system. Thiele postulated that all additions to a conjugated system should take place at the ends, that is 1,4-addition should be the rule. Thiele's own work showed that 1,4-addition was indeed general. However, more extensive experimental work by many workers, including Arthur Michael,[5,6] showed that 1,4-addition was not the invariable rule, and indeed it is not clear why Thiele's ideas would require it.

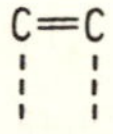

C=C   C=O   C=C—C=C   C=C—C=O

8 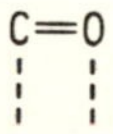   **9** 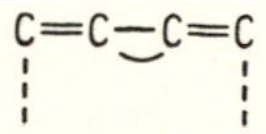   **10** 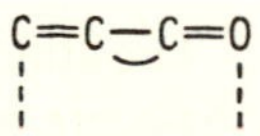   **11**

Thiele's partial valence structure for benzene is shown by **12**; because benzene is completely conjugated, there are no partial valences left over. Michael[5] objected to the lack of symmetry of **12**, and R. Willstätter pointed out that his synthetic cyclooctatetraene[7] **13**, whose structure he established and which by Thiele's views should resemble benzene in properties, was actually completely different from benzene in behavior. Thiele's partial valence theory was thus only partially successful as a guide to the behavior of conjugated systems, but its insight was confirmed by the later quantum mechanical theories of aromatic and conjugated systems.

<u>12</u>   <u>13</u>

The early structural ideas of Berzelius[3] were dualistic; they involved the union of positively and negatively charged atoms or groups, whereas the structural theory in general disregarded the role of charge separations and ions in organic structures. Several factors caused a revival of structural theories based on the electrical nature of bonding and the role of separated charges.

In his Faraday Lecture of 1881 in London, H. von Helmholtz[8] showed that Michael Faraday's work on electrolysis demonstrated that electricity consisted of indivisible units of charge. Helmholtz concluded that chemical bonding involved integral quantities of electricity, and that, contrary to Berzelius, there was not a continuously variable amount of electricity in different bonds or molecules.

This concept of "atoms" of electricity was supported by Arrhenius' theory of ionization[9] in 1887. J. J. Thomson's discovery of the electron in 1897 and his experiments on the charge-to-mass ratio in electrons and positive ions[10] gave an even stronger impetus to the development of new ideas of valence. A number of papers resulted after 1900 in which dualistic formulas with bonds indicated by plus and minus charges were the characteristic feature.[11,12,13] For example, W. A. Noyes[13a] suggested that ammonia could react with chlorine by ionizing in two ways; chlorine likewise could react as positive or negative ions. The implication of this idea, that NCl_3 could

$$Cl^- \qquad\qquad\qquad Cl^+$$
$$+$$
$$N + Cl^- \qquad\qquad \overline{N}| \;\; Cl^+$$
$$+ \qquad\qquad\qquad\quad -$$
$$Cl^- \qquad\qquad\qquad Cl^+$$

<u>14</u>   <u>15</u>

exist in two forms, **14** and **15**, called electron isomers or *electromers*, was extended by others[13b,13c] to many organic compounds, but attempts to isolate electromers were unsuccessful.[13] The quantum mechanical theory of bonding was to show why such electromers could not exist as discrete structures.

This dualistic theory,[13] which developed luxuriantly in America, included the idea that valence was a transfer of electrons, a directed force, and led to structures[13b] such as **16** and **17** for maleic and fumaric acids, assumed to differ only in the relative direction of the carbon valences. Other formulas[13b] like **18** for carbon tetrachloride and **19** for methyl chloride were suggested. Benzene was represented by **20**, which is a composite of six different charge distributions. Analogous structures were written for benzene analogs with two or more aromatic rings.[13c]

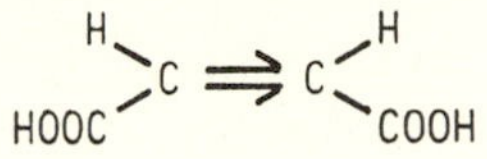

16

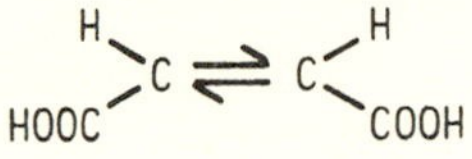

17

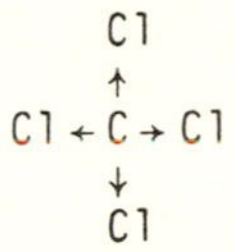

18

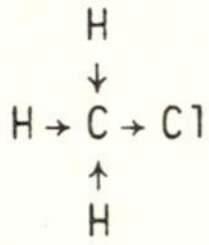

19

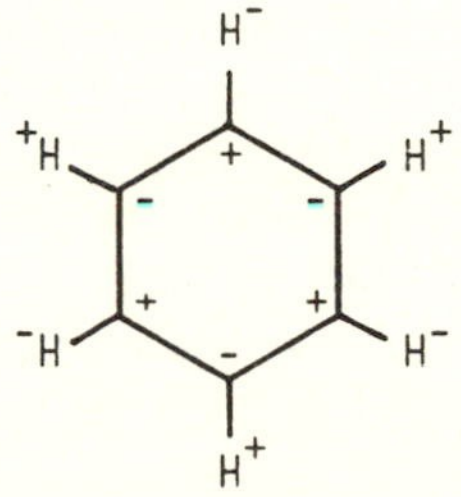

20

An important paper of 1904 by R. Abegg[14] emphasized the tendency of elements at both ends of the first rows of the periodic table to form ions. He pointed out the corollary: carbon, in the middle of the first row, tended to be electrically neutral.[15] He followed Helmholtz in attributing valence to electrons and concluded that atoms had two kinds of valence, normal and contra-valence, the absolute sum of which is eight, as shown in his table.[16] The number eight represented for all atoms the number of "points of attachment" for electrons. (Most writers in English have used the more literal translation of "points of attack" for his term, "Angriffstelle"). His scheme was ambiguous, as he recognized, for the noble gases which could have normal and contra-valences of 0, −8 or 0, +8, depending on their place in Table 1.

Table 1. (Abegg)

Group	1 (Li)	2 (Be)	3 (B)	4 (C)	5 (N)	6 (O)	7 (F)
Normal Valence	+1	+2	+3	±4	-3	-2	-1
Contra-valence	(-7)	(-6)	(-5)		(+5)	(+6)	(+7)

Thomson[17a] suggested that atoms, which he regarded as spheres of positive charge with electrons imbedded, were held together by partial interpenetration of the spheres. When Rutherford's work showed that the positive charge resided in a very small atomic nucleus, Thomson[17b] suggested that atoms were held together by two "tubes of force" each containing an electron.

These and other proposals indicated that valence theory was in a state similar to the structural theory before Kekulé; the time was ripe for a new synthesis. The field was in "the state of imaginative muddled suspense which precedes successful inductive generalization."[18]

The new synthesis was achieved by Gilbert Newton Lewis (1875-1946), the greatest American chemist[19] since Willard Gibbs. Lewis was brought up in Nebraska, received his undergraduate and doctoral training with Richards at Harvard, and spent a year in Nernst's laboratory in Berlin. Harvard proved to be unstimulating intellectually to the highly original young man, Richards being a poor judge of talent, and Lewis soon joined A. A. Noyes at MIT, where the first good school of physical chemistry in this country was being developed. In 1911 Lewis went to Berkeley, where he built up one of the greatest schools of physical chemistry anywhere.

In his fundamental paper of 1916,[20,21] Lewis quoted an unpublished memorandum of his from 1902 in which he suggested that each atom tended to form a stable shell of two inner electrons and eight outer electrons and that these shells were interpenetrable. He placed the electrons at the corners of a cube. His 1902 model explained well the formation of ionic compounds by transfer of electrons to form stable ions with the requisite electronic structures, as in the formation of sodium chloride, Na^+Cl^-, from a sodium and a chlorine atom.

$$Na(2 + 8 + 1) \text{ electrons } + Cl(2 + 8 + 7) \text{ electrons } \longrightarrow$$
$$Na^+(2 + 8) \ Cl^-(2 + 8 + 8)$$

His model did not show how non-ionic bonds, as in organic compounds, were formed, and apparently the ideas were not published at that time for that reason. Evidently influenced somewhat by the speculations[12,22,23] of a young Englishman, A. L. Parson, who was visiting Berkeley, Lewis proposed in 1916 that non-ionic bonds are formed by *sharing* a pair of electrons between atoms. This *covalent* bond (the name due to I. Langmuir) was visualized as two cubes sharing an edge. The simplest molecule, H_2, is held together by one pair of shared electrons, and in its combination with other atoms H is bonded similarly by one shared pair.

Lewis considered that there was a continuous gradation of polar character (or degree of electron sharing) in bonds. One extreme was the complete transfer of electrons to form ions, as in a NaCl molecule or crystal **21**; the other was represented by the chlorine molecule **22**, or by ethane **23**. In these examples, the atoms all have complete octets, by transfer or by sharing. The Na^+ ion has a shell of eight electrons (not written), the Cl_2 molecule has two octets formed from the seven electrons of each Cl atom, and in ethane the H atoms have a stable pair of electrons joining them to the C atoms. The two C atoms in ethane have octets of electrons formed by sharing with each other and with H.

$$Na^+ : \ddot{C}l : ^- \qquad\qquad : \ddot{C}l : \ddot{C}l : \qquad\qquad \begin{matrix} & H & H & \\ H : & C : & C : & H \\ & H & H & \end{matrix}$$

$$\underline{21} \qquad\qquad\qquad \underline{22} \qquad\qquad\qquad \underline{23}$$

$$\begin{matrix} & H & H & \\ H : & C : & C : & \ddot{C}l : \\ & H & H & \end{matrix} \qquad\qquad \begin{matrix} & & Cl^- \\ H & & + \\ H : C : & C + Cl^- \\ H & & + \\ & & Cl^- \end{matrix}$$

$$\underline{24} \qquad\qquad\qquad\qquad \underline{25}$$

Replacement of an H in ethane by Cl to give ethyl chloride **24** distorts the average position of electrons, because Cl tolerates a partial negative charge better than C and the electrons in the C:Cl bond are displaced toward the *electron-attracting* Cl. This means that the C is somewhat more positive, and it tends to attract electrons slightly from the C:C bond next to it toward the Cl. This in turn affects the C:H bonds in **24** by an *inductive effect*, transmitted down the chain. Further substitution of Cl for H in **24** to give CH_3CHCl_2 would distort the electronic arrangement further as described, but there would not be a complete transfer of electrons to form ions, or to justify a dualistic formula like **25**. These inductive electronic displacements can be conveniently indicated, following the convention introduced later by English chemists, by $\delta+$ or $\delta-$ for partial positive or negative charges; thus **25** would be written as $CH_3^{\delta+}C^{\delta-}Cl_3$.

Lewis' ideas readily explained the increasing dissociation constants of halogen-substituted acetic acids. In chloroacetic acid, $ClCH_2COOH$, the electron pair bonding chlorine to carbon is displaced toward chlorine, and hence it requires less energy for the hydrogen to depart for the H_2O molecule from the OH without its electrons, as compared to acetic acid, CH_3COOH.

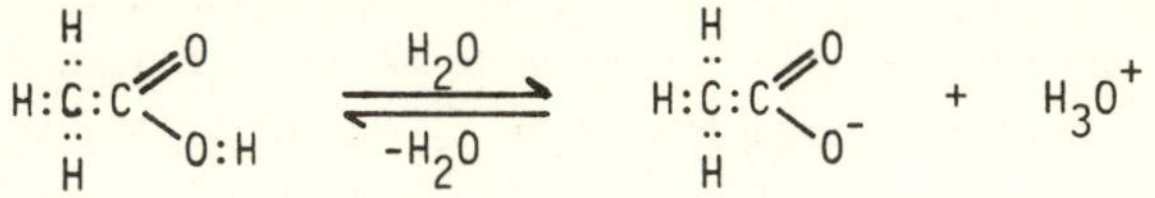

$$Cl:\overset{\overset{H}{\cdot\cdot}}{\underset{\underset{H}{\cdot\cdot}}{C}}:C\overset{\diagup O}{\diagdown O:H} \underset{}{\overset{H_2O}{\rightleftharpoons}} Cl:\overset{\overset{H}{\cdot\cdot}}{\underset{\underset{H}{\cdot\cdot}}{C}}:C\overset{\diagup O}{\diagdown O^-} + H_3O^+$$

$$H:\overset{\overset{H}{\cdot\cdot}}{\underset{\underset{H}{\cdot\cdot}}{C}}:C\overset{\diagup O}{\diagdown O:H} \underset{-H_2O}{\overset{H_2O}{\rightleftharpoons}} H:\overset{\overset{H}{\cdot\cdot}}{\underset{\underset{H}{\cdot\cdot}}{C}}:C\overset{\diagup O}{\diagdown O^-} + H_3O^+$$

The measurements of C. G. Derick (noted in an earlier essay) on the acid dissociation constants of halogen-substituted acids, although not mentioned by Lewis, support the latter's ideas on the rapid decrease of the inductive effect of chlorine as the intervening carbon chain in chlorinated acids increases in length. A corollary would explain the lower dissociation constant of isobutyric acid, $(CH_3)_2CHCOOH$, compared to CH_3COOH, as due to the *electron-repelling* inductive effect of CH_3 groups compared to hydrogen, a property of alkyl groups in accord with many other observations.

Lewis also explained very clearly the basic character of ammonia and its organic derivatives by the combination of the unshared pair of electrons on the nitrogen with a proton to form an ammonium salt such as NH_4Cl. This cleared up at one stroke a long-standing problem of the nature of the valences in these salts: four of the nitrogen valences are covalent and the salt is held together electrostatically by charges between the NH_4^+ and the Cl^- ions.

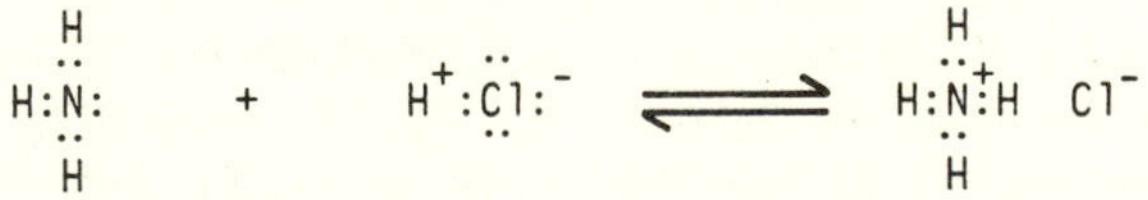

$$H:\overset{\overset{H}{\cdot\cdot}}{\underset{\underset{H}{}}{N}}: + H^+:\overset{\cdot\cdot}{\underset{\cdot\cdot}{C}}l:^- \rightleftharpoons H:\overset{\overset{H}{\cdot\cdot}}{\underset{\underset{H}{}}{N}}:H \quad Cl^-$$

Lewis extended[24] his ideas systematically in 1923, among other things generalizing the acid-base reaction illustrated above in a way which was to become of extraordinary importance in later experimental and theoretical work. His later expression defining[24] "Lewis acids and bases" was that a base was a substance with an unshared pair of electrons, and an acid was a substance deficient in one pair of electrons. By combining, each formed a stable electronic structure with sharing of the electron pair. A Lewis acid, such as boron trifluoride, BF_3, with only six electrons around boron, hence combines readily with a compound having an unshared pair of electrons, for example ethanol, a Lewis base. This combination frequently makes the Lewis acid a very useful catalyst.

$$
\begin{array}{ccc}
\overset{\displaystyle F}{\underset{\displaystyle F}{F:B:F}} + C_2H_5:\overset{..}{\underset{..}{O}}:H & \rightleftharpoons & \overset{\displaystyle \overset{..}{F}}{\underset{\displaystyle}{F:B:F}}\ C_2H_5:\overset{..}{\underset{..}{O}}:H
\end{array}
$$

Another example of a Lewis acid and base combination is the reaction of trimethylboron with ammonia,[25] a process to be studied later in detail with highly important results, particularly by H. C. Brown. Lewis' prophetic vision is shown again in his remarks on the boron hydrides,[26] which were to lead to several major developments in organic and inorganic chemistry.

$$
\begin{array}{ccc}
\overset{\displaystyle CH_3}{\underset{\displaystyle CH_3}{CH_3:B}} + \overset{\displaystyle H}{\underset{\displaystyle H}{:N:H}} & \rightleftharpoons & \overset{\displaystyle CH_3}{\underset{\displaystyle CH_3}{CH_3:B}} : \overset{\displaystyle H}{\underset{\displaystyle H}{N:H}}
\end{array}
$$

A corollary of the importance of the electron *pair* for the structure of stable compounds is that the structures with *unpaired* electrons (free radicals, or odd molecules) are unusually reactive. Such molecules as NO, NO_2, triphenylmethyl, and other organic radicals with unpaired electrons have been investigated with increasing success since Lewis' book was published, particularly by magnetic methods, as Lewis foresaw.[27] Although the idea of electron spin pairing in the covalent bond was several years in the future, Lewis realized that unpaired electrons, as in odd molecules, should be paramagnetic. They behave like a bar magnet, which orients itself in the magnetic field so that its south pole is adjacent to the north pole of the field, and vice versa. Lewis suggested that the paramagnetic properties of O_2 were due to the fact that the ground state of the molecule contained two unpaired electrons. Ordinary molcules with all electrons paired are diamagnetic; they do not orient themselves in the magnetic field, but are repelled from it. In molecules with unpaired electrons, the diamagnetism of the electron pairs is overbalanced by the paramagnetism of the unpaired electron. The development of electron spin resonance spectroscopy (ESR), which allows the quantitative study of unpaired electrons, confirms Lewis' predictions.

W. Kossel in Germany published simultaneously with Lewis' 1916 paper a very long discussion of valence[28] that had many aspects in common with the latter's ideas. Kossel did not explain the formation of covalent bonds as well as Lewis, and his paper did not have the intellectual sweep of Lewis' work. Kossel arranged the octet of electrons on the circumference of a circle; this did not allow as ready a pictorial formation of covalent bonds.

Lewis' book is full of valuable insights for organic chemistry, among others an explanation of the Walden inversion and a clarification of the structure of amine oxides.[29] Lewis' theory had many other ramifications and fruitful suggestions for chemistry as a whole. The pleasure with which one reads his work now is not due solely to his insight, but also to the fact that he was one of the great stylists who has written in English on scientific topics.

An important new concept was developed by Lewis' colleagues at Berkeley between his 1916 paper and his book in 1923. This was the idea that the hydrogen of an OH or an NH group could form a *hydrogen bond*[30] (or bridge) with atoms carrying unshared pairs of electrons, such as O or N. Although hydrogen bonds are weak, with bond strengths varying between 6-10 kcal per mole, roughly ten percent of normal chemical bonds, their cumulative effect in large molecules, such as proteins or nucleic acids, can be decisive in forming and maintaining highly organized structures. They also exercise an important effect on physical and some chemical properties of simple molecules, such as water, ammonia, alcohols, carboxylic acids, and enols. Illustrated are hydrogen-bonded structures for H_2O **26**, the dimer of acetic acid **27**, the internally hydrogen-bonded (*chelated*) structure of enolic methyl acetoacetate **28**, the hydrogen-bonded form of methanol **29**, the hydrogen bonding between peptide linkages as in a protein **30**, and the three hydrogen bonds **31** between the guanine and cytosine ring systems which help bind the two strands of deoxyribonucleic aicd (DNA) into the double helix. The bonds are written with an arrow from the donor atom (N or O, usually) which supplies the pair of electrons to the H for the bond.

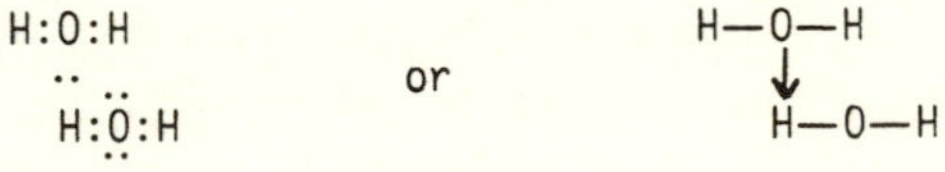

26

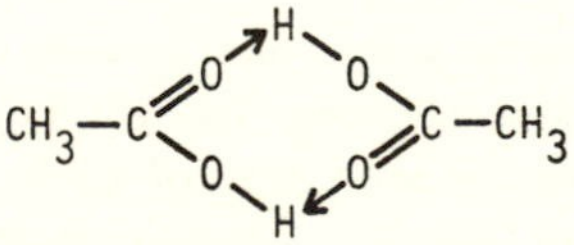

27

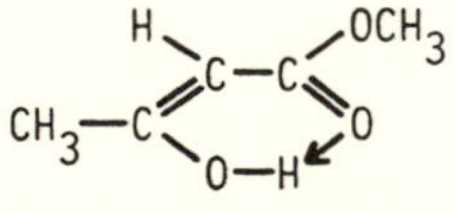

28

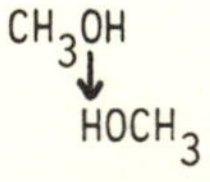

29

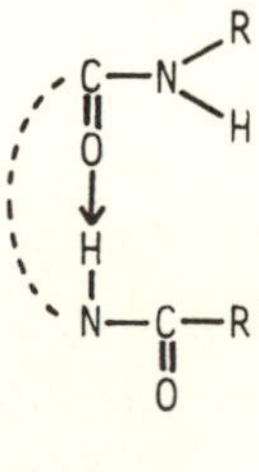

30

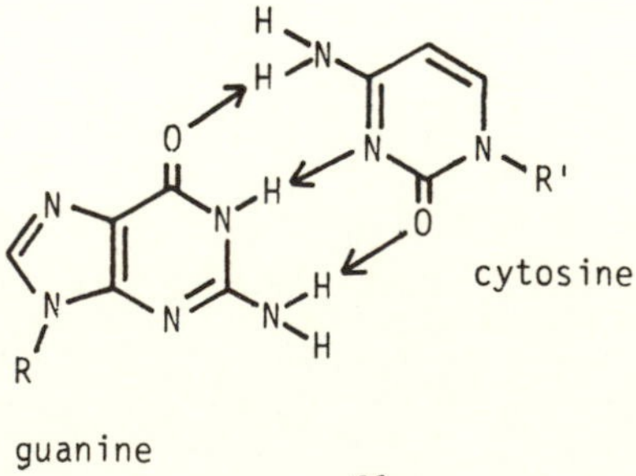

guanine   cytosine

31

Lewis' 1916 paper was extended by Irving Langmuir[31] (1881-1957) of the General Electric Research Laboratories to the heavier elements, and Langmuir introduced the terms *octet* and *covalent.* Although Langmuir's papers may have helped to get the electronic theory known and accepted, Langmuir made few significant additions to the Lewis theory, as Lewis said.[32]

A key point of the Lewis theory is the stability of the noble gas electronic shells, 2 in helium, 8 in neon, 8 in argon, 18 in krypton and xenon. It is an instructive point, and one which would probably have amused Lewis, that this dogma of the electronic theory, which had been unchallenged for many years, was shown to be incorrect by the preparation of many fluorine-containing compounds of noble gases.[33] It is still true, however, that in general the noble gas arrangement of electrons is the most stable one.

In spite of its striking successes, the Lewis valence theory did not give satisfactory answers to a number of fundamental questions. Why do electrons pair to form a bond? Why is a saturated carbon atom tetrahedral? What is the nature of the bonding in simple unsaturated systems like cyclohexene **2**, in conjugated systems like butadiene **4** and benzalacetophenone **5**, and in benzene **1**?

Within the decade following Lewis' book (1923-1933), a more fundamental and detailed theory of chemical bonding developed on the basis of the quantum mechanical theory of atomic structure. This afforded a more satisfactory qualitative and semiquantitative explanation of bonding in organic chemistry. Quantum mechanics is abstract and mathematical, and its modes of thought are somewhat alien to the deep-seated pictorial outlook of organic chemists. The application of quantum mechanics to the structures and reactions of organic compounds is, however, so complex that many simplifying assumptions are required.[34]

The fact that light behaves in some circumstances like a wave phenomenon but in others like a stream of particles has been known for generations. The quantum theory, developed by Max Planck around 1900, postulated that energy could be transferred only in discrete bundles (quanta), the energy being proportional to the frequency associated with the quantum. It followed that a molecule or atom could not contain a continuously variable amount of energy, but could exist only in discrete energy levels differing by definite values associated with the quanta involved in the transition from one energy level to another. In atoms, the possible energy levels are very sharply defined; in most organic molecules, the possible energy levels are subdivided and are no longer sharply separated. The cause for this is understood, although we will not discuss it, but the fundamental tenets of the Planck quantum theory are still valid for organic compounds.

L. de Broglie in 1924 suggested that electrons had both wave and particle character. The wave nature of electrons was demonstrated by diffraction experiments by C. Davisson and L. Germer[35] of the Bell Laboratories in

1927. This wave character is utilized in the electron microscope, which gives far higher magnification than the optical microscope because the wavelength associated with electrons is far shorter than that of visible light.

In 1925 W. Heisenberg and E. Schrödinger independently discovered quantum mechanics. Heisenberg's famous Uncertainty Principle says that the momentum and position of a particle, usually an electron, cannot both be known simultaneously. Schrödinger's wave equation describes the behavior of a system in terms of a wave function ψ (psi), of which the square, ψ^2, represents the probability of finding the particle at any given point. The Schrödinger equation gives ψ for any defined system as a function of the coordinates of the system and of time, but it has not been solved in exact form for any complicated structure. For the hydrogen atom, however, the equation does yield the correct values for the differences in the energy of the electronic levels, as determined by spectroscopic measurement. The wave equation describes *atomic orbitals* for single electrons; orbital is not to be construed literally as a path in which the electron is moving, but rather as expressing the probability of an electron occupying a given part of space, or the electron density.

It is found that the extranuclear electrons of an atom require four *quantum numbers* for their complete description, designated by n (the principal quantum number), ℓ (referring to the angular momentum), m_ℓ (the magnetic quantum number), and m_s (the magnetic spin of the electron). For the last, only two values, $+1/2$ and $-1/2$, are possible. Pauli's Exclusion Principle states that two electrons cannot have all quantum numbers identical. Therefore, in an electron pair bond in which the n, ℓ, and m_ℓ quantum numbers of the two electrons are identical, because these characterize the orbital, the spins must be different (*opposed* or *paired*). If one electron has a spin of $+1/2$, the other, to be paired, must have a spin of $-1/2$; this is usually indicated by the symbol ↑↓ If the m_s values are identical, the two electrons are unpaired and they occupy different orbitals shown as ↑↑. No single orbital can contain more than two electrons, and hence the Lewis shared electron bond has two electrons,[36] differing in their quantum description only in the spin quantum number m_s.

The relationship between the principal quantum numbers for electron orbitals and the more commonly used designations s, p, and d follows.[37] The H atom has one 1s orbital, in which the electron cloud density is distributed symmetrically in a spherical manner around the nucleus. The H_2 molecule, H:H or H_2, has two 1s electrons with opposed spins, the bond being formed by *overlap* of the spherical orbital regions. The isolated C atom in its lowest energy state has two 1s electrons, a pair of 2s electrons with paired spins, and two unpaired electrons, one each in the $2p_x$ and $2p_y$ orbitals. The $2p_x$ orbital shows electron density in the form of two spheres or dumbbells, one sphere on the positive x-axis and one on the negative; at the origin, there is a plane where the electron density is zero (a *nodal plane*). The p_y and p_z orbitals are similarly disposed about the y and z axes.

What happens to the orbitals if C combines with 4H to form methane, CH_4? The electrons of the C in their lowest energy orbitals are arranged with the 1s carbon orbital full. The two 2s electrons and the two 2p electrons of C combine with the four H atoms with their single 1s orbital, to form the *hybridized* molecular orbitals of CH_4, the sp^3 *hybridized* orbitals of saturated carbon. The geometry of the sp^3 paired electrons in CH_4 is that of a regular tetrahedron,[38] instead of the perpendicular arrangement of the $2p_{x,y,z}$ atomic orbitals.

<pre>
                Orbital

                2p_z

                2p_y   ↑

                2p_x   ↑

                 2s   ↑↓

                 1s   ↑↓   ↑

       Element   C   H
</pre>

The tetrahedral arrangement is the one of lowest energy, or maximum overlap of the hybridized orbitals, which is the same thing. It may be regarded as the most stable geometrical arrangement, because there is repulsion between the filled orbitals (shared electron pair). The farthest separation possible for four orbitals is for them to be directed to the corners of a regular tetrahedron. Similar hybridization of atomic orbitals occurs in the formation of other compounds from the first row elements of the periodic system. Normally a compound contains a stable octet formed by sharing.

In ethane the bonds are all sp^3 hybridized; each bond is a sigma, σ, bond, as in other saturated compounds, and the σ bonds are arranged tetrahedrally around each saturated C. The σ nomenclature for the electron pair bond comes from the s atomic orbitals; the σ bonds are cylindrically symmetrical around a line joining the centers of the atoms involved. Other saturated organic compounds have similar bonding.

The formation of unsaturated compounds requires a different type of hybridization of the atomic orbitals of C. The combination of two C and four H in ethylene involves the formation of the framework of ethylene by five sp^2 hybridized orbitals, in which the carbon σ bonds and all six atoms are in one plane; this leaves over one orbital for each carbon, directed above and below the plane of the carbon atoms. These orbitals overlap to form a π *bond* (from p), with its electron density concentrated in two cylinders above and below the plane of the molecule, but with zero density in the plane. The C=C thus is composed of one σ bond in the plane of the molecule, and one π bond with its electron density above and below the plane. This is in accord with the occurrence of *cis-trans* isomerism in suitably substituted C=C compounds and with the resistance to rotation around the C=C, as well as with the planar arrangement of the C=C and the atoms attached directly to it. All this agrees with classical stereochemistry and with physical measurements of very diverse types.

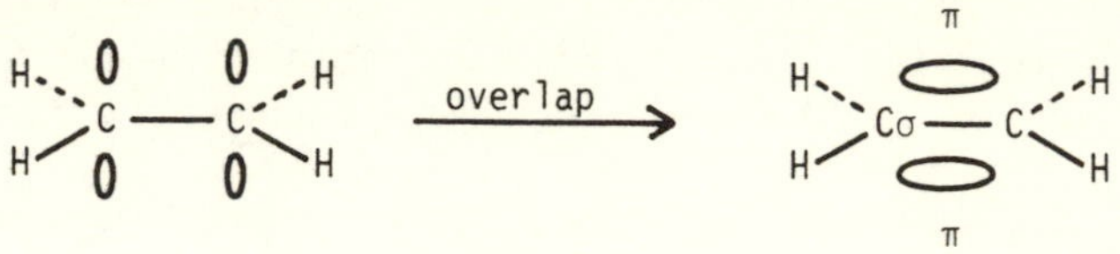

A different type of hybridization is shown in the C≡C of acetylene. This molecule is linear, with three σ bonds (two C-H and one C-C); the remaining two pairs of electrons on the C overlap to form two π bonds between the carbons; these bonds are cylindrically symmetrical around the C-C bond, but with zero electron density in the C-C bond. This *sp* hybridization thus binds the two carbons like a hollow pipe (the two π orbital overlap) with a rod in the center (the σ bond).

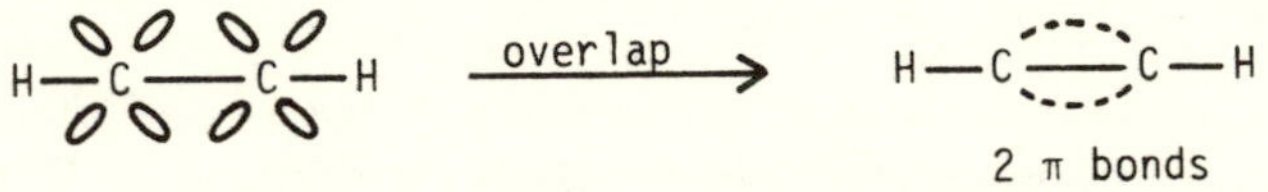

2 π bonds

For a carbonyl compound such as formaldehyde, **32a**, the bonding is similar to ethylene, with sp^2 hybridized C, all atoms in one plane, two C-H and one C-O σ bonds and a π bond with symmetrically situated orbitals above and below the plane of the molecule. Oxygen has six valence electrons and therefore has two unshared pairs, in addition to the two it contributes to the π bond with C in **32a**. There is thus a dipole in the C=O, the oxygen being negative relative to C, as indicated in **32b**. This dipolar nature of the carbonyl group is fundamental in determining the types of reactions which it undergoes.

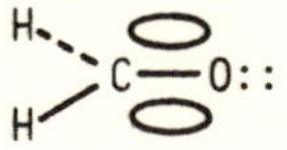

32a 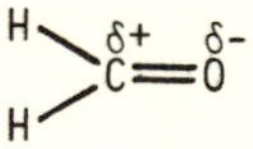   32b

The extension of these ideas of hybridization leads to a satisfactory picture of the structure of benzene and its analogs.[39] Benzene, C_6H_6, has six sp^2 hybridized C atoms connected in a regular hexagon with six H all in the same plane, all connected to C by σ bonds. This leaves six π orbitals with lobes above and below the plane of the ring, and zero density in the plane **33a**. These

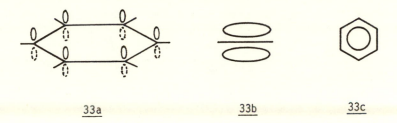

33a   33b   33c

π orbitals overlap to form doughnuts of electron density above and below the plane of the ring, shown in a side view **33b**. This explains the equality of all bond lengths and angles in benzene, with a consequent high degree of symmetry; the symmetry is confirmed by a variety of X-ray diffraction measurements and spectroscopic observations. The π *electron cloud* above and below the plane of the ring, and its zero density in the plane, are confirmed by many modern observations on the nuclear magnetic resonance (NMR) properties of benzene and its derivatives. The sp^2 hybridization and the six π electrons in the benzene ring are normally indicated now by the hexagon and circle **33c**. The symmetrical π electron distribution in benzene makes it a strongly *stabilized* system, with lower energy content than would be calculated for a single Kekulé structure, assuming that the C-C single and double bonds had the same energy content as in unconjugated systems.

The circle in the ring, as in **33c**, is also applied to polycyclic aromatic systems, such as naphthalene, $C_{10}H_8$, and anthracene, $C_{14}H_{10}$. However, in these larger aromatic systems each inscribed circle does *not* represent six π electrons, but a smaller number. Naphthalene has only ten π electrons altogether, and anthracene has only fourteen. If the circle notation for aromatic π electrons is used, this must be remembered.

The exceptional *aromatic* character of benzene requires emphasis. Substitution of a single H on the ring by Br or CH_3, for example **34**, reduces the symmetry of benzene, alters the distribution of the π and σ electrons, and changes the character of the further substitution reactions which can occur. The higher analogs of benzene, such as naphthalene **35** and anthracene **36**, no longer have uniform C-C bond lengths, and the electron density is altered considerably compared to benzene. This results in rates of reactions different from those of benzene and also in preferential attack of reagents at some positions. Naphthalene can form two isomeric monosubstitution products (positions indicated by arrows), and anthracene can form three, while benzene itself has only one. There is similar delocalization of π electrons in other conjugated systems, but owing to lower symmetry, it is not as complete or as effective in reducing the energy of the systems.

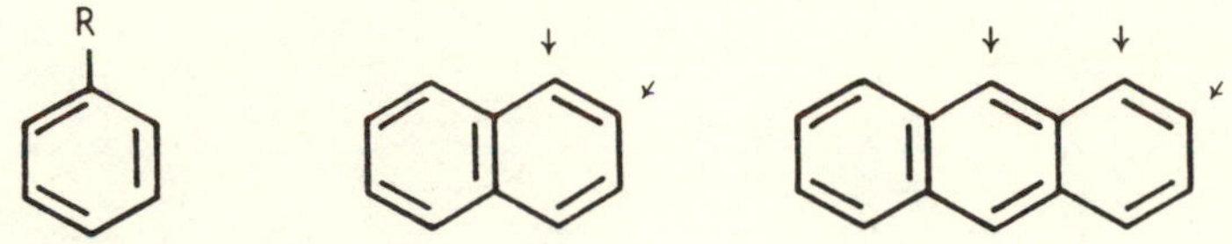

R

34

R = CH$_3$ or Br

35

36

There are two ways of studying the electronic structures of the compounds. The one we have been following roughly, the *molecular orbital (m.o.)* approach, originated with E. Hückel[40] in 1931 and now is the more widely used. His general procedure is to consider the molecule as a whole and to add the valence electrons to occupy the molecular orbitals giving the lowest total energy for the structure. The atomic orbitals must have as high degree of overlap as is consistent with the energy requirement, and this process thus calls for some chemical intuition in many cases. Hückel implied that for cyclic unsaturated systems the best overlap and maximum stability occurred where there were [4n + 2] π electrons in one ring, and when n = 1, there are six π electrons, as in benzene. *Hückel's Rule* applies without qualifications only to benzene, its simple derivatives like **34**, and a few one-ring compounds including some charged species.[41]

The other approach to structure, the *valence bond (v.b.)* procedure, considers that the state of the molecule is intermediate between two or more structures which follow the valence rules.[34a] The true state of the molecule is different from any of the *v.b. limiting forms;* the molecule is said to *resonate* between the limiting v.b. forms, or more correctly, the limiting forms contribute to the *resonance hybrid.* The energy of stabilization of the molecule is called its *resonance energy.* Thus benzene would be described as a resonance hybrid with the two Kekulé forms **37a** and **37b** as the main contributors, with smaller contributions from the Dewar forms, one of which is **37c**. There is no oscillation among the v.b. structures; the double-ended arrow does *not* mean that there is an equilibrium, but that the real state of the molecule is a composite of the v.b. structures. By making suitable assumptions, it is possible to calculate the resonance energy.

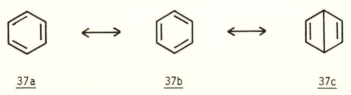

37a   37b   37c

The overall results of the two methods, valence bond and molecular orbital, are similar; the electron delocalization energy of the latter corresponds to the resonance energy of the former, and the assumptions made in quantitative calculations are approximately the same.

The v.b. scheme has the advantage that it is pictorial, and this is sometimes useful. Thus one can say that formaldehyde is a resonance hybrid of the two principal contribution forms, **38a** and **38b**, similar to **32**, and these formulas are useful in elucidating the reactions of formaldehyde and other C=O compounds. With a conjugated ketone such as **39**, one can rationalize 1,4-addition by the contributions of **39b**, the negative part of the addend going to the positive center in **39b**.

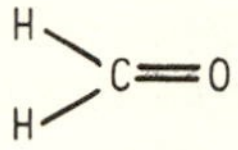

$$\begin{matrix} H \\ \diagdown \\ \diagup \\ H \end{matrix} C=O \qquad\qquad \begin{matrix} H \\ \diagdown \\ \diagup \\ H \end{matrix} \overset{+}{C}-O^-$$

$$\underline{38a} \qquad\qquad\qquad\qquad \underline{38b}$$

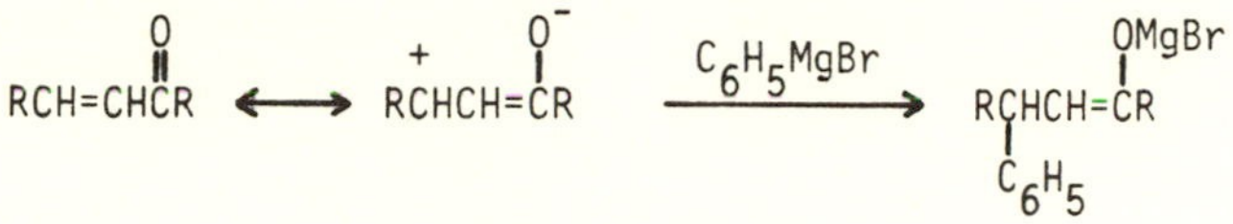

$$\begin{matrix} O \\ \| \\ RCH=CHCR \end{matrix} \longleftrightarrow \begin{matrix} O^- \\ \overset{+}{|} \\ RCHCH=CR \end{matrix} \xrightarrow{C_6H_5MgBr} \begin{matrix} OMgBr \\ | \\ RCHCH=CR \\ | \\ C_6H_5 \end{matrix}$$

$$\underline{39a} \qquad\qquad \underline{39b}$$

$$\xrightarrow[\text{H shift}]{H_2O} \begin{matrix} O \\ \| \\ RCHCH_2CR \\ | \\ C_6H_5 \end{matrix}$$

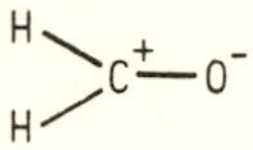

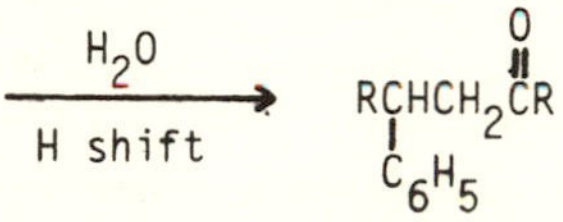

The negative ion from an enolic compound can be written as a hybrid **40**, and the resonance in the peptide bond, which is important in protein structure, is indicated in **41**. As implied in **41b**, the C-N bond in the peptide linkage has a certain amount of C=N character, which means that appreciable energy is necessary for rotation around the C-N bond in proteins.

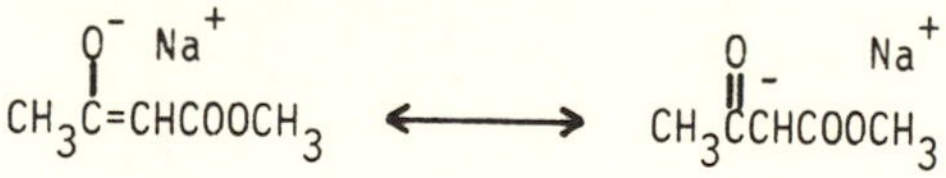

$$\begin{matrix} O^-\ Na^+ \\ | \\ CH_3C=CHCOOCH_3 \end{matrix} \longleftrightarrow \begin{matrix} O\quad\ Na^+ \\ \| \ \ _- \\ CH_3CCHCOOCH_3 \end{matrix}$$

$$\underline{40a} \qquad\qquad\qquad \underline{40b}$$

$$\begin{matrix} O \\ \diagup\!\!\nearrow \\ R-C \\ \diagdown \\ N \\ \diagdown \\ R' \end{matrix}\!\!H \qquad \longleftrightarrow \qquad \begin{matrix} O^- \\ \diagup \\ R-C \\ \diagdown\!\!\!\!_+ \\ N \\ \diagdown \\ R' \end{matrix}\!\!H$$

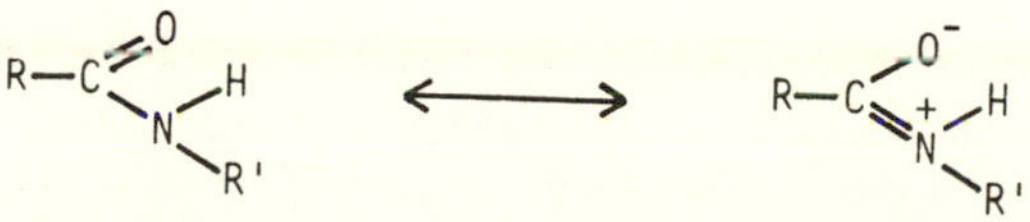

$$\underline{41a} \qquad\qquad\qquad\qquad \underline{41b}$$

The v.b. scheme, however, has been largely supplanted by the m.o. approach. There are serious objections to the use of v.b. structures. One is that they represent what the compound *is not*, rather than what it *is*. The unfortunate term *resonance* implies oscillation or motion; however it is fundamental to the v.b. scheme that there is no change in position of the atomic nuclei involved and no vibration of electrons, at least on any observable time scale. There are philosophical as well as pedagogic objections to describing a molecule by showing what it is not.[42] The m.o. representation in most cases implies no more than it means. In the case of aromatic compounds, circled structures (m.o.) may be preferable to the Kekulé structures **34, 35, 36**, several of which would be required for a suitable v.b. picture of the limiting forms of the resonance hybrid. The m.o. structure for **40** would be **42**, with the delocalized orbital shown by a dotted line, indicating that the O and the CH have most of the electron density. The m.o. picture of conjugated systems such as **39** can be drawn in terms of π orbitals overlapping, as can **42**, and this is sometimes useful.

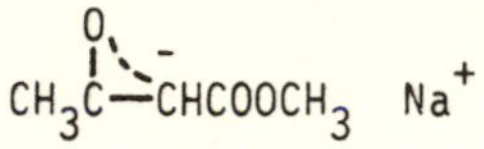

$$CH_3\overset{\overset{\displaystyle O}{\|}}{C}-\overset{-}{CH}COOCH_3 \quad Na^+$$

42

To summarize this account of the contribution of quantum mechanics to the valence theory of organic chemistry, the shared electron pair bond of Lewis now has a satisfactory *a priori* explanation. The tetrahedral sp^3 bonds of saturated carbon, the planar sp^2 bonds of carbon in C=C, and the linear sp bonds in C≡C all have a satisfactory theoretical basis. The electronic nature of aromatic and conjugated systems has been elucidated. Calculations of the stability, and in some cases of the reactivity of systems, can be made on a semiempirical basis. Such calculations may have highly useful implications, as in the case of *carcinogenic* (cancer-producing) polycyclic aromatic compounds[43] such as benzpyrene **43**.

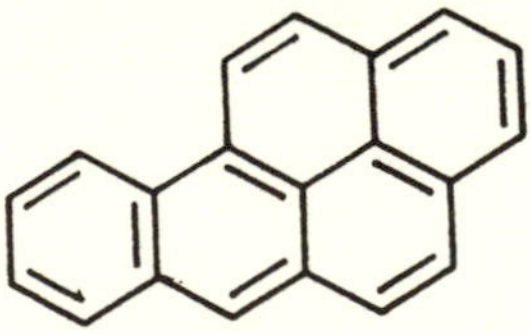

43

The intuition of organic chemists had produced concepts during the 1920s, which, although qualitative, were similar in content to the v.b. and m.o. schemes. The development of quantum mechanics showed why the isolation of electromers (compounds differing only in the position of electrons), such as the two forms of NCl_3 referred to earlier,[13] was impossible.

The v.b. or m.o. picture of a conjugated ketone, such as **39**, is roughly equivalent to Arthur Lapworth's[44] formulation **44** in 1920. Lapworth's scheme and that of many others, including Fry, postulated alternating positive and negative charges (*alternating polarity*) which, however, cannot be reconciled with quantum mechanics.

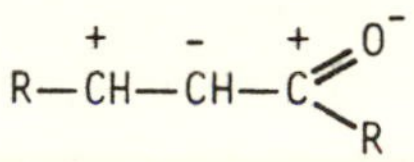

$$
\overset{+}{R-CH}-\overset{-}{CH}-\overset{+}{C}\underset{R}{\overset{O^-}{\diagup}}
$$

44

The electronic theory of the structure and reactions of organic compounds was developed most comprehensively[45] by C. K. Ingold of University College, London; important contributions to some parts of the theory are due to R. Robinson[45b] of Manchester and Oxford, F. Arndt[45c] of Germany, and H. J. Lucas of Caltech. The work on reaction mechanisms by Lucas deserves more credit than it has received.[46] Some of this theory will be discussed later in connection with reaction mechanisms.

To be emphasized now is the anticipation of the quantum mechanical theory of conjugated and aromatic systems by Arndt, Ingold, and others. Arndt suggested a notion similar to v.b. structures of aromatic compounds;[47] Ingold generalized this limited idea greatly and introduced the term *mesomerism*,[48] indicated by curved arrows, which corresponded to the resonance hybrid or the electron delocalization of the quantum mechanical treatment. An example from Ingold,[48b] the mesomeric effect in a conjugated amino ketone **45** is indicated by curved arrows showing displacement of electron pairs; there is an unshared pair on the amino nitrogen, with a partial charge on N and a partial negative one on O. The v.b. notation would be the resonance hybrid **46ab**. Mesomerism in aromatic compounds is indicated by **47**; the influence of Thiele's ideas is obvious.

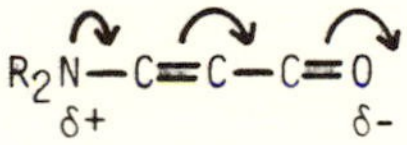

$$
\underset{\delta+}{R_2N}-C\equiv C-\underset{\delta-}{C=O} \qquad R_2N-C=C-C=O \longleftrightarrow R_2\overset{+}{N}=C-C=C-O^-
$$

45 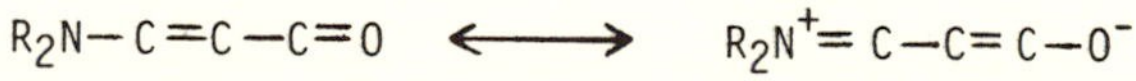   **46a**   **46b**

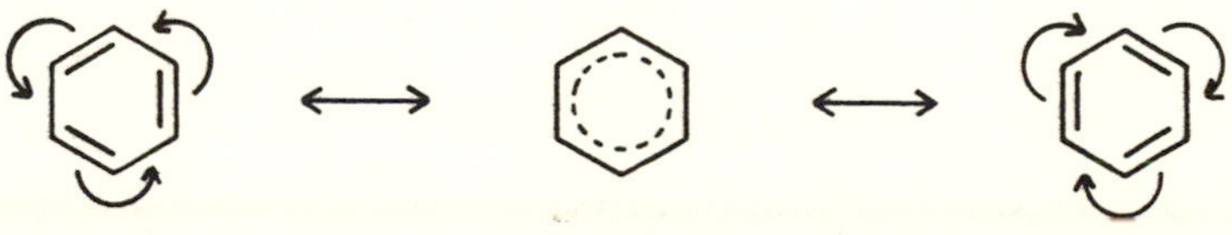

47

Thus by 1935 the quantum mechanical basis for a unified theory of organic structure had been developed, as well as a comprehensive qualitative theory of many, but not all, aspects of organic chemistry. The two were in essential agreement, and their utilization in the classification and rationalization of organic reactions was to be one of the most striking developments of the following decades.[49]

LITERATURE CITED

1. Emil Fischer, Faraday Lecture, "Synthetical Chemistry in its Relation to Biology," JCS, **91**, 1749 (1907).

2. E. Fischer, Ann., **381**, 123 (1911); **386**, 374 (1911); **394**, 350 (1912).

3. C. A. Russell, *The History of Valency*, Leicester, 1971, gives an excellent account of the whole development of valence theory up to the quantum mechanical era. A. N. Stranges, *Electrons and Valence*, College Station, Texas, 1982, discusses in great detail various theories of valence based on separation of charge and the fundamental contributions of G. N. Lewis.

4. J. Thiele, Ann., **306**, 87 (1899), and ten accompanying experimental papers, through p. 266.

5. A. Michael, JPr, **68**, 500 (1903); with V. L. Leighton, p. 521; with P. H. Cobb, **82**, 297 (1910).

6. E. P. Kohler, ACJ, **31**, 243 (1904); E. P. Kohler and Miss M. C. Burnley, ibid., **43**, 412 (1910). A summary of 1,2 and 1,4-additions to conjugated systems is given by W. Hückel, *Theoretische Grundlagen der Organischen Chemie*, Leipzig, 1934, 2nd ed., Vol. 1, pp. 332 ff.

7. R. Willstätter and E. Waser, Ber., **44**, 3423 (1911).

8. H. von Helmholtz, Faraday Lecture, "On the Modern Development of Faraday's Conception of Electricity," JCS, **39**, 277 (1881). This lecture is also notable for its description of the experiments on which the development of the glass electrode rests; cf. M. Cremer, *Z. Biologie*, **47**, 597 ff. (1906). Helmholtz' contribution to this device, so important to later work in chemistry and biology, appears to have been overlooked by workers later than Cremer, including F. Haber and Z. Klemensiewics, Z. Phys. Chem., **67**, 385 (1909). W. M. Clark, *The Determination of Hydrogen Ion*, Baltimore, 1928, p. 431, mentions Helmholtz briefly but none of the monographs on glass electrodes do.

9. S. Arrhenius, Z. Phys. Chem., **1**, 631 (1887).

10. J. J. Thomson, Phil. Mag., **44**, 293 (1897); J. J. Thomson, *Electricity and Matter*, New York, 1904, pp. 70 ff.

11. Described in Russell, also Stranges, ref. 3.

12. An excellent account of the genesis of G. N. Lewis' ideas is by R. E. Kohler, Jr., in *Historical Studies in the Physical Sciences*, Philadelphia, Russell McCormmach, Ed., Vol. 3, p. 343 (1971).

13. Some representative papers were: (a) W. A. Noyes and A. C. Lyon, JACS, **23**, 460 (1901); **35**, 767 (1913); (b) K. G. Falk and J. M. Nelson, ibid., **32**, 1637 (1910); K. G. Falk, **33**, 1140 (1911); the content of these two papers was published in JPr, **88**, 97 (1913), a late example of double publication; (c) H. S. Fry, JACS, **34**, 664 (1912) and other papers; Fry's views were collected in a monograph, *The Electronic Conception of Valence and the Constitution of Benzene*, London, 1921; J. Stieglitz, JACS, **44**, 1293 (1922).

14. R. Abegg, Z. Anorg. Chem., **39**, 330 (1904).

15. A restatement of Michael's and van't Hoff's idea of the "plasticity" of carbon; A. Michael, JPr, **60**, 325ff (1899).

16. Abegg's table does not list the symbols of the elements in the first row.

17. (a) J. J. Thomson, *The Corpuscular Theory of Matter*, London, 1907; (b) Phil. Mag., **27**, 757 (1914).

18. A. N. Whitehead, *Science and the Modern World*, New York, 1948, p. 8.

19. J. H. Hildebrand, *Biog. Mem. Nat. Acad. Scis.*, **31**, 210 (1958). The most vivid accounts of Lewis come from symposium papers, mainly by his collaborators, D. A. Davenport, Ed., in JCE: R. N. Lewis, **61**, 3 (1984): J. W. Servos, ibid., 5; M. Calvin, 14; G. E. K. Branch, 18; G. T. Seaborg, 93; L. Brewer, 101; K. S. Pitzer, 104; J. Biegeleisen, 108; L. Pauling, 201; M. Kasha, 204.

20. G. N. Lewis, JACS, **38**, 762 (1916).

21. G. N. Lewis, *Valence and the Structure of Atoms and Molecules*, New York, 1923; reprinted by Dover, with an Introduction by K. S. Pitzer, 1966, pp. 29 ff.

22. Lewis, *Valence*, pp. 32, 77.

23. A. L. Parson, Smithson. Inst. Publ., **65**, no. 11 (1915).

24. Lewis, *Valence*, p. 142; J. Franklin, Inst., **226**, 293 (1938).

25. Lewis, *Valence*, p. 98.

26. Ibid., pp. 95, 106, 123.

27. Ibid., pp. 80, 148.

28. W. Kossel, Ann. Physik, **49**, 229-362 (1916); Lewis' relationship to Parson, Kossel, and others is discussed by Kohler (ref. 12), and is noted by Lewis.

29. Lewis, *Valence*, p. 113 (inversion); p. 111 (amine oxides).

30. The concept of hydrogen bonding is attributed by Lewis, *Valence*, p. 109, to M. L. Huggins; see M. L. Huggins, JOC, **1**, 407 (1937) for a review and reference to Huggins' Undergraduate Thesis, Berkeley, 1919. W. M. Latimer and W. H. Rodebush, JACS, **42**, 1419 (1920) is the basic paper. For a physical chemical discussion, see G. C. Pimentel and A. L. McClellan, *The Hydrogen Bond*, San Francisco, 1960; M. D. Joesten and L. J. Schaad, *Hydrogen Bonding*, New York, 1974.

31. I. Langmuir, JACS, **38**, 2221 (1916); **41**, 868, 1543 (1919); **42**, 274 (1920). and later papers. See C. C. Suits and M. J. Martin, *Biog. Mem. Nat. Acad. Scis.*, **45**, 215 (1974) for Langmuir. It is an interesting commentary on the vagaries of the Nobel Prize awards that Langmuir received one in 1932 for his work on surface chemistry, but G. N. Lewis never was recognized, either for his work on valence theory or on thermodynamics. Several publications by Langmuir's colleagues at the General Electric Research Laboratories in Schenectady show that the electronic theory of valence was actively pursued in that laboratory.

32. Lewis, *Valence*, p. 87.

33. N. Bartlett, Proc. Chem. Soc., 218 (1962) and many later papers.

34. We shall of necessity rely on secondary sources for our account of the quantum mechanical theory of valence. Among the most useful of these are; (a) Linus Pauling, *The Nature of the Chemical Bond*, Ithaca, 1940; (b) C. A. Coulson, *Valence*, Oxford, 1952; (c) G. W. Wheland, *Resonance in Organic Chemistry*, New York, 1955; (d) C. K. Ingold, *Structure and Mechanism in Organic Chemistry*, Ithaca, 1953, 2nd ed., 1969; (e) L. P. Hammett, *Physical Organic Chemistry*, New York, 1940; 2nd ed., 1970.

35. C. J. Davisson and L. H. Germer, Nature, **119**, 558 (1927); on Davisson, biography by M. J. Kelly, *Biog. Mem. Nat. Acad. Scis.*, **36**, 51 (1962).

36. G. N. Lewis, J. Chem. Phys., **1**, 17 (1933), in a review of the effect of quantum mechanics on his 1923 theory of valence, regards the paired electrons in a bond as a single entity. One paragraph quoted in part earlier is of particular interest for organic chemistry: "Structural organic chemistry, although developed without mathematics except of the most elementary sort, is one of the very greatest scientific achievements. An enormous mass of information was reduced to a well-ordered system through the aid of a few simple principles. It was this system of structural chemistry that served chiefly as my basis when I advanced my theory of valence".

37. Ingold, *Structure and Mechanism*, 2nd ed., p. 13.

38. Pauling, *Chemical Bond*, p. 81. Pauling's book contains leading references to the fundamental work by Heitler and London, Slater, and Pauling himself, on bond formation

by combination and hybridizations of atomic orbitals.

39. Coulson, *Valence*, pp. 223 ff; Ingold, *Structure and Mechanism*, pp. 186 ff.

40. E. Hückel, Z. Physik, **70**, 204 (1931); **76**, 628 (1932); Z. Elektrochemie, **43**, 752 (1937).

41. For a modern discussion of aromatic character and Hückel's Rule, see *Kekulé Centennial*, Washington, 1966, "Nonclassical Aromatic Compounds", by N. C. Rose, pp. 91-110.

42. The use of Kekulé structures in representing the structure of benzene and the v.b. approach in general caused a serious controversy on ideological grounds in the USSR; L. R. Graham, *Science and Philosophy in the Soviet Union*, New York, 1972, chap. 8.

43. A. Pullman and B. Pullman, *Cancérization par les Substances Chimiques et Structure Moléculaire*, Paris, 1955; later books and articles by the Pullmans, Daudel, and Coulson.

44. (a) A. Lapworth, *Mem. Manchester Lit. Philos. Soc.*, **64**, no. 3, 1-16 (1920). (b) An adjacent paper by Sir Robert Robinson, ibid., **64**, no. 4, 1-14 (1920), gives some ideas on conjugated and aromatic systems by a master of organic chemistry; it also contains speculation on the transition state of reactions.

45. C. K. Ingold, Chem. Rev., **15**, 225 (1934), summarizes his earlier speculations, also *Structure and Mechanism*, passim; (b) R. Robinson et al., JCS, 401 (1926) and following papers; also ref. 44; (c) F. Arndt et al, Ber., **57**, 1904 (1924); **63**, 2963 (1930).

46. H. J. Lucas and A. Y. Jameson, JACS, **46**, 2475 (1924) and later papers.

47. It was actually suggested for some heterocyclic compounds with aromatic character, which Arndt described as existing in intermediate stages (Zwischenstufen) differing from the Kekulé structures which could be written.

48. (a) For detailed history, going back to Arthur Michael, see Ingold, *Structure and Mechanism*, 2nd ed., pp. 74 ff; (b) p. 87.

49. One of us, as a graduate student in 1934-1937, remembers clearly the reaction of some classical organic chemists to the "new-fangled theories" of Ingold and others; these were considered not entirely respectable intellectually, and it was said that the Ingold nomenclature was difficult and obscure. This was not confirmed by a reading of Ingold's Chem. Rev. article of 1934, although the presentation was more formalistic than seemed necessary. Ingold's presentations improved with time, the 2nd ed. of *Structure and Mechanism* being the best.

8

General Survey of Organic Chemistry
in the United States, 1914-1939

This interval forms a reasonable unit for considering the changes in organic chemistry in this country over a key twenty-five year period bounded by two World Wars. The present essay surveys some of the striking features of this period, showing the diversity and richness of accomplishment of American chemists. They continued their earlier, promising tradition of research with a highly creditable record of achievement marked by a notable advance in quality of American research. The statement by James B. Conant that the first thirty years of this century were not an outstanding period in the history of American organic chemistry[1] ignores this progress and fails to take into account the historical development of the field, in which Conant himself played an important role from 1918 to 1933. The atmosphere of provincialism and scientific colonialism clearly discernible in American work before 1914 gradually disappeared, and by 1939 the best organic chemistry was plainly equal to the outstanding work in other countries. This trend continued after World War II, when American science in general set the standard for the rest of the world.

The increasing numbers of undergraduate and graduate students, strong departments offering graduate work, and faculty members active in research were the results of the more numerous employment opportunities for chemists in industrial research and production, local and national government research laboratories[2], and private research institutes, as well as college and university teaching. The number[3] of Ph.D.s granted in all branches of chemistry increased from 534 for the five years 1914 to 1918 to 2314 for the five years 1935 to 1939. The total number of bachelor's degrees awarded in all fields in American colleges and universities increased from 31,500 in 1914 to 158,000 in 1939. The membership in the American Chemical Society grew from 7170 in 1914 to 23,519 in 1939. In 1920 it was estimated that there were 300 industrial research laboratories in the United States, and in 1940 the number had grown to 2200.

During this period, students did most of the published university research. Undergraduates and masters candidates did some, but Ph.D. candidates, who usually served as teaching assistants, did the larger part. Research scholarships and fellowships were not generally available in most schools; the institutions bore most of the research costs for supplies and equipment. Few research grants were available, and only a handful of universities, like Yale, Illinois, and Harvard, made postdoctorate appointments, these mainly in the 1930s.

During these years, however, the National Research Council (NRC) offered postdoctoral fellowships awarded by competitive applications, which the recipient could use in this country or abroad.[4] From 1919 to 1951 the NRC awarded 1300 fellowships, of which 229 were in chemistry. The number of new awards in the natural sciences peaked at 88 in 1930-1931, but decreased steadily to 18 in 1940-1941; the stipends varied from $1600 to $2000, reasonable figures for the times. The funds came at first from the Rockefeller Foundation and later from many private sources. Many of the leading organic chemists who emerged in those years had held NRC fellowships. We may mention, for example, M. S. Kharasch, C. F. Koelsch, P. D. Bartlett, F. H. Westheimer, and W. M. Stanley.

American chemistry was not affected as much as physics by the European scientists coming here in the 1930s after the rise of Hitler.[5] The number of chemists was smaller and American chemistry was well-established.

We have tabulated by years the number of papers on organic chemistry in the JACS. We consider a paper to be "organic" if it deals with the compounds of carbon with respect to synthesis; structure of homogeneous natural products; reactivity and reaction mechanisms; relations between chemical structure and physical properties, such as color; and measurements of equilibrium, such as acid-base dissociation constants. Not included under organic papers are those on natural products of uncertain purity; on the preliminary (and usually very laborious) concentration of vitamins, hormones, and similar materials from natural sources; on quantitative methods of analysis, particularly in biochemistry; and on measurements of physical properties of little relevance to reactions of organic compounds. These criteria, although arbitrary in some cases, are clear-cut enough to allow comparisons of trends in publication.

Numbers of Papers Published in
Organic Chemistry in the JACS 1914-1939

5-Year Periods	No. Papers
1914 – 1918	482
1919 – 1923	590
1924 – 1928	975
1929 – 1933	1,785
1934 – 1938	2,056
1939'	*603
Total 26 Years	6,491

*one year only

Notable academic centers for organic research which developed in this period included Illinois, Minnesota, Wisconsin, Caltech, Notre Dame, Duke, Iowa State, Iowa, Pennsylvania, and the University of California at Los Angeles (UCLA). The increased publication from pharmaceutical and other industrial research laboratories is described in a later essay. Expanding research in physical chemistry at Berkeley, Caltech, Chicago, MIT, Princeton, Harvard, and several other centers influenced the development of physical organic chemistry at these and other universities.

The JACS, which absorbed the ACJ in 1914, was the main vehicle of publication for American chemists of all fields from 1914 to 1939. The *Journal of Organic Chemistry* (JOC), established in 1936, rapidly became an important outlet, and the JBC contained many organic papers. We have examined these journals in detail for the 1914-1939 period. The *Journal of Physical Chemistry* continued to attract papers in physical and colloid chemistry, and the *Journal of Chemical Physics* (started in 1931) accommodated papers on the borderline of physics and chemistry; some of this work was to become of fundamental importance to physical organic chemistry.

The Eighth International Congress of Applied Chemistry, which met in the United States in September 1912, furnished a striking picture of the accomplishments of the German chemical industry, the results of sustained cooperative research between industrial and university laboratories. German speakers, exhibits of new synthetics, and a working unit of the newly invented Haber process producing ammonia from hydrogen and nitrogen deepened the impression of German preeminence.

The situation for the American chemical industry and for research in organic chemisty created by the blockade of German chemical imports during World War I was serious. Industrialists, the American Chemical Society, and the universities, however, mobilized research and production to cope reasonably well with the emergency. Although the country did not become completely self-sufficient chemically, as the World War II shortages in rubber and some drugs, such as antimalarial compounds, showed, it was far more independent than in 1914. Strong centers of industrial research and production and of university and college teaching and research developed after 1914. The impetus given to them by the shock of earlier American chemical dependence on Germany sustained the imposing growth after 1920.

An important development for organic research in this country was the accomplishments of Roger Adams and his associates in the Organic Chemicals Manufactures project at Illinois ("preps"); this not only relieved some critical shortages of key organic compounds but led after the war to the founding of the annual publication *Organic Syntheses*. This gave reliable procedures for the preparation of valuable organic compounds and resulted indirectly in the establishment of the Eastman Kodak business of supplying organic chemicals.

We have discussed the significant contributions to physical organic chemistry and reaction mechanisms before 1914 of a number of American chemists. This field became extremely active and productive, particularly after 1930. Leading figures included Conant (Harvard), L. P. Hammett (Columbia), Kharasch (Chicago), H. J. Lucas (Caltech), C. R. Hauser (Duke), and some of their students, particularly Bartlett (Harvard), F. R. Mayo (Chicago), W. G. Young (UCLA), and others for the 1914-1939 period.

Fundamental to any sound development of the understanding of reaction mechanism was G. N. Lewis' proposal of the shared electron bond. Lewis' penetrating insight into the nature of valence was supported by the applications of quantum mechanics to organic bonding by Linus Pauling and others that also justified the postulates of classical stereochemistry (the tetrahedral arrangement around saturated carbon atoms, for example). The quantum mechanical theory of valence was especially useful for conjugated and aromatic systems, avoiding the *ad hoc* nature of earlier explanations. Lewis' more general definition of acids and bases– a "Lewis acid" was an electron-acceptor, and a "Lewis base" an electron-donor– was important not only in the theory of organic chemistry, but also in the reagents which it suggested for experimental work.

Moses Gomberg's key discovery in 1900 of the first free radicals, the triarylmethyls, in which carbon had a valence of three instead of the classical four, led to a whole field of research both in this country and abroad on these "odd molecules" with an unpaired electron, as G. N. Lewis called them. Gomberg and his collaborators, particularly W. E. Bachmann, continued to investigate the triarylmethyl radicals and were joined by Conant, C. S. Marvel (Illinois), and others in America, and by Karl Ziegler in Germany. The realization that trivalent carbon could exist in relatively stable structure made chemists more ready to accept free radicals as intermediates in reactions in solutions when the individual radicals could not then be identified.

Frequently very reactive free radicals occurred in a rapid chain reaction, a series of processes initiated by a free radical and "turning over" many times to yield the product. M. Bodenstein and W. Nernst in Germany demonstrated chain reactions in the gas phase, and the work of Kharasch showed free radicals in solution. His demonstration of the peroxide effect on the addition of HBr to an unsymmetrically substituted ethylenic bond was an experimental and deductive masterpiece; he went on to discover whole families of new reactions which occur by free-radical intermediates. Styrene polymerizing to form the high molecular weight plastic polystyrene, and methyl methacrylate to form the transparent plastic lucite are examples of addition polymerization, another important free-radical chain process first explained in detail by Paul J. Flory (Du Pont) in 1937.

The scientific and technological importance of the field of free-radical chemistry opened up by Kharasch and Flory can scarcely be exaggerated. Free-radical intermediates have been suggested for numerous biochemical processes; the interaction with living systems of light, short wavelength radiation, or high energy particles from radioactive decay undoubtedly involves free-radical intermediates. Although there were many significant contributions from other countries, American work and ideas have dominated the field ever since Gomberg's initial observation. The chemical world recognized Kharasch at the time of his death in 1961 as the leader in free-radical chemistry.

Many organic reactions go through stages involving charged intermediates (ions) in which all electrons are paired. The English chemists, Arthur Lapworth, Sir Robert Robinson, and C. K. Ingold, initiated the development of a coherent and generally consistent body of theory, particularly from 1920 to 1939. This theory, elaborated and formalized primarily by Ingold, dealt successfully with certain ionic reactions such as some replacement reactions at saturated carbon atoms and in aromatic systems, as well as with addition reactions of unsaturated systems. Although the significant achievements of the "English School" later affected American thinking about reaction mechanisms, a scrutiny of the literature indicates that this influence was not marked before 1935. The greatest foreign impact was made by ideas about acids and bases by the Dane, J. N. Brönsted, and his concepts of equilibria and by the work of two Germans, A. Hantzsch and Hans Meerwein.

Brönsted set forth the first quantitative relationship between equilibrium and rate (between thermodynamics and kinetics) in an equation connecting the dissociation constant of an acid (an equilibrium property) with its effect on rates of acid-catalyzed reactions. Hammett demonstrated the wide-spread occurrence of linear free energy relationships. His interest was aroused by reading the paper of 1924 by Brönsted and Pedersen.[6] Following Hammett's pioneer work, the study of linear free energy relations became a major facet of American physical organic chemistry.

Brönsted's influence was also significant in the development of the transition state theory of reactions (the "critical state" of Brönsted), according to which reactants went through an unknown structure (transition state) in which new bonds began to form simultaneously with the breaking of old bonds, the most economical way in terms of energy and entropy by which a reaction could occur. Physical chemists in this country, particularly Henry Eyring (then of Princeton), developed this theory in the early 1930s. Hans Meerwein of Marburg University in Germany showed that BF_3, a Lewis acid, was a valuable acid catalyst; many studies, particularly by J. A. Nieuwland and others at Notre Dame University illustrated the use of this new reagent.

Meerwein, a highly original chemist, also played a major role in the chemistry of molecular rearrangements, a field that has challenged American chemists from before 1914 to the present. The comforting assumption of the classical structural theory that organic reactions always took place with minimal change in structure was almost immediately found to be untrue; deep-seated molecular shifts did occur in many cases. Such rearrangements, caused by acids, bases, or heat, affected the validity of structural proofs, synthetic results, and the understanding of reaction mechanisms and demanded further knowledge of their occurrence and results.

Components of certain essential oils (from "essences", or extracts of plants) such as camphor, menthol, citral, and many others, were found to undergo numerous rearrangements that enormously complicated their structural proof. In the early 1920s, Meerwein's detailed studies concluded that some of these rearrangements occurred through an intermediate carbonium ion in which carbon had three covalent bonds with only six electrons and therefore possessed a positive charge. Conant[7] referred to Meerwein's work on rearrangements in 1929 as "one of the outstanding developments in the past ten years" able to "provide the experimental basis for discussion of the mechanism of a large class of rearrangements". Carbonium ions had been suggested much earlier as intermediates by the American chemists Stieglitz and Norris.

A general theory of molecular rearrangements published by Frank C. Whitmore of Penn State in 1932 correlated a large number of isolated observations on rearrangement reactions. His ideas and extensive experimental results proved extraordinarily useful in predicting the occurrence and products not only in carbonium ion rearrangements, but also in other systems. Indirectly, by accustoming the American chemists to the idea of carbonium ions, the Whitmore theory paved the way for increasing knowledge and acceptance of the English School which had previously postulated such intermediates. Work by Bartlett, Hammett, Saul Winstein (UCLA), and others in the 1930s further familiarized American chemists with carbonium ions as intermediates.

Acid- and base-catalyzed reactions of carbonyl compounds, studied by many workers in this country, were of increasing interest in the 1914-1939 period, here again drawing upon Brönsted's ideas of general acid-base catalysis and the effects of added salts on rates. Many papers dealt mainly with reaction in aqueous solutions and with proton acids, for example the excellent ones by D. R. Merrill and E. Q. Adams (Berkeley) in 1917 on hydrolysis of acetanilide, by Conant and Bartlett on the mechanism of semicarbazone formation in 1932, by A. C. Cope (then at Bryn Mawr) on carbonyl condensations in 1937, and by I. Roberts and H. C. Urey (Columbia) on the mechanism of esterification using labeling by isotopic ^{18}O in 1938.

Supplementing these were steady advances made after 1930 in the mechanism of carbonyl reactions requiring catalysis by very strong bases, like sodamide, $NaNH_2$, and sodium triphenylmethyl, $Na^+C^-(C_6H_5)_3$, which C. R. Hauser used to catalyze carbonyl addition reactions unaffected by weaker bases. M. A. Spielman (Wisconsin) used the highly substituted, very strong base mesitylmagnesium bromide to catalyze similar reactions. Karl Ziegler introduced the valuable strong base lithium diethylamide, $Li^+N^-(C_2H_5)_2$, which is more convenient to use than sodium triphenylmethyl.

The study of the properties of very strong acid solutions by Hantzsch in Germany was extended in this country by Conant and by Hammett and became important synthetically and mechanistically. By 1939 the use of $AlCl_3$ and of HF to bring about reactions of saturated hydrocarbons with unsaturated ones, and thus to produce branched-chain hydrocarbons of enhanced value in high compression engines, had been clearly marked out. The results were a logical development of previous work on strong acids.

Conant pioneered semiquantitative studies on the strengths of very weak acids, mainly hydrocarbons, which opened up an active field of research and had synthetic as well as theoretical implications.

Studies on the mechanism of addition of halogen to the carbon-carbon double bond showed that in polar solvents the addition is a two-step process, usually giving *trans* products.

We have mentioned the role of agricultural and medical research institutes in stimulating investigation of the chemistry of processes taking place in living systems. This means in large part the isolation and study of the mode of action of enzymes. Earlier work on enzymes was necessarily done with impure or mixtures of enzymes. In 1926, however, the isolation of crystalline urease by J. B. Sumner of Cornell and the subsequent preparation of other crystalline enzymes, especially by J. H. Northrop of the Rockefeller Institute, put the chemistry of enzymes on a new level of significance.

On more traditional lines, the high standard of American research in the structural chemistry of vitamins, hormones, carbohydrates, nucleic acid components, and many other natural products had become equal to the best of the German and Swiss work by 1939. This was essential to the development of biochemistry and molecular biology. Leaders in natural product work were W. A. Jacobs and P. A. Levene (Rockefeller Institute), Roger Adams (Illinois), C. S. Hudson (U. S. Public Health Service), Karl Folkers (Merck), and many more.

The dramatic pursuit of carcinogenic compounds, begun in England in the 1920s with the isolation of dibenzanthracene and benzpyrene from coal tar by J. W. Cook, E. L. Kennaway, and others, sparked intense and successful synthetic investigations of polycyclic aromatic compounds in this country, principally by L. F. Fieser (Harvard), M. S. Newman (Ohio State), and W. E. Bachmann (Michigan). The goal of these studies– the elucidation of the mechanism of carcinogenesis by pure organic compounds– proved far more elusive than was suspected in the optimistic early days of this work in the 1930s. This research not only laid the necessary foundation for the significant advances of the next fifty years, but led to notable new knowledge and synthetic procedures for polycyclic carbon ring compounds, particularly the sterols, sex hormones, and related compounds. The elegant total synthesis of the sex hormone equilenin by Bachmann, Wayne Cole, and A. L. Wilds in 1939 at Ann Arbor marked the opening of the era of total synthesis of complex natural products which burgeoned in the following decades.

American production of plastics and polymers, such as Bakelite, predated 1914, but the work of Wallace H. Carothers at Du Pont from 1928 to 1937 on the nature and formation of polymers led to the invention of nylon and neoprene. Carothers' achievements resulted in large research programs on synthetic polymers, stimulating notable advances in the fundamental theory and investigation of the chemistry and the technology of polymers in many industrial and university laboratories. Contemporary with this increased knowledge of synthetic polymers was greater understanding of natural polymers, such as cellulose, starch, proteins, and nucleic acids. The interaction of the two fields resulted in mutual progress.

Basic studies on heterogeneous catalysis (causing reactions between gases and liquids to take place at a solid catalytic surface), processes of moderate importance in academic chemistry but vital to industries producing synthetic ammonia, methanol, motor fuel, and countless organic compounds, were carried out in many university and industrial laboratories, as well as in the Fixed Nitrogen Laboratory of the U. S. Department of Agriculture during and after World War I. Among the prominent investigators of the mechanism of reactions at solid surfaces were Irving Langmuir at General Electric, H. S. Taylor at Princeton, and W. A. Lazier at Du Pont. Homer Adkins at Wisconsin was a leader in the study of catalytic addition of hydrogen to organic compounds.

The rediscovery and experimental proof of the earlier Sachse-Mohr theory that the cyclohexane ring did not have all six carbons in one plane but in bi-planar "chair" or in "boat" forms were supported experimentally by W. Hückel in Germany and J. Böeseken in Holland in the 1920s, and by R. G. Dickinson and C. Bilicke in 1928 at Caltech. Their X-ray crystallographic examination of the hexachloro- and hexabromocyclohexanes showed clearly the "chair" form. This non-planar state was essential in interpreting the theory and chemical behavior of the important fused cyclohexane ring systems in steroids and other natural products, worked out after 1939 by new chemical evidence and new physical methods.

American chemists in this period illuminated the stereochemistry of molecules exhibiting optical activity without asymmetric carbon atoms. Roger Adams' resolutions and studies of properly substituted biphenyls demonstrated the degree of steric hindrance and size of groups blocking free rotation around the ring-joining carbons, a phenomenon first recognized by J. Kenner in England. E. P. Kohler of Harvard confirmed van't Hoff's classical prediction of molecular asymmetry due to hindered rotation in allenic compounds by his resolution of a highly substituted allene.

A fundamental stereochemical problem was posed by the Walden inversion, named after the German-Baltic chemist. He showed that some sequences of reactions led to a product with one sign of rotation, while other sequences gave the same product with the opposite rotation, hence a mirror image of the first. This was a process whose occurrence and mechanism must be predictable, if configurations of organic compounds were to be ascertained.

The 1914-1939 period saw an almost complete solution of this question. Considered by Emil Fischer and others, it was G. N. Lewis, with his usual insight and clear exposition, who expressed in words what was found to be the most satisfactory mechanism. Because he drew no diagrams, Lewis' scheme was sometimes overlooked by later workers. A. R. Olson of Berkeley confirmed the Lewis path.

Levene, investigating the structures of fats, sugars present in nucleic acids, and other compounds attempted to determine the relationship between the sign of rotation and the configuration of optically active compounds, a question answered earlier with some success in sugar chemistry by C. S. Hudson and others. Levene's laborious and careful studies on chosen series of configurationally related compounds, each containing one asymmetric carbon atom, failed to prove a reliable correlation. Nevertheless, they were to be of great value later in establishing the stereochemical relationships of complicated molecules, particularly natural products, after the absolute configuration of key compounds had been revealed by X-ray methods.

A perennial quest in organic chemistry has been the relationship between structure and rate (including mode) of reaction usually called "relative reactivity". Because the classical structural theory does not give *a priori* information about these relationships, the prediction of reactivity requires a foundation of extensive experimental studies on varied reactions applied to varied structures, using the behavior of some prototype compound as a standard.

In the 1914-1939 period many organic chemists carried out studies of relative reactivities; these inquiries usually differed from those prior to 1914 in having more clear-cut objectives. In addition to the collection of reliable data useful in synthetic work, theories of structure, valence, reaction mechanisms, and catalysis could be tested by suitably designed experiments. W. A. Noyes examined the role of ortho substitution in the kinetics of esterification of benzoic acids; Conant measured reaction rates of sodium iodide with many halides in acetone; J. F. Norris examined the relative reactivities of alcohols, his work harmonizing with Ingold's theories; C. R. Noller (Stanford) showed the effect of structure on displacement reactions in alkyl halide-pyridine systems, and many more investigations could be cited.

In this period notable advances in the theory of the structure of aromatic compounds and of aromatic (electrophilic) substitution took place. The application of quantum mechanics to the benzene problem by Pauling and others pointed to a highly symmetrical structure with all C-C bonds identical in length and arranged in the form of a regular hexagon. This agreed with the X-ray crystallographic analysis of hexamethylbenzene and *m*-dinitrobenzene and with dipole measurements on a number of substituted benzenes.

The directing influence of various groups on aromatic substitution, whether to ortho, meta, and/or para positions, had received much attention over the years and had generated a vast literature on electrophilic substitution. Theories of electron displacement, the increased or decreased electron density of positions in the benzene ring occasioned by the directing groups, were in general agreement with the groups' electron-attracting or -donating properties, as also shown by their effects on acid dissociation constants and dipole moments. In addition, these theories were supported by approximate quantum mechanical calculations by Pauling and G. W. Wheland.

The 1914-1939 period, particularly the last ten years, saw the beginning of the Instrumental Revolution, including the use of isotopic tracers and of powerful methods of separation; these completely changed the conduct of research in organic chemistry and in related fields like biochemistry. A practical pH meter for measuring acidity with a glass electrode was commercially available by 1936. The Instrumental Revolution made striking progress after 1939, but the trend was clearly discernible before that. Column chromatography, used by the Russian botanist M. Twsett in 1906, was introduced into organic chemistry in Germany and Switzerland in the 1930s, and then into this country. Many modifications were to become indispensable, particularly in work on natural products and in biochemistry.

The discovery and concentration of the stable isotopes of hydrogen (deuterium, ^{2}H), carbon (^{13}C), oxygen (^{18}O), and nitrogen (^{15}N) by H. C. Urey and his collaborators at Johns Hopkins and Columbia introduced extremely powerful techniques for study of reaction mechanisms in organic chemistry and biochemistry. The radioactive ^{14}C and ^{3}H (tritium) became generally available only after the work on the atomic piles in the Manhattan Project during World War II.

Among instrumental methods, the measurement of ultraviolet and visible spectra (electronic spectra) of organic compounds started as early as 1870, and such measurements were reported with increasing frequency up to 1939. The infrared spectra of organic compounds, which show the frequency of atomic vibrations, were reported occasionally in the 1920s. These spectral measurements represented specialized skills and techniques, as well as a large amount of time. The availability of commercial spectrometers, simple to operate and relatively easy to maintain, did not come until after 1940. The Beckman DU ultraviolet and visible spectrophotometer was the first widely useful instrument for measuring electronic spectra.

Mass spectroscopy, first used effectively by F. W. Aston and by A. J. Dempster in 1918-1919, was utilized particularly in petroleum research laboratories before 1929, but the extraordinary importance of the technique for general structural use came after 1945. Another presently indispensable instrument, the nuclear magnetic resonance spectrometer (NMR), was unknown until after 1945. The measurement of optical rotatory dispersion (ORD), the change in optical rotation with the wave length of the light used, and related measurements were made in the 19th century. Important results required better instrumentation, which became available after 1945.

X-ray diffraction by crystals was observed in 1912, and the first determination of an organic structure was that of the highly symmetrical hexamethylenetetramine by R. G. Dickinson and A. L. Raymond at Caltech. X-ray crystallographic determination of structure for compounds with a low degree of symmetry became possible only when computers were developed after 1945.

All these useful methods of determining structures of organic compounds have become available for routine use since 1945 and have revolutionized the determination of organic structures.

The symbiosis between instrumentation and advance of a science leads to questions about the relationship between technology and advance in fundamental knowledge. There seems to be a feedback mechanism; new ideas or observations result eventually in a new technology (as with the laser), which leads in turn to experiments impossible before and to new observations requiring new theoretical ideas.

The 1914-1939 period thus shows development of the fundamental theory of molecular structure. Based in part on this, there is increasing knowledge of reaction mechanism and of methods of synthesis. The Instrumental Revolution is foreshadowed, and in general American organic chemistry comes of age in international science.

LITERATURE CITED

(This section, as a survey, is less completely documented than usual. Later essays contain more discussion and relevant references.)

1. James B. Conant, *My Several Lives*, New York, 1970, p. 34. Although this book contains some surprising internal contradictions, it is useful as a picture of the chemical career of one of the key figures of this period.

2. An excellent account of the growth of federal laboratories: A. Hunter Dupree, *Science in the Federal Government*, Cambridge, 1957.

3. Figures on doctorates and bachelor degrees are from A. Thackray, J. L. Sturchio, P. T. Carroll, and R. Bud, *Chemistry in America, 1876-1976: An Historical Application of Science Indicators*, Dordrecht, 1985, pp. 257-266. We are indebted to Dr. Sturchio for giving us this data prior to publication. The figures on doctorates by L. R. Harmon and H. Soldz, *Doctorate Production in United States Universities 1920-1962*, National Academy of Sciences - National Research Council, Washington, 1963, pp. 10, 20, start only with 1920 and differ slightly from those quoted. Figures on American Chemical Society membership from H. Skolnik and K. M. Reese, Eds., *A Century of Chemistry*, Washington, 1976, p. 456. For research laboratories in the U. S., *Research - A National Resource*, 3 Vol., National Resources Planning Board, Washington, 1938-1941, Vol. 2: *Industrial Research*, R. Stevens, p. 37 and F. S. Cooper, p. 130; also K. A. Birr, in *Science and Society in the United States*, D. Van Tassel and M. Hall, Eds., Homewood, Ill., 1966, p. 69.

4. M. Rand, *Scientific Monthly*, **73**, 71(1951).

5. D. J. Kevles, *The Physicists*, New York, 1978.

6. L. P. Hammett, Foreword, p.vii, to N. B. Chapman and J. Shorter, Eds., *Advances in Linear Free Energy Relationships*, New York, 1972.

7. J. B. Conant. JACS, **51**, 1619 (1929), a review of C. W. Porter's monograph, *Molecular Rearrangements*, New York, 1928.

ROGER ADAMS AND CARL S. MARVEL

9
Structural and Synthetic Studies, 1914-1939

The papers in this division number in the thousands and only those of originality, special usefulness, or mechanistic interest can be mentioned. Categories and times often overlap, and a completely consistent or satisfactory scheme of organization is impossible. Should a given procedure be discussed under Lewis acids or under synthetic methods? Furthermore, the structural and synthetic work involved in this period for such specific topics such as natural products, aromatic chemistry, organometallic compounds, and numerous reactions of well-defined mechanistic type which follow a logical combination of theory and synthesis will be discussed in later essays.

In general the changes in the character of synthetic organic chemistry follow the changes in American research over this period. The work becomes less provincial, more original, and more significant. Although much uninspired compound-making and pursuing the vagaries of homologous series took place, some of this work was valuable, particularly in medicinal chemistry and in areas such as the preparation of monomers for polymerization experiments. Furthermore, the behavior of higher homologs frequently shows an interesting contrast with that of lower homologs, facts of fundamental mechanistic interest.

The growing interest in reaction mechanisms frequently required synthesis and purification of novel compounds. Many of the accomplished workers in this field, Lucas, Kharasch, Conant, and Bartlett, for example, became expert synthetic chemists in order to prepare compounds for mechanistic study. The field of polycyclic aromatic compounds and the study of the stereochemical course of reactions required ingenious synthetic work in many cases.

J. F. Thorpe (1872-1940, professor at Imperial College, University of London, 1914-1938) had postulated a novel type of structure[1] to explain many observations on the reactions of glutaconic esters, which Thorpe believed to

$$ROOCCH=CHCH_2COOR$$

Normal form

$$ROOC\overset{|}{C}HCH_2\overset{|}{C}HCOOR$$

Labile form

Esters of glutaconic acid

exist in "normal" and "labile" forms, the latter being written as a symmetrical diradical. Kohler and Reid showed that Thorpe's main argument for the labile form— its failure to give a Michael addition product— was incorrect; under suitable experimental conditions methyl cyanoacetate added to both forms, the so-called normal and labile, to give the same product.[2] Kohler's work showed that the classical structural formula for glutaconic acid is satisfactory, provided possible *cis-trans* isomerism and tautomerism are considered.

$$ROOCCH=CHCH_2COOR \quad + \quad NCCH_2COOCH_3 \xrightarrow[\text{base}]{\text{trace}} ROOCCH_2\underset{|}{C}HCH_2COOR$$
$$NCCHCOOCH_3$$

A minor masterpiece of structural determination by F. J. Moore and Ruth Thomas of MIT was the elucidation of xanthogallol,[3] a yellow compound

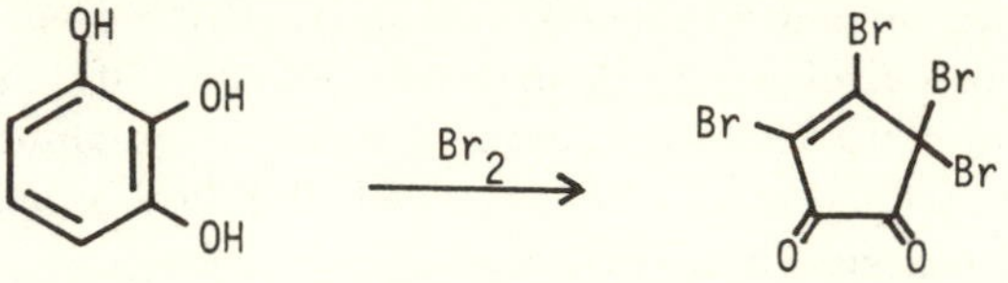

pyrogallol   xanthogallol

obtained from bromine and pyrogallol. A completely erroneous structure had been proposed by German workers many years earlier.[4] It is clear from Moore's incisive paper that American chemistry was the poorer for his failure to carry out more researches on structural problems. Moore (1867-1926, A.B. 1889, Amherst; Ph.D. 1893, Heidelberg) taught at MIT from 1894 to 1926. He was primarily a teacher and wrote a useful *History of Chemistry* (1918), which went through three editions. With Edgar F. Smith he was one of the founders in 1921 of the Division of the History of Chemistry of the American Chemical Society.

The extensive study and clinical applications of the "nitrogen mustards" in cancer chemotherapy made the first report on this series significant;[5] K. Ward of the Hercules Powder Co. prepared several examples of these compounds and found that they were reactive toward nucleophiles and were also strong vesicants (blister producers), especially the tertiary amine. Detailed research during World War II on the nitrogen mustards as possible war gases showed that they were unsuitable for this purpose, but quantitative and mechanistic information on their reactions with nucleophiles was obtained.[6] Pharmacological tests demonstrated their anticancer activity which led after the war to the synthesis and testing of thousands of compounds of this type, called *alkylating agents*.[6] Several alkylating agents are now used routinely in conjunction with other compounds in treating cancer.

$$(ClCH_2CH_2)_2S \qquad ClCH_2CH_2NH_2 \quad (ClCH_2CH_2)_2NH \quad (ClCH_2CH_2)_3N$$

Sulfur mustard   Nitrogen mustards
("mustard gas")

Organic fluorine compounds were studied intensively in this country, following the discovery that Freon 12, CCl_2F_2, could be used in refrigeration systems. Methods of synthesis involved exchange of fluorine for chlorine by inorganic fluorides.[7] Continued research on organofluorine compounds at Du Pont led to the synthesis of tetrafluoroethylene and its polymerization to Teflon, a very stable, tough, self-lubricating polymer.[8] Because of its chemical inertness, Teflon was used in the Oak Ridge diffusion plant for separating ^{238}U in the atomic bomb project. It was described scientifically only in 1946 in a paper by Hanford and Joyce.

$$CCl_4 \xrightarrow[SbCl_5]{SbF_3} CCl_2F_2$$

$$HCCl_3 \xrightarrow[SbCl_5]{SbF_3} HCClF_2 \xrightarrow{650\text{-}800°} CF_2{=}CF_2 \xrightarrow[\text{polymerization}]{\text{addition}} (CF_2CF_2)_n$$
$$\text{Teflon}$$

William E. Hanford (1908- , B.S. 1930, Philadelphia College of Pharmacy; Ph.D. 1935, Illinois with Roger Adams) was a research chemist and research executive at Du Pont, General Aniline and Film, M. W. Kellogg, and Olin from 1935 to 1974. Robert M. Joyce (1915- , B.S. 1935, Nebraska; Ph.D. 1938, Illinois, also with Adams) worked on polymers, fluorine compounds, films, and pharmaceuticals at Du Pont from 1938 to 1978 as a chemist and later as executive in several divisions of the company.

The direct fluorination of organic compounds by elementary fluorine, a process of experimental difficulty and some danger, was studied by L.A. Bigelow during a long career at Duke University.[9] This work eventually yielded results of practical as well as scientific importance. Lucius A. Bigelow (1892-1971, B.S. 1915, MIT; Ph.D. 1919, Yale) taught at several schools including Brown from 1920 to 1924 and at Duke from 1924 to 1961. Nearly all his research was on the action of elementary fluorine on organic compounds.

Marvel observed the isomerization of substituted acetylenes to allenes in numerous cases.[10] Carl S. (Speed) Marvel (1894- , A.B., A.M. 1915, Illinois Wesleyan; Ph.D. 1920, Illinois) taught at Illinois from 1920 to 1961 and at Arizona from 1961 on. Marvel worked in many areas of synthetic organic chemistry on organometallic compounds, free radicals, and acetylenes. He

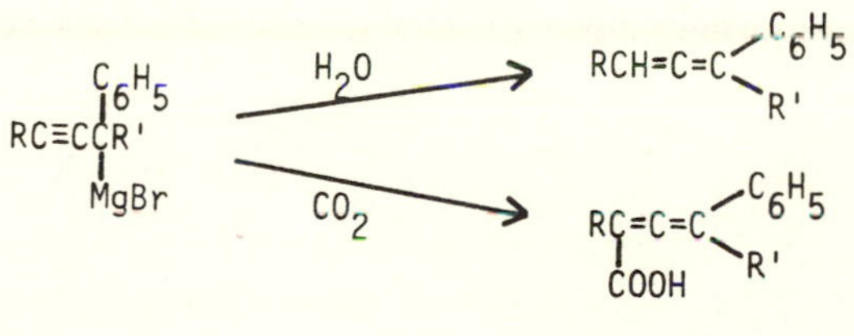

$$RC{\equiv}CR' \text{ (with } C_6H_5 \text{ and MgBr)} \xrightarrow{H_2O} RCH{=}C\overset{C_6H_5}{\underset{R'}{\diagup\diagdown}}$$

$$\xrightarrow{CO_2} RC{=}C\overset{C_6H_5}{\underset{R'}{\diagup\diagdown}} \text{ with COOH}$$

was the first university chemist in this country to study the structure and mechanism of formation of synthetic polymers. A consultant to Du Pont for over fifty years, he exercised great influence through his teaching and research. He continued research at Arizona for many years on polymers resistant to high temperatures; his productive career in research has been one of the longest in American chemistry.

Charles D. Hurd investigated the chemistry of ketenes in detail and worked out the pyrolysis of acetone to ketene.[11]

$$CH_3\overset{\overset{\text{O}}{\|}}{C}CH_3 \quad \xrightarrow{\text{heat}} \quad CH_2=C=O \quad + \quad CH_4$$

Hurd (1897- , B.S. 1918, Syracuse; Ph.D. 1921, Princeton with L. W. Jones) taught at Northwestern from 1924 to 1961. He studied carbohydrates, heterocyclic compounds, and nitrogen compounds and published a comprehensive monograph, *The Pyrolysis of Carbon Compounds* (1929). He proposed the accepted mechanism for the Claisen rearrangement.

The acetal of ketene, $CH_2=C(OC_2H_5)_2$, had been reported by the German chemist Scheibler to be formed from sodium ethoxide and ethyl acetate. Attempts by S. M. McElvain at Wisconsin to repeat this were uniformly unsuccessful, and McElvain finally clarified the problem by an unambiguous synthesis of ketene acetal.[12]

$$ICH_2CH(OC_2H_5)_2 \quad + \quad (CH_3)_3COK \quad \xrightarrow{(CH_3)_3COH} \quad CH_2=C(OC_2H_5)_2 \quad + \quad KI$$

This led to the development of a new field of chemistry,[13] the study of ketene acetals; Scheibler's materials had not contained the acetal. Further, McElvain's synthesis is an early example of the use of *tert*-butyl alcohol as a solvent for a base-catalyzed reaction; the tertiary alcohol does not react with the ketene acetal, in contrast to ethyl alcohol which adds to ketene acetals.[13] In this respect *tert*-butyl alcohol has some of the advantages of a "dipolar aprotic solvent", similar to dimethyl sulfoxide, $CH_3S(\rightarrow O)CH_3$, (DMSO).

$$CH_2=C(OC_2H_5)_2 \quad + \quad C_2H_5OH \quad \longrightarrow \quad CH_3C(OC_2H_5)_3$$

The latter type of solvent was to be very widely used in later years for displacement and base-catalyzed reactions. The double bond in the ketene

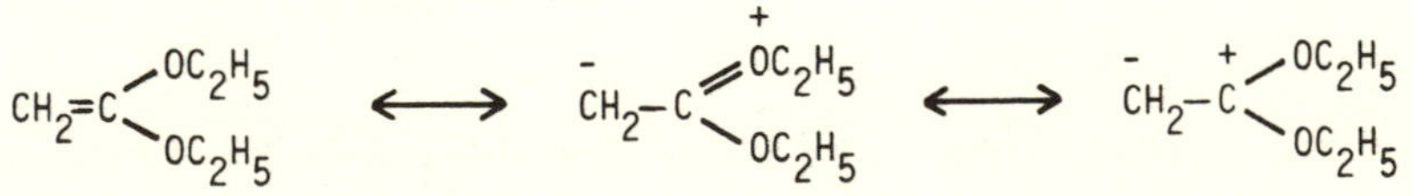

$$CH_2=C\begin{smallmatrix} OC_2H_5 \\ OC_2H_5 \end{smallmatrix} \quad \longleftrightarrow \quad \overset{-}{C}H_2-C\overset{+}{\begin{smallmatrix} OC_2H_5 \\ OC_2H_5 \end{smallmatrix}} \quad \longleftrightarrow \quad \overset{-}{C}H_2-\overset{+}{C}\begin{smallmatrix} OC_2H_5 \\ OC_2H_5 \end{smallmatrix}$$

acetals behaves as if it had large contributions from resonance forms. Its many reactions follow the pattern expected from these structures, with electrophiles adding to the electron-rich carbon.

Samuel M. McElvain (1897-1973, B.S. 1919, Washington University at St. Louis; Ph.D. 1923, Illinois with Roger Adams) was a member of the Wisconsin faculty from 1923 to 1961. His research included nitrogen heterocycles, reaction mechanisms (particularly the ester condensation), and structural elucidation of the active principles in catnip.

An original synthesis of ethylenic hydrocarbons of wide applicability was devised by C. E. Boord of Ohio State.[14] This was used to synthesize ten of the thirteen possible hexenes,[15] among other applications. Cecil E. Boord (1884-1964, B. S. Wabash; Ph.D. 1912, Ohio State) taught at Ohio State from 1912 to 1964. He directed a research project on the synthesis and properties of hydrocarbons.

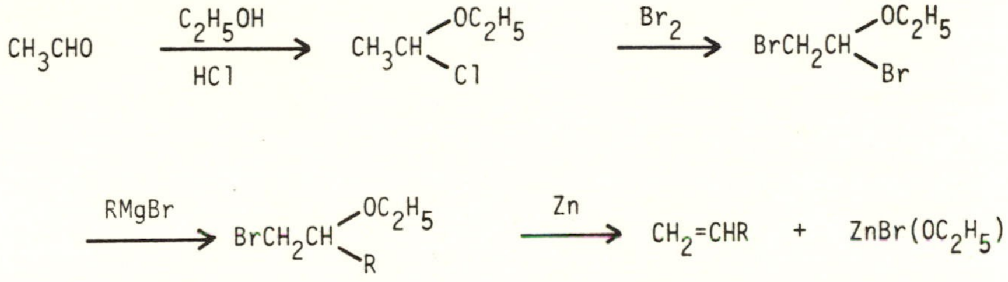

$$CH_3CHO \xrightarrow[HCl]{C_2H_5OH} CH_3CH\begin{smallmatrix}OC_2H_5\\Cl\end{smallmatrix} \xrightarrow{Br_2} BrCH_2CH\begin{smallmatrix}OC_2H_5\\Br\end{smallmatrix}$$

$$\xrightarrow{RMgBr} BrCH_2CH\begin{smallmatrix}OC_2H_5\\R\end{smallmatrix} \xrightarrow{Zn} CH_2=CHR + ZnBr(OC_2H_5)$$

The structures of an interesting series of compounds, the rubrenes, were established after synthetic work by Marvel and by C. F. Koelsch. Rubrene itself was prepared in France by Charles Dufraisse by heating the chloride from diphenylphenylethynylcarbinol.[16] The rubrenes took up oxygen in the presence of light to form photooxides which evolved oxygen on heating. This reversible oxygenation of many substituted rubrenes[16] was investigated because of the apparent similarity to the behavior of hemoglobin in the blood. The structures postulated by Dufraisse had not been proved, however.

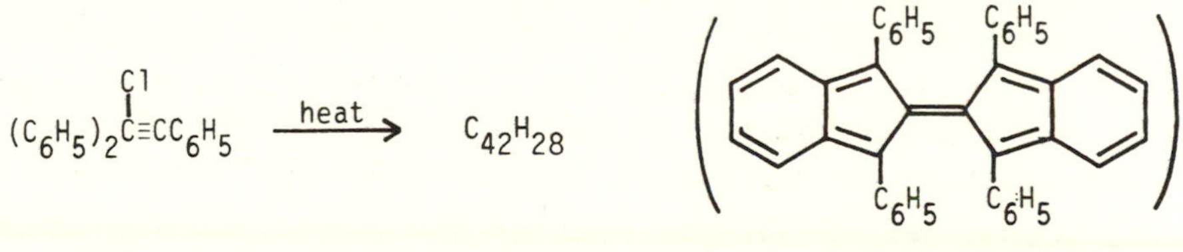

$$(C_6H_5)_2\overset{Cl}{\underset{|}{C}}CC_6H_5 \xrightarrow{heat} C_{42}H_{28}$$

Rubrene (old structure)

Marvel, in connection with work on possible free-radical precursors, synthesized a dihydrodihydroxy derivative of the supposed rubrene structure and found that it showed none of the properties of the rubrene family.[17] Koelsch synthesized the same derivative by a different procedure and found the two synthetic samples to be identical.[18] The French group, suspicious of the old structure, proposed a new structure, tetraphenylnaphthacene, which was shown to be correct by synthesis.[19] The formation of photooxides was found to be a general reaction of anthracene and naphthacene derivatives.[16,19]

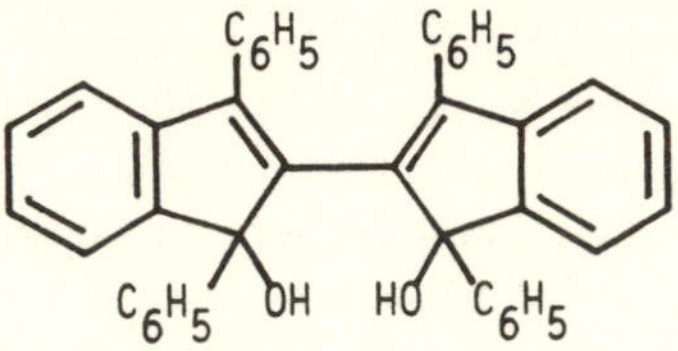

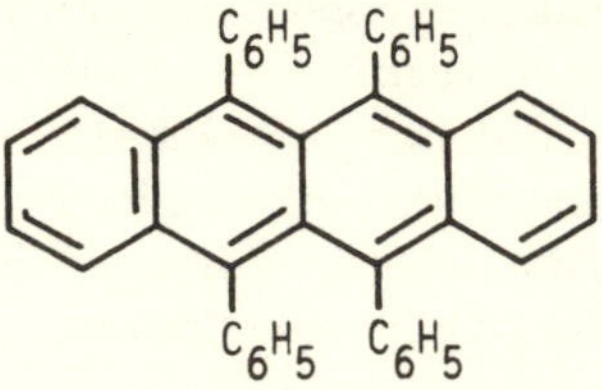

Rubrene (correct structure)

C. Frederick Koelsch[20] (1907- , B.S. 1928, Wisconsin; Ph.D. 1931 with McElvain) taught at Minnesota from 1932 to 1973. He received the Langmuir Award (precursor of the ACS Award in Pure Chemistry) in 1934. Koelsch was a brilliant experimentalist with a remarkable knowledge of chemistry who trained some outstanding Ph.Ds. He prepared a free radical in 1932, but due to a referee's objections his work was not published until 1957, when electron spin resonance measurements on the original sample showed its character.

The determination of the enediol structure of ascorbic acid (vitamin C)

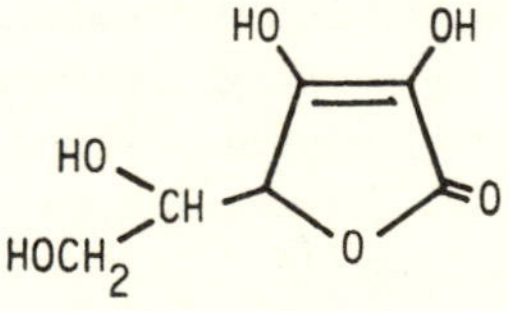

Vitamin C
(Ascorbic acid)

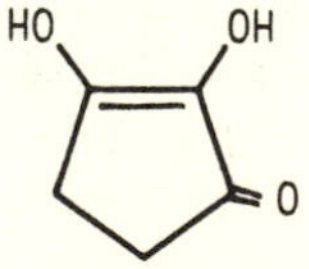

Reductic acid

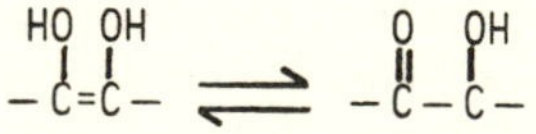

enediol   hydroxy
                   ketone

and of the related reductic acid[21] led to a search for other stable enediol systems. Kohler and R. B. Thompson obtained an enediol diacetate from a hydroxy ketone, although the free enediol was not stable.[22]

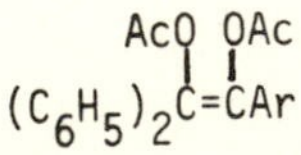

After Kohler's death Thompson reported on a relatively stable enediol from catalytic reduction of dimesityl diketone.[22] Fuson and J. Corse reported earlier on a stereoisomeric enediol obtained by action of the Bachmann-Gomberg reagent ($Mg\text{-}MgI_2$) on the same diketone[23] and a more stable enediol from 2,4,6-triethylphenyl homolog. Enediols had been found by R. P. Barnes and Leila Green of Howard University[24] as derivatives of $C_6H_5C(OH){=}C(OH)C({=}O)Ar$, with Ar being $C_6H_2(CH_3)_3(2,4,6)$.

R. Percy Barnes (1898- , A.B. 1921, Amherst; Ph.D. 1933, Harvard with Kohler) taught at Howard University from 1922 to 1967 and published research on α-diketones; he was a member of the National Science Board from 1950 to 1958.

The stabilizing effect of the mesityl group on an attached carbonyl group was strikingly shown by Fuson's preparation of tri- and tetraketones with terminal mesityl groups.[25] The tetraketone readily lost carbon monoxide on heating to form the triketone.

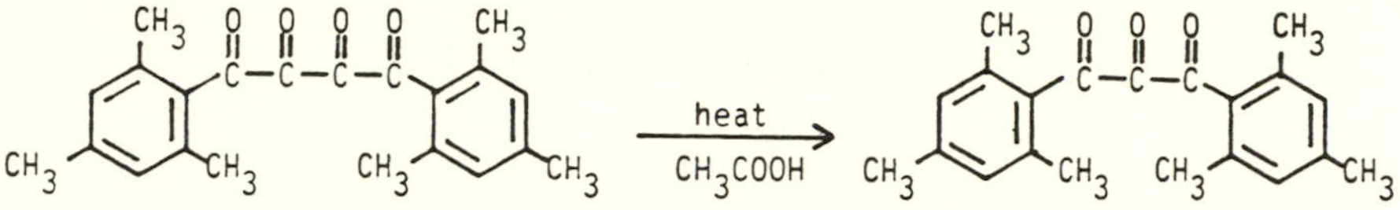

Reynold C. Fuson (1895-1979, A.B. 1920, Montana; A.M. 1921, Berkeley; Ph.D. 1924, Minnesota) taught at Illinois from 1927 to 1963, training many chemists who became prominent in industry and academic life. An inveterate traveler and excellent linguist, he was a leader in the development of the Illinois graduate program in organic chemistry through his teaching, textbooks, and research. His work dealt with the structure of benzene, conjugated systems, and steric hindrance.

J. R. Johnson[26] with H. R. Snyder and others at Cornell planned a study of organic boron compounds to give information about electron-deficient structure, because trivalent boron has only six electrons around boron, is a Lewis acid, and is somewhat similar to a carbonium ion. The results are

$$C_4H_9MgBr \xrightarrow[\text{or } BF_3]{(CH_3O)_3B} (C_4H_9)_3B \xrightarrow[C_6H_5CO_3H]{H_2O_2 \text{ or}} (C_4H_9O)_3B$$

$$\xrightarrow{NaOH} C_4H_9OH + Na_3BO_3$$

interesting in themselves, and also for comparison with the extended later studies of Herbert C. Brown on organoboron compounds as synthetic intermediates. Johnson prepared his boron compounds by action of a Grignard reagent on a borate ester or boron trifluoride. He found that tri-*n*-butylboron was oxidized quantitatively by hydrogen peroxide or organic peracids to tributyl borate, which was hydrolyzed by base to *n*-butyl alcohol. Johnson also found that carbon-boron bonds could be cleaved by bromine and other reagents and attributed these cleavages correctly to coordination with the boron to complete its octet of electrons, followed by cleavage of the carbon-boron bond by rearrangement or by displacement. The peroxide rearrangement and hydrolysis is the final stage in Brown's hydroboration reaction; however, Brown could prepare trialkylboranes from boron hydrides and ethylenic hydrocarbons. The boron hydrides were not available at the time of Johnson's work, and his procedure involved the isolation of the trialkylboranes. The formation of "salts" from trivalent boron compounds (acting as Lewis acids) and electron donors (such as ammonia) is discussed elsewhere. Johnson's work was the most comprehensive study on organoboron compounds in the country prior to Brown's.

John R. Johnson (1900-1983, B.S. 1920, Illinois; Ph.D. 1922, Illinois with Roger Adams) was a faculty member at Cornell University from 1927 to 1970. He did research on arsenic, boron, and germanium compounds; furan derivatives; ketenes; mold metabolites, such as gliotoxin; and reaction mechanisms. He and Adams wrote a laboratory manual for elementary organic chemistry, which was widely used for more than fifty years. Harold R. Snyder (1910- , B.S. 1931, Illinois; Ph.D. 1935, Cornell with J. R. Johnson) was on the Illinois faculty from 1936 to 1975. His research included practical syntheses of amino acids, chemistry of indole, and use of polyphosphoric acid in organic synthesis.

A novel method of ring closure, good for four-, five-, and six-carbon rings, was discovered by R. C. Fuson.[27] The mechanism proposed involved a 1,4-diradical, but ionic displacement reactions seem more probable.

$$
\begin{array}{ccc}
\text{COOR} & \text{COOR} & \text{COOR} \\
| & | & | \\
\text{CH}_2\text{—CHBr} & \text{CH}_2\text{—CHCN} & \text{CH}_2\text{—C—CN} \\
| \quad \xrightarrow{\text{NaCN}} & | & | \quad | \\
\text{CH}_2\text{—CHBr} & \text{CH}_2\text{—CHBr} \longrightarrow & \text{CH}_2\text{—C—H} \\
| & | & | \\
\text{COOR} & \text{COOR} & \text{COOR}
\end{array}
$$

The action of diazomethane on cyclic ketones was shown by Mosettig and Burger at Virginia to lead to ring enlargement of the ketones, a useful synthetic procedure.[28] Alfred Burger (1905- , Ph.D. 1928, Vienna) was associated with the chemistry department at Virginia from 1929 to 1970. He was an active research worker in many fields of medicinal chemistry, wrote and edited valuable monographs on medicinal chemistry, and edited the *Journal of Medicinal Chemistry* from 1959 to 1971. Erich Mosettig (1892-1962, Ph.D. 1923, Vienna) was a staff member at Virginia from 1922 to 1939 and chemist at the National Institutes of Health from 1939 to 1957. He did research on morphine alkaloids and morphine substitutes, steroids, antimalarials, and other types of pharmacologically active compounds.

By generating the diazomethane *in situ*, Kohler's group[29] prepared large quantities of seven-, eight-, nine-, and ten–membered ring ketones, and of cycloheptene and cyclooctene for measurements of heats of hydrogenation.

$$CH_3N(NO)COOC_2H_5 + Na_2CO_3 \longrightarrow CH_2N_2 \text{ (Meerwein)}$$

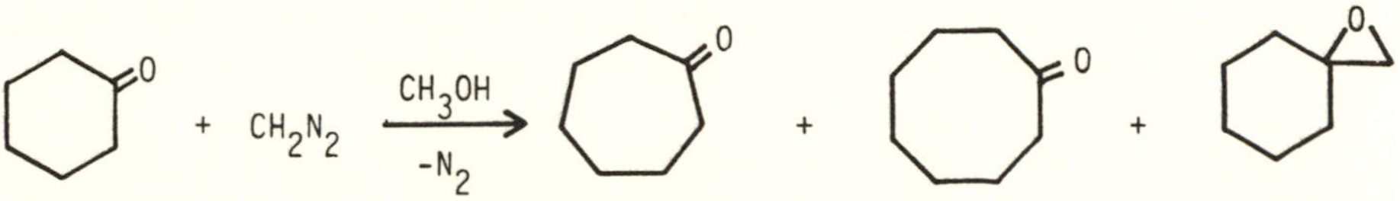

Kohler's work was also directed toward a synthesis of cyclooctatetraene (COT), because there was a widespread feeling in the late 1930s that Richard Willstätter's synthesis of this compound was not valid and that he had probably obtained styrene. This would account for the non-aromatic behavior of Willstätter's product.[30,31] Willstätter[30] had reduced his product catalytically with uptake of $4H_2$ to cyclooctane, under conditions where benzene took up no hydrogen, and oxidized this to suberic acid, observations not explained by his critics but proving the presence of the eight-membered ring. Cope and Overberger's synthesis of COT and proof of its identity with the material obtained by W. Reppe from acetylene finally confirmed Willstätter's work and the non-aromatic character of COT.[32]

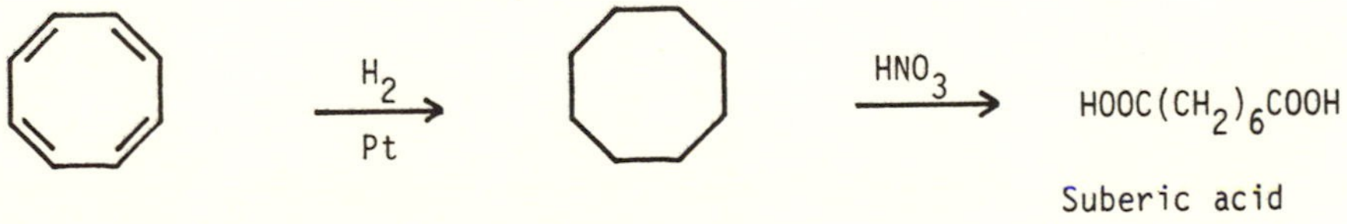

Suberic acid

$$4 \text{ HC}\equiv\text{CN} \xrightarrow{\text{Ni catalysts}} (CH=CH)_4 \text{ (COT, Reppe)}$$

Arthur C. Cope (1909-1966, A.B. 1929, Butler; Ph.D. 1932, Wisconsin with McElvain) taught at Bryn Mawr, Columbia, and MIT, where, as chairman, he built up an outstanding chemistry department. He served the American Chemical Society and other scientific organizations in many positions of policy, in which he was closely associated with Roger Adams. He discovered a rearrangement of a carbon system (the Cope rearrangement) analogous to the Claisen rearrangement; studied carbon ring compounds of medium size, including cross-ring effects; and worked on amine oxide rearrangements and platinum complexes of unsaturated hydrocarbons. Many of his students have had outstanding research careers.

Charles G. Overberger (1920- , B.S. 1941, Penn State; Ph.D. 1944, Illinois with C. S. Marvel) taught at Brooklyn Polytechnic Institute until 1967 and then at Michigan where he was vice-president for research from 1972 to 1982. His research included many aspects of polymer chemistry, including polymers as models for enzymes. His professional activities included the presidency of the American Chemical Society.

Studies on oxidation by Nicholas A. Milas included a method of *cis*-hydroxylation of double bonds with hydrogen peroxide in *tert*-butyl alcohol with a trace of osmium tetroxide as catalyst,[33] useful for water insoluble compounds. Milas, born in Candia, Crete, in 1897 (B.S. 1922, Coe College; Ph.D. 1926, Chicago with Ethel M. Terry) taught at MIT from 1928 to 1965. In addition to studies on peroxides, he did synthetic work on vitamins A and D.

Workers at Mathieson Alkali in Niagara Falls[34] demonstrated the preparation, oxidizing properties, and usefulness of ethyl hypochlorite; they noted that *tert*-amyl hypochlorite was particularly stable. This type of reagent was studied particularly in later years as a source of chlorine atoms.[35]

An entirely different type of oxidation was the "Hooker oxidation" discovered by Samuel C. Hooker, a man whose career was unusual by any standards.[36,37] As the example below shows, the oxidation shortens the side chain by one carbon, and the quinone ring apparently opens and recyclizes in the opposite direction because the alkyl and hydroxyl groups change places; this is made clear by marking the benzenoid ring with bromine. Hooker suggested this sequence, which has some further support. The hydroxynaphthoquinones of this type were found to have promising antimalarial activity, and the intensive investigation of the field by Fieser and chemists at Abbott Laboratories has already been noted.

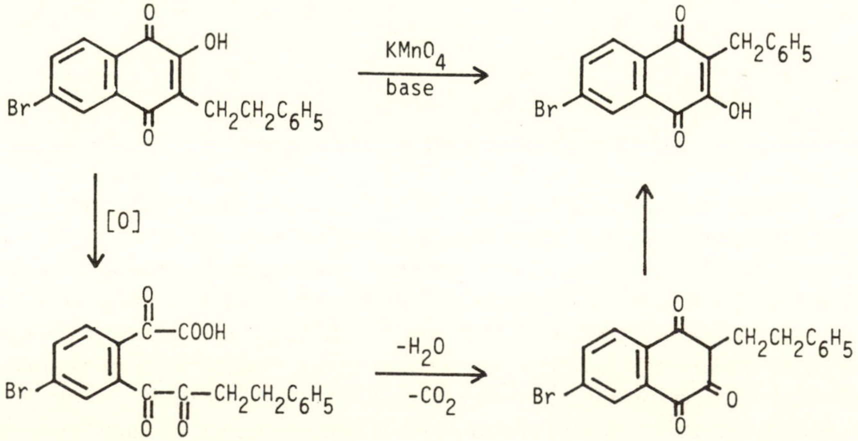

Samuel C. Hooker (1864-1935) was born in England, received his Ph.D.

in 1885 from Munich, and became chief chemist of a Philadelphia sugar company. He carried out researches on lapachol, a hydroxynaphthoquinone from tropical wood, publishing in ACJ and JCS. His managerial ability carried him to responsible posts in the American Sugar Refining Company, from which he retired in 1915. Thereafter he resumed work on lapachol and related compounds in his private laboratory, located, appropriately, on Remsen Street in Brooklyn. His voluminous research records were prepared for publication by Fieser and appeared in eleven papers[36] in 1936 in JACS. Hooker collected one of the most complete chemical libraries in the world, now at Wayne State University in Detroit and known as the Kresge-Hooker Library. Hooker was also an accomplished amateur magician.

An early example of a silicon compound as a synthetic reagent is the use of SiF_4 as a reagent for esterifying alcohols; this presumably functions by forming an acyl silicon fluoride.[38] Silicon compounds of many types were to become widely used in synthesis.

$$RCOOH \quad + \quad R'OH \quad \xrightarrow{\quad SiF_4 \quad} \quad RCOOR'$$

An extremely useful synthetic procedure for methylating amines is the reductive methylation with formaldehyde and formic acid developed by H. T. Clarke. It has the notable advantage over methylation with methyl iodide or similar reagents in that it yields no quaternary ammonium salts.[39] The mechanism apparently involves formation of a methylol amine, $RNHCH_2OH$, which is reduced by formic acid to the methyl amine.

$$RNH_2 \quad + \quad CH_2O \quad \xrightarrow{\quad HCOOH \quad} \quad RNHCH_3 \quad + \quad RN(CH_3)_2 \quad (no \ \overset{+}{RN}(CH_3)_3 \ X^-)$$

Hans T. Clarke (1887-1972) was born in England and studied at University College and with Emil Fischer at Berlin, where he had first hand experience with the toxicity of "mustard", $(ClCH_2CH_2)_2S$. After a D.Sc. from London, he joined Eastman Kodak in Rochester, New York, pioneered in setting up their organic chemicals department, and was a co-founder of *Organic Syntheses*. As head of the department of biological chemistry at the College of Physicians and Surgeons at Columbia from 1928 to 1956, he built up a vigorous research group and by his strong support made possible the fundamental discoveries of R. Schoenheimer and D. Rittenberg on metabolic pathways using deuterium and other isotopes as labels.

Another reaction of amines was the demonstration by W. M. Lauer at Minnesota of the mechanism of alkylation of an enamine on carbon.[40] This supplemented earlier work by Robinson, but its potentialities as an impor-

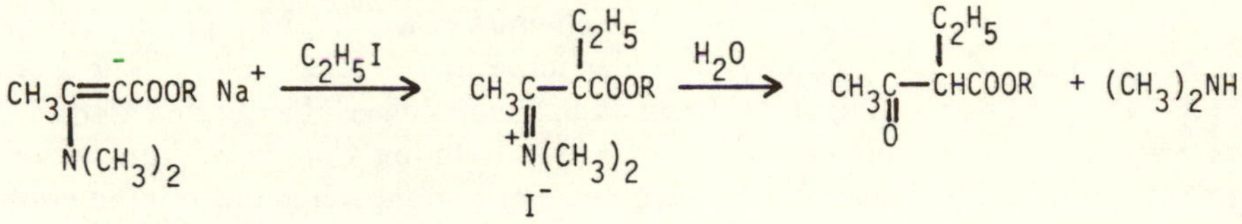

$$CH_3\overset{-}{\underset{N(CH_3)_2}{C=CCOOR}} Na^+ \xrightarrow{C_2H_5I} CH_3\overset{C_2H_5}{\underset{\overset{+}{N(CH_3)_2}}{\underset{I^-}{C-CCOOR}}} \xrightarrow{H_2O} CH_3\overset{C_2H_5}{\underset{O}{C-CHCOOR}} + (CH_3)_2NH$$

tant synthetic procedure were not realized until Gilbert Stork at Columbia developed the enamine synthesis.[41]

Walter M. Lauer (1895-1976, A.B. 1913, Ursinus; Ph.D. 1924, Minnesota) taught at Minnesota from 1920 to 1964. He introduced Pregl's methods of microcombustion to this country and did excellent work on the structure of bisulfite compounds, the Claisen rearrangement, antimalarial drugs, hydrogen isotope exchange reactions, and long-chain fatty acids.

Another very useful synthetic reaction was foreshadowed by D. D. Coffman and Marvel[42] who treated phosphonium compounds with strong bases and obtained phosphonium ylides. These reacted with water to form phosphine oxides. If the ylides had been treated with a carbonyl compound, RCOR', the reaction later known as the Wittig reaction would have resulted, with $(C_6H_5)_3PO$ and the unsaturated compound $RR'C=CHCH_3$ as products. Marvel wrote the structures of the ylides as shown; he stated, however, that they probably did not contain pentavalent phosphorus but rather resembled the amine oxides and phosphine oxides in bonding.

$$(C_6H_5)_3\overset{+}{P}C_2H_5 \ I^- \xrightarrow{LiC_4H_9} (C_6H_5)_3P=CHCH_3 \xrightarrow{H_2O} (C_6H_5)_2\overset{O}{\overset{\|}{P}}C_2H_5 + C_6H_6$$

The discovery by G. Domagk in Germany and the E. Fourneau group in France of the clinical value of sulfanilamide and related compounds in treating bacterial infections was the most important advance in the chemotherapy of bacterial disease up to that time.[43] A burst of synthetic activity in many countries ensued, and many hundreds of analogs of sulfanilamide were synthesized and tested.[44] The synthetic chemistry involved was straightforward; hence many of the useful sulfanilamide derivatives were prepared by several groups simultaneously.[45]

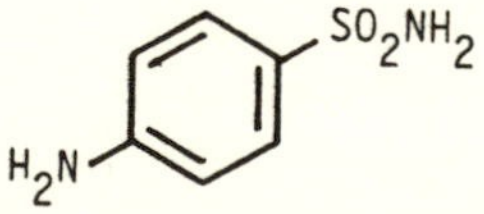

Sulfanilamide

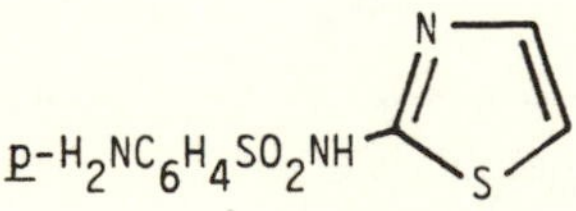

Sulfathiazine

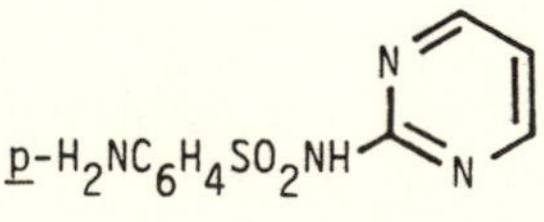

Sulfadiazine

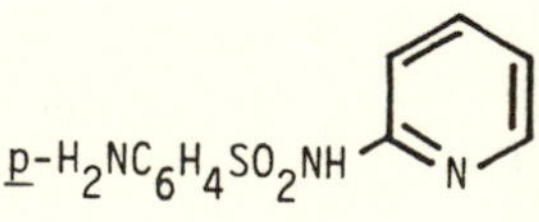

Sulfapyridine

The speed of development of the sulfa drug field was striking; it was also significant because of the correct suggestion by D. D. Woods[46] and by P. Fildes in Britain that the sulfa drugs acted as antagonists or antimetabolites to p-aminobenzoic acid. It was not known at that time why the latter is required for bacterial growth; the later isolation and structural proof of the

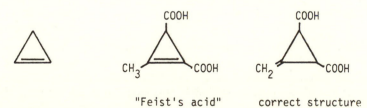

p-Aminobenzoic acid   Folic acid (pteroylmonoglutamic acid)

several folic acids,[47] however, showed that p-aminobenzoic acid is required for folic acid biosynthesis, and the folic acids in turn are required in the biosynthesis of nucleic acids. This was one of the first steps in rationalizing the hitherto completely empirical nature of chemotherapy. P. H. Bell and R. O. Roblin of the American Cyanamid Company in Stamford, Connecticut, extended the idea of antagonism of sulfa drugs to p-aminobenzoic acid.[48]

Richard O. Roblin (1907- , A. B. 1930, Rochester; Ph.D. 1934, Columbia with Marston T. Bogert) was active in medicinal chemical research at the Stamford laboratories, working on sulfa drugs, diuretics, antibiotics, and antimetabolites. He became vice-president of Cyanamide and president of Cyanamide Europe.

The structure of cyclopropene is clearly significant for theories of bonding

"Feist's acid"   correct structure

and strain. N. J. Demjanov had reported its preparation with reasonable evidence for its structure.[49] M. J. Schlatter of Caltech improved the preparation and made the compound available for physical studies. F. Feist had early attributed the cyclopropene structure to an acid, but the correct structure was proved by nuclear magnetic resonance measurements to be a methylenecyclopropane by A. T. Bottini and J. D. Roberts.[50] Kohler in 1930 synthesized a highly substituted cyclopropene by elimination of nitrous acid from a nitrocyclopropane; the structure was established conclusively.[51] In the 1950s and 1960s cyclopropenes were prepared in variety by the addition of carbenes, RCH, (from diazoacetic esters, $N_2CHCOOR$, or diazoalkanes, $RCHN_2$) to acetylenes, using light as a catalyst.[52]

The inadequacy of traditional ideas of strain was shown most dramatically by the preparation of many bicyclobutanes; some were formed from cyclopropenes and carbenes.[52]

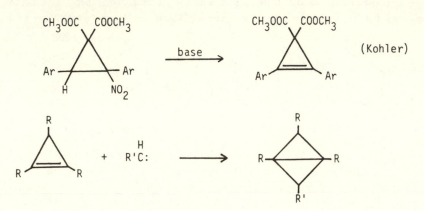

(Kohler)

A general reaction of great promise was the conversion by chlorine-water of S-alkylisothioureas to the corresponding sulfonyl chlorides by Treat B. Johnson and J. M. Sprague at Yale.[53] The S-alkyl compounds are readily prepared from thiourea and alkyl halides, or alcohols plus acid. It was found later that this reaction can cause serious explosions,[54] so it must be used cautiously.

$$S=C(NH_2)_2 \quad + \quad RX$$

$$S=C(NH_2)_2 \quad + \quad ROH \quad + \quad HCl \longrightarrow RS-C{\overset{NH}{\underset{NH_2}{\diagup}}} \quad \xrightarrow[H_2O]{Cl_2} \quad RSO_2Cl$$

A synthesis of α-amino acids[57] was developed by M. S. Dunn at UCLA;

$$CH_2(COOC_2H_5)_2 \xrightarrow{C_4H_9ONO} HON=C(COOC_2H_5)_2 \xrightarrow[\text{Raney Ni}]{H_2} H_2NCH(COOC_2H_5)_2$$

$$\xrightarrow{C_6H_5COCl} C_6H_5CONHCH(COOC_2H_5)_2 \xrightarrow[\text{2. RX}]{\text{1. Na}} C_6H_5CONH\overset{R}{\underset{\mid}{C}}(COOC_2H_5)_2$$

$$\xrightarrow[HCl]{H_2O} H_2N\overset{R}{\underset{\mid}{C}}HCOOH$$

he employed the steps shown, some being modifications of earlier methods. In turn modifications produced better yields, wider applicability, and ease of operation. Such syntheses made amino acids readily available for feeding experiments and for the synthetic work on polypeptides which began around 1950.

Structural work which was fundamental to the understanding of disease caused by pneumococci was done at Rockefeller during this period by M. Heidelberger and W. F. Goebel in O. T. Avery's laboratory.[56] They showed that the antigenic properties of the pneumococcus were due to the polysaccharide coat or capsule of the bacillus. They found that types III and VIII pneumococcus had similar immunological properties because they contained the same aldobionic acid grouping in the capsular polysaccharide. Goebel showed by a combination of degradation and synthesis that the type III polysaccharide was made up of aldobionic acid units as shown.[59] The type VIII structure was similar but not identical. This work had implications in immunology far beyond the group of organisms under study.

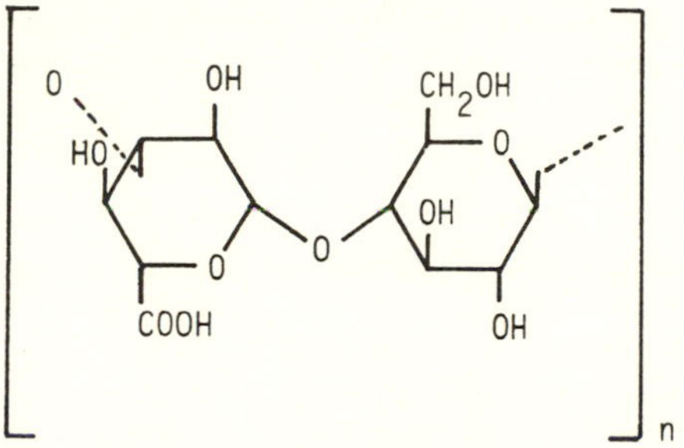

LITERATURE CITED

1. J. F. Thorpe, JCS, **87**, 1669 (1905); Thorpe and A. S. Wood, ibid., **103**, 1579 (1913).
2. E. P. Kohler and G. H. Reid, JACS, **47**, 2803 (1925), and refs. therein.
3. F. J. Moore and Ruth M. Thomas, ibid., **39**, 974 (1917). Mrs. Moore established a scholarship in memory of her husband, which assisted one of us (DST) during two years of graduate work.
4. C. A. Theurer, Ann., **245**, 327 (1888).
5. K. Ward, Jr., JACS, **57**, 914 (1935).
6. G. Golumbic, J. S. Fruton and M. Bergmann, JOC, **11**, 518 (1946) and following papers; P. D. Bartlett et al., JACS, **69**, 2971 (1947); use of nitrogen mustards in cancer, A. Gilman and F. S. Philips, Science, **103**, 409 (1946); *Clinical and Biological Effects of Alkylating Agents*, Ann. N. Y. Acad. Scis., **68**, 657-1266 (1958).
7. Many papers; e.g., A. L. Henne et al., JACS, **58**, 402, 887 (1936).
8. W. E. Hanford and R. M. Joyce, ibid., **68**, 2082 (1946); M. M. Renfrew and E. E. Lewis, Ind. Eng. Chem., **38**, 870 (1946), and refs. therein; R. J. Plunkett, U. S. Patent 2,230,654 (Feb. 4, 1941); Chem. Abstr., **35**, 3365 (1941).
9. L. A. Bigelow et al., JACS, **55**, 4614 (1933) and many later papers.
10. C. S. Marvel et al., ibid., **57**, 2619 (1935); J. R. Johnson, et al., ibid., **60**, 1885 (1938).
11. C. D. Hurd and P. B. Cochran, ibid., **45**, 515 (1923); J. H. Williams and C. D. Hurd, JOC, **5**, 122 (1940); Hurd, *Organic Syntheses*, Coll. Vol. 1, 2nd ed., p. 330.

12. F. Beyerstedt and S. M. McElvain, JACS, **58**, 528 (1936); **59**, 226 (1937).
13. S. M. McElvain, Chem. Rev., **45**, 453 (1950). See biography of McElvain by Gilbert Stork, *Biog. Mem. Nat. Acad. Scis.*, **54**, 221 (1983).
14. C. E. Boord, et al., **52**, 3396 (1930) and later papers.
15. C. G. Schmitt and C. E. Boord, ibid., **54**, 751 (1932).
16. Review by C. Dufraisse, Bull. Soc. Chim., [4] **53**, 789 (1933).
17. J. C. Eck and C. S. Marvel, JACS, **57**, 1898 (1935).
18. C. F. Koelsch and H. J. Richter, ibid., p. 2010.
19. C. Dufraisse et al., Bull. Soc. Chim., **2**, 1546 (1935); **3**, 1847 (1936) and following; C. F. H. Allen and L. Gilman, JACS, **58**, 937 (1936).
20. Biography, W. E. Noland, *Minnesota Chemist*, **27**, No. 7, p. 7 (1975).
21. T. Reichstein and R. Oppenauer, Helv., **17**, 390 (1934).
22. E. P. Kohler and R. B. Thompson, JACS, **59**, 887 (1937); Thompson, ibid., **61**, 1281 (1939).
23. R. C. Fuson et al., ibid., **61**, 975, 2010 (1939).
24. R. P. Barnes and Leila S. Green, ibid., **60**, 1549 (1938).
25. A. R. Gray and R. C. Fuson, ibid., **56**, 1367 (1934); Fuson, et al., ibid., 2099.
26. J. R. Johnson, H. R. Snyder et al., ibid., **60**, 105, 111, 115, 121 (1938).
27. R. C. Fuson and T. Y. Kao, JACS, **51**, 1536 (1929), and later papers.
28. E. Mosettig and A. Burger, ibid., **52**, 3456 (1930), and refs. therein to action of diazomethane on open-chain aldehydes and ketones by Arndt and by Meerwein.
29. E. P. Kohler, M. Tishler, et al., ibid., **61**, 1057 (1939), using a modified method of Meerwein.
30. R. Willstätter and E. Waser, Ber., **44**, 3423 (1911); Willstätter and M. Heidelberger, ibid., **46**, 517 (1913).
31. S. Goldwasser and H. S. Taylor, JACS, **61**, 1260 (1939); C. D. Hurd and L. R. Drake, ibid., **61**, 1943 (1939).
32. A. C. Cope and C. G. Overberger, ibid., **69**, 976 (1947).
33. N. A. Milas and S. Sussman, ibid., **58**, 1302 (1936) and many other papers.
34. M. C. Taylor et al., ibid., **47**, 395 (1925); cf. R. D. Chattaway and O. G. Backeberg, JCS, **123**, 2999 (1923).
35. Review of organic hypohalites, M. Anbar and D. Ginsburg, Chem. Rev., **54**, 925 (1954).
36. S. C. Hooker, JACS, **58**, 1168, 1174, 1179 (1936); L. F. Fieser, et al., ibid., 1223.
37. C. A. Browne in *Dictionary of American Biography*, Suppl. Vol. 1.
38. J. A. Gierut, F. J. Sowa and J. A. Nieuwland, JACS, **58**, 786 (1936).
39. H. T. Clarke, et al., ibid., **55**, 4571 (1933) and refs. therein. Biography of Clarke by H. B. Vickery, *Biog. Mem. Nat. Acad. Scis.*, **46**, 3 (1975). For Clarke's share in the isotopic tracer studies, see the excellent account of R. E. Kohler, Jr., *Historical Studies in Physical Sciences*, **8**, 257 (1977).
40. W. M. Lauer and G. W. Lones, JACS, **59**, 232 (1937); R. Robinson, JCS, **109**, 1038 (1916); Robinson was then at the University of Sydney.
41. G. Stork, et al., JACS, **76**, 2029 (1954) and later papers.
42. D. D. Coffman and C. S. Marvel, ibid., **51**, 3496 (1929); review of Wittig reaction, A. Maercker, *Organic Reactions*, **14**, 270 (1965).
43. G. Domagk, Deut. Med. Wochschr., **61**, 250 (1935) and later papers (Chem. Abstr., **29**, 4831 (1935)); E. Fourneau, et al., Compt. Rend. Soc. de Biol., **122**, 258 (1936) (Chem. Abstr., **30**, 6061 (1936)).
44. E. H. Northey, Chem. Rev., **27**, 85 (1940); E. H. Northey, *The Sulfonamides and Related Compounds*, New York, 1948.
45. Some key American papers: M. L. Crossley, et al., JACS, **60**, 2217 (1938); R. J. Fosbinder and L. A. Walter, ibid., **61**, 2032 (1939); W. A. Lott and F. H. Bergeim, ibid.,

61, 3593 (1939) (sulfathiazole); R. O. Roblin, et al., ibid., **62**, 2002 (1940) (sulfadiazine). Extensive clinical work on sulfanilamide, sulfathiazole, and sulfapyridine, but not on sulfadiazine, was reported in L. Goodman and A. Gilman, *The Pharmacological Basis of Therapeutics*, First ed., New York, 1941.

46. D. D. Woods, Brit. J. Exptl. Path., **21**, 74 (1940); P. Fildes, Lancet, **238**, I, 955 (1940).

47. So many large groups were involved in the isolation and synthesis of the members of the folic acid complex that equitable citations are difficult; however, several laboratories of the American Cyanamide Company were primarily responsible; e.g., E. L. R. Stokstad et al., JACS, **70**, 5, 10, 14, 19, 23, 25, 27 (1948), and earlier papers.

48. P. H. Bell and R. O. Roblin, ibid., **64**, 2905 (1942).

49. N. J. Demjanov and Marie Dogarenko, Ber., **56**, 2200 (1923); M. J. Schlatter, JACS, **63**, 1733 (1941).

50. F. Feist, Ber., **44**, 136 (1911); A. T. Bottini and J. D. Roberts, JOC, **21**, 1169 (1956); M. D. Ettlinger, JACS, **74**, 5805 (1952) and refs. therein to X-ray crystallography and IR evidence.

51. E. P. Kohler and S. F. Darling, JACS, **52**, 1174 (1930); S. F. Darling and E. W. Spanagel, ibid., **53**, 1117 (1931). Kohler's first publications on cyclopropanes were James B. Conant's Ph.D. Thesis: Kohler and Conant, ibid., **39**, 1404, 1699 (1917).

52. For review of reactions with diazoacetic ester, V. Dave and E. W. Warnhoff, *Organic Reactions*, **18**, 217 (1970).

53. T. B. Johnson and J. M. Sprague, JACS, **58**, 1348 (1936); **59**, 1837 (1937).

54. K. Folkers, et al., ibid., **63**, 3530 (1941).

55. M. S. Dunn et al., JBC, **94**, 599 (1932); **130**, 341 (1939).

56. An interesting account of Avery's group and its fundamental discovery that DNA was the carrier of genetic information in the pneumococcus is *The Professor, The Institute and DNA*, by René J. Dubos, New York, 1976.

57. W. F. Goebel, et al., JBC, **121**, 195 (1937); **139**, 511 (1941); **110**, 391 (1935); JACS, **59**, 2745 (1937); and with M. Heidelberger, JBC, **70**, 613 (1925); **74**, 613 (1927).

WERNER E. BACHMANN

10

The Structure and Reactions of Aromatic Compounds, 1914-1939

In earlier essays the unusual structural problems posed by benzene and its analogs ("aromatic compounds") have been discussed. In the period under review, the structure of aromatic compounds was satisfactorily solved on both an experimental and a theoretical basis, and the reactions of aromatic compounds began to be reasonably well understood.

The development of the quantum mechanical theory of electron bonding, by Pauling and Slater in this country and by E. Hückel and others abroad, led to the picture of benzene by 1930 as a symmetrical molecule with its hydrogens in the plane of the ring and a π electron cloud above and below the plane of the ring.[1] Pauling's "valence bond" symbolism looked on benzene

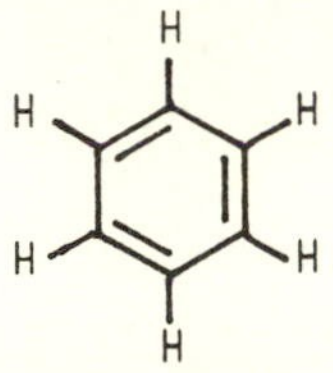

Valence bond structures

as a "resonance hybrid", with contributions from the two Kekulé structures as well as from other structures of higher energy. This was indicated by a double-ended arrow, which did *not* mean a rapid oscillation between the two Kekulé structures, a perennial misunderstanding; it did mean that the actual structure was a composite of the Kekulé structures. In Hückel's molecular orbital picture, the six π electrons (the aromatic sextet) were likewise considered to be disposed symmetrically; many years later the circle inside the hexagon was used to indicate the aromatic system. The great stability of benzene, meaning by stability its lower rate of many reactions and lower heat of combustion compared to a structure containing three carbon-carbon double bonds, is satisfactorily explained by the quantum mechanical theory.

Of the many early hypotheses, two are interesting enough to mention. In 1922 E. C. Crocker, an MIT research associate, discussed the benzene structure in terms of Lewis bonds, suggesting that electron pairing was due to magnetic forces and emphasizing the high degree of symmetry of benzene and the symmetrical distribution of electrons outside the ring.[2] He considered Kekulé structures to be unstable and preferred a symmetrical electron distribution. He extended his ideas to aromatic heterocycles and to explaining the phenomena of orientation in substituted benzenes. Although he had no theoretical basis, Crocker obviously anticipated the conclusions of later work.

The second speculation was less successful; M. L. Huggins at Berkeley suggested that the benzene ring was composed of tetrahedral carbons and was puckered, without a high degree of symmetry.[3] Some clever experiments by Reynold C. Fuson showed that Huggins' ideas were untenable.[4]

The planar, regular hexagonal structure of benzene was postulated by Pauling in a theoretical paper[5] in 1926; C. P. Smyth at Princeton supported this by dipole moment measurements on benzene derivatives.[6] The X-ray diffraction experiments of Kathleen Lonsdale in 1929 in England on hexamethylbenzene[7] were conclusive in establishing the highly symmetrical structure of benzene. The higher aromatic compounds, such as naphthalene, $C_{10}H_8$, anthracene and phenanthrene, both $C_{14}H_{10}$, do not have all C-C bonds in the rings of the same length in contrast to benzene, because the π electron distribution is not as symmetrical as in benzene.

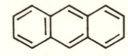
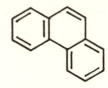

Naphthalene   Anthracene   Phenanthrene

German chemists intensively studied the chemistry of aromatic compounds after 1850; this was partly because important naturally occurring dyes, such as alizarin and indigo, were found to be aromatic derivatives. Their synthesis and sale proved to be the foundation of the German dye industry that strengthened German research by providing employment for Ph.D.s, grants for university research, and industrial consultantships for German professors.[8] Great numbers of synthetic dyes were prepared, some of which became profitable, and these were almost all derived from aromatic systems. Some German chemists therefore tried to rationalize and classify the reactions of aromatic compounds, although in general they were more interested in making new compounds and proving their structures than in speculating about how and why they were formed.[9]

A fundamental problem in aromatic chemistry was that of the directive effect of groups already in benzene on the position taken by an entering substituent, the *orientation of aromatic substitution*. It was recognized experimentally in the 19th century that substituents could be divided into two general groups: (1) those which direct entering substituents *ortho* and *para* (*o-* and *p-*), the most notable of these being OH, NH_2, CH_3 (and other alkyl groups), NHCOR, OCH_3, and NR_2; (2) those groups which direct entering substituents to the *meta* (*m-*) position, such as NO_2, COOR, COR, SO_3H, and N^+. These directing influences apply to reagents which introduce the "typical" aromatic substituents, such as NO_2, Br, SO_3H, and N=NAr. It was not clearly recognized until the 1920s that the "typical" aromatic substitution reaction involved an attack on the aromatic carbon by an electrophilic reagent, that is, one which reacted preferentially at a position with a high electron density. The diagrams show a further important point, that *o,p*-directing groups increase the rate of further substitution, while *m*-directing groups decrease it.

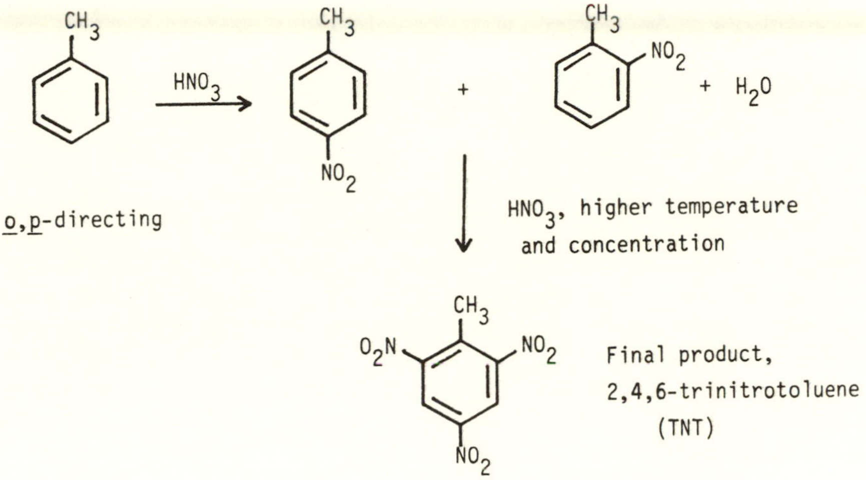
CH3
HNO3
NO2
+ H2O
o,p-directing
HNO3, higher temperature
and concentration
O2N
Final product,
2,4,6-trinitrotoluene
(TNT)

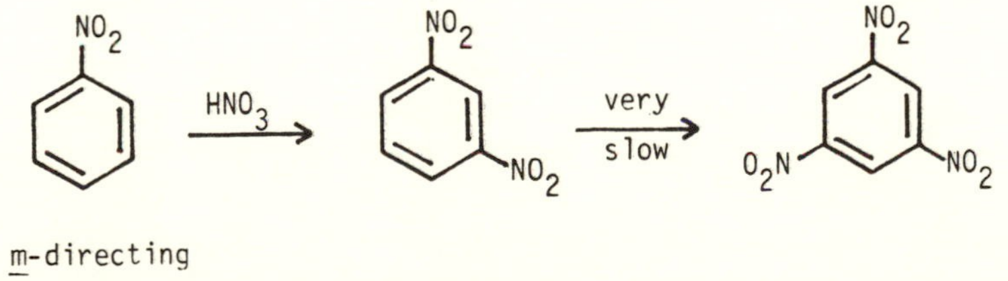
NO2
HNO3
very
slow
O2N
m-directing

Various empirical correlations between orienting effect and chemical nature of groups were made early, such as the fact that unsaturated groups are generally *m*-directing, while saturated groups are *o,p*-directing. Arnold F. Holleman, a Dutch chemist, in a monograph[10] in 1910, summarized the experimental observations, furnishing much of the quantitative data from his own work. He believed that the aromatic substitution took place by addition to a double bond, as in the Kekulé structure, followed by elimination. This idea had some serious flaws which required *ad hoc* explanations. Theories of polarity, especially those of H. Shipley Fry, were current for several years[11] and were an outgrowth of ideas of K. G. Falk and J. M. Nelson, J. Steiglitz, and W. A. Noyes in this country, Vorländer in Germany, and A. Lapworth and R. Robinson in England.[12] These theories assumed an alternation of charges around the benzene ring and led to correct identification of *o,p*- and *m*-directing groups. The alternating polarity theory was, as discussed elsewhere, applied to saturated carbon chains but was disproved experimentally by H. J. Lucas and by J. B. Conant.

Harry Shipley Fry (1878-1949, A.B. 1901; Ph.D. 1905, both at Cincinnati) spent his whole career teaching at Cincinnati (1902-1945). He is best known for his ideas on valence and structure of benzene, summarized in a monograph in 1921[11] based on the dualistic theories of the chemists listed above. Fry's monograph appeared before Lewis' book on valence and marked the high point of the pre-Lewis theories.

The mechanism of orientation utilizing Lewis' electronic theory of molecular structure was sketched out in 1926 by H. J. Lucas,[13] who pointed out that the electron-attracting or -withdrawing effects of substituents on the aromatic ring could best be determined by their effect on the dissociation constants of p-substituted benzoic acids. Thus p-nitrobenzoic acid is a stronger acid than benzoic acid, and the nitro group in nitrobenzene is a meta directing group, because it decreases electron density at the o- and p-positions, so that the m-positions are relatively electron-rich. Thus electrophilic substitution attacks the m-positions, although more slowly than the o,p-substitution, due to general deactivation of the ring by electron withdrawal by m-directing groups. Lucas' ideas were not worked out in detail, although he measured dissociation constants of several substituted benzoic acids to determine the electronic nature of several groups.

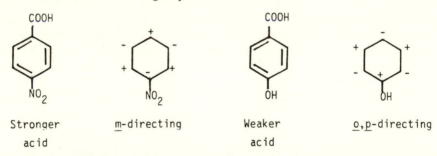

Stronger   m-directing   Weaker   o,p-directing
acid   acid

A more detailed analysis of the electronic effects of groups by C. K. Ingold, Robinson, and others showed that two effects were needed, one a direct inductive electrostatic effect of the Lewis-Lucas type, and the second a result of unshared pairs of electrons on the substituent which interacted with the benzene ring by an "electromeric" effect (Ingold's term)[12] or a "resonance" effect[14] in the Pauling terminology. This dual effect explained some anomalies such as the behavior of the halogens, which are o,p-directing but which decrease the rate of substitution. Although Ingold's system was described in several publications, a systematic exposition was not made by Ingold until 1934 in a review article.[15] This was highly formalistic, lacked specific experimental examples, and led to the unjustified view that the ideas of the "English School" were obscure and their papers hard to read. For this reason the ideas of the English school did not have the influence on American physical organic chemistry which perhaps they should have had. An extensive reading of the American literature leads to the conclusion that American physical organic chemistry was an indigenous parallel development, not a derivative of the British school. The work of Bartlett, Branch, Conant, Hammett, Hauser, Kharasch, Kraus, Lucas, Whitmore, Winstein, Young (arranged alphabetically), and others, during the period under review, supports this view.

The determination of the position of attack in substituted benzenes was difficult in all of the early work because of the limited methods of separation and purification available. The determination of relative reaction rates by competitive experiments was used for very fast reactions, such as the bromination of phenols and anilines. Extended studies by A. W. Francis at Yale and MIT furnished useful experimental data, but little new in fundamental understanding.[16]

A kinetic study of an important technical dye-forming reaction, the coupling of a diazonium compound with a phenol, showed a bimolecular process

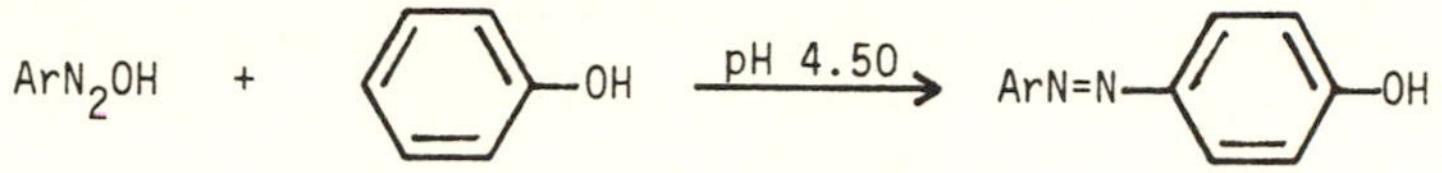

$$ArN_2OH + \text{\langle benzene \rangle}-OH \xrightarrow{pH\ 4.50} ArN=N-\text{\langle benzene \rangle}-OH$$

which was pH dependent and which was assumed to be the reaction of the diazonium compound with the undissociated phenol.[17] A further study over a wider pH range on coupling with amines and phenols showed that the reaction really involved the unprotonated amine or the phenoxide anion, which would be expected *a priori* to be attacked much more rapidly by the electrophilic diazonium group than the undissociated phenol because of the high electron density in the phenol anion.[18]

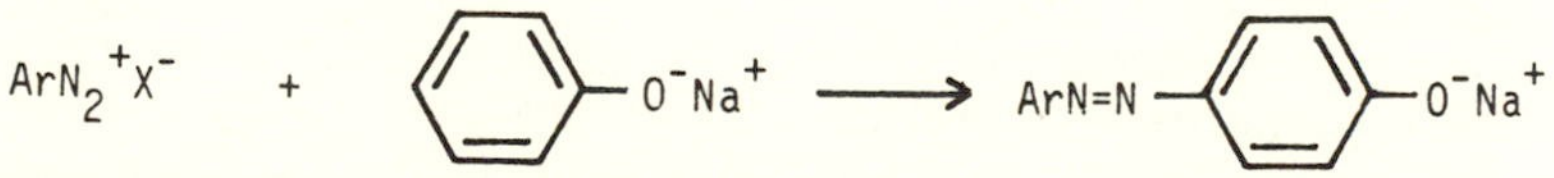

$$ArN_2^+X^- + \text{\langle benzene \rangle}-O^-Na^+ \longrightarrow ArN=N-\text{\langle benzene \rangle}-O^-Na^+$$

Nucleophilic aromatic substitution, worked on extensively by C. Loring Jackson before 1912, is not a common reaction in this period; however, one example should be mentioned.[19] In agreement with later work, F was more rapidly replaced by the nucleophile than Cl was.

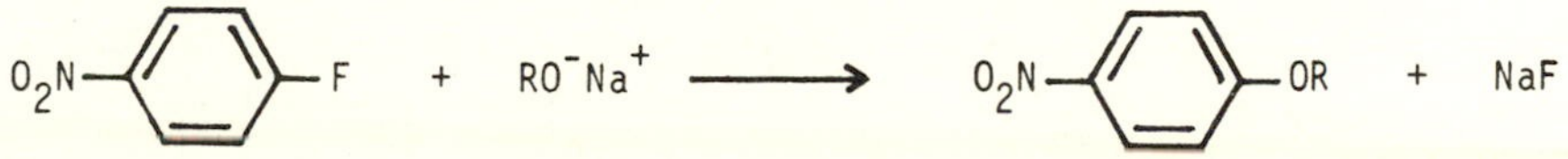

$$O_2N-\text{\langle benzene \rangle}-F + RO^-Na^+ \longrightarrow O_2N-\text{\langle benzene \rangle}-OR + NaF$$

This reaction brings to mind F. Sanger's use of 2,4-dinitrofluorobenzene to tag the terminal amino group of polypeptides during the determination of the amino acid sequence in insulin.[20]

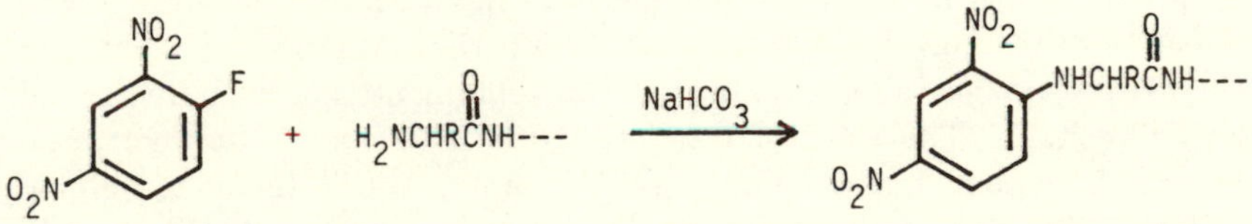

Several cases of replacement of an aromatic halide by a strong base show the formation of a rearranged product, where the basic group is on a carbon adjacent to the position occupied by the halogen.[21] These almost certainly proceed by elimination of halogen acid to form a "benzyne", which then adds a molecle to reform the aromatic structure. The intermediacy of benzynes in such reactions was shown conclusively by J. D. Roberts in 1953.[22]

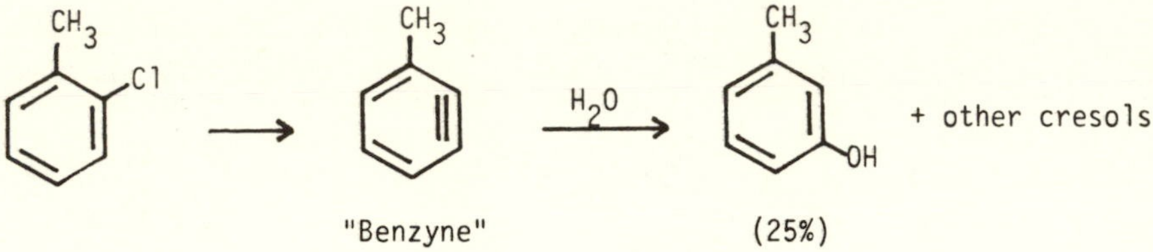

The action of sodium on toluene in liquid ammonia with some water forms a liquid highly reactive toward bromine; this is probably an early example of the Birch reduction to yield a dihydrobenzene[23] by addition of sodium atoms or the equivalent in electrons to an aromatic system.

The formation of biaryls by decomposition of diazonium compounds in an aromatic solvent, the Gomberg-Bachmann reaction,[24] although more complex than the condensed equation shows, is probably a free-radical substitution. This is supported by its failure to follow the rules of orientation for electrophilic aromatic substitution; thus in nitrobenzene, the entering aryl group goes mainly o- and p-, instead of m. Although the yields are not high, this is a useful method of synthesis in some cases.

$$ArH \ + \ Ar'N_2OH \ \xrightarrow{NaOH} \ Ar-Ar' \ + \ H_2O \ + \ N_2$$

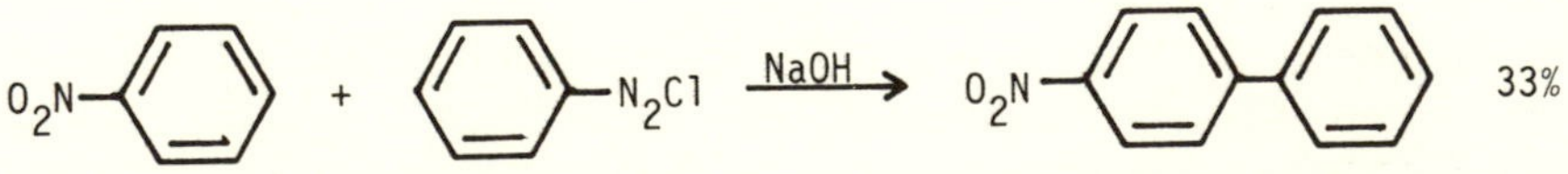

A. A. Morton of MIT noted that bromobenzene promoted the reaction of metallic sodium with triphenylmethyl chloride and other halogen compounds.[25] It seems probable that, in contrast to Morton's explanation, a free-radical reaction is involved, with a phenyl radical formed from sodium and bromobenzene.[26]

During the 1914-1939 period, synthesis of aromatic derivatives was active everywhere that organic chemistry was pursued, and it is impossible to describe even the American work in detail. Dyes continued to be synthesized in large numbers for textiles and for specialized uses, such as the cyanine dyes for photographic sensitizers.[27] The synthesis of aromatic medicinal compounds, of compounds important in the study of natural products, including heterocyclic compounds, and of aromatic compounds for theoretical study, continued at an increasing rate. Several new factors not present in 1914 increased the research interest, particularly in polycyclic aromatic compounds.

Dehydrogenation by sulfur or selenium of many natural products containing reduced, fused-ring systems yielded aromatic compounds which required synthesis for structure identification. Such natural products include polyisoprenoids such as the steroids, triterpene acids, and some alkaloids.

A second factor was the discovery in England that some polycyclic aromatic hydrocarbons were carcinogenic (cancer-producing), some of these were isolated from coal tar.

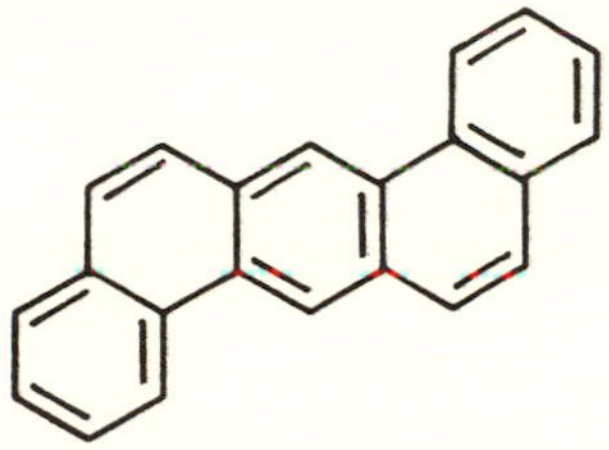

Dibenzanthracene 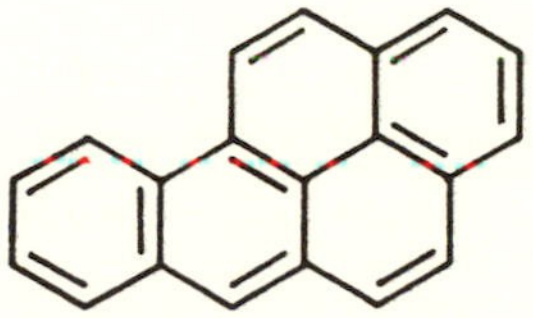   Benzo[a]pyrene

Sir Percival Pott (1714-1788), an English physician, reported in 1775 that chimney sweeps had a high incidence of skin cancer, and in the early 20th century it was noted that workers in the coal tar industry were particularly susceptible to cancer. In 1930 the synthetic compound dibenzanthracene was shown to produce skin cancer in mice,[28] the first pure chemical compound found to be carcinogenic. Careful examination of coal tar by the British chemist James W. Cook led to the isolation and identification by synthesis of benzo[a]pyrene.[29] This carcinogen has since been identified in soot, cigarette smoke, and internal combustion engine exhaust fumes. Testing of other available polycyclic compounds soon showed that methylcholanthrene, which had been prepared in Munich by the German chemist Heinrich Wieland in several laboratory steps from the naturally occurring bile acid, desoxycholic acid, was a carcinogen as potent as benzo[a]pyrene. There is some similarity in structure between dibenzanthracene and methylcholanthrene.

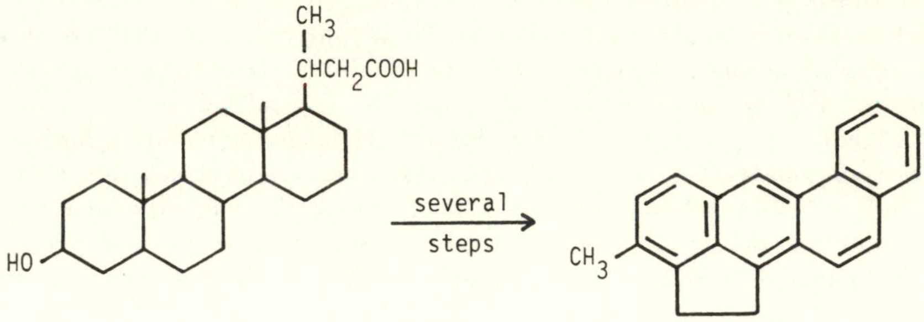

Desoxycholic acid   Methylcholanthrene

The synthesis and testing of many isomers, homologs, and derivatives of these carcinogens were taken up by Cook with great energy and ingenuity, with the hope that knowledge of the effect of structure on carcinogenic properties would give evidence about the mechanism of cancer initiation. Although the work resulted in notable increase in knowledge of chemical behavior and synthetic methods, information about the mechanism of carcinogenesis accumulated very slowly. Only after several decades, with the vastly more effective techniques of separation, labeling, and structure determination made possible by the Instrumental Revolution, was notable progress made.

The leading academic worker in the United States in aromatic chemistry was Louis F. Fieser of Harvard. He took up the synthesis of carcinogens with gusto, and with A. M. Seligman, soon synthesized methylcholanthrene, making it readily available for biological study.[30] Fieser's research on aromatic compounds continued with a large group of collaborators until World War II. His work on vitamin K is mentioned elsewhere.

Arnold M. Seligman (1912-1976, A.B. 1934; M.D. 1939, both from Harvard) carried out the synthesis of methylcholanthrene the year after he graduated from Harvard and later became professor of surgery at Johns Hopkins. He was a gifted scientist and published many papers in areas related to cancer, drawing on both his medical and chemical background. He died of cancer.

Fieser published several papers which strengthened his belief in the validity of the Kekulé structures for aromatic compounds and in the fixation of the Kekulé double bonds in polycyclic aromatics. 2-Naphthol couples rapidly

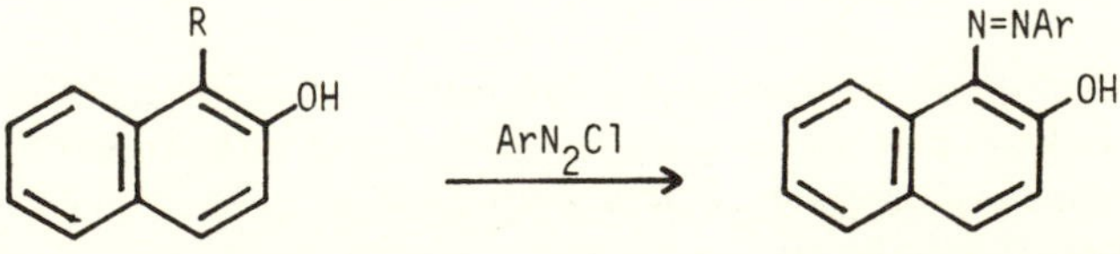

R = H
R = CH$_3$, no reaction

with diazonium compounds to form the 1-arylazo dyes. However, 1-methyl-2-naphthol does not couple, indicating not enough C=C character in the 2,3-bond.[31] Although the experiments were clear cut, the results were explained just as well by the resonance picture.

One of Fieser's postdoctorates, Melvin S. Newman, had a highly productive research career at Ohio State University from 1936 on. He developed novel methods for the synthesis of aromatics, including a rational synthesis of coronene, a hydrocarbon with six benzene rings around a hole, like a doughnut. Newman also worked on steric hindrance and on other problems of importance to physical organic chemistry.[32]

M. T. Bogert at Columbia developed some useful syntheses for bicyclic and tricyclic aromatic compounds which were important in natural product work. His procedure was an intramolecular Friedel-Crafts cyclization, presumably involving a carbonium ion attack on an aromatic ring.[33]

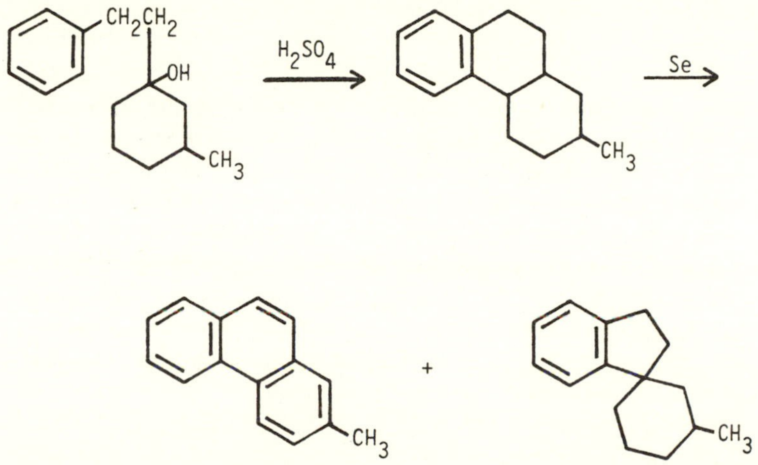

Marston T. Bogert (1868-1954, A.B. 1890, Columbia; Ph.B. 1894, Columbia School of Mines) was self-taught in organic chemistry and spent his career at Columbia. He trained many research students who became leaders in various fields of chemistry and was active in scientific organizations. A courtly personage, he was president of ACS in 1907 and 1908. He worked extensively on heterocyclic compounds, but his research on aromatic syntheses was his best.

W. E. Bachmann of Ann Arbor spent a year in J. W. Cook's laboratory in England and became active in the synthesis of polycyclic aromatics.[34] His work was interrupted early by World War II research.

Highly branched alkylated aromatics, usually derivatives of phenols, are formed by a Friedel-Crafts type of reaction of phenols with tertiary alcohols or unsaturated hydrocarbons, with H_2SO_4 (Niederl), BF_3, or anhydrous HF

as catalyst.[35] A compound of great technical importance as an intermediate
in the formation of various condensation polymers is bis-phenol A, prepared
by condensation of phenol with acetone. It is used in making epoxy resins[36]
by treating with epichlorohydrin.

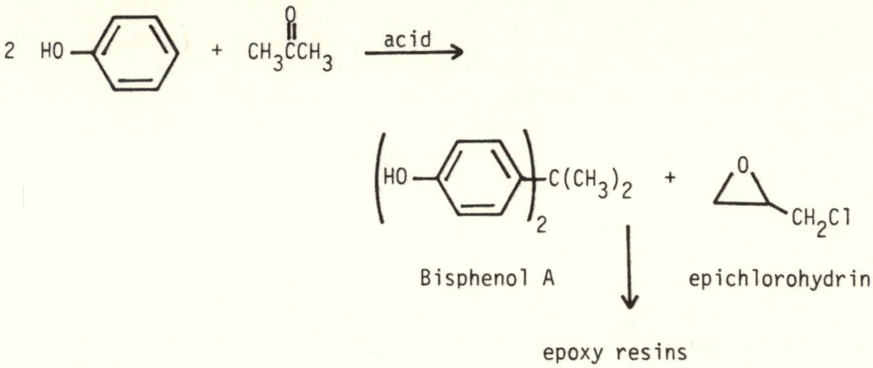

Bisphenol A   epichlorohydrin

epoxy resins

A long study of the preparation and separation of polymethylbenzenes
by the Friedel-Crafts reaction by Lee Irvin Smith of Minnesota made these
compounds readily available[37] and was important in the elucidation of the
structures of the vitamin E (tocopherol) series. Smith (1891-1973, B.S. 1913,
Ohio State; Ph.D. 1920, Harvard with Kohler) built up an outstanding school
of organic chemistry at Minnesota, where he taught from 1920 to 1960. His
work centered on reactions of quinones and of aromatic compounds.

Studying the less common aromatic substituents, Kharasch showed that
mercuration of nitrobenzene gave the *o, p* instead of the expected *m*-product.[38]

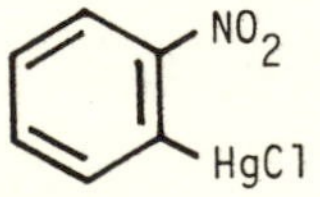

Aromatic arsenic derivatives were explored in many papers by Cliff S.
Hamilton at Nebraska;[39] some of these were important as drugs and agricultural chemicals.

Hamilton (1889-1973, B.Sc. 1912, Monmouth; Ph.D. 1922, Northwestern
with W. Lee Lewis) taught at Lincoln, Nebraska, from 1923 to 1958. He
was notably successful at training undergraduates and Masters' candidates in
research, and many of his students made excellent records as Ph.D. students
at Illinois in particular, and in their later careers in industrial research or
teaching. The new chemical laboratory on the Lincoln campus is named for
Hamilton.

LITERATURE CITED

1. For reviews, see L. Pauling, *The Nature of the Chemical Bond*, Ithaca, 1939; Pauling in
   Organic Chemistry, H. Gilman, Ed., 1st ed., Chap. 22, 1938; E. Hückel, Z. physik, **70**,

204 (1931) and later papers.

2. E. C. Crocker, JACS, **44**, 1618 (1922); Crocker, B. S. 1914, MIT, worked for the consulting laboratory Arthur D. Little for many years and became an expert on flavors and odors.

3. M. L. Huggins, ibid., **44**, 1607 (1922); **45**, 264 (1923); **53**, 1182 (1931). Huggins (1897-1981) was an able X-ray crystallographer and structural chemist and was one of the early proponents of the hydrogen bond (or bridge, as he called it).

4. R. C. Fuson, ibid., **46**, 2779 (1924); **47**, 2018 (1925).

5. L. Pauling, ibid., **48**, 1132 (1926).

6. C. P. Smyth, et al., ibid., **49**, 1030 (1927).

7. K. Lonsdale, Proc. Royal Soc. A, **123**, 494 (1929).

8. L. F. Haber, *The Chemical Industry During the Nineteenth Century*, 1958, 1969, passim.

9. D. S. Tarbell, JCE, **51**, 7 (1974).

10. A. F. Holleman, *Die Direkte Einführung von Substituenten in den Benzolkern*, Leipzig, 1910; Chem. Rev. **1**, 187 (1924).

11. H. S. Fry, *The Electonic Theory of Valence and the Constitution of Benzene*, New York, 1921.

12. C. K. Ingold, *Structure and Mechanism in Organic Chemistry*, 2nd ed., 1969, p. 266 ff., gives an account of the development of ideas about aromatic substitution.

13. H. J. Lucas, JACS, **48**, 1827 (1926). It should be pointed out that Hammett's sigma values, which were based on acid dissociation constants, could be used as an index of o, p- or m-directing activity. Substituents with a negative sigma value (compared to H) in the p-position are o, p-directing groups and increase the rate of electrophilic substitution; groups with a positive sigma value are m-directing and decrease the rate of substitution: Hammett, *Physical Organic Chemistry*, p. 194. Neither Hammett nor Ingold quotes Lucas' paper on aromatic substitution, although Ingold quotes his work on electronic effects of alkyl groups on olefin reactions.

14. G. W. Wheland and L. Pauling, JACS, **57**, 2086 (1935); with semiquantitative calculations on the resonance energy and mechanism of substitution.

15. C. K. Ingold, Chem. Rev., **15**, 225 (1934). This alleged obscurity was apparently one reason why L. F. Fieser, a distinguished experimentalist in aromatic chemistry, rejected the newer ideas on aromatic chemistry (L. F. Fieser, in *Organic Chemistry*, H. Gilman, Ed., 1938, Chap. 2.) The English views were explained lucidly by J. R. Johnson in Chap. 19 of the same work.

16. A. W. Francis et al., JACS, **47**, 2211, 2340 (1925); **48**, 655, 1631 (1926).

17. J. B. Conant and W. D. Peterson, ibid., **52**, 1220 (1930).

18. R. Wistar and P. D. Bartlett, ibid., **63**, 413 (1941).

19. M. J. Rarick, R. Q. Brewster and F. B. Dains, ibid., **55**, 1289 (1933).

20. F. Sanger, Biochem. J., **39**, 507 (1945); *Nobel Lectures Chemistry, 1942-1962*, Amsterdam 1964, p. 545.

21. V. E. Meharg and I. Allen, Jr., ibid., **54**, 2920 (1932); H. Gilman and S. Avakian, ibid., **67**, 349 (1945); for a comprehensive review see J. F. Bunnett and R. E. Zahler, Chem. Rev., **49**, 382 (1951).

22. J. D. Roberts, et al., ibid., **75**, 3290 (1953) and later papers.

23. C. B. Wooster and K. L. Godfrey, ibid., **59**, 596 (1937); other examples: H. Gilman and A. L. Jacoby, JOC, **3**, 108 (1938); N. O. Kappel and W. C. Fernelius, ibid., **5**, 40 (1940); W. E. Bachmann and L. H. Pence, JACS, **59**, 2339 (1937). For a summary of the work of Arthur J. Birch, later professor at the Australian National University, Canberra, who made this reduction a useful synthetic method, see Birch, Quart. Rev., **4**, 69 (1950).

24. M. Gomberg and W. E. Bachmann, JACS, **46**, 2339 (1924); also Gomberg and J. C. Pernert, ibid., **48**, 1372 (1926); Bachmann and R. A. Hoffman, Org. Reactions, **2**, 224 (1944).

25. A. A. Morton and J. R. Stevens, JACS, **54**, 1919 (1932).

26. Cf. J. R. Johnson, in Gilman, op. cit., p. 1633. J. F. Bunnett and J. K. Kim, JACS, **92**, 7463, 7464 (1970) and later papers, strengthen this view.

27. E. Q. Adams and H. L. Haller, JACS, **42**, 2389 (1920). L. G. S. Brooker et al., ibid., **57**, 2488 (1935); 59, 67, 74 (1937), and many other papers and patents from Brooker's group at the Eastman Kodak Co.

28. E. H. Kennaway, Biochem. J., **24**, 497 (1930). On the general subject, see Louis F. Fieser, *The Chemistry of Natural Products Related to Phenanthrene*, New York 1936, Chap. 3. For a more recent review, *Chemical Carcinogenesis, Part A*, P. Ts'o and J. A. DiPaolo, Eds., 1974.

29. J. W. Cook, et al., JCS, 395 (1933). The nomenclature of these polycyclic compounds is troublesome because over the years several numbering systems have been used. The benzopyrene derivative shown was first called 1,2-benzpyrene, then 3,4-benzpyrene and most recently benzo[a]pyrene, which is now used in all publications dealing with the carcinogenically active metabolites of benzpyrene.

30. L. F. Fieser and A. M. Seligman, JACS, **57**, 228, 942 (1935); 58, 2482 (1936).

31. L. F. Fieser and W. C. Lothrop, ibid., **57**, 1459 (1935); 58, 749 (1936).

32. Polycyclic syntheses, M. S. Newman and L. M. Joshel, ibid., **60**, 485 (1938); Coronene, Newman, ibid., **62**, 1683 (1940).

33. M. T. Bogert, et al., ibid., **56**, 248 (1934) and later papers. See biography of Bogert by L. P. Hammett, *Biog. Mem. Nat. Acad. Scis.*, **45**, 96 (1974).

34. W. E. Bachmann and Charlotte H. Boatner, JACS, **58**, 2194 (1936).

35. For phenol alkylation with H_2SO_4, see J. B. Niederl et al., ibid., **55**, 284, 2571 (1933); for BF_3, F. J. Sowa, et al., ibid., **55**, 3402 (1933); for HF, J. H. Simons, S. Archer et al., ibid., **60**, 2952, 2953, (1938).

36. Kirk-Othmer, *Encyclopedia of Chemical Technology*, 3rd ed., 1980, 9, 267.

37. Some papers of many: L. I. Smith and F. J. Dobrovolny, JACS, **48**, 1413, 1420, 1693 (1926), describe the preparation of durene (1,2,4,5-tetramethylbenzene), 1,2,3,4,5-pentamethylbenzene (prehnitene), and hexamethylbenzene.

38. M. S. Kharasch and L. Chalkley, Jr., ibid., **43**, 607 (1921). The anomalous behavior of the usual mercuration reaction was explained by W. J. Klapproth and F. H. Westheimer, ibid., **72**, 4461 (1950), who showed that in strong acid mercuration involved electrophilic attack.

39. A typical paper: H. P. Brown and C. S. Hamilton, JACS, **56**, 151 (1935).

11

The Origin of "Organic Syntheses"

The annual publication *Organic Syntheses* played such a key role in the development of organic research in this country after 1920 that it deserves special mention.[1]

Clarence G. Derick, Roger Adams' predecessor in organic chemistry at Illinois (later an industrial chemist and consultant), hired about five graduate students during the summer of 1914 to prepare organic compounds needed in teaching and research. The outbreak of World War I in August of that year immediately exposed America's critical lack of specialized organic chemicals and gave new significance to this "summer preps", or "Organic Chemical Manufactures" project as it was known officially. The urgent national situation caused a marked expansion of the Illinois prep labs in 1915.

Ernest H. Volwiler worked in "preps" during the summer of 1915 and was in charge of the program in 1916 and 1917; Carl S. Marvel worked from June of 1916 until August of 1919. In September of 1917 Marvel and another graduate student were given "manufacturing scholarships" instead of teaching assistantships for the academic year. The workers received twenty-five cents per hour and eventually some academic credit. As every experienced synthetic chemist knows, the fact that procedures for preparing a certain compound have been published in a journal or described in a patent does not guarantee that a repetition of the published procedures in another laboratory will lead to a reasonable yield, or, indeed, to any yield at all of the desired product. Marvel and his associates, therefore, had to do considerable research on conditions and methods to develop efficient and economical procedures for the preparation of key compounds on a sizeable laboratory scale. The prep lab workers accomplished prodigies of synthetic work, aided by their ingenuity in devising equipment out of materials at hand.[2]

Roger Adams, who joined the Illinois department in September 1916 as the replacement for Derick, took charge of the preps operation with vigor. Adams (1889-1971, A.B. 1909, Ph.D. 1912, both at Harvard with H. A. Torrey, T. W. Richards, Latham Clarke, and C. L. Jackson) was the leading figure in American organic chemistry for a generation.[3] A contemporary of J. B. Conant, F. C. Whitmore, Henry Gilman, and J. B. Sumner at Harvard, he spent a year in Germany on a traveling fellowship, returned to Harvard as instructor, and moved to Illinois in 1916. He served as head of the chemistry department from 1926 to 1954, succeeding W. A. Noyes, and trained about 250 research and postdoctoral students. The important problems he worked on included the development of platinum oxide for catalytic hydrogenation (Adams catalyst), chaulmoogric acid, stereochemistry of compounds with restricted rotation and of deuterium compounds, the structure of gossypol, chemistry of compounds isolated from marihuana, alkaloid chemistry, and quinone imines. He had a varied and influential career as industrial consultant, officer of national scientific societies, and scientific worker for the federal government during and after two World Wars. He and his colleagues made Illinois a leading international center for research in organic chemistry. We discuss various aspects of his work later.

The preps lab scheme appealed strongly to Adams: it was work of national importance, it was good training for the students, it could be a potent asset to the research program and the national reputation of the Illinois department, and the strong practical side of Adams' nature found it congenial. He introduced into the preps operation a systematic scheme of cost accounting for labor and materials, so that the cost per gram of each batch of each compound was known. Each worker had the cost records of his predecessors on each prep, and it became a mark of distinction for the worker to cut the cost of the prep. This accounting scheme gave the preps program an impressive financial basis and was excellent training in chemical economics for participating graduate students. The preps scheme would have been much less important without this exercise of Adams' business sense, which validated it to the university administration and to outsiders.

Adams wrote a detailed balance sheet covering the period June 1, 1917-February 1, 1918, for the trustees,[4] the figures showing a net profit of $2800. He recommended that this credit be kept to cover future deficits or overhead costs. A rough balance sheet, covering the years 1917-1920, indicated a surplus of about $3000.

The preps operation was successful pedagogically, scientifically, and financially. The list of chemicals increased, and thirty university and many industrial laboratories bought reagents. The preps lab sold over sixty lbs. of dimethylglyoxime, a reagent then used for nickel analysis in steel, and by September 1917 had prepared about eighty different chemicals.[5] In 1918 about sixty chemicals were in stock and several scores of additional ones were available on order.[6] The *Boston Transcript* carried the story of the Illinois prep labs' contribution to the war effort on its editorial page at the time of the Boston meeting of the American Chemical Society in September 1917;[7] evidently Adams seized the opportunity to present a paper describing them. The Illinois venture was so valuable and successful that after the war much of the synthetic work was transferred to the Eastman Kodak Co. in Rochester, New York, under the director of Dr. Hans T. Clarke (later professor of biochemistry at Columbia), who spent several weeks in Urbana studying the operation. This was the origin of the Eastman Line of chemicals for research and teaching use. The Illinois work continued for many years in the summer preps, in which a number of graduate students were paid to prepare chemicals needed for the research program at Illinois and for the course in qualitative organic analysis, providing the laboratory with an unmatched variety of chemicals. In later years the prep labs synthesized amino acids, used by Dr. W. C. Rose of the Illinois department in his classic feeding studies on mice to determine the essential amino acids. In 1942, for example, R. L. Shriner directed the summer program, in which twenty-six workers produced 200 different organic compounds, an increase of thirty-four over the summer before.

Some of the tested procedures of the Illinois prep laboratory for specific compounds were published in JACS[8] and later in four *University of Illinois Bulletins* on *Organic Chemical Reagents*.[9] Adams considered organizing a group of university laboratories in a cooperative program of preparing needed compounds. This idea proved too cumbersome, and instead, in 1921 he launched the plan of publishing an annual volume of methods for preparing useful organic compounds with Volume 1 of *Organic Syntheses*[10] with Adams as editor-in-chief and Conant, Clarke, and Oliver Kamm (Parke, Davis and Co.), as associate editors. This volume contained seventeen procedures, and the plan for the series, as set out in the preface, called for an annual publication of preparations each of which would be checked in a laboratory different from that of the submitter. Exact details of procedures, reagents, yields, physical properties, alternative methods of preparation, and the names of submitters and checkers were to be given. It was planned to have about thirty procedures in each volume: Volume 2 had twenty-five, Volume 3, twenty-nine, and thereafter about thirty appeared in each volume. The preface asked for the cooperation of chemists in submitting procedures for compounds which they had studied in detail.

In the fiftieth annual volume of *Organic Syntheses*, Adams described his difficulties in getting a publisher for the first volume,[11] because all the publishers he approached at first used the same referee, who repeated his negative recommendation. However, John Wiley and Sons finally published the first volume and had no reason to regret it, publishing over fifty annual volumes since that time.

Many organic chemists cooperated in submitting preparations, and they regarded service on the editorial board as an honor. The success and continued vitality of *Organic Syntheses* in its sixth decade shows that it filled a real need. The annual red-backed volumes and the five, green, collective volumes, containing the carefully edited contents of the nine previous annual issues instead of the five proposed in the first preface, are a remarkable record of the development of synthetic organic chemistry since 1921. New reactions, new reagents, new instrumentation, new ideas on reaction mechanisms, new editors, and new submitters have kept the series an active force in organic chemistry. The first volume, modestly styled a "pamphlet" in the preface, and its many successors are a fitting monument to Adams, his associates, their successors, and to the Illinois prep labs. In a letter written for Adams' retirement dinner in 1954, Conant recalled the discussion of the idea of *Or-*

ganic Syntheses in 1918 during their war service in Washington and said that it should be called the "Adams Annual" because he was the moving spirit.[12] This series was close to Adams' heart, and he took a leading part in it up to his last years. *Organic Syntheses* was the model for *Inorganic Syntheses* and *Biochemical Preparations*, which have been successful and useful in their fields.

Although the records of *Organic Syntheses* prior to 1937 apparently are lost, the editorial board probably met as in later years for an afternoon of work and dinner at each national ACS meeting. Charles F. H. Allen, one of E. P. Kohler's students then teaching at McGill, served as secretary to the board from 1928 to 1937 (Volumes 9-17). In 1938 A. Harold Blatt of Queen's College, another of Kohler's Ph.D.s, became secretary and was associated with *Organic Syntheses* or *Organic Reactions* for many years. Until 1939 the enterprise proceeded very informally with Adams keeping the accounts; no one received any salary, and the editors invested the royalties.[1]

A change in the income tax laws in 1939 led to the incorporation of *Organic Syntheses* as a nonprofit membership corporation in New York. The officers were Adams, president; W. W. Hartman (Eastman Kodak), treasurer; Blatt, secretary; L. F. Fieser (Harvard) and J. R. Johnson (Cornell), directors. Thereafter the corporation held its official meetings separately from the board of editors. The latter usually numbered about six, and each member was responsible for assembling at least one of the annual volumes. By 1976 about sixty chemists had served on the board of editors.

Adams and his associates discussed another innovation[13] in chemical publication as early as 1939, but the first volume of *Organic Reactions* appeared only in 1942. Subsequent volumes have appeared periodically, Volume 25 in 1977. Chemists with first hand experience with some reaction wrote a comprehensive survey of the reaction, usually with extensive tables. The articles discussed the scope, limitations, and mechanism of the reaction with experimental conditions for specific cases. The series was thus a valuable source of information and ideas, which mirrored the new methods in organic chemistry and frequently suggested good problems in reaction mechanisms.

LITERATURE CITED

1. This is described in D. S. and A. T. Tarbell, *Roger Adams*, Washington, 1981, pp. 52-56, 104-105. From a somewhat different viewpoint, R. L. and R. H. Shriner tell the story in *Cumulative Indices, Organic Syntheses*, New York, 1976, pp. 423-427. The present account is adapted from our book by permission of the American Chemical Society.
2. C. S. Marvel to DST, January 23, 1976; taped interview with Marvel by DST, March 1, 1977.
3. D. S. and A. T. Tarbell, op. cit. ref. 1, passim; idem., *Biog. Mem. Nat. Acad. Scis.*,53, 2 (1982).
4. Statement signed by Adams, Illinois Archives, Dean's Office, Department and Subject File, 15/1/1, 17, chemistry; undated but after February 1, 1918.
5. R. Adams, *Science*, , 47 225 (1918); J. Ind. Eng. Chem., 9, 685 (1917).
6. R. Adams, JACS, 40, 869 (1918).

7. R. Adams Archive, University of Illinois.

8. R. Adams and Oliver Kamm, JACS, 40, 1281 (1918); R. Adams, O. Kamm and C. S. Marvel, ibid., **41**, 276, 789 (1919); ibid. **42**, 299, 310 (1920).

9. The first was Adams, Kamm and Marvel, *University of Illinois Bulletin*, XVI, no. 43; reviewed in JACS, **42**, 1074 (1920) by Moses Gomberg with complimentary remarks on the value of the Illinois preps program. Gomberg reviewed the second issue, ibid., **43**, 1748 (1921). The four bulletins were dated June 23, 1919; October 11, 1920; October 9, 1921; and October 23, 1922 (information from the Du Pont Lavoisier Library, through the kindness of Dr. R. M. Joyce).

10. *Organic Syntheses*, Vol. 1, New York, 1921.

11. R. Adams, *Organic Syntheses*, 50, vii (1970).

12. J. B. Conant to R. M. Joyce, July 13, 1954.

13. Ref. 1, pp. 185-6.

HENRY GILMAN

Organometallic Compounds, 1914-1939

This class is defined as compounds with an organic group attached to a metal through a carbon atom. Its most important members during this period are the Grignard reagents, organomagnesium halides, RMgX, (named for Victor Grignard, French chemist and Nobel Prize winner who discovered them in 1901) and the lithium compounds, RLi. Although the history of organometallic compounds goes back to the 19th century,[1] significant advances were made by American workers in the 1914-1939 period.

The leading figure for many decades was Henry Gilman (1893- , A.B. 1916, Ph.D. 1918 with E. P. Kohler, both at Harvard); he had the distinction of being the first student to publish research with Roger Adams. He built up a flourishing graduate department at Iowa State University, where he was professor from 1919. His ninetieth birthday was celebrated there by a symposium in 1983. He published many hundreds of research papers, over sixty appearing in the Netherlands journal *Rec. trav. chim. Pays Bas* during the years 1924-1936, the largest block of American papers published in an overseas journal since 1914. He edited the valuable two-volume cooperative work, *Organic Chemistry An Advanced Treatise* (first edition 1938, revised 1943, 1947), generally known as "Gilman", which was a great boon to all research workers and to students in particular.

Gilman's contribution to knowledge of organometallic compounds covered much of the periodic table and can only be characterized as monumental. He described new or improved methods of preparation and of analysis, studied the reactions, and developed new synthetic procedures utilizing organometallic compounds. Some of his work was of great mechanistic interest. A few examples follow, selected from the work of Gilman and of other chemists.

Gilman described valuable quantitative analyses and qualitative tests for Grignard reagents and organolithium compounds.[2] A quantitative method, showing the amount of CH_3MgI adding to an organic compound and the amount reacting with NH, COOH, and other "active hydrogens" to liberate methane, CH_4, was elaborated by Kohler, J. F. Stone, and Fuson.[3] Many of Gilman's early papers gave detailed directions for optimum yields of Grignard reagents, such as $tert\text{-}C_4H_9MgCl$.[4]

Grignard reagents were known from Kohler's work at Bryn Mawr to add 1,4 to conjugated systems. Gilman found that with a highly substituted system, 1,4-addition could involve the unsaturation of the benzene ring.[5] A useful synthetic procedure is the reaction of a Grignard reagent with an alkyl ester of a sulfonic acid; the sulfonate group, $ArSO_3$, is displaced, with a coupling of the group from the Grignard and the alkyl ester.[6] A serviceable synthetic example is due to Marvel.

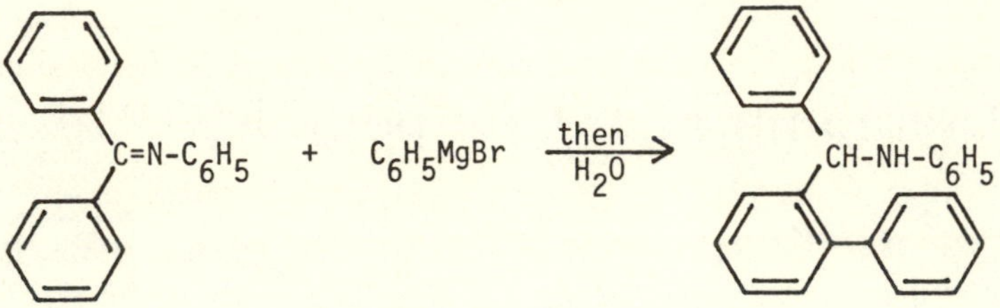

$$ArSO_3(CH_2)_3Cl \; + \; C_2H_5MgBr \; \longrightarrow \; C_2H_5(CH_2)_3Cl \; + \; ArSO_3^-MgBr^+$$

An unpredicted result was obtained in an attempt to make a highly branched alcohol by a Grignard reaction; reduction, instead of addition, ensued.[7] Other highly branched compounds gave similar results,[7,8] and a

$$\underline{i}\text{-}C_3H_7\overset{\overset{\textstyle O}{\|}}{C}C_3H_7\text{-}\underline{i} \; + \; \underline{i}\text{-}C_3H_7MgCl \; \longrightarrow \; \underline{i}\text{-}C_3H_7\overset{\overset{\textstyle OH}{|}}{C}HC_3H_7\text{-}\underline{i} \; + \; C_3H_6$$

reasonable explanation suggests that if addition of the Grignard to the carbonyl group, $C=O$, is retarded by a high degree of substitution, slower possible reactions such as reduction can occur. Much evidence indicates that reduction involves a cyclic transition state with transfer of a β-hydrogen; the

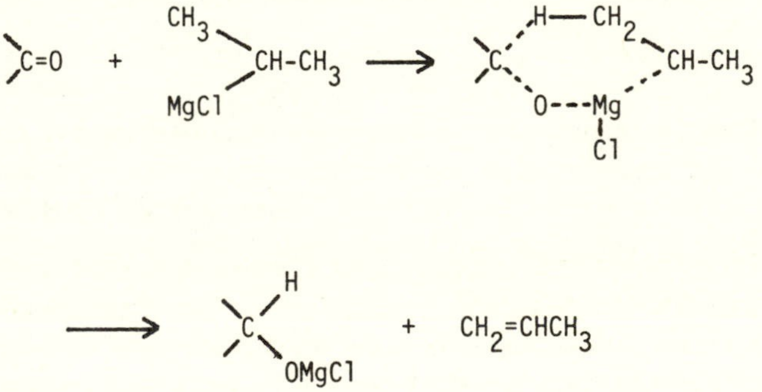

Grignard reduction is thus similar to reduction by other reagents, such as alkoxides. The organolithium compounds, valuable as synthetic reagents and as strong bases, had been prepared from organomercury compounds by a troublesome procedure. K. Ziegler in Germany[9] showed that they could be made from metallic lithium and alkyl halides, and Gilman demonstrated that they could be prepared as readily as Grignard reagents in ordinary laboratory equipment,[10] important findings because lithium compounds are useful for reactions in numerous cases where the corresponding Grignards are not.

Lithium compounds can be made readily from some halides, such as *p*-bromo-N-dimethylaniline, p-$(CH_3)_2NC_6H_4Br$, which do not add Grignards. Some highly hindered carbonyl groups which do not add Grignards react with lithium compounds; thus the tertiary, highly substituted alcohol is obtained from isopropyllithium and diisopropyl ketone.[11]

$$\underline{i}\text{-}C_3H_7\overset{\overset{\displaystyle O}{\|}}{C}C_3H_7\text{-}\underline{i} \quad + \quad \underline{i}\text{-}C_3H_7Li \quad \longrightarrow \quad (\underline{i}\text{-}C_3H_7)_3COH$$

Lithium compounds are strong enough bases to replace hydrogen (*to metalate*) in some substituted aromatic rings; usually the hydrogen *ortho* to the substituents is replaced.[12]

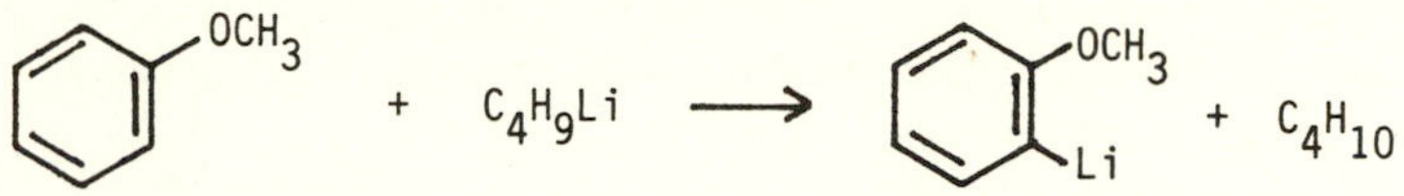

Aromatic halogens can be exchanged for lithium by RLi,[13] a most useful synthetic procedure.

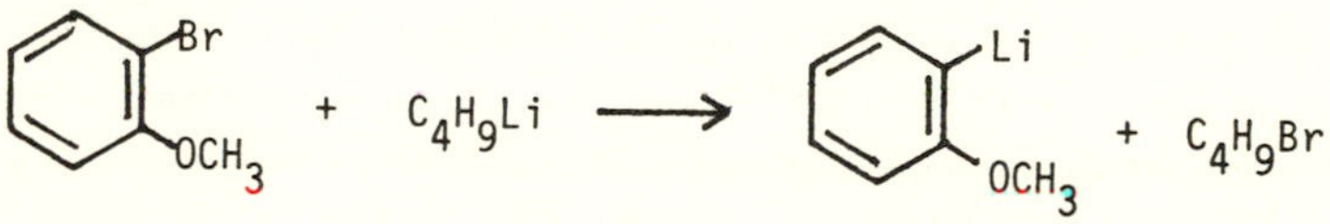

Lithium compounds give 1,2-addition to conjugated systems,[14] but Grignards generally add 1,4.

$$C_6H_5CH{=}CH\overset{\overset{\displaystyle O}{\|}}{C}C_6H_5 \quad + \quad C_6H_5Li \quad \longrightarrow \quad C_6H_5CH{=}CH\overset{\overset{\displaystyle OH}{|}}{C}(C_6H_5)_2$$

89% 1,2 addition

Lithium compounds give good yields of ketones with carboxylic acids, in a widely applicable process.[15,16]

The examples above, and many others which might be quoted, indicate the significance of lithium compounds, whose applications increased steadily after 1939. The structure of organometallic compounds and particularly the effect of solvent on their reactions were worked out gradually and will be discussed later.

The reactions of organosodium compounds were also explored, particularly by A. A. Morton. He obtained a high yield of triarylcarbinols from chlorobenzene, metallic sodium, and diethyl carbonate.[17] The action of sodium

$$ArCl \quad + \quad Na \quad + \quad (C_2H_5O)_2CO \quad \longrightarrow \quad Ar_3COH$$

on 1-chloropentane followed by carbonation gave a mixture of caproic acid and butylmalonic acid. Although removal of a second proton from $C_5H_{11}^-Na^+$ would hardly have been predicted, this was shown to be the most likely source of the malonic acid.[18] Dicarbanions and polycarbanions are useful synthetic intermediates, but all other ones known have carbonyl groups to stabilize the negative charge, whereas $C_5H_{10}Na_2$ does not.

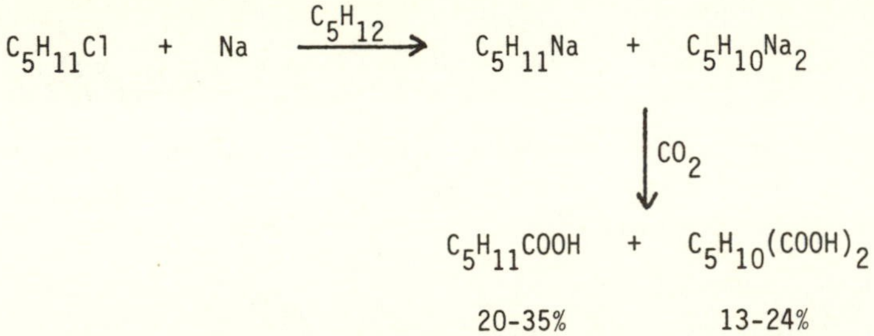

$$C_5H_{11}Cl \quad + \quad Na \quad \xrightarrow{C_5H_{12}} \quad C_5H_{11}Na \quad + \quad C_5H_{10}Na_2$$

$$\downarrow CO_2$$

$$C_5H_{11}COOH \quad + \quad C_5H_{10}(COOH)_2$$

$$20\text{-}35\% \qquad\qquad 13\text{-}24\%$$

Avery A. Morton (1892-   , A.B. 1913, Cotner; Ph.D. 1924, MIT with J. F. Norris) taught at MIT from 1919 to 1946. He worked on polymerization, organometallic compounds, and heterocycles. His originality has been undervalued because his papers are not clearly written.

W. E. Bachmann and H. T. Clarke showed that chlorobenzene and sodium gave the products below;[19] these may have been formed from phenylsodium by halogen-metal interchange.

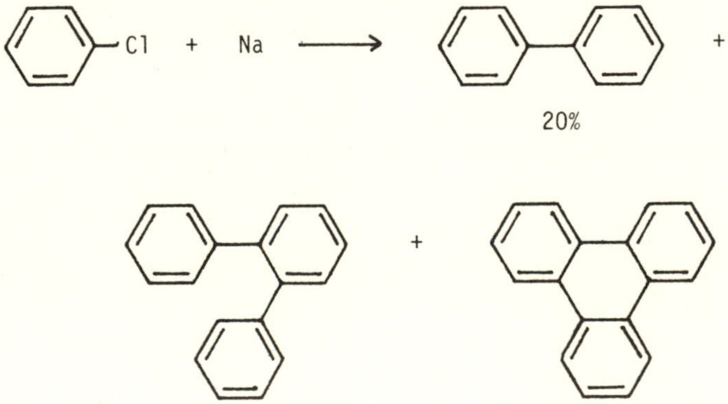

20%

triphenylene

Werner E. Bachmann (1901-1951, A. B. 1923, Ph.D. 1926, with Moses Gomberg, both at University of Michigan, Ann Arbor) was one of the most gifted synthetic chemists of his time. His work with Gomberg resulted in two reactions known by their joint names, and he taught at Michigan from 1925 until his death, training many Ph.D. students who had outstanding research careers. He worked on reduction of ketones, synthesis of steroids, the war-time explosive RDX, carcinogenic hydrocarbons, and the structure of penicillin. His work was marked by elegance, economy, and insight.[20]

The use of organometallic compounds derived from acetylenes for synthesis goes back beyond J. U. Nef in 1899;[21] the Grignard, sodium, and lithium acetylides were the most useful.[21] A typical example uses sodium acetylide for preparing a secondary acetylenic alcohol.[22] The metal acetylides became very important in synthesis, particularly for production of polyunsaturated compounds, such as Vitamin A, and unsaturated fatty acids.[21b]

$$Na + NH_3 \quad \xrightarrow{Fe(NO_3)_3} \quad NaNH_2 \quad \xrightarrow{HC\equiv CH} \quad NaC\equiv CH$$
$$(liq.)$$

$$\xrightarrow{C_2H_5CHO} \quad C_2H_5\overset{OH}{\underset{}{C}}HC\equiv CH$$

The question of the constitution of the Grignard reagent RMgX, usually in diethyl ether solution, had been the subject of extensive experiment and debate ever since its discovery.[23] W. S. Schlenk at Berlin found that dioxan precipitated halide and some magnesium from an ether solution of a Grignard; he suggested the existence of the equilibrium shown and thought that the position of this equilibrium could be determined by dioxan treatment and analysis.[24]

$$2C_6H_5MgBr \rightleftharpoons (C_6H_5)_2Mg + MgBr_2$$

precipitated by dioxan

P. D. Bartlett and C. M. Berry at Minnesota showed that a solution of diethylmagnesium, $(C_2H_5)_2Mg$ prepared by Schlenk's dioxan procedure, gave normal addition to cyclohexene oxide, whereas the ordinary Grignard reagent

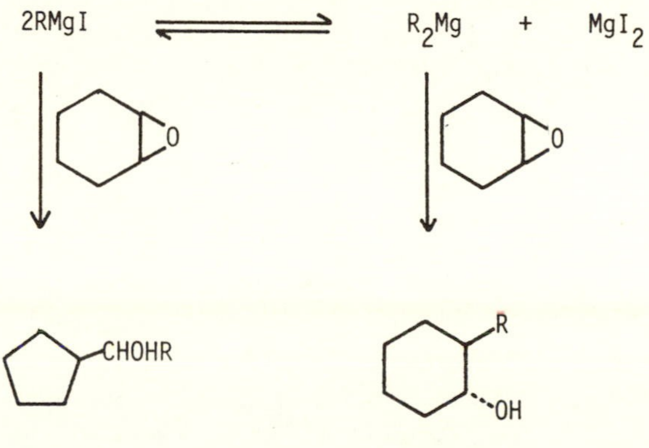

$$2RMgI \rightleftharpoons R_2Mg + MgI_2$$

R = CH₃, C₂H₅   trans

gave a rearrangement of the epoxide ring, probably due to the action of MgI_2 as a Lewis acid.[25] Cope noted a difference in rate of reaction of C_6H_5MgBr and $(C_6H_5)_2Mg$ with dimethyl sulfate.[26] Further work showed that while there was undoubtedly an equilibrium of the Schlenk type, the apparent position of equilibrium varied with time and method of determination.[27]

Extended studies on the electrolysis and conductivity of Grignard solutions by W. V. Evans and Ralph Pearson at Northwestern[28] gave a more complete picture of the reagent's behavior in ether solution. The electron-donating nature of diethyl ether,$C_2H_5OC_2H_5$, (a Lewis base) makes it bond with Mg, and complex anions may form; larger anions are also possible, held

$$\left[\begin{array}{c} R \\ R:\overset{..}{Mg}:O(C_2H_5)_2 \\ \overset{..}{X} \end{array}\right]^{-} \qquad \left[\begin{array}{c} X \\ (C_2H_5)_2O:\overset{..}{Mg}:O(C_2H_5)_2 \\ OC_2H_5 \end{array}\right]^{+}$$

together by coordination between X and Mg. Most of the current is carried by the anions, because they do not combine further with the electron-donating solvent and hence move more rapidly. The cations are strong Lewis acids, bind strongly with ether, and do not migrate rapidly because of their solvation. The Grignard reagent can dissociate according to the following equations in ether solution.

$$RMgX \rightleftharpoons R^{-} + MgX^{+}$$
$$R_2Mg \rightleftharpoons R^{-} + RMg^{+}$$

The nucleophilic character of the R group is confirmed by the reactions of Grignards and lithium compounds; the R group attacks the positive end of a C=O, for example, while the positive MgX goes to the negative oxygen. This paper of Evans and Pearson deserves careful study; it shows the value of the Lewis acid-base concepts and contains the germs of Pearson's later more detailed ideas on hard and soft acids and bases.

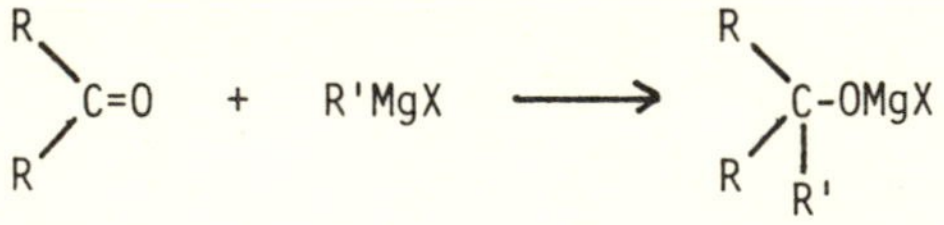

$$\underset{R}{\overset{R}{>}}C=O \ + \ R'MgX \ \longrightarrow \ \underset{R}{\overset{R}{>}}\underset{R'}{\overset{|}{C}}-OMgX$$

Ralph G. Pearson (1919-   , B.S. 1940, Lewis Institute; Ph.D. 1943, Northwestern with Ward V. Evans) has taught since 1943 at Northwestern and the University of California at Santa Barbara. His research has been on kinetics of reactions of both organic and inorganic compounds, on which he published a standard book (with A. A. Frost). He has worked on reactions of both organic and inorganic compounds. He has proposed very useful general categories of hard and soft acids and bases, an extension of the G. N. Lewis ideas on acids and bases.

The importance of solvent on preparation and properties of organometallic compounds began to emerge before 1939 and was demonstrated in detail after that time. The usual methods of determining molecular weight showed that lithium compounds were associated in solution, dimers and hexamers being indicated in several cases. More exact information from nuclear magnetic resonance studies on ^1H and ^7Li showed that in ether solution CH_3Li and C_2H_5Li were associated to $(RLi)_4$; in the presence of sodium alkoxide, $NaOR$, Li_4R_3OR is formed.[29] Coordination of RLi with basic donors, such as $(CH_3)_2NCH_2CH_2N(CH_3)_2$, was found to increase greatly the ability of RLi to remove protons (i.e., the basic properties of RLi).[30] It became clear that solvent had a profound effect on the reactions and properties of RLi, because of coordination of RLi with the solvent and changes in the degree of association of RLi with itself.

A key observation was that of N. D. Scott at Du Pont,[31] who showed that the addition of sodium to naphthalene and biphenyl occurred in dimethyl ether or, more conveniently, in ethylene glycol dimethyl ether. In diethyl ether such addition did not occur.

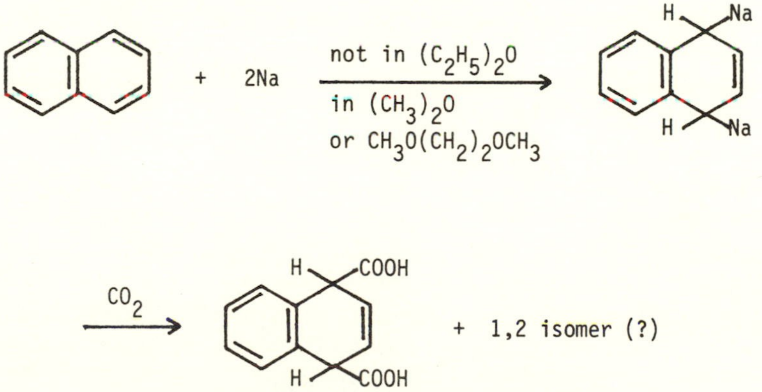

naphthalene + 2Na $\xrightarrow[\text{in } (CH_3)_2O \text{ or } CH_3O(CH_2)_2OCH_3]{\text{not in } (C_2H_5)_2O}$ 1,4-dihydronaphthalene-1,4-diyl-disodium (H, Na / H, Na)

$\xrightarrow{CO_2}$ 1,4-dihydronaphthalene-1,4-dicarboxylic acid (H, COOH / H, COOH) + 1,2 isomer (?)

Norman D. Scott (1894-1948, B.S. 1916, Drake; Rhodes Scholar, B.A., M.S., B.Sc. 1921, Oxford; Ph.D. 1924, Wisconsin with The Svedberg) taught at Wisconsin and Harvard. From 1927 he was associated with research at Du Pont.

Allene Jeanes and Adams suggested[32] that the effect of $(CH_3OCH_2)_2$ on sodium addition was due to formation of a chelate with the Na^+, and the essentials of this idea have been widely applied. Later work has shown that tetrahydrofuran is frequently a much better solvent for organometallic compounds than diethyl ether, probably because the oxygen in the former is able to form more stable solvates of organometallic compounds than the latter.

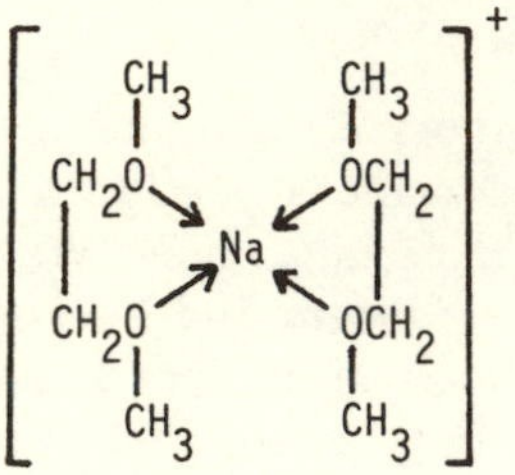

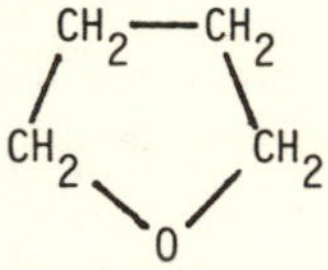

tetrahydrofuran   diethyl ether

Some discussion has been given previously of reactions of sodium in liquid ammonia with organic compounds. Addition reactions of sodium to unsaturated or aromatic compounds to form organometallic substances had been described by Schlenk and E. Bergmann,[33] and C. F. Koelsch,[34] among others.[35] The use of the ethylene glycol dimethyl ether solvent made this reaction more general and reproducible. The addition of lithium, sodium, and potassium to phenanthrene was found to give 9,10-addition,[32] not 1,4 as previously thought.[33] Jeanes found that carbonation gave the *trans*-9,10-dihydro-9,10-dicarboxyphenanthrene. Another product previously reported was found to be a 9-fluorenecarboxylic acid,[32] caused by the presence of the hydrocarbon fluorene as an impurity in the phenanthrene used.[33]

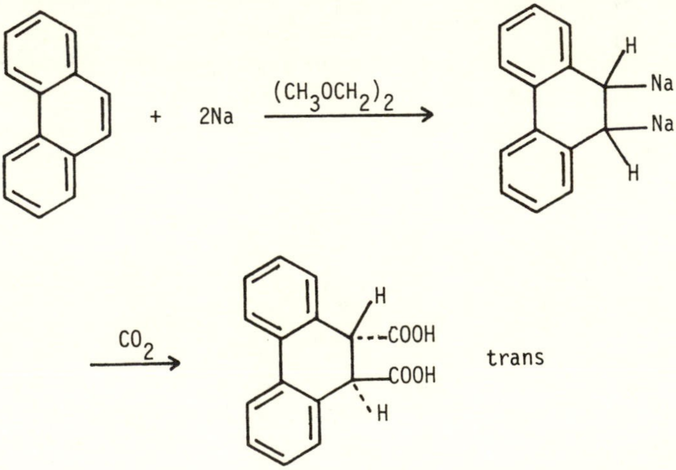

The addition of alkali metals to *cis* or *trans* stilbene followed by carbonation was found by G. F. Wright in Toronto to give mainly the *meso* dicarboxylic acid;[36] the yields were much better here with the glycol ether solvent. This addition of sodium, lithium, or potassium was accompanied by isomerization of *trans* stilbene, either at the addition step, or at the carbonation step. The former seems more likely.

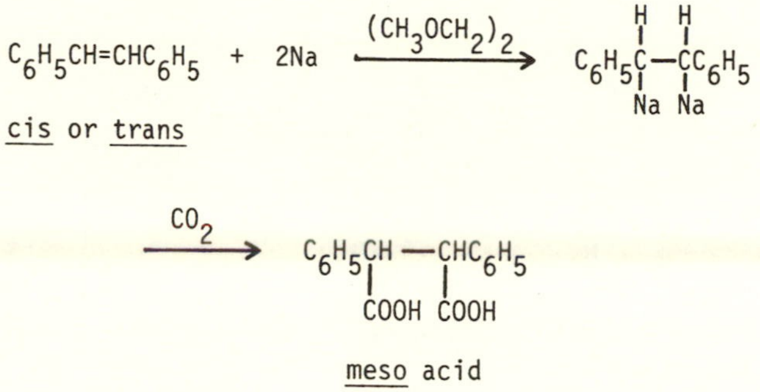

$$C_6H_5CH=CHC_6H_5 \ + \ 2Na \ \xrightarrow{(CH_3OCH_2)_2} \ C_6H_5\overset{\overset{H}{|}}{\underset{\underset{Na}{|}}{C}}-\overset{\overset{H}{|}}{\underset{\underset{Na}{|}}{C}}C_6H_5$$

cis or trans

$$\xrightarrow{CO_2} \ C_6H_5\overset{}{\underset{\underset{COOH}{|}}{CH}}-\overset{}{\underset{\underset{COOH}{|}}{CH}}C_6H_5$$

meso acid

George F. Wright (1904-   , B.S. 1919, Iowa State; Ph.D. 1932, with Henry Gilman) taught at Toronto from 1938 on. His research dealt with organometallic compounds, especially organomercurials, dipole moments, and stereochemistry.

Certain organoalkali compounds, such as phenylisopropylpotassium, $C_6H_5CKCH(CH_3)_2$, can initiate polymerization of dienes like butadiene to give rubber-like materials. This is a chain reaction, with carbanions as intermediates in some cases and free radicals in others. Butadiene and isoprene (2-methylbutadiene) can be polymerized by lithium metal or butyllithium to the *cis*-polymers; the product from isoprene is very similar to natural rubber.[37] A. A. Morton's alfin catalyst, composed of an organosodium compound (allyl- or benzyl-sodium) and sodium isopropoxide, polymerizes butadiene, isoprene, and styrene to rubber-like materials.[38] Most of the synthetic rubber produced during and since WW II was prepared, however, by free-radical polymerization using other catalysts than organometallic compounds.

The central carbon-carbon bond in highly substituted ethane derivatives

$$(C_6H_5)_3CCH(C_6H_4CH_3-\underline{p})_2 \ + \ 2Na \ \longrightarrow \ (C_6H_5)_3CNa \ +$$

$$+ \ NaCH(C_6H_4CH_3-\underline{p})_2$$

is cleaved to give organometallic compounds. Thus pentaphenylethanes are cleaved by forty percent sodium amalgam or by liquid sodium-potassium alloy.[39] This C-C bond cleavage yields carbanions rather than free radicals.[40] It was used to correlate structure with ease of C-C bond cleavage in cases where heterocyclic groups or highly unsaturated groups were present instead of aromatic groups.

Conant studied the free-radical dissociation of substituted dixanthyls as a function of the saturated R group;[41] the secondary alkyldixanthyls showed dissociation to a free radical more rapidly than those with primary alkyl groups. The dixanthyls were cleaved by alkali metals, even where there was no visible dissociation to free radicals.

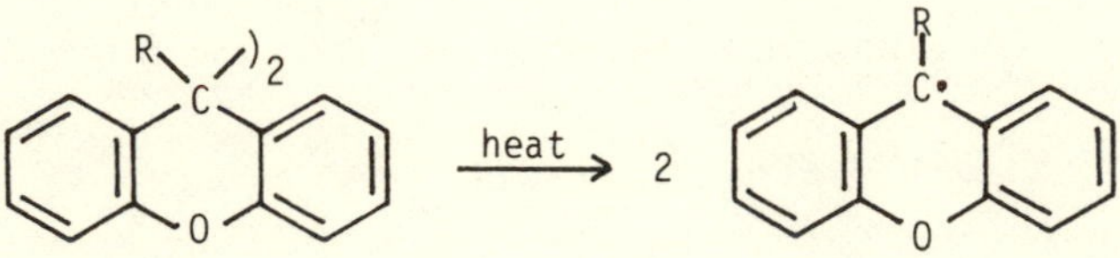

Acetylenic derivatives such as that below showed cleavage by alkali metals, although they did not show free-radical dissociation.[42] Cleavage of other highly arylated C-C bonds, accompanied by rearrangement, was observed by Koelsch.[34]

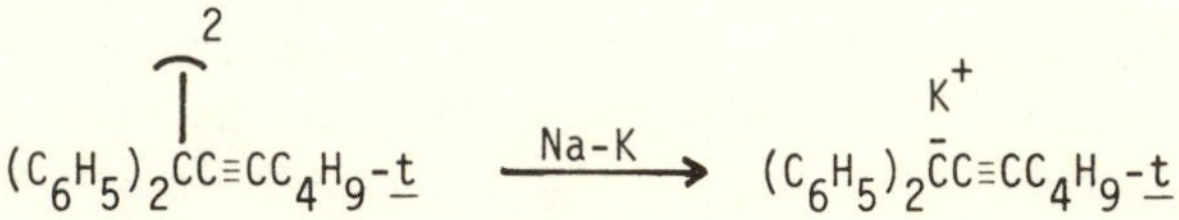

It should be emphasized that thermal dissociation of these compounds, when it occurs, gives free radicals, each with an unpaired electron. The alkali metal cleavage yields carbanions; in both cases the odd electron or the electron pair is stabilized by resonance in the aromatic-unsaturated system. The alkali metal cleavage is thus due to the acidity of the corresponding C-H bond; the structural features stabilizing the unpaired electron are similar to those stabilizing the electron pair.

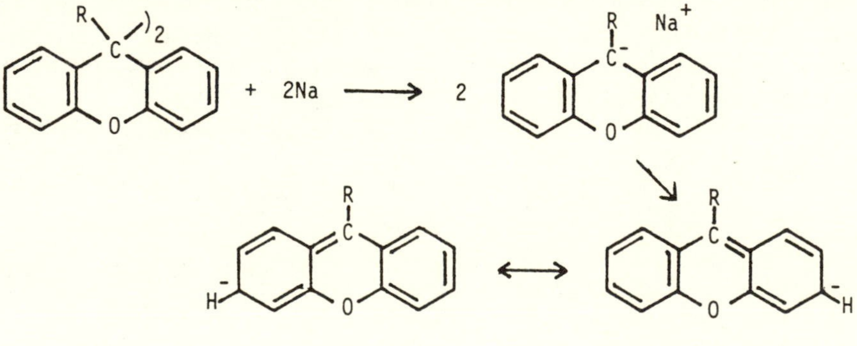

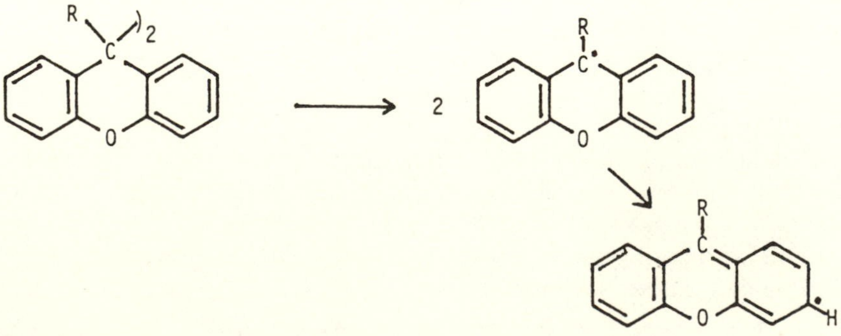

LITERATURE CITED

1. For history of early organometallic work, see H. Gilman and R. E. Brown, JACS, **52**, 1181, 5045 (1930); for review of organometallics in general, see Gilman's chapter in *Organic Chemistry*, 1938 ed., pp. 406-488.

2. Quantitative method: H. Gilman et al., JACS, **45**, 150 (1923); qualitative test: H. Gilman and F. Schulze, ibid., **47**, 2002 (1925); H. Gilman and J. Swiss, ibid., **62**, 1847 (1940).

3. E. P. Kohler, J. F. Stone, and R. C. Fuson, ibid., **49**, 3181 (1927).

4. H. Gilman and H. H. Parker, ibid., **46**, 2816 (1924).

5. H. Gilman, et al., ibid., **51**, 2252 (1929); a similar case was found by E. P. Kohler and E. M. Nygaard, ibid., **52**, 4128 (1930).

6. H. Gilman and N. J. Beaber, ibid., **47**, 518 (1925); S. S. Rossander and C. S. Marvel, ibid., **50**, 1491 (1928); H. Gilman and L. H. Heck, ibid. **50**, 2223 (1928); H. Gilman and J. Robinson, *Organic Syntheses*, **10**, 4 (1930).

7. J. B. Conant and A. H. Blatt, JACS, **51**, 1227 (1929).

8. F. C. Whitmore et al., ibid., **60**, 2028 (1938), for example. The extensive literature on reduction by Grignards was reviewed by M. S. Kharasch and O. Reinmuth, *Grignard Reactions of Nonmetallic Substances*, New York, 1954, pp. 147-160.

9. K. Ziegler and H. Colonius, Ann., **479**, 135 (1930).

10. H. Gilman, et al., JACS, **54**, 1957 (1932); **55**, 1252 (1933), and later papers.

11. W. G. Young and J. D. Roberts, ibid., **66**, 1444 (1944); cf. P. D. Bartlett and A. Schneider, ibid., **67**, 141 (1945) for the synthesis of tri-*t*-butylcarbinol from Na, $(CH_3)_3CCl$, and $(CH_3)_3CCOOCH_3$.

12. H. Gilman and R. L. Bebb, ibid., **61**, 109 (1939); H. Gilman and J. W. Morton, *Organic Reactions*, **8**, 258 (1954).

13. H. Gilman, et al., JACS, **61**, 106 (1939); JOC, **3**, 108 (1938); G. Wittig et al., Ber., **71**, 1903 (1938); R. G. Jones and H. Gilman, *Organic Reactions*, **6**, 339 (1951); W. E. Parham, C. K. Bradsher and D. A. Hunt, JOC, **43**, 1606 (1978).

14. A. Luttringhaus, Ber., **67**, 1602 (1934).

15. H. Gilman and P. R. Van Ness, JACS, **55**, 1258 (1933).

16. Margaret J. Jorgenson, *Organic Reactions*, **18**, 1 (1970).

17. A. A. Morton and J. R. Stevens, JACS, **53**, 2244 (1931); A. A. Morton and W. S. Emerson, ibid., **59**, 1947 (1937).

18. A. A. Morton and I. Hechenbleickner, ibid., **58**, 1697 (1936); A. A. Morton, et al., ibid., **60**, 1426, 1924 (1938).

19. W. E. Bachmann and H. T. Clarke, ibid., **49**, 2089 (1927).

20. Biography by R. C. Elderfield, *Biog. Mem. Nat. Acad. Scis.*, **34**, 1 (1960).

21. J. U. Nef, Ann., **308**, 274 (1899) and refs. therein; J. A. Nieuwland and R. R. Vogt, *The Chemistry of Acetylene*, New York, 1945, pp. 74-94; R. A. Raphael, *Acetylenic Compounds in Organic Synthesis*, London, 1955, passim.

22. F. C. McGrew and Roger Adams, JACS, **59**, 1497 (1937); K. N. Campbell, Barbara K. Campbell, and L. T. Eby, ibid., **60**, 2882 (1938).

23. M. S. Kharasch and O. Reinmuth, op. cit., pp. 99-115 for detailed discussion.

24. W. S. Schlenk and W. Schlenk, Jr., Ber., **62**, 920 (1929); **64**, 734 (1931).

25. P. D. Bartlett and C. M. Berry, JACS, **56**, 2683 (1934); they used Noller's method of dioxan precipitation, C. R. Noller, ibid., **53**, 635 (1931).

26. A. C. Cope, ibid., **56**, 1578 (1934).

27. C. R. Noller and W. R. White, ibid., **59**, 1354 (1937); Cope, ibid., **60**, 2215 (1938). The presence of oxygen further complicates the picture by preventing precipitation of $MgCl_2$: C. R. Noller and A. J. Castro, ibid., **64**, 2509 (1942).

28. W. V. Evans and Ralph Pearson, ibid., **64**, 2865 (1942). The detailed electronic picture is from Kharasch and Reinmuth, op. cit., p. 111.

29. L. M. Seitz and T. L. Brown, JACS, 88, 2174 (1966); T. L. Brown et al., J. Organometallic Chem., **77**, 299 (1974), and refs. therein.

30. A. W. Langer Jr., Trans. New York Acad. Sci., ser. 2, **27**, 741 (1965).

31. N. D. Scott, J. R. Walker and V. L. Hansley, JACS, 58, 2442 (1936) and refs. therein; Walker and Scott, ibid., 60, 951 (1938).

32. Allene Jeanes and Roger Adams, ibid., 59, 2608 (1937) and refs. therein.

33. W. Schlenk and E. Bergmann, Ann., 463, 1 (1928); later workers have made extensive corrections to this paper, eg., refs. 32, 34, 35, 36.

34. C. F. Koelsch, JACS, 56, 480 (1934).

35. Review by C. B. Wooster on organoalkali compounds, Chem. Rev., 11, 1 (1932); K. Ziegler on organopotassium compounds, Angew. Chem., 49, 455, 499 (1936).

36. G. F. Wright, JACS, **61**, 2106 (1939).

37. J. K. Stille, *Introduction to Polymer Chemistry*, New York, 1962, pp. 73 ff, 183 ff.

38. A. A. Morton, et al., JACS, 69, 950 (1947) and later papers; Stille, pp. 79-80.

39. W. E. Bachmann and F. Y. Wiselogle, JOC, 1, 354 (1936).

40. K. Ziegler and B. Schnell, Ann., **437**, 227 (1924); K. Ziegler, ref. 35, cleavage of R_3COCH_3 by potassium to give R_3CK.

41. J. B. Conant, et al. JACS, 48, 1743 (1926); Conant and Mildred W. Evans, ibid., **51**, 1925 (1929).

42. P. L. Salzberg and C. S. Marvel, ibid., 50, 1737 (1928); D. W. Davis and Marvel, ibid., **53**, 3840 (1931), and other papers by Marvel.

13

Catalytic Hydrogenation and
Other Catalytic Processes

The addition of hydrogen to carbon-carbon multiple linkages and to other unsaturated groups in the presence of a solid catalyst was developed in Europe around 1900.

In 1897 P. Sabatier and J. B. Senderens at Toulouse, France, reduced ethylene to ethane with hydrogen and solid nickel catalyst,[1] and in a long series of papers they extended their method using high temperatures. Their nickel catalysts were prepared by reduction of nickel oxides with hydrogen, or by pyrolysis of salts such as nickel formate. The catalysts consisted of metallic nickel mixed with variable amounts of oxides. Sabatier received the Nobel Prize for chemistry in 1912 for his work on catalytic reduction. Paul Sabatier (1854-1941) took his doctorate in the physical sciences at the Collège de France (1880) and was professor at Toulouse from 1884 to 1929. He was not interested in the great practical applications of his discoveries.

V. I. Ipatieff developed catalytic reactions under high pressure, including hydrogenations, in Russia around 1900 and later in this country. Ipatieff (1867-1952) was a graduate (1892) of the Mikhail Military Academy of St. Petersburg (now Leningrad). He was largely self-taught as a chemist, and he became professor of chemistry and explosives at the Academy from 1899 to 1930. Around 1900 he found that at high pressures and temperatures solid catalysts can direct reactions to form specific products.[2] Ipatieff moved to the United States in 1930 and made fundamental discoveries in the application of catalysis to petroleum technology, particularly in the formation of branched-chain hydrocarbons for high octane fuels.

In 1902-1903 the German chemist W. Normann (1870-1939) was granted patents on the hydrogenation (hardening) of vegetable oils to (solid) fats with nickel catalysts, but the process did not become commercially useful for nearly a decade.[3] Then the need to produce solid fats for soaps and shortenings fostered a growing industry in Europe and America. Procter and Gamble's "Crisco", the first all-hydrogenated, purely vegetable shortening (from cottonseed oil), was marketed in 1911,[4] but their basic patent for hydrogenation was in litigation from 1915 to 1920, when it was declared invalid by the U.S. Supreme Court.[5] By 1922 about sixty firms were engaged in hydrogenating vegetable oils in this country.[6]

It was in this context that the unique contribution of Murray Raney (1885-1965) was made.[7] Raney graduated in mechanical engineering at the University of Kentucky (1909) without ever completing high school and worked in Chattanooga, Tennessee, from 1913 on. There, with electric power available from the Hales Bar Dam on the Tennessee River and abundant cotton-seed oil from the crops of Georgia and Alabama, Raney became engrossed in the study of nickel catalysts for use in hydrogenating cottonseed oil. In the 1920s he developed a novel method of preparing catalytically active nickel by treating a nickel-silicon[8] or nickel-aluminum[9] alloy with aqueous sodium hydroxide, as illustrated.

$$NiAl_2 \ + \ 6 \ NaOH \ \xrightarrow{\ H_2O\ } \ Ni \ + \ 2 \ Na_3AlO_3 \ + \ 3 \ H_2$$

"Raney nickel", prepared from the nickel-aluminum alloy, became well known from Homer Adkins' publication described below and has become the most widely used hydrogenation catalyst. Much cheaper than the platinum or other noble metal catalysts and less subject to poisons, it does not require recovery of the nickel after use. It is widely used in industrial processes and is still manufactured in a plant designed and built by Raney, later bought by W. R. Grace and Company.[10]

> Raney was almost unique, among chemist-inventors of the 20th cen-tury, in making his invention in almost complete isolation from other technical people; he was literally almost the last of the basement in-ventors in chemistry and chemical technology. Practically all other major discoveries in chemistry in this century of practical impor-tance, such as nylon, teflon, insulin, penicillin, polyethylene, and silicones, to name a few, have been made by workers in university laboratories, research institutes, or industrial research laboratories with supporting staffs and research facilities of some magnitude. The same is true of practically all discoveries of fundamental prin-ciples in chemistry and other sciences. Although individual intelli-gence and originality are indispensable for an important discovery, there is usually required in addition a setting of the necessary tech-nical equipment and also of knowledgeable colleagues, to produce a stimulating atmosphere in which originality can flower and be productive... The striking thing about Raney's discoveries is that they represented an individual effort of intelligence and originality, without the institutional environment of equipment and collabora-tion, which has been present in nearly all comparable cases.[7]

In addition to its use as a hydrogenation catalyst, Raney nickel has the novel property of removing sulfur from organic carbons. Carbon-hydrogen bonds replace carbon-sulfur bonds, the hydrogen adsorbed on the nickel sur-face being the active agent. This reaction has been employed to establish structures, as with biotin and penicillin, and for laboratory scale syntheses.[11]

$$RSR' \ \xrightarrow{\ Raney \ Ni-H_2\ } \ RH \ + \ R'H \ + \ H_2S$$

Starting in 1930, Homer Adkins at the University of Wisconsin described hydrogenations at temperatures up to 250° and pressures up to 400 atmospheres using three catalysts: Raney nickel, nickel on kieselguhr (prepared by reduction of nickel oxide), and copper chromite developed by Adkins.[12,13,14,15] Adkins' monograph[13] and some later publications described in detail the methods of hydrogenation, purification, and yields of products for most classes of unsaturated organic compounds. In general, the nickel catalysts are more useful for carbon-carbon multiple bonds, and copper chromite for carbon-oxygen bonds, although these specifications are by no means absolute. We make no attempt to summarize Adkins' monograph which is fundamental for the field, but we note a few points.

Hydrogenolysis (cleavage by hydrogen) of carbon-oxygen bonds and some other bonds, particularly by nickel, is described in detail. Benzyl alcohol gives toluene almost quantitatively at 100°, for example. Ethers, acetals, esters, and anhydrides show some hydrogenolysis of carbon-oxygen bonds. Carbon-carbon and carbon-nitrogen bonds are also hydrogenolyzed in some cases. However, a good degree of control can be exercised over hydrogenations by varying catalyst, solvent, temperature, and pressure.

$$C_6H_5CH_2OH \xrightarrow[100°]{H_2, Ni} C_6H_5CH_3 + H_2O$$

Adkins' work, and the extensive industrial work on high pressure reactions of various types in addition to catalytic reduction, popularized the high pressure technique, which was widely used in university laboratories by 1940. It was useful in that sizeable batches could be run with acceptable yields. The high pressure technique had disadvantages: the initial equipment costs were high; the conditions normally used in most university laboratories involved a certain hazard; and the maintenance of the high pressure equipment was troublesome, particularly if it was used only intermittently.

Homer Adkins (1892-1949, B.S. 1915, Denison; Ph.D. 1918, Ohio State with W. L. Evans) was a faculty member at Wisconsin from 1919 to 1949. He developed an excellent graduate department in organic chemistry and trained many outstanding students.[16] In addition to his fundamental work on hydrogenation, he investigated various synthetic procedures, and rates and equilibria of carbonyl reactions.[15]

The development after 1945 of the complex metal hydrides,[17] such as lithium aluminum hydride, $LiAlH_4$, and its congeners, gave more convenient and specific reducing agents which can bring about most of the reductions hitherto done under high pressure, using ordinary glass equipment at atmospheric pressure. High pressure hydrogenation on a laboratory scale thus became almost obsolete after 1950, but is still important on an industrial scale.

The application of complex metal hydrides for reduction is due primarily to Herbert C. Brown. His book on boranes[17] summarizes this research.

The use of noble metal catalysts, particularly platinum and palladium, for catalytic reductions at room temperature and low pressure (one-three atmospheres) was developed at about the same time as the first nickel catalysts.[18] Platinum black (finely divided platinum metal) prepared by reducing platinic chloride with various reagents[19] (best with formaldehyde by Loew in 1890)[20] was widely used by Willstätter and others.[21] There was also much work on "colloidal" platinum and palladium, prepared by a very involved procedure using partially hydrolyzed egg albumin to stabilize the colloidal metal.[22] The number of modifications proposed and the frequent reports of unreproducible results indicate how unsatisfactory these platinum and palladium catalysts really were. The studies of Roger Adams at Illinois were to be as important in this field as Murray Raney's work on nickel catalysts.

The discovery of the platinum oxide catalyst was Adams' first major research contribution and was reported as usual in formal publications.[23] It did, however, have an element of serendipity which became clear from later accounts.[24] After a year or two of tedious experimentation to determine the best conditions for the Willstätter-Waldschmidt-Leitz method[20] of reducing platinum chloride to platinum black with formaldehyde, a student at Illinois was making a large run using one ounce of chloroplatinic acid to prepare a supply of catalyst. A broken casserole cascaded platinum black over an old, wooden table top. The valuable mass, when scraped up, included wood splinters, dust, and debris. Solution in aqua regia did not destroy all the foreign matter, and Adams suggested fusion with sodium nitrite to remove the organic material. This yielded a brown platinum oxide which was reduced readily by hydrogen, giving a highly active platinum black catalyst.[24] The nitrite fusion was standardized,[23] and the product was found to be an excellent catalyst for reducing a variety of organic compounds.

Adams' group studied the effects of promoters, such as ferric chloride, in great detail,[25] and "Adams catalyst" became a very valuable reagent in organic chemistry. Although palladium and rhodium are more useful in some cases, the Adams catalyst still holds its place.[26] By 1940 the Adams catalyst, the palladium catalyst described below, and Raney nickel allowed the reduction of a wide variety of unsaturated groups with good selectivity and good yields. Platinum and palladium catalysts are particularly useful for reducing carbon-carbon multiple bonds, although they have many other applications.[26] They can be used on a preparative, diagnostic, or microquantitative scale,[27] a feature particularly useful in determining the amount of carbon-carbon unsaturation in small amounts of natural products.

W. H. Hartung, working in the laboratories of the pharmaceutical firm Sharp and Dohme, found palladium on charcoal to be a valuable tool. It was the catalyst of choice[28] for the removal of protecting benzyl groups from oxygen or nitrogen by catalytic hydrogenolysis. This reagent also reduced nitriles and oximes to primary amines; in the example illustrated biologically active ethanol amines are formed.[29]

$$\underset{\underset{ArCCH=NOH}{\overset{O}{\overset{\parallel}{}}}}{} \quad \xrightarrow[\substack{\text{absolute } C_2H_5OH, \\ \text{dry HCl}}]{H_2, \ Pd\text{-}C} \quad \underset{\underset{ArCCH_2NH_2 \cdot HCl}{}}{\overset{OH}{\overset{\mid}{}}}$$

Walter H. Hartung (1895-1961, B.A. 1918, Minnesota; Ph.D. 1926, Wisconsin with Homer Adkins) was a research chemist at Sharp and Dohme and then taught pharmaceutical chemistry at Maryland, North Carolina, and the Medical College of Virginia. His research included amino acids, peptides, amino alcohols, and derivatives of the tropane ring system.

A common problem with the noble metal catalyst, and in some instances with nickel, was the poisoning of the catalyst by sulfur. For example, the reductive removal of a protecting benzyl group from sulfur could not be done with palladium or nickel, but it could be readily accomplished by V. du Vigneaud's procedure using sodium-liquid ammonia.[30]

Continuing studies on Raney nickel produced reductions that were valuable and different from those obtained by conventional hydrogenation procedures.[31] E. Schwenk and others at the Schering Corporation showed that the combination of Raney alloy and aqueous sodium hydroxide had useful reducing properties, giving hydrogenolysis of aromatic ketones to the hydrocarbon, reducing other carbonyl groups to alcohols, and replacing halogen, methoxy, and sulfonate by hydrogen.

Papers by Adkins, Reid, and other university chemists illustrated catalytic processes other than hydrogenation brought about by solid catalysts and liquid or gaseous reagents (heterogeneous catalysis).[32] Primary or secondary amines react with alcohols and Raney nickel to give alkylation. Many variations are possible.[33] In 1914 L. P. Kyriakides studied conversion of diols to dienes[34] over several catalysts at 300-450°.

$$\underset{\underset{C_6H_5CH_2SCH_2CHCONHR}{}}{\overset{NH_2}{\overset{\mid}{}}} \quad \xrightarrow[\text{liq. } NH_3]{Na} \quad C_6H_5CH_3 \ + \ \underset{\underset{HSCH_2CHCONHR}{}}{\overset{NH_2}{\overset{\mid}{}}}$$

$$C_6H_{11}NH_2 \ + \ C_2H_5OH \quad \xrightarrow{Ni, \ 140°} \quad C_6H_{11}NHC_2H_5 \ + \ C_6H_{11}N(C_2H_5)_2$$
$$\qquad\qquad\qquad\qquad\qquad\qquad\qquad (46\%) \qquad\quad (14\%)$$

$$\underset{(CH_3)_2\overset{\overset{OH}{\mid}}{C}-\overset{\overset{OH}{\mid}}{C}(CH_3)_2}{} \quad \xrightarrow{Cu, \ 430°} \quad CH_2=\overset{\overset{CH_3}{\mid}}{C}-\overset{\overset{CH_3}{\mid}}{C}=CH_2$$

E. Emmett Reid demonstrated reductions involving ammonia and hydrogen sulfide to synthesize alkyl cyanides and mercaptans,[35] respectively. He also observed hydrogen exchanges,[36] using a ceria catalyst.

$$CH_3COOH + NH_3 \xrightarrow[500°]{Al_2O_3 \text{ on thoria}} CH_3CN + 2 H_2O \quad (35\%)$$

$$C_4H_9OH + H_2S \xrightarrow[380°]{thoria} C_4H_9SH + H_2O + C_3H_7CHO$$
$$(53\%) \qquad\qquad (17\%)$$

$$RCHO + C_2H_5OH \xrightarrow[300-380°]{ceria} RCH_2OH + CH_3CHO$$

Numerous studies from American university laboratories in the 1920s defined some of the problems of heterogeneous catalysis and suggested possible mechanisms. Adkins, particularly with W. A. Lazier[37] who was to become a leader in the field at Du Pont, studied dehydration and dehydrogenation of alcohols and found that the resulting reactions depended on the way the catalyst, such as alumina, was prepared. They suggested that the type of reaction, when there were several competing ones, depended on the spacing of the active sites on the catalyst.

Further suggestions on mechanisms came from H. S. Taylor's group at Princeton. R. N. Pease (a student of Taylor) showed that carbon monoxide was an effective poison for the catalytic activity of copper in hydrogenating ethylene.[38] This and other observations led Taylor to propose the active site theory, according to which the products were determined by the active sites, atoms on the surface of the catalyst which had only part of their bonding capacity satisfied by neighboring atoms in the surface. These active sites thus adsorbed reactants strongly, were attacked preferentially by poisons, and determined the course of the catalytic reaction by the intensity of their valence forces.[39] Taylor showed that some adsorption on catalytic surfaces had a measurable activation energy (activated adsorption), and some required very small energies (chemisorption).[40] Activated adsorption was postulated as an intermediate stage in reactions on a catalytic surface.

Hugh S. Taylor (1890-1974, B.Sc. 1909; D.Sc. 1914, both at Liverpool) taught at Princeton from 1914 to 1958, where he built up an outstanding school of physical chemistry. R. N. Pease (1895-1964) and Henry Eyring (1901-1981) were among his colleagues. Taylor retained his British citizenship and was eventually knighted.

C. R. Hoover at Wesleyan (Connecticut) showed that a mixture of CO and H_2 over nickel-palladium catalyst on pumice at 100° produced small amounts of ethylene.[41] Methane (20-25 percent) was formed at 250°. Such work was to be pursued with great success in industrial laboratories. Hoover (1885-1942) was killed in a blimp crash while doing war research for the National Defense Research Committee.

Two catalytic gas phase oxidation reactions, developed in government laboratories during World War I, provided new means of readily obtaining organic intermediates and became important industrial processes. One was the oxidation of naphthalene by air over a vanadium-molybdenum oxide catalyst giving a good yield of phthalic anhydride, worked out by H. D. Gibbs (1872-1934) in the Bureau of Chemistry of the U. S. Department of Agriculture.[42] Details of the process were made available to industrial firms who wished to utilize it.

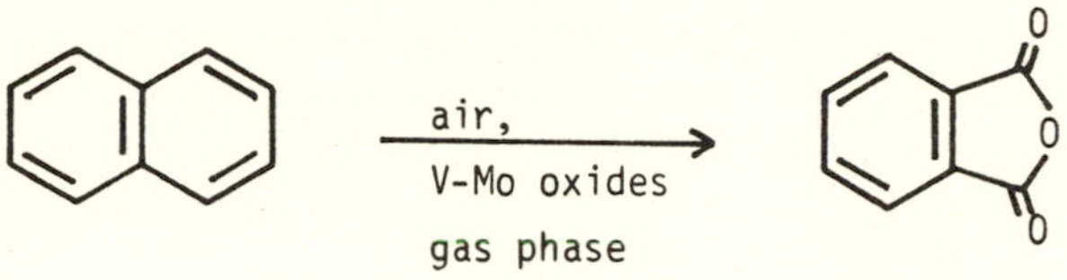

Phthalic anhydride

The second reaction was the vapor phase oxidation of benzene to maleic anhydride using a similar catalyst, worked out by C. M. Weiss and C. R. Downs of the Barrett Company in New York City.[43]

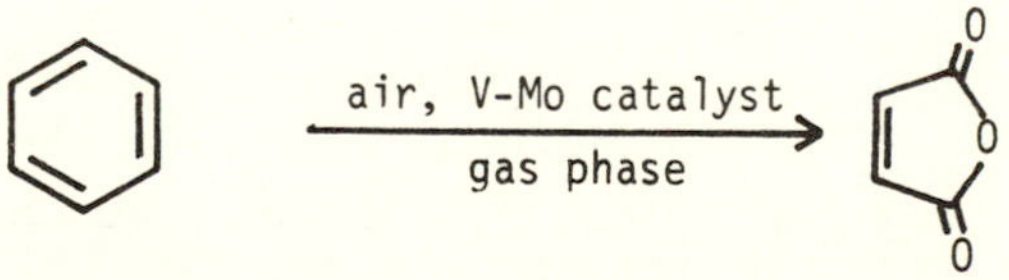

Maleic anhydride

The gas phase oxidation of ethylene to ethylene oxide by air over silver-containing catalysts was developed both in this country and abroad. It led to further important products, such as ethylene glycol and many more. The ready reaction of ethylene oxide with many types of compounds, water, alcohols, amines, and carboxylic acids gives it a key position.[44]

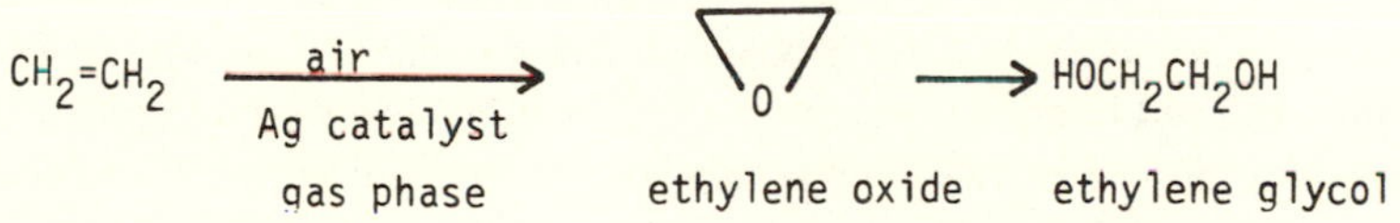

ethylene oxide   ethylene glycol

Of the host of industrial catalytic processes, we mention only two sets, those involved in the manufacture of nylon and those used in making high octane gasoline from petroleum.

The development of a commercial process for 6,6 nylon, the polyamide[45] from adipic acid, $HOOC(CH_2)_4COOH$, and the diamine, $H_2N(CH_2)_6NH_2$, was due to Elmer K. Bolton of Du Pont. The laboratory preparation, as described elsewhere, was discovered by W. H. Carothers and his group at Du Pont. Bolton (1886-1968, A.B. 1908, Bucknell; Ph.D. 1913, Harvard with C. L. Jackson) spent two years in Willstätter's laboratory, where his close friend and classmate, Roger Adams, had worked the year before. Bolton worked at Du Pont from 1916 to 1951 becoming essentially director of research. He played a key role (if not highly publicized) in the development of nylon, Neoprene rubber, dyes, and many other products.[46]

The first commercial process for the nylon intermediates apparently involved the stages indicated below.[47] The ingenuity of chemists over the years led to many modifications, among them oxidation of cyclohexane catalytically to adipic acid, and new ways of converting adipic acid to the diamine.[48] As the world knows, the importance of this breakthrough in industrial production of synthetic fibers cannot be overestimated.

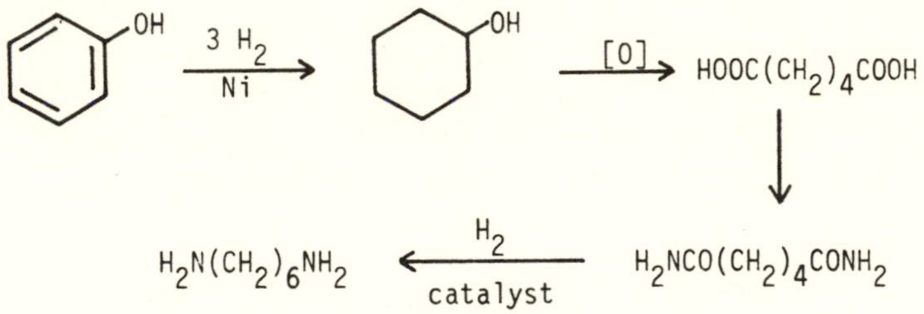

The petroleum industry faced a double challenge. Not only was more and more fuel needed for the burgeoning numbers of internal combustion engines, but its quality required adjustment for the increased working pressure inside the cylinders. In answer, the industry developed a variety of catalytic processes for synthesizing more branched-chain hydrocarbons for better performance as fuels. The reactions involved included branched-chain cracking, hydrogenation, dehydrogenation, isomerization of straight to branched-chain hydrocarbons, reactions of alkenes with alkanes, cyclization, and many others. Every major refining company developed its own set of processes applied on an enormous scale. Independent laboratories like Universal Oil Products and independent inventors like Eugene Houdry made valuable contributions.[49,50]

The discovery of ortho and para hydrogen and, more strikingly, of deuterium, allowed highly informative experiments to be carried out on the mechanism of action of heterogeneous catalysis. The availability of radioactive carbon, ^{14}C, and other isotopes after 1945 contributed a new dimension to these difficult problems. The experimental work of Otto Beeck at the Shell Development Laboratory in California on heterogeneous catalysis, which began before 1940, was outstanding.[51]

Adkins, Adams, Ipatieff, Taylor, and other American academic chemists, and Raney, Carothers, and many more American industrial chemists made contributions to catalytic processes which were of world-wide significance.

LITERATURE CITED

1. P. Sabatier and J. B. Senderens, Compt. Rendus, **124**, 1358 (1897), and many later papers; P. Sabatier, *Catalysis in Organic Chemistry*, tr. by E. E. Reid, New York, 1922.

2. V. I. Ipatieff, *Catalytic Reactions at High Pressure and Temperatures*, New York, 1937; see also Ipatieff's fascinating autobiography concerning life and chemistry in Russia before and after the fall of the Tsarist government in 1917: *The Life of a Chemist: Memoirs of V. I. Ipatieff*, translated by V. Haensel and Mrs. R. H. Lusher, Stanford, 1946.

3. Carleton Ellis, *Hydrogenation of Organic Substances*, 3rd. ed., New York, 1930, pp. 150, 316. This book is a huge, uncritical catalog of publications and patents; Leprince and Siveke, German Patent 141,029 (1902); W. Normann, British Patent 1515 (1903). See also the account of Normann in *The Sources of Inventions*, by John Jewkes, David Sawers, and Richard Stillerman, New York, 1959, p. 306; Sabatier-Reid, op. cit., p. 190.

4. W. Haynes, Ed., *American Chemical Industry*, Vol. VI, 1949, p. 344.

5. Anon., Chem. Met. Eng., **23**, 1145 (1920); Anon., J. Ind. and Eng. Chem., **13**, 172 (1921).

6. Sabatier-Reid, op. cit., p.. 349.

7. For Raney's career, D. S. Tarbell and A. T. Tarbell, JCE, **54**, 26 (1977), based on unpublished material.

8. U. S. Patent 1,563,587 issued December 1, 1925, to "Murray Raney of Chattanooga, Tennessee."

9. U. S. Patent 1,628,190 issued May 10, 1927, to Raney as above.

10. At South Pittsburgh, Tennessee; the manufacture of the alloy is a spectacular process, involving mixing of molten nickel and aluminum.

11. J. Bougault et al., Bull. soc. chim., **7**, 781 (1940); review, G. R. Pettit and E. E. Van Tamelen, Organic Reactions, **12**, 356-529 (1962).

12. H. Adkins and H. I. Cramer, JACS, **52**, 4349 (1930); and refs. therein.

13. For a summary of his work to 1937, see H. Adkins, *Reactions of Hydrogen with Organic Compounds over Copper-Chromium Oxide and Nickel Catalysts*, Madison, 1937; among individual papers, the following are of special interest: with R. Connor, JACS, **54**, 4678 (1932), with R. Connor and K. Folkers, ibid., 1138. The first publication in a scientific journal on Raney nickel was by L. W. Covert and H. Adkins, ibid., **54**, 4116 (1932), which gave an improved procedure for the preparation of the catalyst from the nickel-aluminum alloy. Preparation of the catalyst from the alloy was described briefly in Raney's patent,[9] in more detail in Adkin's monograph (p. 20) and by R. Mozingo, *Organic Syntheses*, **21**, 15 (1941); Coll. Vol. III, 181. Procedures for making more active Raney nickel catalysts from the alloy are given by H. R. Billica and Adkins, ibid., p. 176. The later review by Adkins and R. L. Shriner in Gilman's *Organic Chemistry*, 2nd. ed., 1943, pp. 779-835, covers high pressure and low pressure hydrogenations with platinum as well as nickel and copper chromite catalysts.

14. Preparation of catalysts: copper chromite, Adkins, R. Connor and K. Folkers, JACS, **54**, 1138 (1932); nickel on kieselguhr, Adkins et al., ibid., p. 1651.

15. Adkins, monograph, pp. 19-20.

16. Biography of Adkins by Farrington Daniels, *Biog. Mem. Nat. Acad. Scis.*, **27**, 293 (1952).

17. Lithium aluminum hydride was first prepared by A. E. Finholt, A. C. Bond, Jr., and H. I. Schlesinger, JACS, 69, 1199 (1947). Applications of complex hydride reductions were exhaustively reviewed by N. G. Gaylord, *Reduction with Complex Metal Hydrides*, New York, 1956; H. C. Brown, *Boranes in Organic Chemistry*, Ithaca, 1972, pp. 209 ff.

18. Adkins and Shriner, op. cit.; P. N. Rylander, *Catalytic Hydrogenation over Platinum Metals*, New York, 1967; Morris Freifelder, *Practical Catalytic Hydrogenation*, New York, 1971; R. L. Augustine, *Catalytic Hydrogenation*, New York, 1965.

19. For early methods, see Sabatier-Reid, op. cit., pp. 53 ff.

20. O. Loew, Ber., 23, 289 (1890); improved procedure by R. Willstätter and E. Waldschmidt Leitz, ibid., 54, 121 (1921).

21. R. Willstätter and D. Hatt, ibid., 45, 1471 (1912); with E. Waser, 44, 3426 (1911); R. Feulgen, ibid., 54, 360 (1921).

22. C. Paal, and C. Amberger, Ber., 37, 125 (1904) and later papers.

23. Roger Adams, V. Voorhees, W. H. Carothers and R. L. Shriner, JACS, 44, 1397 (1922); 45, 1071, 2171 (1923); *Organic Syntheses*, Coll. Vol. 1, 463 (1932).

24. N. J. Leonard, JACS, 91, a-d (1969) written for Adams' eightieth birthday; the manuscript was authenticated by Adams. Supplementary accounts are in an unpublished manuscript by Adams, "A Sketch of Research Achievements," in the Adams papers in the Illinois library, and J. H. Wolfenden, JCE, 44, 299 (1967). A detailed account of his career is given by D. S. and A. T. Tarbell, *Roger Adams, Scientist and Statesman*, Washington, 1981; a briefer account by the same authors is *Biog. Mem. Nat. Acad. Scis.*, 53, 1-47 (1982), which includes a complete list of his scientific publications.

25. W. H. Carothers and Roger Adams, JACS, 47, 1047 (1925), and nearly 20 other papers, as well as many other examples of applications.

26. See the Rylander and Freifelder monographs, Ref. 18.

27. J. F. Hyde and H. W. Scherp, JACS, 52, 3359 (1930).

28. Removal of benzyl groups attached to nitrogen, oxygen, or sulfur by hydrogenolysis was reviewed by W. H. Hartung and R. Simonoff, *Organic Reactions*, 7, 263 (1953).

29. W. H. Hartung, JACS, 50, 3370 (1928); 53, 2248 (1931). The palladium charcoal used was prepared, following E. Ott and Schröter, Ber., 60, 633 (1927), by reducing $PdCl_2$ with H_2 in presence of animal charcoal.

30. V. du Vigneaud, L. F. Audrieth, and H. S. Loring, JACS, 52, 4500 (1930); R. H. Sifferd and du Vigneaud, JBC, 108, 753 (1935) and later papers.

31. B. Whitman, E. Schwenk et al., JBC, 118, 789 (1937); JOC, 7, 587 (1942); 9, 1 (1944).

32. For general surveys, for the period to 1939, see Ipatieff, *Catalytic Reactions*, Ref. 2 above; H. Adkins, Ind. Eng. Chem. 32, 1189 (1940); G. Schwab, H. S. Taylor and R. Spence, *Catalysis*, New York, 1937. For later summaries, see the series *Advances in Catalysis*, Vol. 1, 1948, New York; *Catalysis*, 8 vol., Vol. 1, 1954, New York.

33. H. Adkins and C. F. Winans, JACS, 54, 306 (1932) and later papers.

34. L. P. Kyriakides, ibid., 36, 980 (1914).

35. G. D. Van Epps and E. E. Reid, ibid., 38, 2128 (1916); R. L. Kramer, and E. E. Reid, ibid., 43, 880 (1921).

36. C. H. Milligan and E. E. Reid, ibid., 44, 202 (1922).

37. H. Adkins and W. A. Lazier, ibid., 46, 2291 (1924); 47, 1719 (1925); 48, 1671 (1926); H. Adkins and P. E. Millington, ibid., 51, 2449 (1929).

38. R. N. Pease and L. Stewart, ibid., 47, 1235 (1925); H. S. Taylor and R. N. Pease, ibid., 44, 1637 (1922).

39. H. S. Taylor and R. M. Burns, ibid., 43, 1273 (1921); H. S. Taylor, Proc. Roy. Soc., 108A, 105 (1925); *Advances in Catalysis*, Vol. 1, 1 (1948).

40. H. S. Taylor, JACS, 52, 5298 (1930); Schwab, Taylor and Pence, *Catalysis*, New York, 1937, pp. 204-218.

41. C. R. Hoover, *et al.*, JACS, **49**, 796 (1927); this work was started in 1913 in an effort to make glycol catalytically.
42. Preliminary paper, H. D. Gibbs, Ind. Eng. Chem., **11**, 1031 (1919); review of subsequent work, J. K. Dixon and J. C. Longfield, in *Catalysis*, P. H. Emmett, Ed., Vol. 7, 1960, pp. 196 ff.
43. C. M. Weiss and C. R. Downs, Ind. Eng. Chem., **12**, 228 (1920); review by Dixon and Longfield, op. cit., p. 186.
44. Review id., ibid., p. 237.
45. The designation 6,6 means that each component contained six carbon atoms. There was also a 5,10 nylon, derived from $HOOC(CH_2)_8COOH$ and the five-carbon diamine, but this acid was prepared from castor oil, and Bolton decided to produce the 6,6 form commercially because both components could be made from benzene.
46. For Bolton, see R. M. Joyce, *Biog. Mem. Nat. Acad. Scis.*, **54**, 50 (1983).
47. E. K. Bolton, Ind. Eng. Chem., **34**, 53 (1942); ibid., **37** (2), 106 (1945).
48. Examples are given by W. S. Emerson, *Guide to the Chemical Industry*, 1983, New York, pp. 50-51; J. M. Tedder, A. Nechvatal and A. H. Jubb, *Basic Organic Chemistry, Part 5. Industrial Products*, Chichester, 1975, pp. 122-125.
49. Universal Oil Products originally belonged to a group of oil companies; they conveyed it to a trustee in 1944, with the American Chemical Society as beneficiary of the trust. The sizable income from this source goes to the Petroleum Research Fund of the ACS, which makes grants to individual investigators for advanced research and education in the petroleum field, as very broadly interpreted.
50. Houdry (1892-1962) was born in France, but his most important work was done in this country; see C. G. Moseley, JCE, **61**, 655 (1984).
51. Review; Otto Beeck, *Advances in Catalysis*, Vol. II, 1950, p. 151; Beeck, A. E. Smith and A. Wheeler, Proc. Roy. Soc., **A177**, 62 (1940).

ELMER P. KOHLER

14

Stereochemistry, 1914-1939

Before considering work in stereochemistry as such, some calculations of the numbers of possible isomers and stereoisomers of compounds containing a given number of carbons should be mentioned. This problem has had perennial attention ever since 1875.

H. R. Henze and C. M. Blair at Texas calculated the number of saturated hydrocarbons, saturated alcohols, and ethylenic hydrocarbons.[1] The number of possible isomers with no rings of formula $C_{25}H_{52}$ was found to be 36,797,588. They also calculated the number of possible stereoisomers and non-stereoisomeric saturated hydrocarbons and checked their figures by writing structures through $C_{14}H_{30}$. Clearly no one is going to check their figures by trying to synthesize 6500 compounds of the $C_{14}H_{30}$ formula. E. E. Reid at Johns Hopkins synthesized twenty-two isomeric octyl alcohols, $C_8H_{17}OH$, out of the eighty-nine possible ones.[2] Branch and Hill at Berkeley also published methods of calculating numbers of possible stereoisomers of open chain compounds.[3]

The possibility that stereoisomers might exist differing only in the degree of rotation (*conformational isomers*) around a carbon-carbon single bond was considered by several German chemists around 1890.[4] C. A. Bischoff, professor at Riga, claimed erroneously to have isolated conformational isomers derived from dialkylsuccinic acids.[5] Methods of denoting different conformations similar to the much later Newman projections were used.[4,5] Opinions about the freedom of rotation around carbon-carbon single bonds wavered, but it seemed to be finally agreed that conformational isomers as shown were not stable enough to be isolated.

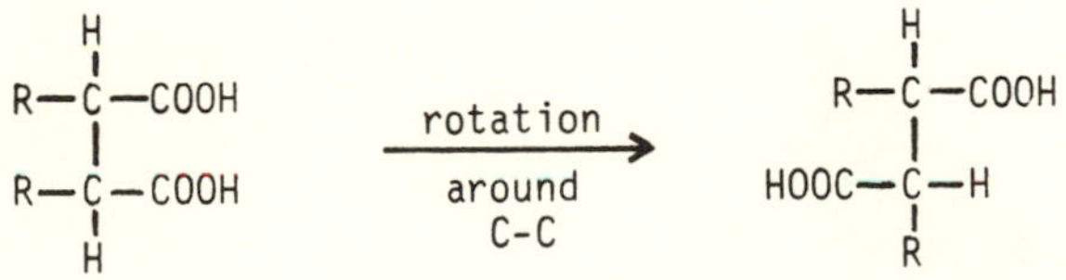

$$
\begin{array}{ccc}
\underset{\underset{H}{|}}{\overset{\overset{H}{|}}{R-C-COOH}} & \xrightarrow[\substack{around \\ C-C}]{rotation} & \underset{\underset{R}{|}}{\overset{\overset{H}{|}}{R-C-COOH}} \\
\underset{\underset{H}{|}}{R-C-COOH} & & HOOC-C-H
\end{array}
$$

This very old problem was advanced conclusively by American work in the 1930s. H. Eyring at Princeton, in a 1932 study of energy barriers, concluded[6] that there was a small barrier of about 0.36 kcal. to rotation of the two methyls in ethane, due to repulsion of the hydrogens when opposite each other, the *eclipsed* position as opposed to the *staggered* position. Experimental evidence on the height of the energy barrier was obtained from G. B. Kistiakowsky's calorimetric measurements on the heat of hydrogenation of ethylene.[7] Calculations of the equilibrium between C_2H_6, C_2H_4, and H_2, using the experimental heat of hydrogenation and statistical calculations of entropies and heat capacities, showed a discrepancy which could be explained in two ways: (a) the measured heat of hydrogenation of ethylene was too high by 0.9-1.5 kcal,[8] (b) the energy barrier of free rotation in ethane was larger than previously believed.[9] Further experiments ruled out alternative (a), verifying the reliability of the heat of hydrogenation measurement,[10] and additional experimental and theoretical work by Kistiakowsky and by K. S. Pitzer at Berkeley showed that the barrier to rotation of the methyl groups in ethane was actually about three kcal.[11]

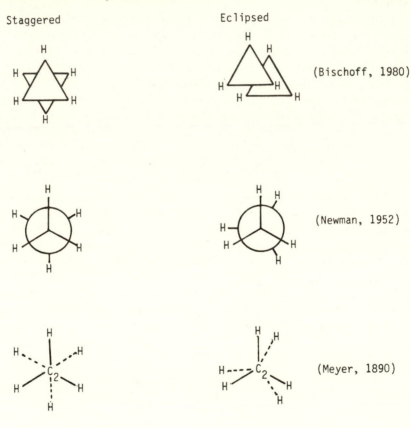

Staggered

Eclipsed

(Bischoff, 1980)

(Newman, 1952)

(Meyer, 1890)

This result showed that, although the barrier to free rotation was small and would not allow isolation of conformational isomers in saturated open chain systems at room temperature, there was a repulsion in the eclipsed conformation which made the staggered one more stable. This result is interesting in a more general context, that of intellectual inertia in accepting new ideas. As long as the accepted idea was that the barrier to free rotation was very small, there was reluctance on the part of E. Teller and B. Topley to consider the possibility of a much larger barrier, but Kistiakowsky was willing to consider it, probably because of his justified confidence in the accuracy of his experimental measurement. Once the validity of the three kcal barrier to free rotation in ethane had been established, there was no further reluctance to accept it as a general phenomenon. The work of Pitzer and others was extended to estimate repulsions of eclipsed forms in saturated rings, particularly in cyclohexane rings, as discussed later. Many barriers to rotation were measured later by nuclear magnetic resonance and microwave spectroscopy.

Another type of restricted rotation around carbon-carbon bonds which has a sufficiently high energy barrier to allow resolution into the stable optically active forms occurrs in the optically active biphenyl derivatives, such as dinitrodiphenic acid. If the two benzene rings cannot be coplanar, the whole molecule is asymmetric because it has no plane or other elements of symmetry, and it can exist in two optically active forms. This requires that enough bulky groups are present, usually at least three, in the ortho positions of the

two aromatic rings to prevent them from becoming coplanar or from passing through the coplanar position. This type of optical activity, in which there is no asymmetric carbon but an asymmetric molecule, was first demonstrated by the English chemists G. H. Christie and J. Kenner.[12] Roger Adams and his students at Illinois greatly extended this field.[13] The work was originally undertaken with the assumption that biphenyl had a folded structure with the two rings over each other (the Kaufler structure) like two pages of a partially opened book. Dipole moment measurements on 4,4'-disubstituted biphenyls showed that the folded structure was wrong, and experimental work which seemed to support the Kaufler formula by presumed ring closure between the 4 and 4'-positions was corrected.

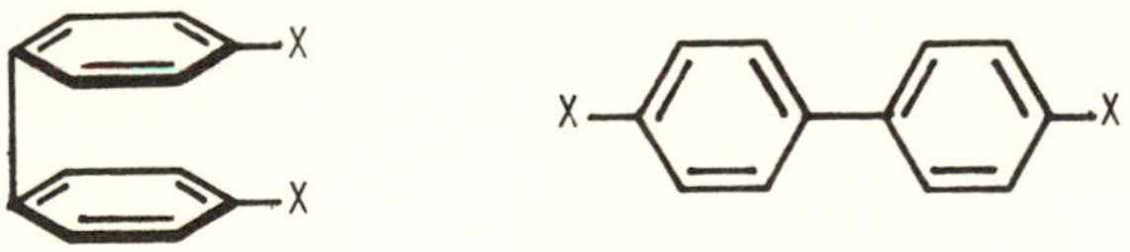

Adams[13] showed that the size and number of the groups in the ortho position determined the resolvability of biphenyls and their rate of racemization when resolved, the latter showing the rate of rotation of the two rings through a coplanar form. The size of atoms or groups as determined by X-ray data correlated well with the effectiveness of groups in preventing coplanarity and thus preserving optical activity.

This explanation of the optical activity as due to hindered rotation through a coplanar structure was supported by physical measurements. X-ray crystallographic measurements showed that the rings in resolvable biphenyls were at angles to each other of not more than 45°. Ultraviolet absorption spectra showed that with resolvable biphenyls, where coplanarity of rings was prevented, there was no electronic interaction between the two rings, and the ultraviolet spectrum was the sum of those of the two individual rings. Where coplanarity is possible, there was electronic interaction between the rings with some double bond character between them; the ultraviolet absorption was therefore at longer wavelengths, compared to each ring by itself.[14] This is an accurate indication of coplanarity.

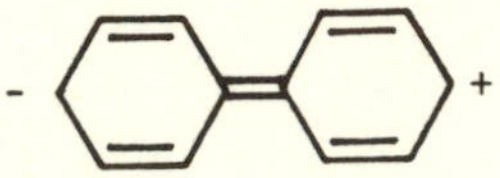

coplanar; interaction 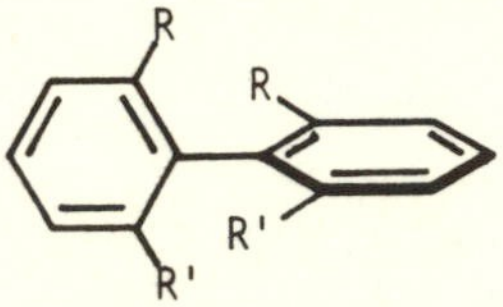   noncoplanar; no interaction

Adams found that fluorine, which is no larger than hydrogen, did not hinder rotation sufficiently to allow resolution. The rates of racemization were roughly correlated inversely with the size of groups preventing rotation. More exact kinetic studies by Kistiakowsky and later by F. H. Westheimer[15] showed an energy barrier to rotation of about 20-27 kcal depending on the substituents. This is to be compared with the three kcal. barrier for ethane; the higher value for these biphenyl derivatives is due to the large groups and the more rigid aromatic system.

Adams investigated other types of hindered rotation and the resulting optical activity in terphenyls, for example, where there were two centers of hindered rotation, and in heterocyclic compounds. N. Kornblum synthesized and resolved the biphenyls below, where rotation was hindered by a third

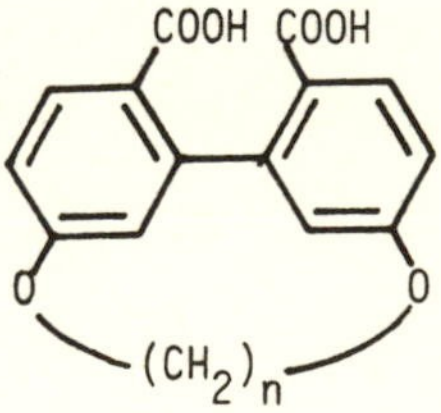

n = 8, 10

ring and not by ortho substituents;[16] the half life for racemization was 170 minutes (at 43° in dioxan) for n=8 and 120 minutes for n=10. In 1933 E. Mack at Ohio State suggested that 2,2,3-trimethylbutane and hexamethylethane should show hindered rotation like the biphenyls; this was based on scale models derived from collision areas and viscosity measurements. The barrier to rotation is undoubtedly larger than in ethane or butane, but is smaller than in the resolvable biphenyls.[17]

The barriers to rotation for open chain compounds described above were useful in considering the stereochemistry of saturated carbocyclic rings, particularly cyclohexane derivatives. Most of the key ideas in this field were developed elsewhere, but some significant contributions were made in the United States.

In a famous paper Adolph Baeyer[18] assumed that in saturated carbocyclic rings, all the carbon atoms lay in one plane. The normal bond angle for a regular tetrahedron is 109° 28′, and Baeyer assumed that in rings larger or smaller than a five-membered ring, a "strain" was involved in spreading the tetrahedral bond angles in larger rings, or in compressing them for rings smaller than five carbons. Strain would manifest itself by increased reactivity or by thermodynamic instability compared to some standard, such as the cyclopentane ring which is almost strain-free with the normal tetrahedral angles. Although the idea of strain has had some qualitative usefulness in dealing with reactivity of three and four-ring compounds, Baeyer's notions were taken more seriously than they deserved, and his assumption of coplanarity in cyclohexane and larger rings turned out to be erroneous. In Baeyer's view the tetrahedral angle for a planar cyclohexane would be spread (or strained) by 5° 16′, for a coplanar cycloheptane by 9° 51′, and by an increasing amount for larger coplanar rings.

Baeyer's assumption of coplanarity for carbocyclic rings was challenged in 1890 by the young German chemist Herman Sachse (1862-1893) of Charlottenburg,[19] who pointed out that a strain-free structure with the normal tetrahedral angles could be written for cyclohexane if the ring carbons were arranged in *two parallel* planes. Sachse's ideas were adopted and given a clearer presentation in 1918 by E. Mohr (1873-1926), professor at Heidelberg. Sketches of the resulting "Sachse-Mohr" structures for cyclohexanes

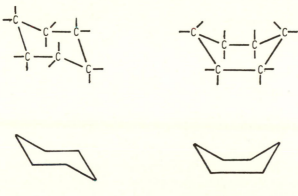

Chair, or rigid, form   Boat, or movable, form

are shown above,[20] with their modern designations of *chair* and *boat* forms. The earlier designation of the *rigid* form and the *movable* form were superseded by the more descriptive terms of chair and boat respectively. The positions of hydrogens or other substituents are indicated by dashes. The chair and boat forms are conventionally written, as shown, as projections on a plane at an angle to the main plane of the carbocyclic ring.

These projections do not show clearly the high degree of symmetry of the chair form, which is apparent from an inspection of models. In the chair form, the alternate carbon atoms lie in the same plane, so that the six carbon atoms lie in two parallel planes each containing three carbons. The hydrogen atoms are also symmetrically placed, six of them with their bonds perpendicular to the planes of the carbons, the *axial* (a) hydrogens, alternately above and below the carbon planes. The other six hydrogens, the *equatorial* (e) hydrogens, are on the periphery of the ring and are roughly in the planes of the ring carbons. (The terms axial and equatorial were introduced much later by workers in the field).

The boat form is less symmetrical, having four carbons in one plane, and two carbons, the *bowsprit* carbons, out of the plane of the other four. The boat form has higher energy than the chair, because the former has partially eclipsed hydrogens with the resulting repulsive forces.

Mohr pointed out that by rotation around the C-C bonds a chair form could *flip* into another chair form; a substituent which occupied the axial (a) position before the flip would assume the equatorial (e) position after flip, and vice versa.

The Sachse-Mohr structures soon received experimental support. Researches by H. G. Derx, working in Böeseken's laboratory at Delft, on the reaction of acetone with cis-1,2-cyclohexanediol to form a cyclic ketal were interpreted in terms of Sachse-Mohr structures, although this did not constitute a proof of these structures.[21]

At Göttingen in 1925 W. Hückel isolated *cis* and *trans* decalins; these molecules were of comparable stability and could be explained best by the fusion of two chair form strainless rings.[22]

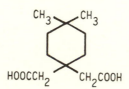

trans-Decalin   cis-Decalin

The synthesis of large ring compounds by L. Ruzicka (1887-1976) in the Federal Technical Institute at Zurich in the 1920s disproved Baeyer's strain theory for large ring compounds.[23] Carothers suggested that large rings probably had two long, parallel chains with loops at the end.[24]

The most conclusive evidence for the Sachse-Mohr chair form in cyclohexane rings was the X-ray crystallographic structure determination[25] of hexabromo- and hexachlorocyclohexane by R. G. Dickinson and C. Bilicke at Caltech in 1928. This showed the equatorial halogens in the plane of the carbocyclic ring, with the carbons in the two planes of the chair form. Although the significance of these observations was not emphasized in the original paper, the diagram was published in Lucas' text of 1935 with a clear statement of its meaning.[26]

In the early 1930s there were several reports of the isolation of stable boat and chair forms of substituted cyclohexanes; these were found to be based on faulty observations. A thorough attempt by R. F. Miller and Adams to show the presence of isomers in a cyclohexane with no asymmetric carbon atoms was unsuccessful; there was no evidence for more than one compound.[27]

$$CH_3 \qquad CH_3$$
$$HOOCCH_2 \qquad CH_2COOH$$

It was concluded that interconversion of boat and chair forms, or flips of chair forms, required so little energy that stereoisomers of this type would not be isolable at room temperature.

The English sugar chemist W. N. Haworth introduced the term *conformation* and considered chair structures for the six-membered pyranose rings of sugars.[28] D. H. Brauns of the U.S. Department of Agriculture applied conformational ideas to problems in the sugar field.[29] Two workers at the Bell Laboratories discussed the stability of the chair and other possible forms of the cyclohexane rings.[30]

By 1950 the time was ripe for a major extension of the structural theory to describe the effects of conformation, properties, and reactions for both open chain and cyclic structures. The key papers that integrated the various lines of research and established conformational thinking on a solid basis,[31] were by D. H. R. Barton in 1951 and 1953. It does not lessen the importance of the justly celebrated papers to point out some of the earlier work on which these were based: the determination of the stereochemistry of the steroid and triterpene series, the electron diffraction experiments of O. Hassel,[32] infrared and Raman spectroscopy, the estimation of barriers to rotation in open chain compounds, the mechanism of elimination reactions, and many more. The general ideas of conformational analysis have become part of the working tools of the organic chemist,[33] just as the ideas of classical stereochemistry are. Conformational phenomena have been a major field of research in this and other countries for many years.

Sir Derek Barton (1918-   , Ph.D. 1942, Imperial College), the outstanding English organic chemist of his generation, was professor at Glasgow and at Imperial College, and since 1978 has been Directeur, Institut de Chimie des Substances Naturelles at Gif-sur-Yvete, France. He received the Nobel Prize in 1969 with Hassel of Norway for his work on conformational analysis. Barton's work has been outstanding in structure, synthesis, and biogenesis of natural products, and on the mechanism of organic reactions. He was a leader in developing photochemistry as a synthetic tool, particularly for the partial synthesis of aldosterone, and his imaginative work on phenol oxidation has been very influential.

Other aspects of stereochemistry received attention during the 1914-1939 period. Van't Hoff had recognized in his first publications on stereochemistry that a properly substituted allene, although it had no asymmetric carbon atom, would be an asymmetric molecule and should be obtainable in optically active form. The asymmetry of the allene is caused by the arrangement of the two carbon-carbon double bonds in planes at right angles to each other; it is in principle analogous to the asymmetry of the hindered biphenyls. Kohler, J. T. Walker, and M. Tishler confirmed this by resolving the substituted allene shown into optically active forms.[34] An asymmetric synthesis forming an optically active allene was carried out simultaneously in England by W. H. Mills and P. Maitland.[35]

$$\underset{C_6H_5}{\overset{HOOCCH_2O_2C}{>}} C{=}C{=}C \underset{C_{10}H_7{-}1}{\overset{C_6H_5}{<}}$$

A stereochemical problem which had not been solved conclusively by the 1930s was the possibility of resolving a trivalent nitrogen compound. If the unshared pair of electrons occupied the apex of a tetrahedron **1, 2**, and if the molecule did not undergo the umbrella-like vibrations through a plane as shown, a tertiary amine should be resolvable. Workers in many countries had failed many times to effect such a resolution.[36] It was generally thought from spectroscopic evidence[37] that the rate of vibration was too great to allow a resolution at room temperature. T. D. Stewart at Berkeley failed to resolve the amine **3** at -70° using optically active acids.[38] This and attempts to resolve N-acyl- or N-sulfonylamides may have been doomed to failure because of the delocalization of the unshared pair of electrons on the nitrogen by the aromatic ring or by the acyl group. A more promising attempt by Adams and T. L. Cairns was to resolve ethylenimine derivatives **4**, but this gave no conclusive result.[39]

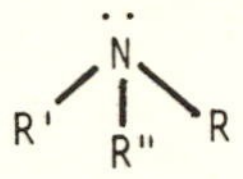

$$C_6H_5CH_2\overset{..}{N}C_2H_5$$
$$C_6H_5$$

$$\underline{1} \qquad \underline{2} \qquad \underline{3} \qquad \underline{4}$$

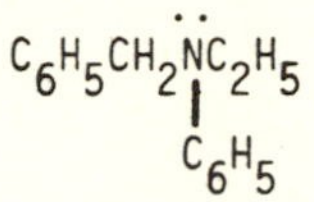

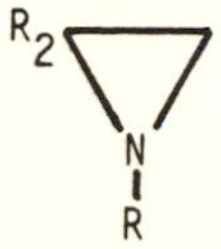

The discovery of deuterium D, the heavy isotope of hydrogen, by H. C. Urey in 1932,[40] and the subsequent availability of D_2O of high purity suggested a novel stereochemical problem. Would symmetrically placed deu-

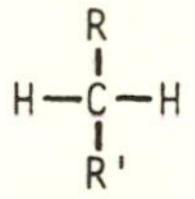

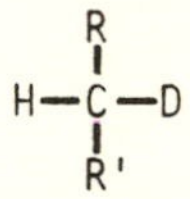

$$\overset{R}{\underset{R'}{H-C-H}} \qquad \overset{R}{\underset{R'}{H-C-D}}$$

terium and hydrogen introduce enough asymmetry so that optically active compounds of appreciable rotation could be obtained? The simplest case is shown here. Adams investigated this problem systematically at Illinois in the 1930s and no cases of optical activity were found. In one example, optically active ethylethynylcarbinol was reduced catalytically with deuterium; the saturated ethyl-ethyl-D_4-carbinol was inactive.[41] Likewise a repetition of the reported resolution of $C_6H_5CH(NH_2)C_6D_5$ gave only optically inactive products.[42] By 1953, however, several examples of optically active deuterium compounds, such as $C_6H_5CHDCH_3$, had been discovered.[43]

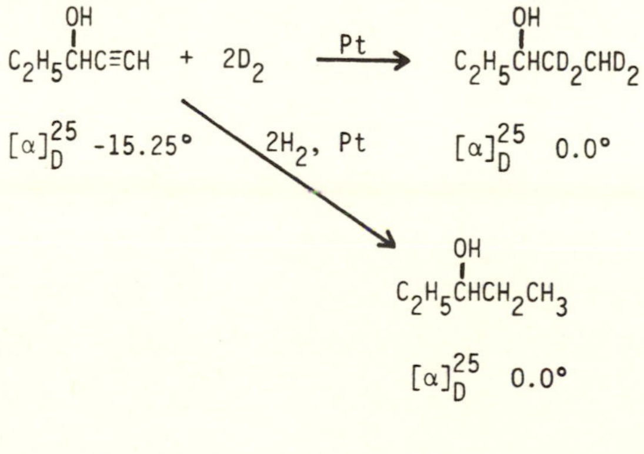

$$\underset{[\alpha]_D^{25} \ -15.25°}{\overset{OH}{C_2H_5\overset{|}{C}HC\equiv CH}} + 2D_2 \xrightarrow{Pt} \underset{[\alpha]_D^{25} \ 0.0°}{\overset{OH}{C_2H_5\overset{|}{C}HCD_2CHD_2}}$$

$$\xrightarrow{2H_2, \ Pt}$$

$$\underset{[\alpha]_D^{25} \ 0.0°}{\overset{OH}{C_2H_5\overset{|}{C}HCH_2CH_3}}$$

The use of stereoisomerism to establish the structure of mercury compounds by Marvel has already been mentioned. W. A. Noyes attempted to obtain diazo compounds in optically active forms to support his structure for these compounds, if the charges were stable as shown. Numerous attempts to obtain optically active forms were inconclusive,[44] and revised resonance

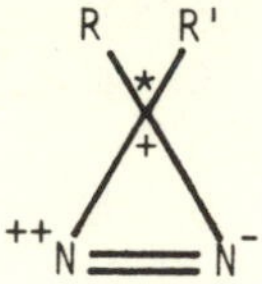

structures for diazo compounds indicated that the carbon was not asymmetric.[45] A successful use of optical activity to establish structure was Marvel's demonstration that azoxy compounds could be obtained optically active;[46] this ruled out the symmetrical ring structure, which would have a plane of symmetry.

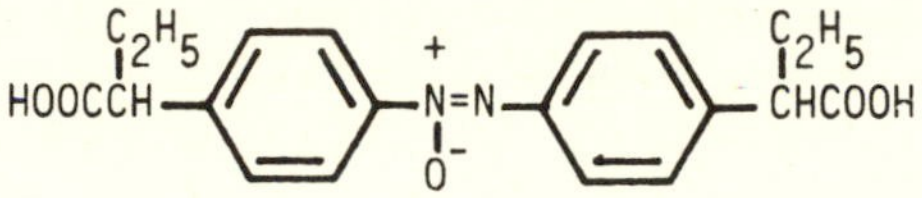

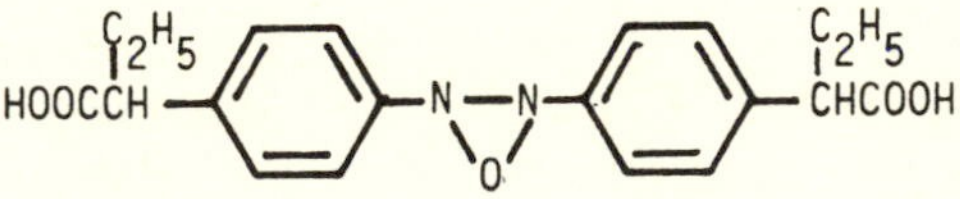

Another stereochemical problem, whose real significance lay more in the future, was the question of the preferential absorption of one form of a *dl* dye by wool, cotton, or silk. It might be thought that because these natural textile fibers are optically active with many asymmetric centers, they might adsorb one mirror image of a dye preferentially. Experiments failed to give definite support for this idea.[47] The general question, however, underlay the action of enzymes in forming optically active products and the design of reagents for asymmetric syntheses. V. Prelog's resolution of Troeger's base[48] by preferential adsorption on a lactose column was an example of preferential adsorption by an optically active adsorbent.[49] Many other examples are now known.

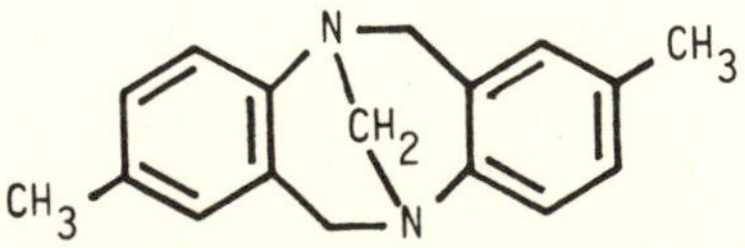

Troeger's base

The relation of configurations of natural products, such as sugars, amino acids, steroids, and alkaloids to each other and to a common reference compound is important in establishing biogenetic relationships and in structural work. Such relations are also useful in mechanistic studies and in correlating structures with sign, magnitude of rotation, and effects on the latter of changes in wavelength of light used to measure optical activity (the optical rotatory dispersion or ORD). This requires a comprehensive set of relationships between series of compounds of known configuration which are related to a standard compound of known or assigned configuration. Such configurationally related series, based on the known absolute configuration of tartaric acid, allow assignments of absolute configuration to all asymmetric centers in complex molecules.

Much of the key experimental work on which these webs of related configurations are based was done by P. A. Levene[50] at the Rockefeller Institute between 1925 and 1937. For example, Levene and H. L. Haller showed that

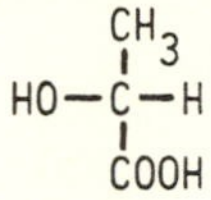

(-) Lactic acid

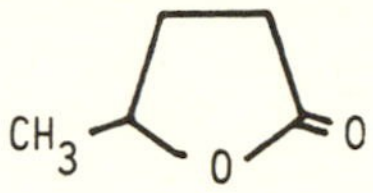

(+) Valerolactone

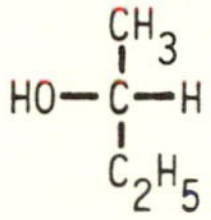

(-) 2-Butanol

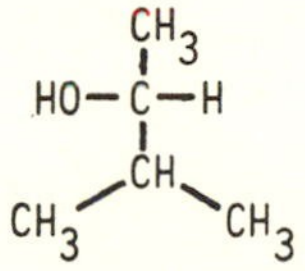

(-) 3-Methyl-2-butanol

(−)-lactic acid, (+)-valerolactone, (−) 2-butanol, and (−) 3-methyl-2-butanol all had the same configuration.[51,52] P. D. Bartlett in Levene's laboratory[53] related the configuration of α-hydroxyacids to those of alcohols below.

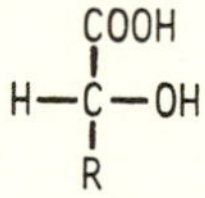

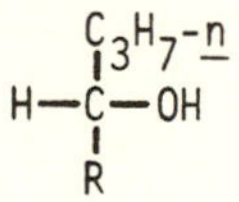

$$R = CH_3C(CH_3)_2 \quad \text{or} \quad (CH_3)_2CHCH_2$$

Levene's work was started originally to help elucidate the structures and configurations of sugars present in nucleic acids. The periodic acid oxidation of sugars by C. S. Hudson and others discussed later gave indispensable links in correlating configurations of sugars with other classes.[54]

In the decades following 1939, stereochemistry developed a comprehensive and satisfactory nomenclature for configurations and for symmetry properties of molecules; absolute and relative configurations were established for most classes of compounds; applications of rotatory dispersion measurements to configurations were developed; the stereochemistry of atoms other than carbon was unravelled; and the stereochemical course of many reactions was elucidated. In addition, as described above, nuclear magnetic resonance and X-ray crystallography became indispensable tools in studying conformation and configuration. The use of enzymes as stereospecific reagents, and the probing of enzymatic reactions by stereospecific deuterium labeling, proved to be subtle and highly illuminating techniques.

The classical stereochemical exposition of Adams, Shriner, and Marvel in Gilman's *Organic Chemistry* was superseded by books by E. L. Eliel and by Kurt Mislow,[55] and by several serial publications.[54] American chemists played their full share in the increase of stereochemical understanding.

LITERATURE CITED

1. H. R. Henze and C. M. Blair, JACS, **53**, 3042, 3077, (1931) and later papers; D. Perry, ibid., **54**, 2913 (1932), for saturated alcohols, saturated hydrocarbons, and ethylenic hydrocarbons; D. D. Coffman, C. M. Blair, and H. R. Henze, ibid., **55**, 252 (1933), for acetylenic hydrocarbons; refs. to earlier work in these papers.

2. G. L. Dorough, E. E. Reid et al., ibid., **63**, 3100 (1941).

3. G. E. K. Branch and T. L. Hill, JOC, **5**, 86 (1940); they do not refer to Henze's work. Special cases are discussed by J. K. Senior of Chicago, Ber., **60**, 73 (1927).

4. V. Meyer, Ber., **23**, 567 (1890); K. Auwers and V. Meyer, ibid., 2082; A. Eiloart, ACJ, **12**, 231 (1890) (from J. Wislicenus' laboratory at Leipzig); JPr, **43**, 124 (1891) (from Cornell University).

5. C. A. Bischoff, Ber., **23**, 623 (1890) and many accompanying papers, especially ibid., **24**, 1085 (1891). Examination of Bischoff's papers and other contemporary ones[4] shows that Bischoff's ideas were not as novel as claimed by G. V. Bykov, in *van't Hoff-LeBel Centennial*, O. B. Ramsay, Ed., Washington, 1975, p. 114.

6. H. Eyring, JACS, **54**, 3191 (1932).

7. G. B. Kistiakowsky et al., ibid., **57**, 65 (1935).

8. E. Teller and B. Topley, JCS, 876 (1935).

9. H. A. Smith and W. E. Vaughan, J. Chem. Phys., **3**, 341 (1935).

10. G. B. Kistiakowsky et al., JACS, **58**, 137 (1936).

11. J. D. Kemp and K. S. Pitzer, ibid., **59**, 276 (1937); J. Chem. Phys., **4**, 749 (1936); J. B. Howard, ibid., **5**, 451 (1937); K. S. Pitzer, ibid., 473; G. B. Kistiakowsky et al., ibid., **6**, 407 (1938).

12. G. H. Christie and J. Kenner, JCS, **121**, 614 (1922).

13. Reviews of the field: R. Adams and H. C. Yuan, Chem. Rev., **12**, 261 (1933); R. L. Shriner and R. Adams, in *Organic Chemistry*, H. Gilman, Ed., New York, 1938, p. 259-303.

14. X-ray: G. W. Clark and Lucy W. Pickett, JACS, **53**, 167 (1931); Ultraviolet absorption: M. T. O'Shaughnessy and W. H. Rodebush, ibid., **62**, 2906 (1940); M. Calvin, JOC, **4**, 256 (1939).

15. G. B. Kistiakowsky and W. R. Smith, JACS, **58**, 1043 (1936). Later important theoretical and experimental work was done by F. H. Westheimer at Chicago on this problem; Westheimer in *Steric Effects in Organic Chemistry*, M. S. Newmann, Ed., New York, 1956, pp. 543-555.

16. R. Adams and N. Kornblum, JACS, **63**, 188 (1941); and refs. therein.

17. W. A. Everhart, W. A. Hare and E. Mack, Jr., ibid., **55**, 4894 (1933).

18. A. Baeyer, Ber., **18**, 2277 (1885).

19. H. Sachse, ibid., **23**, 1363 (1890), C. A. Russell, *van't Hoff-LeBel Centennial*, pp. 159-178, discusses the development of conformational analysis.

20. E. Mohr, JPr, **98**, 315 (1918); **103**, 316 (1922).

21. H. G. Derx, Rec. Trav. Chim., **41**, 312 (1922) and later papers.

22. W. Hückel, Ann., **441**, 1 (1925) and later papers; W. A. Wightman, JCS, 1421 (1925) in Ingold's laboratory at Leeds discussed ring flips in the decalin series.

23. L. Ruzicka, A. Stoll and H. Schinz, Helv. Chim. Acta, **9**, 249 (1926) and many other papers.

24. A review by Carothers of W. Hückel's book of 1927 on the present state of the strain theory, W. H. Carothers, JACS, **50**, 2852 (1928), shows Carothers' broad interests in organic chemistry; he commented on the new vistas in stereochemistry opened up by the evidence for the nonplanar strainless rings in sugar, polysaccharide, and petroleum chemistry.

25. R. G. Dickinson and C. Bilicke, ibid., **50**, 764 (1928).

26. H. J. Lucas, *Organic Chemistry*, New York, 1935, p. 89.

27. R. F. Miller and R. Adams, JACS, **58**, 787 (1936).

28. W. N. Haworth, *The Constitution of Sugars*, London, 1929, p. 90; H. S. Isbell, Chem. Soc. Rev., **3**, 1 (1974).

29. D. H. Brauns, JACS, **51**, 1820 (1929).

30. A. H. White and S. O. Morgan, ibid., **57**, 2078 (1935).

31. D. H. R. Barton, JCS, 1027 (1953); Experientia, **6**, 316 (1950). For an interview with Barton, see P. Farago, JCE, **50**, 234 (1973).

32. O. Hassel, Tidsskr. Kjemi, **3**, 32 (1943) and later papers.

33. Detailed reviews: E. L. Eliel, N. A. Allinger, S. J. Angyal and G. A. Morrison, *Conformational Analysis*, New York, 1965, W. G. Dauben and K. S. Pitzer, in *Steric Influences in Organic Chemistry*, M. S. Newman, Ed., New York, 1956, 1-60.

34. E. P. Kohler, J. T. Walker and M. Tishler, JACS, **57**, 1743 (1935).

35. W. H. Mills and P. Maitland, Nature, **135**, 994 (1935).

36. R. L. Shriner, R. Adams, and C. S. Marvel, in Gilman. op. cit., pp. 328-338.

37. R. M. Badger, Phys. Rev., **35**, 1038 (1930).

38. T. D. Stewart and C. Allen, JACS, **54**, 4027 (1932).

39. R. Adams and T. L. Cairns, ibid., **61**, 2464 (1939); Cairns, ibid., **63**, 871 (1941). J. F. Kincaid and F. C. Henriques, Jr., ibid., **62**, 1474 (1940), calculated the energy barrier to vibration in ethylenimines to be thirty-eight kcal, which is ample to allow resolution at room temperature. They calculated for a tertiary amine 15 kcal, almost large enough to allow resolution at -80°. A. T. Bottini and J. D. Roberts, ibid., **78**, 5126 (1956) concluded from NMR measurements on N-ethyl-ethyleneimine that the barrier to inversion was much smaller than thirty-eight kcal, and hence resolution would only be possible at low temperature.

40. H. C. Urey, F. G. Brickwedde and G. M. Murphy. Phys. Rev., **39**, 864 (1932); ibid., **40**, 1 (1932).

41. F. C. McGrew and R. Adams, JACS, **59**, 1497 (1937).

42. R. Adams and D. S. Tarbell, ibid., **60**, 1260 (1938).

43. E. L. Eliel, ibid., **71**, 3970 (1949); E. R. Alexander and A. G. Pinkus, ibid., 1786; A. Streitweiser, ibid., **75**, 5014 (1953) and later papers; F. A. Loewus, F. H. Westheimer and B. Vennesland, ibid., **75**, 5018 (1953).

44. C. S. Marvel and W. A. Noyes, ibid., **42**, 2259 (1920); A. Weissberger, Ber., **66**, 559 (1933).

45. Structure of diazo compounds: N. V. Sidgwick, L. E. Sutton and W. Thomas, JCS, 406 (1933).

46. T. T. Chu and C. S. Marvel, JACS, 55, 2841 (1933).

47. A. W. Ingersoll and R. Adams, ibid., 44, 2930 (1922); C. W. Porter and H. K. Ihrig, ibid., 45, 1990 (1923); W. R. Brode and R. Adams, 48, 2193 (1926).

48. M. A. Spielman, ibid., 57, 583 (1935).

49. V. Prelog and P. Wieland, Helv. Chim. Acta, 27, 1127 (1944).

50. For Levene's career, see earlier essay. Summaries of his stereochemical work: P. A. Levene and A. Rothen, JOC, 1, 76 (1936); idem., in Gilman's Organic Chemistry, 1938, pp. 1779-1849.

51. P. A. Levene and H. L. Haller, JBC, 69, 165 (1926); and refs. therein.

52. Levene, A. Walti, and H. L. Haller, ibid., 71, 465 (1927); Levene and S. A. Harris, ibid., 113, 55 (1937). Possible a priori objections to the stereochemical validity of Levene's results were removed by W. E. Doering and R. W. Young, JACS, 74, 2997 (1952); K. B. Wiberg, ibid., 74, 3891 (1952).

53. P. D. Bartlett, M. Kuna, and P. A. Levene, JBC, 118, 503, 513 (1937).

54. Excellent accounts of correlation of configuration and of the value of Levene's contribution are by J. A. Mills and W. Klyne in Progess in Stereochemistry, W. Klyne, Ed., 1, 1954, p. 182; W. Klyne and P. M. Scopes, ibid., 4, 1969, p. 97; O. B. Ramsey, Stereochemistry, London, 1981.

55. E. L. Eliel, Stereochemistry of Carbon Compounds, New York, 1962; Kurt Mislow, Introduction to Stereochemistry, New York, 1965.

15
Natural Products, 1914-1930

An examination of the organic chemical literature on the structure of natural products for this period in the JACS, the JBC, and *Science* shows that, with a few notable exceptions, there was little in this country to compare with the best work abroad. The outstanding researches of Richard Willstätter on the coloring materials of flowers and on chlorophyll; of A. Windaus on the sterols and colchicine; of H. Wieland on the bile acids; of Hans Fischer on porphyrins; of L. Ruzicka on the large-ring natural products, civetone and muscone; of W. H. Perkin, Jr., and Robert Robinson on a variety of alkaloids, plant pigments, and other natural products had few counterparts in this country. Up to 1930 the JACS had a larger number of significant papers in physical organic chemistry than in natural product chemistry. This situation is neatly exemplified by the work of J. B. Conant, who published during the 1920s a series of original papers on oxidation-reduction potentials of quinones, irreversible oxidations, free-radical formation by one electron transfer, reactivity of organic halides with sodium iodide in acetone, mechanism of semicarbazone formation, and other fundamental problems in physical organic chemistry. He did publish during the 1920s some work on hemoglobin and its derivatives, mainly physical organic in character. In 1929 his first paper on chlorophyll derivatives[1] showed a shift in his interests toward natural products and the chemistry of living systems; his selection as president of Harvard in 1933, however, finished his chemical career and ended his contributions to the chlorophyll problem.

In many cases the distinction between the organic chemistry of natural products and biochemistry is somewhat arbitrary. Work on the structure or synthesis of a pure natural product obviously is organic chemistry, but the long processes of concentration and separation of vitamins and hormones from natural sources, and operations on ill-characterized materials of natural origin are not considered here as natural product chemistry, unless a pure compound has been isolated and studied. The fact that many of these products were mixtures was no reproach to the chemists or biochemists involved; it merely demonstrates the complexities of the problems, the limited methods of separation, and the uncertain criteria of purity available at the time.

In the 1930s, however, the output in natural product chemistry rose in this country, both in university and industrial laboratories (principally pharmaceutical firms), and by the outbreak of World War II American work was extensive and of a technical standard comparable to the best foreign work. Up to 1930 the American physical organic research compared favorably with most of that abroad. Germany, long prominent in synthesis and structure of natural products, was the last major center to develop good research in physical organic chemistry. With some significant exceptions, such as A. Hantzsch, Karl Ziegler, F. Arndt, and Hans Meerwein, physical organic chemistry in Germany played a minor role on the international scene until after World War II.

The earlier research on natural products in this country has been discussed in previous essays. Although there were some excellent studies prior to 1914, the American university system of graduate training and research in chemistry was not designed to cope with difficult, long-term problems requiring a high degree of technical skill. The research laboratories of the pharmaceutical firms were developing in 1914, but their outstanding chemical achievements were to become imposing only in the 1930s. The biochemistry departments, normally located in medical schools but in some notable cases as at Illinois, part of the chemistry department, carried out some excellent work on isolation and characterization of natural products. In general, the medical school departments of biochemistry had only one professor, on the European model, and subordinate staff members investigated the problems of the professor. These departments thus had more continuity of experienced assistants and could attack long range problems more effectively than those divisions organized for graduate training. As will be seen, the Rockefeller Institute for Medical Research, with its continuity of senior staff and considerable stability of junior staff members (mainly postdoctorates), was able to compete with the monolithic European departments in probing difficult problems for many years with highly skilled personnel. Until about 1930, the Rockefeller Institute had few rivals in this country for the high caliber of its research on natural products, the studies of P. A. Levene, Walter A. Jacobs, and their respective groups being outstanding. Most of the Rockefeller results were published in the JBC rather than in the JACS.

Probably the most significant, single American accomplishment in natural products between the two World Wars was the first isolation in 1926 of an enzyme, urease, as a crystalline solid by James B. Sumner of Cornell.[2] Enzymes at that time were considered to be colloidal materials of indefinite composition carrying an active group. Sumner's discovery and his steadfast contention that enzymes were definite chemical compounds thus clashed with orthodox biochemical thought of the day, championed particularly by the great German chemist Richard Willstätter. The latter accepted Sumner's views with serious reservations[3] in 1933, and, although Sumner, J. H. Northrop, and H. C. Sherman soon obtained additional enzymes in crystalline form, biochemists as a group did not recognize the significance of Sumner's pioneer discovery for many years. Sumner received the Nobel Prize in 1946 with Northrop and W. M. Stanley[4] for work on crystalline enzymes and viruses, but he was not elected to the National Academy of Sciences until 1948. Sumner's discovery led directly to the study of the chemical structure of pure enzymes and to the isolation of crystalline tobacco mosaic virus in

1935 by Stanley.[5] Much of modern biochemistry is a development of Sumner's discovery.

James B. Sumner (1887-1955, B.S. 1910, A.M. 1913, Ph.D. 1914, with Otto Folin, all at Harvard) taught at Cornell from 1914 to 1955, first in the Medical College in New York and later at Ithaca. In addition to urease, he isolated several other enzymes in crystalline form, including catalase and peroxidase. John H. Northrop (1891- , B.S., M.A., Ph.D., all at Columbia) worked at the Rockefeller Institute for Medical Research from 1916 to 1961, developing methods for isolating and studying crystalline enzymes and crystalline viruses. Wendell M. Stanley (1904-1971, B.S. 1926, Earlham; Ph.D. 1929, Illinois with Roger Adams) did research at Rockefeller from 1931 to 1947, then directed the virus laboratory at Berkeley until he retired in 1969. He showed that viruses were nucleoproteins and was an early proponent of the idea that viruses caused cancer.

Henry C. Sherman (1875-1955, B. S. 1893, Maryland; A.M. 1896, Ph.D. 1897, both at Columbia) taught at Columbia from 1899 to 1950. Starting as an analytical chemist, he investigated foods, nutrition, vitamins, and enzymes (particularly those acting on carbohydrates) and published many books and papers on nutrition.

The chemistry of natural products of biochemical importance was the life work of P. A. Levene (1869-1940) of the Rockefeller Institute. Levene,[6] born in Russia, studied in the Classical Gymnasium in St. Petersburg and received medical training at the Imperial Military Medical Academy, being one of the few Jews admitted. Among his teachers were Alexander Borodin, the composer (and organic chemist by profession), and Ivan Pavlov, the physiologist and later the discoverer of the conditioned reflex. Levene and his family came to New York in 1891 to escape Russian anti-Semitism. During the next fifteen years he practiced medicine, worked in laboratories of physiological chemistry both in New York and Germany, and recovered from an attack of tuberculosis in the Adirondacks. He spent one summer working with Emil Fischer in Berlin and in 1905 he joined the Rockefeller Institute, where he remained the rest of his life.

Some of Levene's work later played an important role in relating absolute configurations of natural products to simple compounds of known absolute configurations, such as tartaric acid[6]. However, most of his research dealt with the chemistry of the nucleic acids and of their constituents: sugars, particularly sugar phosphates; and the organic bases purines and pyrimidines present in nucleic acids. Levene was versatile and prolific, although some of his apparently great productivity was due to his propensity to publish a large number of very short papers, rather than a smaller number of more comprehensive accounts.

Levene and Jacobs[7] had identified the five-carbon sugar (pentose) from yeast nucleic acid as D-ribose **1a** in 1909. Levene later isolated and identified[7]

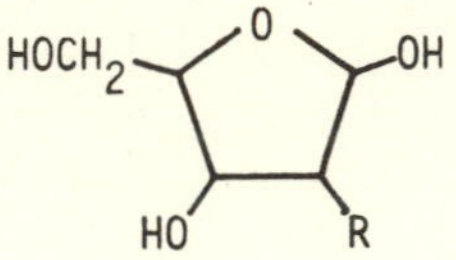

1a, R = OH
1b, R = H

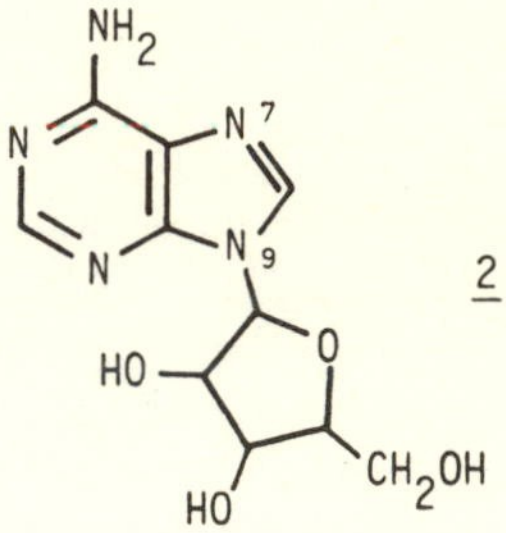

the sugar present in thymonucleic acid (DNA, deoxyribonucleic acid from the thymus gland) as D-2-deoxyribose 1b. He and Tipson[8] showed that adenosine, present in ribonucleic acid (RNA) and in numerous important coenzymes such as adenosine triphosphate (ATP), was adenine combined with D-ribose, the sugar being in a five-membered ring (furanose) form 2. At this time, Levene believed that the sugar was attached to the 7-nitrogen of adenine, instead of the 9-nitrogen, as was shown later to be correct. Levene had early recognized[9] that nucleic acids were composed of organic bases (purines or pyrimidines) combined with pentose sugars, as in 2, and that these were joined by phosphate ester linkages from one sugar unit to another, these combinations (sugar, phosphate, and base) being "nucleotide" units. The true molecular size of the nucleic acids was not realized until the 1940s, and the "tetranucleotides" of Levene and his contemporaries were thought to represent the true molecular complexity of the nucleic acids. A "dinucleotide" from ribonucleic acid (RNA) can be represented by 3, in which "B" stands for a purine or pyrimidine base attached to the pentose sugar, and the phosphate ester groups connect the 3'- and 5'- positions of the sugar units.

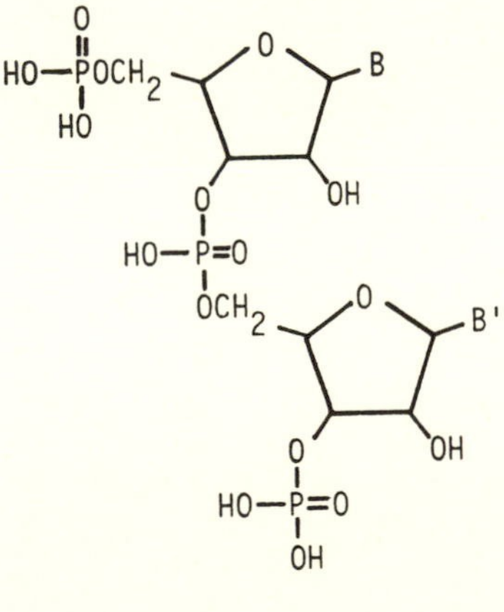

Levene made a start on the synthesis of nucleosides from purines and a sugar derivative,[10] worked on important amino sugars and their derivatives,[11] and made useful contributions to the chemistry of brain constituents[12] and to protein chemistry. His work as a whole is of fundamental significance to organic and biochemistry.

Brilliant investigations of the structures of natural products were also carried out by Walter A. Jacobs (1883-1967) at the Rockefeller Institute.[13] After his graduation from Columbia, Jacobs took his Ph.D. at Berlin in 1907 with Emil Fischer, the acknowledged master of organic chemistry. By this time, few American students traveled abroad for Ph.D.s in organic chemistry, although some still went, mainly to Germany, for postdoctoral experience. Jacobs' subsequent career (1907-1959) was all spent at Rockefeller, where he worked for some years with Levene. Later he worked on various mercury and arsenic compounds as chemotherapeutic agents[14] against diseases caused by parasites, such as trypanosomiasis ("African sleeping sickness"), but the medical usefulness of his compounds was limited. The wide-spread interest in mercury and arsenic derivatives after 1914 in this country was caused by Paul Ehrlich's discovery of the effectiveness of Salvarsan (arsphenamine), an arsenic compound, in syphilis. The blockade cut off supplies of this drug from Germany during World War I, and the American literature from 1915 into the 1920s contains many papers on organic compounds containing arsenic, mercury, and antimony.[15]

Jacobs, with Michael Heidelberger (later a distinguished immunochemist at Rockefeller and at Columbia), began studies in 1922 on the family of cardiac drugs,[16] derived in part from African arrow poisons and from the common foxglove plant (*Digitalis*). Both are valuable agents in treating heart disorders, having a common characteristic reaction on the heart. This complicated and difficult class of compounds all occurred in nature as glysosides (connected to one or more sugar units), and Jacobs' investigations focused on the sugar-free molecules (the aglycones). Jacobs' work, like Levene's, was done by himself personally and by postdoctorates, including the key men E. L. Gustus and R. C. Elderfield. At this time, the Rockefeller workers had no teaching responsibilities, and the Jacobs group could compete successfully with the German university laboratories, particularly with Windaus and R. Tschesche at Göttingen, where pursuit of closely related compounds progressed. The elucidation of the structures of these drugs required nearly fifteen years. In the closing stages the relationship of these compounds to the sterol and bile acid structures was established, but during most of the inquiries, each product obtained by degradation had to be purified by crystalization, its empirical formula determined by elemental analysis, and its relation to its precursors deduced from its chemical properties and further transformations. No spectroscopic methods were used until the work was nearly concluded, and almost none of the present techniques produced by the Instrumental Revolution was then available.[17]

In 1934 Jacobs and Elderfield and Tschesche in Göttingen showed the carbon skeleton of the cardiac aglycones and the steroids to be the same.[18] Jacobs had proposed in 1933 a partial structure for strophanthidin, the most complex of the aglycones, to which he assigned an incorrect ring system, although the correct functional groups were given. The correct structures for strophanthidin and digitoxigenin (one of the aglycones from *Digitalis*) deciphered by Jacobs and by Elderfield[19] are 4 and 5.

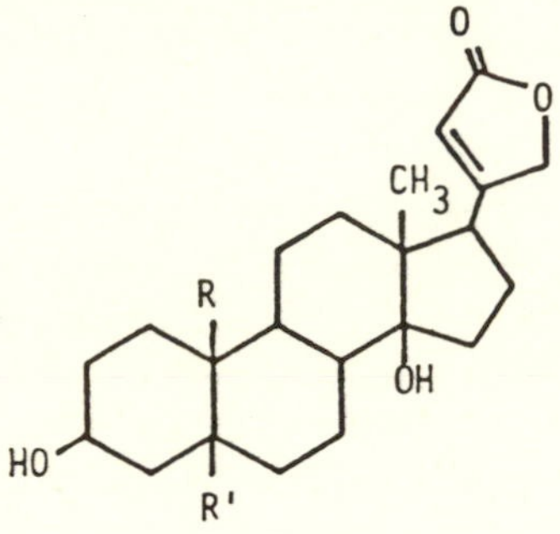

<u>4</u>   Strophanthidin, R = CHO; R' = OH

<u>5</u>   Digitoxigenin, R = CH₃; R' = H

Robert C. Elderfield (1904-1979, A.B. 1926, Williams; Ph.D. 1930, MIT, with Tenney B. Davis) was a research chemist at Rockefeller from 1930 to 1936, then taught at Columbia until 1952 and at Michigan until 1970. His work included syntheses in the field of heterocyclic compounds (on which he edited a valuable nine-volume series of monographs), chemotherapy, cardiac drugs, and alkaloids.

Jacobs' fundamental work with Lyman C. Craig during the 1930s on the important ergot alkaloids, which contain lysergic acid (the parent of lysergic acid diethylamide or LSD), will be discussed later.

At the university level, the program at Yale initiated by H. L. Wheeler and Treat B. Johnson around 1900 was continued after Wheeler's death in 1914 by Johnson. Their investigations covered purines and pyrimidines, known to be components of nucleic acids and (later) vitamins and coenzymes. The Yale chemists published nearly 200 papers on purines and pyrimidines, as well as many studies on amino acids and on other heterocyclic ring systems. The Johnson group, although its work was not as clearly focused as that of Levene, continued to make useful contributions to synthetic methods and to the fundamental chemistry of these compounds. An example is an improved synthesis of orotic acid **6**, a pyrimidine derivative which became of great consequence in biochemistry and cancer chemotherapy.[20] Johnson's greatest contribution was a method for synthesis of a uracil glycoside **7** (a "nucleoside" of a type that occurs in nucleic acids). This valuable general procedure pointed the way to other methods of preparing this important class of compounds in later years.[21]

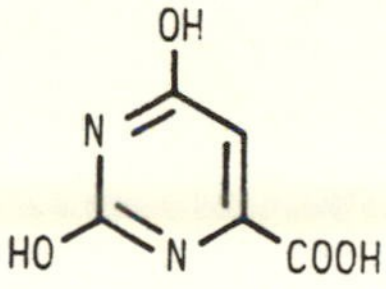

6, Orotic acid 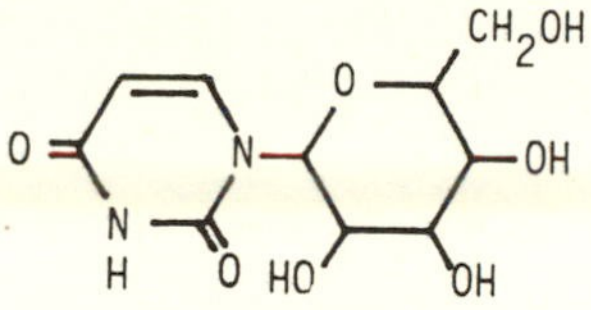   **7**, Uracil nucleoside

Roger Adams' best known work on natural products during the 1920s was the structural determination and synthesis of chaulmoogric and hydnocarpic acids **8**, constituents of chaulmoogra oil used at that time in treating leprosy.[22] A large number of homologs and analogs was produced in Adams' laboratory. Adams also worked on syntheses[23] for various naturally occurring anthraquinones such as morindone **9**; these compounds are similiar in structure to the natural dye alizarin, which had been synthesized by the German chemists Graebe and Lieberman in 1869.

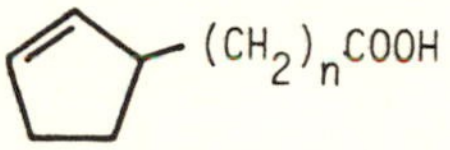

8a, Chaulmoogric acid, n = 12   **9**, Morindone
8b, Hydnocarpic acid, n = 10

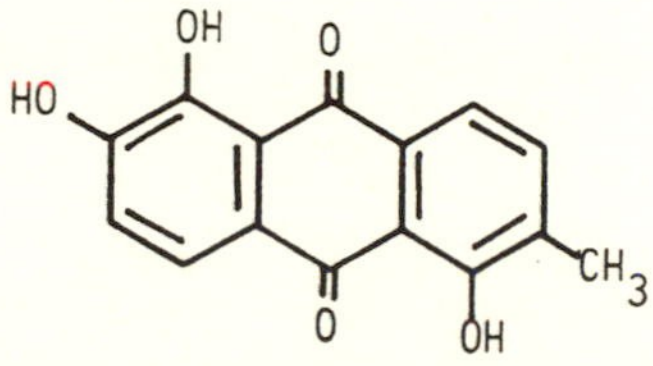

L. F. Fieser at Bryn Mawr prepared another naturally occurring quinone, lapachol **10**, occurring in various tropical woods.[24] The antimalarial activity of lapachol analogs is noted in an earlier essay.

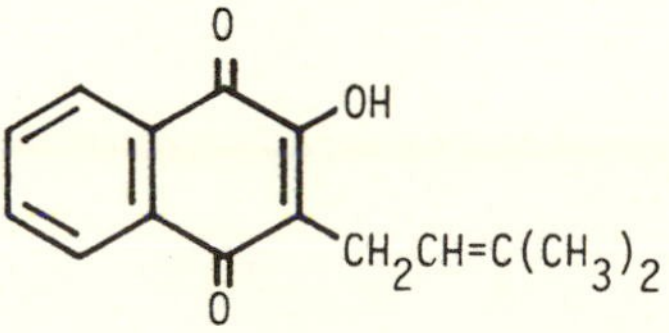

10, Lapachol

Louis F. Fieser (1899-1978, A.B. 1920, Williams; Ph.D. 1924, Harvard with Conant) taught at Bryn Mawr for several years and at Harvard from 1930 to 1968. He and his wife, Mary, with a large research group, did research on quinones, aromatic compounds, carcinogenic compounds, and steroids. They were coauthors of valuable monographs on steroids, phenanthrene derivatives, several very successful textbooks, and a series of volumes on organic chemical reagents.

A number of chemists in government laboratories, mainly in the Department of Agriculture, published results on the structures of complex natural products, particularly those having insecticidal or other marked biological activity. Among these investigators were E. K. Nelson, E. P. Clark, F. B. La Forge, and H. L. Haller; some had been colleagues of P. A. Levene. The determination of the structure of rotenone **11**, a potent insecticide in derris root from South America, was successfully accomplished.[25]

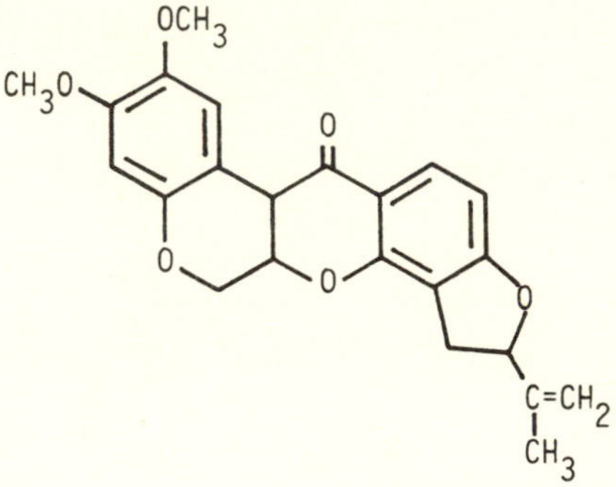

11, Rotenone

E. C. Kendall[26] at the Mayo Clinic in Rochester, Minnesota, isolated the active principle of the thyroid gland, thyroxin, and although Kendall suggested a completely erroneous structure, the correct structure was established several years later in England by Harington as **12**.[27] About three decades in the future, Kendall was to receive world-wide acclaim and the Nobel Prize for research on steroids of the adrenal cortex and his share in the discovery of the anti-inflammatory properties of cortisone.

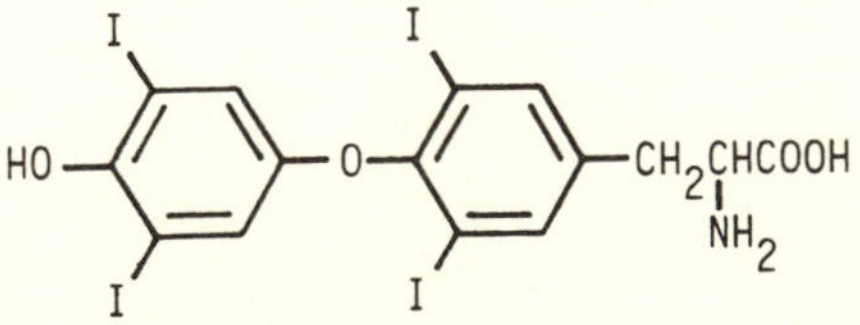

12, Thyroxin

Claude S. Hudson, mentioned in an earlier essay, continued his inquiries in sugar chemistry from 1914 to 1930 in several government laboratories with numerous collaborators. Hudson extended his work on quantitative correlation of optical rotation of sugars with their configuration by the "Hudson isorotation rules". These rules were shown by the English chemist, W. N. Haworth, to break down in some cases.[28] Hudson, however, carried on a variety of isolation and structural studies on carbohydrates, trained some excellent collaborators in the field, and must be regarded as one of the international masters of sugar chemistry.[29] His most elegant and fundamental work was to come after 1930, with his application of periodic acid as a specific oxidizing agent in sugar chemistry.

Melville L. Wolfrom (1900-1969, A.B. 1924, Ohio State; Ph.D. 1927, Northwestern with W. Lee Lewis) worked with Hudson and then with Levene. He taught at Ohio State from 1929, training nearly one hundred Ph.D.s and many post-doctorates.[30] Wolfrom in 1929 reported a derivative of the free aldehyde form of glucose **13**, a notable advance, and a problem which he developed in detail later.[31] Wolfrom later did important work on polysaccharides, particularly on cellulose, starch, and the naturally occurring agent heparin, used medically to prevent blood clotting.

$$
\begin{array}{l}
HC{=}O \\
| \\
(CHOCOCH_3)_4 \\
| \\
CH_2OCOCH_3
\end{array}
$$

13, Aldehyde form of D-glucose as the pentaacetate

Princeton University was the last of the major northeastern universities to develop graduate training in organic chemistry. Eugene Pacsu (1891-   ), who came from Hungary to Hudson's laboratory and then taught at Princeton from 1930 to 1960, built up an active research program in sugar chemistry.[32] Karl Paul Link at Wisconsin began in carbohydrate chemistry[33] but later branched out to elucidate the structure of the important, naturally occurring anticoagulant dicoumarol. Hudson, however, remained the dean of American sugar chemists.

In 1931 Lyndon F. Small reported the first of a series of papers from the University of Virginia in which he studied in detail the transformation products of morphine and other opium alkaloids.[34] The project at Virginia was supported at first by funds from the Bureau of Social Hygiene, Inc., under the aegis of the Committee on Drug Addiction of the National Research Council. It was directed toward finding derivatives of the opium alkaloids which would retain the analgesic activity of morphine without the addicting properties. Included in the project were pharmacologists, such as Dr. Nathan Eddy of Michigan, and representatives of other medical disciplines. This represents a notable early example of a concerted attack on a serious medical and social problem, drug addiction. Many years later after Small's death it appeared that analgesic and addicting properties could be separated by changes in the morphine stucture.[35] Although Small and his group did not reach their goal, they laid the foundations for significant advances in analgesics.

L. F. Small (1897-1957, B.S. 1920, Dartmouth; Ph.D. 1926, Harvard with Conant) worked with Heinrich Wieland in Munich for two years and then taught at Virginia from 1928 to 1939. He joined the U. S. Public Health Service, the forerunner of the National Institutes of Health, where he did research until his death. He was the first editor of JOC.[36] Small's collaborator on much of the Virginia research and on the authoritative monograph on the opium alkaloids[37] was Robert C. Lutz (1900- , B.A. 1921, Ph.D. 1925, Harvard, also with J. B. Conant) who taught at Virginia from 1928 to 1970. He made extensive studies on conjugated systems.

During this period many chemists and biochemists continued the successful development of the chemistry of amino acids and polypeptides. H. D. Dakin and R. West discovered the useful new reaction of amino acids shown.[38]

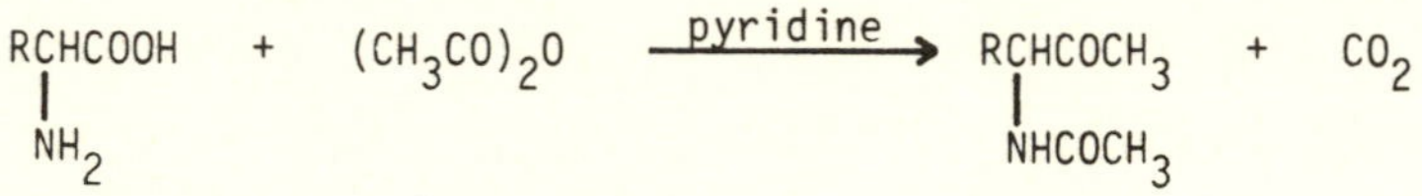

$$\underset{\underset{NH_2}{|}}{RCHCOOH} + (CH_3CO)_2O \xrightarrow{\text{pyridine}} \underset{\underset{NHCOCH_3}{|}}{RCHCOCH_3} + CO_2$$

Dakin also studied further the racemization of hydantoins and polypeptides by base. The essential amino acid methionine **14** was isolated from the hydrolysis of casein by J. H. Mueller of Columbia,[39] and Marvel at Illinois developed a convenient synthesis.[40]

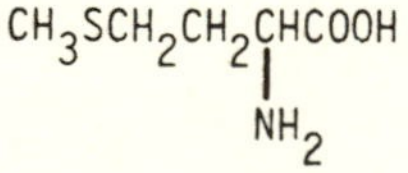

$$\underset{\underset{NH_2}{|}}{CH_3SCH_2CH_2CHCOOH} \qquad \underset{\underset{NH_2}{|}}{HSCH_2CHCOOH} \qquad \underset{\underset{NH_2}{|} \quad \underset{NH}{|}}{HOOCCHCH_2SSCH_2CHCOOH}$$

<u>14</u>, Methionine 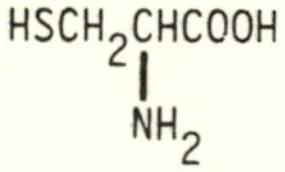   <u>15</u>, Cysteine 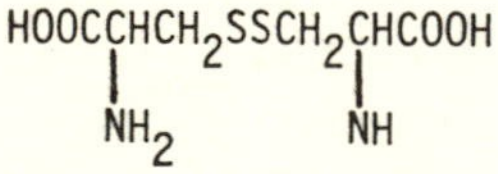   <u>16</u>, Cystine

The chemistry of another important sulfur-containing amino acid, cysteine **15**, and its disulfide, cystine **16**, was explored in detail in numerous papers in the JBC in the 1920s. Cysteine and cystine are essential amino acids occurring in many polypeptides and proteins. Many workers investigated the action of base on cysteine and its derivatives; the usual result is the elimination of sulfur from the molecule.[41]

Vincent du Vigneaud (1901-1978) became the master of the chemistry of organic sulfur compounds of biochemical significance. The first Ph.D. from the University of Rochester Medical School, he worked with John R. Murlin on the sulfur components (later found to be cystine) of the newly discovered antidiabetic hormone insulin. Du Vigneaud studied this problem further with

J. J. Abel of the Johns Hopkins Medical School. He then spent several years at Illinois, where he discovered the valuable method of protecting SH groups by forming the $SCH_2C_6H_5$ derivative (the S-benzyl compounds). The benzyl group was removed by reductive cleavage with sodium in liquid ammonia.[42]. This procedure became so important for later work that it is illustrated here. Du Vigneaud was professor of biochemistry at George Washington University and later at Cornell Medical College. His outstanding work on a variety of problems to be mentioned later brought him the Nobel Prize in 1955.[43]

$$RSSR \text{ or } RSH \xrightarrow[\text{liq. } NH_3]{Na} RSNa \xrightarrow{C_6H_5CH_2Cl} RSCH_2C_6H_5$$

$$\xrightarrow[\text{liq. } NH_3]{Na} RSNa + CH_3C_6H_5$$

The structure of the important tripeptide glutathione, first isolated by Sir F. G. Hopkins at Cambridge University, was established by several American groups simultaneously[44] and was synthesized elegantly by du Vigneaud, using his sodium-liquid ammonia technique.[45]

In a study of fundamental significance, J. T. Edsall at the Harvard Medical School showed, by measuring the Raman spectra, that amino acids in neutral water solution existed as dipolar (doubly charged) ions, not as the neutral molecules.[46]

$$\underset{NH_2}{RCHCOOH} \rightleftharpoons \underset{^+NH_3}{RCHCOO^-} \qquad \text{dipolar ion}$$

Although we have not tried to cover the papers on concentration and purification of hormones and vitamins, it is worth pointing out here that in 1928 Oliver Kamm at Parke Davis reported the separation and concentration of two active principles from the posterior lobe of the pituitary gland. One, named oxytocin, caused contraction of smooth muscle, including the uterus, and the other, vasopressin, as its name suggests, increased blood pressure.[47] The complete purification and structural elucidation of these polypeptide hormones was far beyond the level of knowledge and technique of 1928. However, they were studied by du Vigneaud immediately before and after World War II with resounding success. These researches were as important a milestone in the development of the organic chemistry of life processes as Sumner's discovery of the first crystalline enzyme in 1926.

LITERATURE CITED

1. J. B. Conant and J. F. Hyde, JACS, 51, 3668 (1929), Chlorophyll(I); last paper with Conant's name, Chlorophyll(XIV), with B. F. Chow and Emma M. Dietz, ibid., 56, 2185 (1934). On Conant, see P. D. Bartlett, *Biog. Mem. Nat. Acad. Scis.*, 54, 91 (1983).

2. J. B. Sumner, JBC, 69, 435 (1926); J. B. Sumner and D. B. Hand, ibid., 76, 149 (1927). J. J. Abel, Proc. Nat. Acad. Sci., 12, 132 (1926), isolated the hormone insulin in crystalline form, but it is not an enzyme and is not a large enough molecule to be considered a protein.

3. Richard Willstätter, *Aus Meinem Leben*, Weinheim, 1949, p. 363.

4. See biography of Sumner by L. A. Maynard, *Biog. Mem. Nat. Acad. Scis.*, 31, 376 (1958).

5. W. M. Stanley, Science, 81, 644 (1935) and later papers; J. H. Northrop, *Crystalline Enzymes*, New York, 1939; H. C. Sherman, Mary L. Caldwell and L. E. Booher, Science, 74, 37 (1931).

6. D. D. Van Slyke and W. A. Jacobs, *Biog. Mem. Nat. Acad. Scis.*, 23, 75 (1945).

7. P. A. Levene and W. A. Jacobs, Ber., 42, 3247 (1909); P. A. Levene, Mikeska, and T. Mori, JBC, 85, 785 (1929). With L. W. Bass, Levene summarized the current knowledge in *Nucleic Acids*, New York, 1931.

8. P. A. Levene and R. S. Tipson, JBC, 94, 809 (1931).

9. P. A. Levene and Ida P. Rolf, ibid., 40, 415 (1919) and refs. therein.

10. P. A. Levene and H. Sobotka, ibid., 65, 463 (1925).

11. P. A. Levene and F. B. Laforge, ibid., 21, 345, 351 (1915).

12. P. A. Levene, ibid., 24, 69 (1916). These references are merely illustrative of Levene's great output.

13. Biography is by R. C. Elderfield, *Biog. Mem. Nat. Acad. Scis.*, 51, 246 (1980).

14. W. A. Jacobs and M. Heidelberger, JBC, 20, 513 (1915), and many later papers.

15. For reviews, see G. W. Raiziss and J. L. Gavron, *Organic Arsenical Compounds*, New York, 1923; F. C. Whitmore, *Organic Compounds of Mercury*, New York, 1921; W. G. Christiansen *Organic Derivatives of Antimony*, New York, 1925.

16. W. A. Jacobs and M. Heidelberger, JBC, 54, 253 (1922), the first of several score papers on the cardiac aglycones.

17. Examination of the original papers in the JBC is instructive; overall summaries are given by R. C. Elderfield, Chem. Rev., 17, 187 (1935); L. F. Fieser, *Chemistry of Natural Products Related to Phenanthrene*, 1st ed., New York, 1936; L. F. Fieser and Mary Fieser, *Steroids*, New York, 1959.

18. W. A. Jacobs and R. C. Elderfield, JBC, 108, 497 (1935); Science, 80, 434 (1934); R. Tschesche, and H. Knick, Z. Physiol. Chem., 222, 58 (1933).

19. W. D. Paist, E. R. Blout, F. C. Uhle and R. C. Elderfield, JOC, 6, 273 (1941).

20. T. B. Johnson and E. F. Schroeder, JACS, 53, 1989 (1931).

21. G. E. Hilbert and T. B. Johnson, ibid., 52, 4489 (1930); paper CXVII on pyrimidines.

22. R. L. Shriner and R. Adams, ibid., 47, 2727 (1925); C. R. Noller and R. Adams, ibid., 48, 1080 (1926) and later papers; Adams published over 20 papers in all on this problem.

23. R. A. Jacobson and R. Adams, ibid., 47, 283, 2011 (1925).

24. L. F. Fieser, JACS., 49,, 857 (1927).

25. L. B. La Forge and H. L. Haller, ibid., 54, 810 (1932).

26. E. C. Kendall and A. E. Ostertag, JBC, 40, 265 (1919) and earlier papers.

27. C. H. Harington, Biochem. J., 20, 300 (1926).

28. W. N. Haworth, JACS, 52, 4168 (1930).

29. Hudson's work, published mainly in JACS, is conveniently available in *Claude S. Hudson, Collected Papers*, edited by R. M. Hann and N. K. Richtmyer, 2 vol., New York, 1946.

Hudson's isorotation rules are summarized by him in Sci. Papers. Nat. Bureau of Standards, **21**, 241 (1926).

30. Biography by D. Horton and W. Z. Hassid, *Biog. Mem. Nat. Acad. Scis.*, **47**, 487 (1975).
31. M. L. Wolfrom, JACS, **51**, 2188 (1929).
32. E. Pacsu, ibid., **53**, 3099 (1931) and later papers from Princeton.
33. K. P. Link, ibid., **52**, 2091 (1930).
34. L. F. Small and F. L. Cohen, ibid., **53**, 2214, 2227 (1931); L. F. Small and R. E. Lutz, ibid., **54**, 4715 (1932).
35. Marshall Gates, Scientific American, November 1966, p. 131.
36. For Small's career, see Erich Mosettig, *Biog. Mem. Nat. Acad. Scis.*, **33**, 397 (1959).
37. L. F. Small and R. E. Lutz, *Chemistry of the Opium Alkaloids*, Washington, 1932.
38. H. D. Dakin and R. West, JBC, **78**, 91, 745, 757 (1927). For Dakin's career, see JCS, 3319 (1952); further on Dakin-West reaction, G. H. Cleland and C. Niemann, JACS, **71**, 841 (1949).
39. J. H. Mueller, JBC, **56**, 157 (1923).
40. W. Windus and C. S. Marvel, JACS, **52**, 2575 (1930).
41. A more detailed review of this problem is given by D. S. Tarbell and D. P. Harnish, Chem. Rev., **49**, 1 (1951).
42. V. du Vigneaud, L. F. Audrieth and H. S. Loring, JACS, **52**, 4500 (1930); R. H. Sifferd and du Vigneaud, JBC, **108**, 753 (1935).
43. Interview with du Vigneaud by R. A. Plane, J. Chem. Ed., **53**, 8 (1976); V. du Vigneaud, *A Trail of Research* , Ithaca, 1952.
44. E. C. Kendall et al., JBC, **84**, 657 (1929); B. H. Nicolet, ibid., **88**, 389 (1930).
45. V. du Vigneaud and G. L. Miller, ibid., **116**, 469 (1936).
46. J. T. Edsall, J. Chem. Phys., **4**, 1 (1936); **5**, 225, 508 (1937).
47. O. Kamm et al., JACS, **50**, 573 (1928).

WALTER A. JACOBS

16
Work on the Structure of Natural Products in the United States, 1930-1940

As the preceding essay shows, there was excellent work carried out in this country during the 1914-1930 period, some of which was of international significance. Nevertheless, the 1930-1940 decade differed from the earlier period both qualitatively and quantitatively; American research which was competitive and even superior to much overseas study increased in amount and proportion of the whole.

The 1930-1940 decade brought a rich harvest of important results, both here and abroad, in the isolation and structural determination of vitamins (A, B_1, B_2, B_6, C, D, E, K), sex hormones (male and female), adrenal cortex compounds, alkaloids of many classes, particularly the ergot series, carbohydrates of both low and high molecular weight, and miscellaneous natural products such as gossypol (from cottonseed meal). In many of these cases, the triumphs of the 1930s were the culmination of long series of isolation experiments, some lasting for decades. In many cases also, isolation and structural determinations were carried out simultaneously here and abroad, and it is scarcely important to decide in each case who was "first".

The isolation and structure proof of the estrogenic hormone estrone by E. A. Doisy and his collaborators at St. Louis University and at Washington University in St. Louis went back to experiments of 1924.[1] The Doisy group reported[2] the isolation of crystalline estrone, which they called "theelin", in 1929. It was shown to be identical with material isolated in Germany by Adolph Butenandt (1903-). The structure of estrone was worked out by Butenandt and others,[3] and Doisy later isolated the more potent estrus-producing hormone, estradiol, from 3000 lb. of hog ovaries.[4]

Edward A. Doisy (1893- , A.B. 1914, Illinois; Ph.D. 1920, Harvard) taught at Washington and St. Louis University Medical Schools and made fundamental contributions in the vitamin K series discussed below.

Progesterone, the hormone that controls pregnancy, was isolated simultaneously by the physiologists Willard M. Allen (1904-) at Rochester, Frederick L. Hisaw (1891-1972) at Harvard, and K. H. Slotta at Breslau.[5] Slotta proposed the correct structure.[6] Progesterone is the hormone whose production is prevented by the contraceptive pill. The isolation, structure proof, and partial synthesis of the other important steroid sex hormones, equilenin, androsterone, and testosterone were largely done abroad.

The *total* synthesis of the sex hormone equilenin, by W. E. Bachmann, Wayne Cole, and A. L. Wilds at Ann Arbor, was a landmark in the synthesis of complex natural products.[7] Equilenin contains two asymmetric carbons (marked by asterisks) and hence it can exist in four stereoisomeric forms. There had been much unsuccessful work by several groups overseas on this problem. Bachmann's beautiful synthesis, in which he isolated all four stereoisomers, and in which all steps were developed to give high yields, instantly became a classic. In this country it encouraged the total synthesis of more complex steroids in later years and of many other natural products with many asymmetric centers. Bachmann played a key role in chemical research during World War II, and, although he died at a tragically early age, his synthetic accomplishments pointed the way to a major synthetic development in natural products, not only in America but also in organic chemistry worldwide.[8]

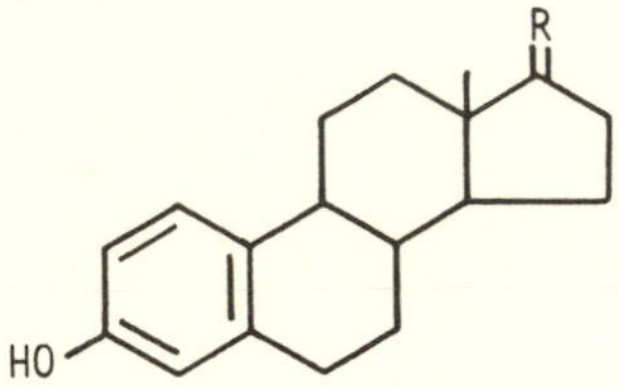

Estrone (R = O)
Estradiol (R = H, OH)

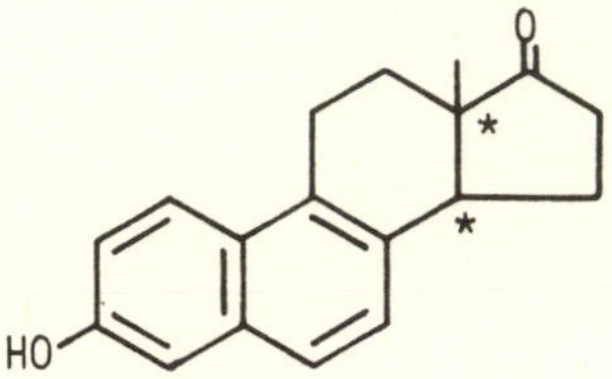

Equilenin

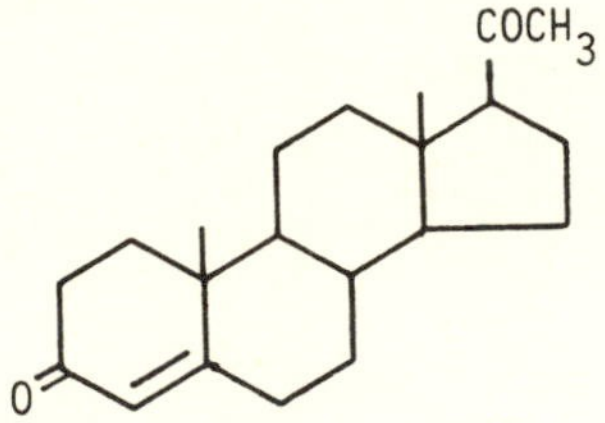

Progesterone

The existence of "accessory food factors" or "vitamines", as they were first called by Casimir Funk[9] in 1912, was recognized by 1900. Feeding experiments showed that diets composed only of fat, protein, and carbohydrate led to pathological conditions or deficiency diseases in animals and, as was recognized later, in humans. Beriberi, pellagra, and scurvy are examples of human diseases caused by lack of vitamins.

The chemical work on the isolation of pure vitamin A and its structure elucidation was done mainly by Paul Karrer[10] in the University of Zurich, Switzerland. There had been many investigations in the United States of the concentration of the fat-soluble vitamins, and it was recognized that there were more than one. Several approaches in this country to the synthesis of vitamin A[11] led to methods which were eventually successful after 1940. Workers at Distillation Products, Inc., in Rochester, New York, first obtained Vitamin A in crystalline form and accomplished its synthesis by several routes.[12]

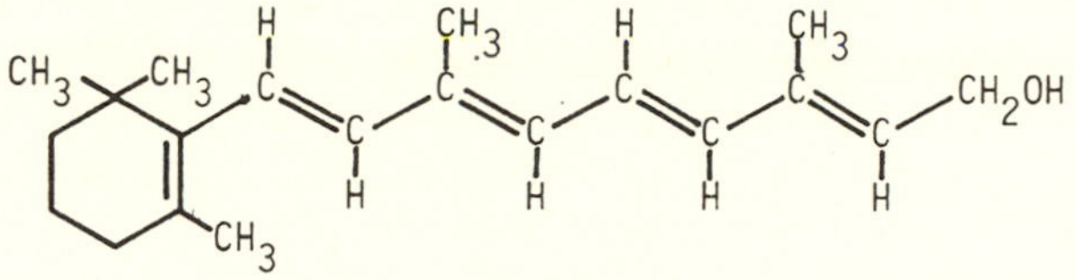

Vitamin A

The isolation of crystalline vitamin B_1 or thiamine, the establishment of its structure, and its synthesis by Robert R. Williams (1886-1965) were the culmination of a unique twenty-five year quest. Robert R. Williams, brother of Roger J. Williams, an outstanding biochemist, was born in India, the son of American Baptist missionaries.[13] After obtaining the B.S. and M.S. degrees at Chicago, Williams worked in the Bureau of Science in Manila, Philippine Islands, on the problem of isolating the material from rice bran which prevents the deficiency disease beriberi. This disease was first recognized by Chinese and was observed[14] by European doctors in Java in 1645. By 1915 he had obtained concentrates from rice hulls which had a favorable effect on human beriberi, and he realized that the active principle was a nitrogen-containing base. Williams returned to this country in 1915, and worked in government laboratories and after 1925 as chemical director of the Bell Laboratories. He continued his own research at night, aided by a grant from the Carnegie Corporation and the use of the chemical laboratories at Columbia.

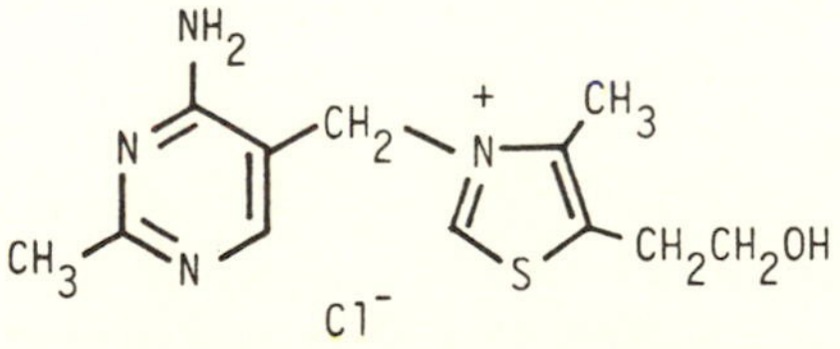

Thiamine (Vitamin B_1) 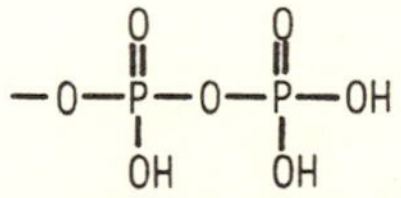   Pyrophosphate ester

Several other groups were working on the problem, and crystalline vitamin B_1 was first obtained by Jansen and Donath (1927) in Holland. Williams greatly improved the isolation procedure in 1933; in 1935-1936 he established the correct structure and developed a practical synthesis.[15] Williams had valuable collaboration from chemists at Columbia, Merck, and Bell Laboratories, and with commendable public spirit he assigned the patents to the Research Corporation, a non-profit organization which used the proceeds of the "Williams-Waterman Fund" to support basic research in chemistry. The structural work on thiamine is notable for the major contribution made by ultraviolet spectroscopy,[16] a portent of the Instrumental Revolution, which was in full swing by 1940.

The biochemically active form of thiamine (the pyrophosphate ester instead of H in structure) has several roles as a coenzyme; it causes loss of carbon dioxide from α-keto acids accompanied by oxidation, and it is involved in the transfer of units containing two carbon atoms in sugar metabolism. The chemical basis for its action, which involves the thiazolium nucleus (the five-atom ring containing S and N), is reasonably well understood.[17]

The isolation, structure proof, and synthesis of vitamin B_2, riboflavine, were carried out mainly by Karrer in Zurich and Richard Kuhn in Heidelberg during the 1930s. Max Tishler's group at Merck greatly improved the syntheses of riboflavine.[18]

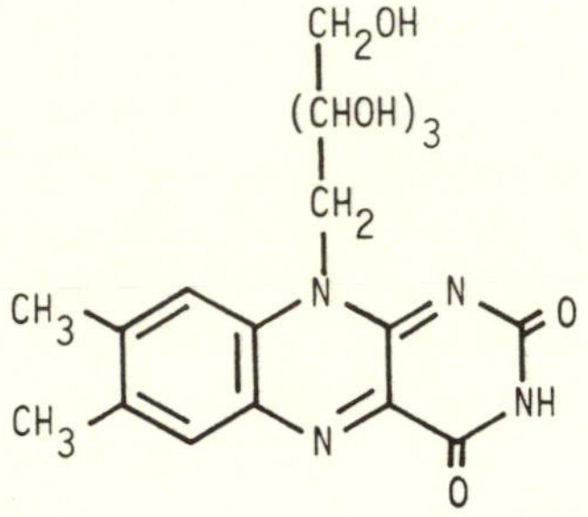

$$CH_2OH$$
$$(CHOH)_3$$
$$CH_2$$

Riboflavine

Max Tishler (1906- , B.S. 1928, Tufts; Ph.D. 1934, Harvard with E. P. Kohler) did research at Merck from 1937 to 1970, rising rapidly from research chemist to the presidency of Merck, Sharp, and Dohme Research Laboratories. He later served as university professor of science at Wesleyan University in Connecticut. At Merck he made great contributions to the synthesis of many compounds of medical importance, including cortisone.

The vitamin B_6 family, designated pyridoxine, was isolated from rice bran and from yeast by five groups of workers in 1938, including three teams in this country.[19] The Merck group established the structure of pyridoxine[20] and S. A. Harris and K. Folkers there synthesized the compound.[21] The biochemical role of the pyridoxine family in transamination, decarboxylation, and other transformations of amino acids was elegantly established by Esmond E. Snell at Texas.[22]

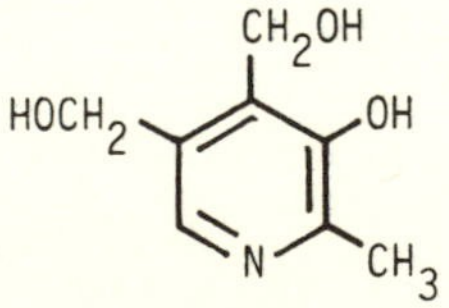

$$CH_2OH$$

Pyridoxine

Karl Folkers (1906- , B.S. 1928, Illinois; Ph.D. 1931, Wisconsin with

Homer Adkins, postdoctorate at Yale) made his career at Merck from 1933 to 1963, first in research and later as vice-president. After a few years as president of Stanford Research Institute, he moved to the University of Texas at Austin. His unusual record of structure determinations and syntheses of vitamins and other compounds of basic biochemical importance includes the ubiquinones (coenzyme Q) and polypeptide hormones.

The isolation and synthesis of vitamin C (ascorbic acid) promised freedom from the menace of scurvy, a deficiency disease which had taken a tragic toll of sailors, soldiers, pioneer settlers, and others subsisting on a diet devoid of fresh meat or fruits and vegetables. The Scottish naval physician Dr. James Lind[23] (1716-1794) published a treatise on scurvy in 1754 and, based on his own observations, recommended oranges, lemons, green food, and onions as antiscorbutics. It had been known for over 150 years at that time that lemon juice was a specific against scurvy. This knowledge had not been used, and in Baron Anson's voyage around the world in 1740-1744 with a British naval squadron, seventy-five percent of the crews died of scurvy.[23] Although he probably had not seen Lind's treatise, it was rightly regarded as a personal triumph for Captain James Cook that he kept his crews free of scurvy by use of fresh food when available during his three extended voyages of discovery in 1768-1780.[24] Only in 1795 did the British Admiralty require the provision of lime juice for protection of naval crews.[23]

Albert Szent-Györgyi (1893-), a Hungarian biochemist who had a long and varied career, isolated a "hexuronic acid" from oranges and cabbages and from the cortex of adrenal glands; he easily oxidized and reduced this material and obtained the empirical formula.[25] Szent-Györgyi appears to have been interested in only its oxidizing properties and not in any antiscorbutic activity.

Charles G. King (1896-), a biochemist at the University of Pittsburg, concentrated the antiscorbutic principle from lemon juice[26] and reported in 1932 that crystalline vitamin C from lemon juice was identical to the hexuronic acid from adrenal glands, and that both samples showed the same strong antiscorbutic activity in animal tests.[27] At about the same time, J. L. Svirbely and Szent-Györgyi, now working in the University of Szeged, Hungary, reached the same conclusion.[28]

The structure of vitamin C was rapidly established in 1933 by the English sugar chemists, W. N. Haworth and E. L. Hirst, and excellent, low-cost syntheses were soon devised,[29] so that vitamin C, which is required in much larger quantities for human nutrition than any other vitamin, became an inexpensive compound. Although vitamin C was the first vitamin to be obtained crystalline, its biochemical function is still not completely worked out. It is necessary for the biosynthesis of collagen,[30] material present in connective tissue, and arthritis is associated with malfunction of this system.

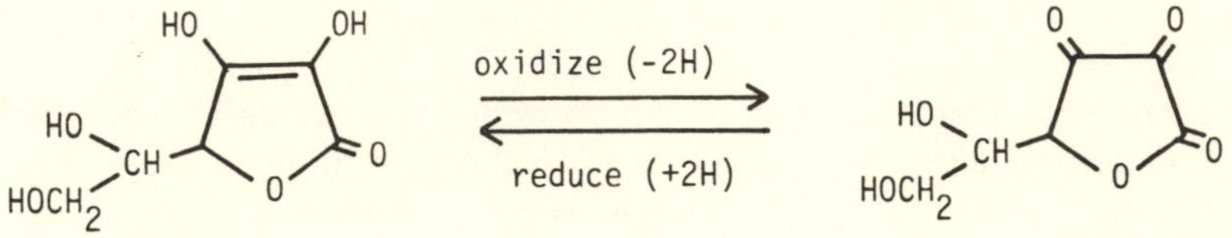

Vitamin C   Dehydroascorbic acid
(Ascorbic acid)

The recognition of a fat-soluble factor essential for proper calcification of bone and hence for the prevention of the deficiency disease rickets was due to extended feeding experiments[31] by E. V. McCollum (1879-1967) at Wisconsin and Johns Hopkins. These led to the famous observations by Harry Steenbock (1886-1967) at Wisconsin, which showed that exposure of animals or foodstuffs to sunlight prevented rickets.[32,33] The chemical elucidation of "vitamin D", the antirachitic materials, as derivatives of sterols isomerized by light was a problem of great complexity, solved mainly in overseas laboratories.[34] Light acts upon 7-dehydrocholesterol to give the antirachitic compound cholecalciferol. However, more recent work in this country by H. F. DeLuca at Wisconsin has shown that the active product is not cholecalciferol itself, but a product formed by *in vivo* oxidation, with OH groups replacing H at positions 1 and 25 (indicated by * in the formula).[33,35]

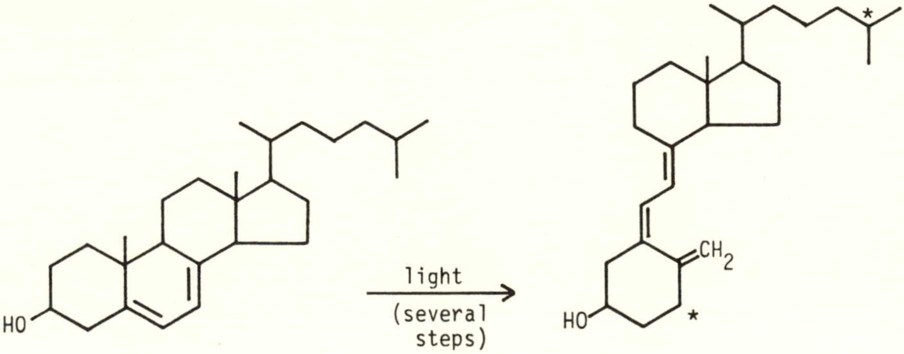

7-Dehydrocholesterol   Vitamin D$_3$
(cholecalciferol)

A series of feeding experiments by H. M. Evans (1882-1971) at Berkeley showed in 1922 that a factor present in lettuce and more abundantly in wheat germ oil was necessary for reproduction in rats.[36] The factor, named vitamin E, next in the series after the antirachitic vitamin D, was concentrated and isolated in pure form in 1936 with the collaboration of O. H. Emerson and G. A. Emerson of Merck.[37] It proved to be a member of a homologous series and was called α-tocopherol; the other members have one or two fewer methyl groups on the aromatic ring.[38] The structure was established by key degradations by E. Fernholz at Merck.[39] Karrer in Switzerland, A. R. Todd in England, and L. I. Smith at Minnesota synthesized α-tocopherol.[40] The twenty-carbon chain (including the carbons in the oxygen-containing ring) is made up of four isoprene units, $CH_2=CHCH(CH_3)=CH_2$, these are synthesized biologically from acetate units, CH_3COOH, and represent one of the key building blocks of natural products, including the steroids and the vitamin K series.

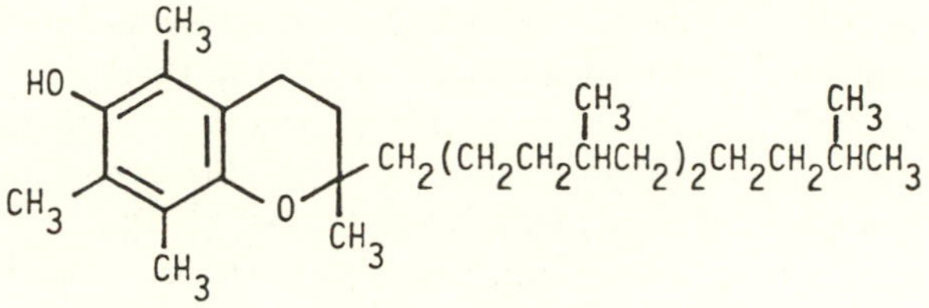

α-Tocopherol

The biochemical role of the tocopherols in human nutrition is not entirely clear, although the very rapid oxidation of the tocopherols suggest that they may be antioxidants or parts of oxidation-reduction systems *in vivo*.[33] The antioxidant activity may prevent oxidation of sensitive unsaturated fats and of SH groups by oxygen, leading to cross-linked chains of unsaturated fats. Although the efficacy of the vitamin E compounds in treating various human diseases has been asserted, there is no general agreement on this.

H. Dam in Copenhagen demonstrated the existence of the vitamin K family; he observed in 1929 that a fat-soluble factor was necessary for proper blood clotting in chicks. This factor, called "Koagulations-vitamin" (hence vitamin K) could be assayed by measuring the decrease in blood-clotting time of chicks deficient in prothrombin.[41] Pure vitamin K_1 was isolated nearly simultaneously by Dam and Karrer at Copenhagen and Zurich respectively, and by Doisy's group in St. Louis University.[42] Doisy obtained a crystalline derivative, from which the yellow oily vitamin could be regenerated unchanged.[42] Vitamin K_1 occurs in hog liver fat, in leaves of vegetables, and particularly in alfalfa. The ultraviolet absorption spectrum was useful in checking the purity and helping to determine the structure of the vitamin. The already active problem was still further enlivened by publications from several other groups, including H. J. Almquist and A. A. Klose at Berkeley, Fernholz at Merck, and L. Fieser of Harvard, some of whom had been working on the problem for a long time. The July 1939 issue of the JACS contains six preliminary communications on vitamin K compounds, and the September 1939 issue has seven.

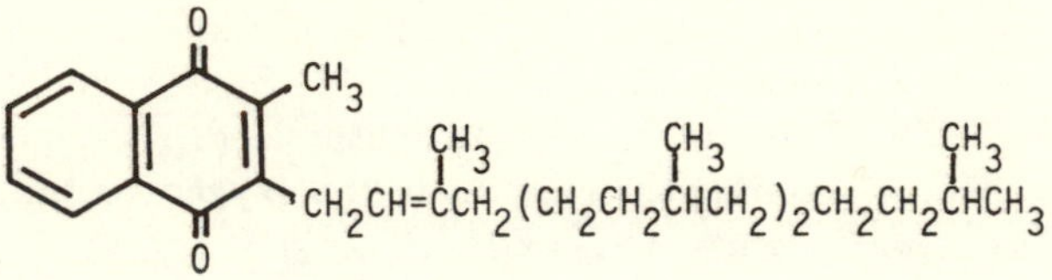

Vitamin K_1
(phylloquinone or 2-methyl-3-phytyl-1,4-naphthoquinone)

Degradations of K_1 quickly established its structure as 2-methyl-3-phytyl-1,4-naphthoquinone.[42] It was synthesized in 1939 in three American laboratories simultaneously by the Doisy, Almquist, and Fieser groups, by condensation of 2-methyl-1,4-naphthohydroquinone with the twenty-carbon alcohol phytol (four isoprene units) or derivatives thereof.[43] The Doisy group also isolated vitamin K_2 from putrified fish meal; its structure was later found to be similar in structure to K_1, but to have a side chain of seven isoprene units (thirty-five carbons).

Testing of the long-known compound 2-methyl-1,4-naphthoquinone, menadione, revealed vitamin K activity,[41] and several water-soluble derivatives of the corresponding hydroquinone show clinically useful activity.

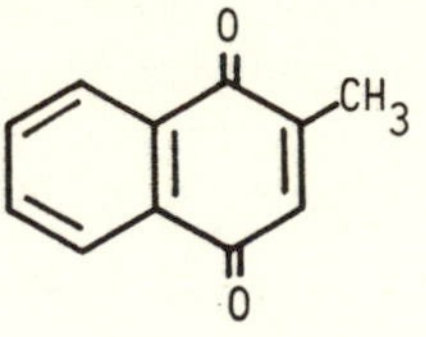

menadione 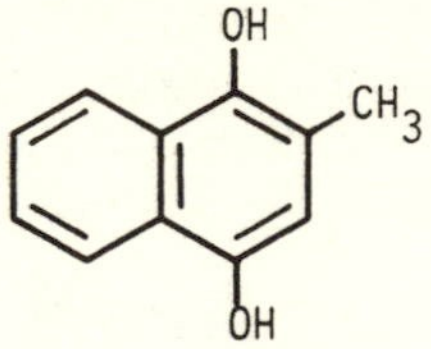   menadione hydroquinone

The striking biological action of the vitamin K series is due to its participation in the complex process of blood clotting; it does not prevent bleeding in hemophiliacs but is used routinely in childbirth to prevent hemorrhages in babies, some of whom do not have sufficient vitamin K at birth to give normal blood clotting. The nature of the blood clotting process and the complete mechanism of the antihemorrhagic action of vitamin K compounds is not yet understood.[33] Even in the rapidly moving field of natural products of the 1930s, the speed of isolation of pure vitamin K_1, proof of structure, synthesis, and the introduction of useful derivatives for therapy by several independent groups is striking. Dam and Doisy shared the Nobel Prize in medicine in 1943 for their work on vitamin K.

In later years numerous quinones, the ubiquinones and plastoquinones with polyisoprene side chains, were isolated and found to have essential biochemical functions in oxidation-reduction systems.

The fascinating story of the dicoumarol-warfarin compounds, which are blood anticoagulant agents and antagonize the blood coagulant (antihemorrhagic) action of the vitamin K compounds, is appropriate here. Karl Paul Link (1901-1978), trained as a biochemist and sugar chemist at Wisconsin and abroad, spent his career at Wisconsin as a biochemist in the School of Agriculture and in a courageous battle against illness. In the 1930s he was asked to work on the "sweet clover disease" of cattle; this sickness caused increased blood clotting time and usually fatal hemorrhage in cattle who had eaten spoiled sweet clover.[44] After developing a quantitative bioassay for prothrombin time (a measure of anticoagulant activity), Link's group isolated the active material from spoiled clover and proved its structure by degradation and synthesis to be the methylene-bis-coumarin (dicoumarol) below.[45] They went on to synthesize many analogs and derived compounds and proved that the anticoagulant activity of this series was antagonized by vitamin K compounds. One analog, warfarin, is an excellent rat poison, causing fatal hemorrhaging in rats. A pharmaceutical preparation of warfarin, Coumadin, is used in humans to prevent unwanted blood clots and control the prothrombin time. Link's work on the sweet clover disease is not only outstanding scientifically, but also it illustrates again how research and synthesis on natural products can lead to very valuable results. Dicoumarol had been reported by the German chemist Richard Anschütz in 1903 in impure form, but he had not examined its biological properties.

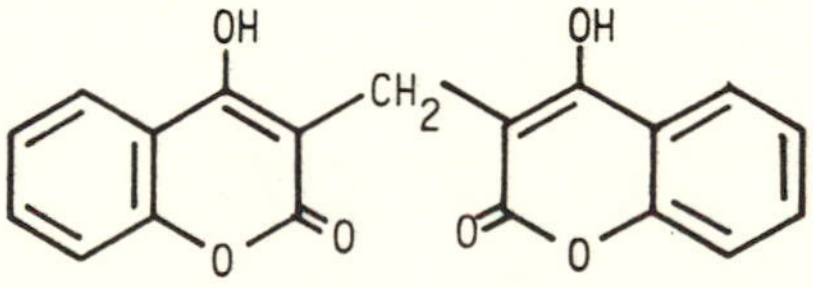

Dicoumarol

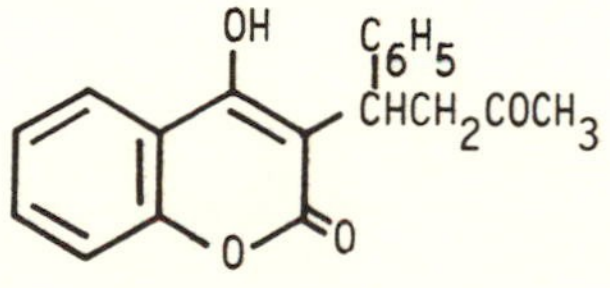

Warfarin

To the 1930s also belonged the isolation of nicotinamide (niacin) from yeast by Conrad Elvehjem[46] at Wisconsin; this compound cures the human deficiency disease pellagra, shown by J. Goldberger of the U. S. Public Health Service to be cured by yeast.[47] Niacin functions as part of a coenzyme required for biochemical oxidation-reduction.

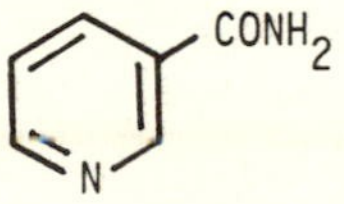

Nicotinamide
(niacin)

During the 1930s several groups isolated pure compounds from the outer layer (cortex) of the adrenal glands, the corticosteroids or corticoids. It had been known that adrenalectomy (removal of the adrenal glands, two small glands near the kidneys) was followed by death, and workers at Princeton showed that extracts from the adrenal cortex would prolong the life of adrenalectomized animals.[48] Isolation of several pure compounds, all inactive in adrenalectomized dogs, was reported by Oskar Wintersteiner and J. J. Pfiffner at Columbia.[49]

During 1935-1936, Edward C. Kendall's group at the Mayo Clinic[50] separated five pure compounds;[51] one, compound E (later called cortisone), showed activity in the muscle work assay procedure for adrenal activity developed in late 1935 by D. J. Ingle in Kendall's laboratory. Kendall's group proposed a partial steroid structure for compound E that proved correct in its main features. Workers in T. Reichstein's laboratory in Basel, Switzerland, were isolating corticoids at the same time,[52] and the early literature is complicated by three different systems of indicating each compound at the three laboratories. By 1943 the three groups had isolated twenty-eight steroids from adrenal cortex. The varied designations of the individual compounds required a glossary for reference.[53]

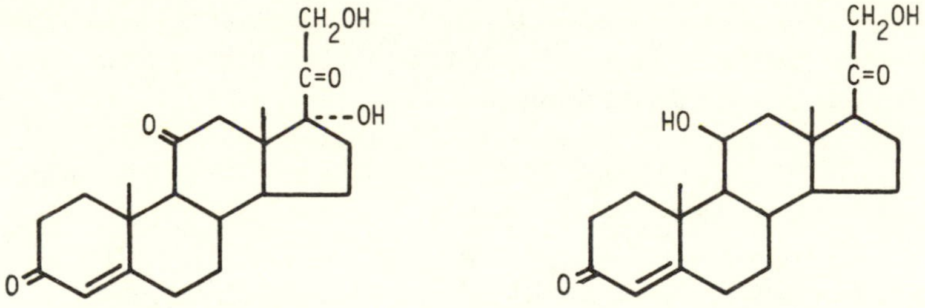

Cortisone
(Kendall's E)

Corticosterone
(Kendall's B)

The dramatic importance of the adrenal cortex compounds became clear in the following decades with the discovery of the anti-inflammatory (antiarthritic) activity of cortisone, for which Kendall, Philip S. Hench of Mayo Clinic, and Reichstein shared the Nobel Prize in medicine in 1950. The isolation and synthesis of the most active naturally occurring corticoid, aldosterone, by British, Swiss, and American chemists brought to a climax these long years of complex investigations.

During this time R. E. Marker, supported by Parke Davis and Company, published voluminous research in steroid chemistry from Pennsylvania State College.[54] Much of Marker's research was rather casually done and described, but he made a fundamental contribution to the structure of the plant sapogenins by establishing the presence of the spiroketal group, as in diosgenin below. The transformation of this material, readily available from yams (*Dioscorea* roots) in Mexico, to the natural hormone progesterone was worked out by Marker[55] and put into production by Syntex in Mexico City. Progesterone thus became the most readily available steroid hormone, and the Upjohn Company in Kalamazoo, Michigan, developed methods for its conversion to cortisone. Progesterone has various medical and veterinary use and can also be converted to antiprogestational compounds used in the Pill (oral contraceptives).[56] The work on the sapogenins by Marker and others shows once again how the basic research on structural organic chemistry may have very important and direct practical applications.

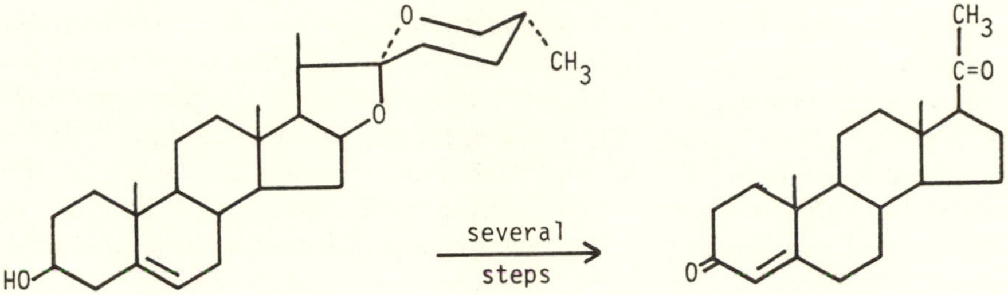

Diosgenin → several steps → Progesterone

The biochemist William C. Rose (1887-1985) of Illinois isolated the amino acid threonine, another discovery of vital importance. This is one of the amino acids essential for animal nutrition and was isolated after many controlled feeding experiments, in which mixtures of purified amino acids were fed instead of protein.[57] The preps program at Illinois that made pure amino acids available in quantity made these experiments possible. Threonine was synthesized, along with the three other stereoisomers, by Herbert E. Carter (1910-), also of Illinois;[58] only the naturally occurring stereoisomer is effective in maintaining growth.

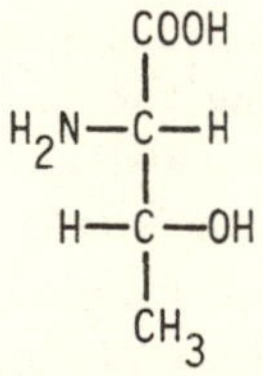

L(+) Threonine

Small's work on the opium alkaloids, mentioned earlier, continued through the 1930s, and Roger Adams began a study of the Senecio alkaloids, a large class of compounds widely distributed in plants, of which monocrotaline is a representative.[59]

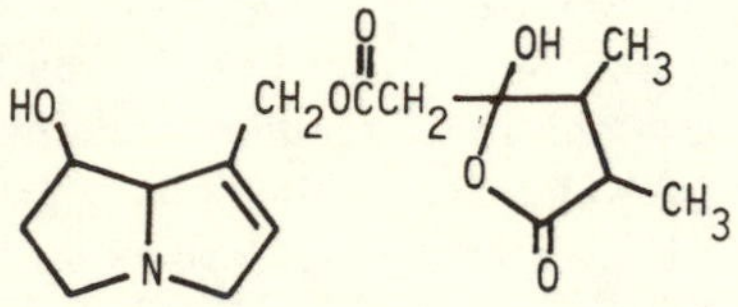

Monocrotaline

The alkaloid physostigmine (eserine) from Calabar beans, used in treating glaucoma, was synthesized in 1935 by Percy L. Julian and J. Pikl at De-Pauw University.[60] This work is noteworthy not only for its intrinsic chemical elegance, but also because Julian was one of the very first black chemists to achieve a prominent place in organic chemical research. Julian (1899-1975, A.B. 1920, DePauw; A.M. 1923, Harvard; Ph.D. 1931, Vienna, with E. Späth) taught at Fisk, West Virginia State College for Negroes, Howard, and, after taking his Ph.D., at DePauw from 1932 to 1936. From 1936 to 1953 he was a director of research at the Glidden Company with outstanding accomplishments in developing soya products and in the preparation of steroids from soybeans. In 1953 he established a consulting firm with a branch in Mexico which produced steroids from *Dioscorea*. He later sold the Mexican laboratory to Smith, Kline, and French. Julian took part in many public service activities; he was one of the first black members of the National Academy of Sciences and had numerous publications and patents in organic chemistry.

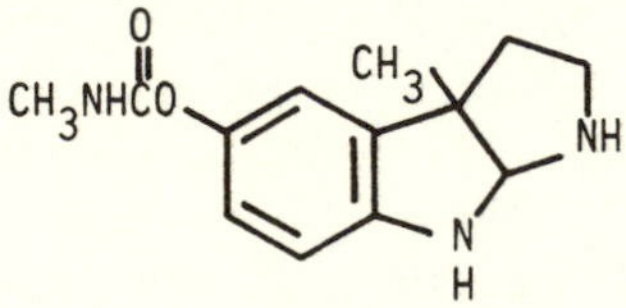

Physostigmine

Significant advances in solving the complex problem of the ergot alkaloids, important chemically, medically, and socially, were made by W. A. Jacobs and Lyman C. Craig at Rockefeller. Craig (1906-1974, B.S. 1928; Ph.D. 1931 with R. M. Hixon, both degrees at Iowa State) was associated with Jacobs for some years and later had his independent research program at Rockefeller. He was a masterly experimentalist and an ingenious designer of apparatus; he developed the powerful countercurrent distribution technique for separating closely related compounds. He worked on several families of alkaloids with Jacobs, but his later research investigated polypeptides, proteins, antibiotics, and methods of dialysis.[61]

Ergot is a fungus which grows on rye; it has been used crude for centuries to induce childbirth, but toxic symptoms (ergotism) can result from eating bread made from contaminated rye. Ergot contains a mixture of alkaloids and the first pure compound was isolated in crystalline form in 1875. The materials responsible for the oxytocic action (uterus-contracting) were isolated later but found to be a complex mixture. Craig and Jacobs, W. A. Stoll in Switzerland, and S. Smith and G. M. Timmis in England solved the structures of the ergot alkaloids.[62] The alkaloids occur in stereoisomeric pairs and Craig and Jacobs showed that hydrolysis gave lysergic acid, derived from a new ring system, for which they established the structure shown by degradation and synthesis.[63] Stoll proved both the position of the double bond and the difference between the two series (differing configuration at position 8). His laboratory also pointed out that the diethylamide of lysergic acid (LSD) induced abnormal psychic states, and this has led to serious social problems.

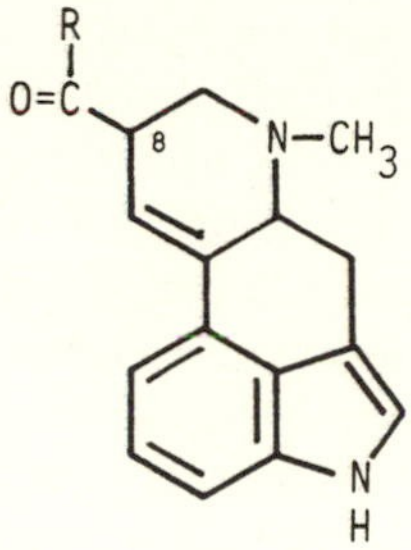

R = OH, lysergic acid

R = $N(C_2H_5)_2$, lysergic acid diethylamide (LSD)

R = $NHCH(CH_3)CH_2OH$, ergonovine

R = 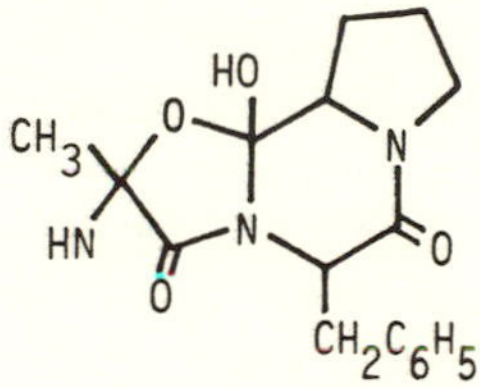 , ergotamine

Jacobs and Craig also verified that most of the ergot alkaloids gave, in addition to lysergic acid, phenylalanine, proline, an α-keto acid, and ammonia, all bound to the COOH by peptide linkages. The structure of lysergic acid was confirmed by an elegant synthesis in 1954 by many workers.[64] These complex structures were established by classical degradations and the piecing together of links of evidence with little use of instrumental methods, which were slowly becoming available by 1940 from the Instrumental Revolution.

For centuries the hallucinogenic activity of marihuana (hashish, hemp, pot, grass) has been known; the active material comes from the hemp plant, *Cannabis sativa* or *C. indica*. Only sporadic work had been done on the chemical nature of marihuana until Roger Adams at Illinois and A. R. Todd in England in the late 1930s took up the problem.[65] The active constituents of marihuana were not obtained pure until after World War II, when better procedures for separation were available. However, Todd and Adams isolated and proved the structures of several compounds from marihuana, showing them to be derived from dibenzopyran, or readily convertible to that system.

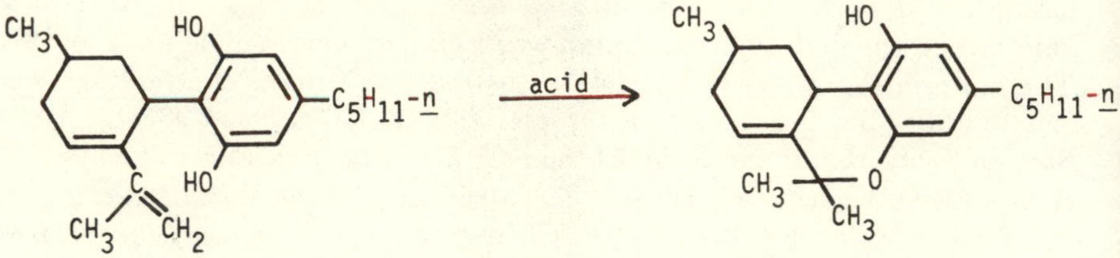

A cannabidiol

A tetrahydrocannabinol

The elucidation of the structure of gossypol, the toxic yellow pigment in cottonseed meal, by Adams' group[66] was another example of the high calibre of structural work being done during this period. The structure established was confirmed by synthesis of the degradation products, apogossypolic acid and desapogossypolone tetramethyl ether, by Adams and B. R. Baker.[67] The large number of functional groups in gossypol and the sensitivity of the compound to oxidation in air made its chemistry complex and baffling; it is not surprising that in spite of the accessibility of the compound (0.5 percent obtained from cottonseed meal), the structural problem had discouraged several American and European chemists.[68] The gossypol question overlapped with the marihuana and Senecio alkaloid problems in Adams' laboratory, which became a leading center for natural product studies before 1941. Thereafter Adams was obliged to devote full time to war research for the Office of Scientific Research and Development during the war years. Ultraviolet and visible absorption spectra proved useful in the gossypol work.

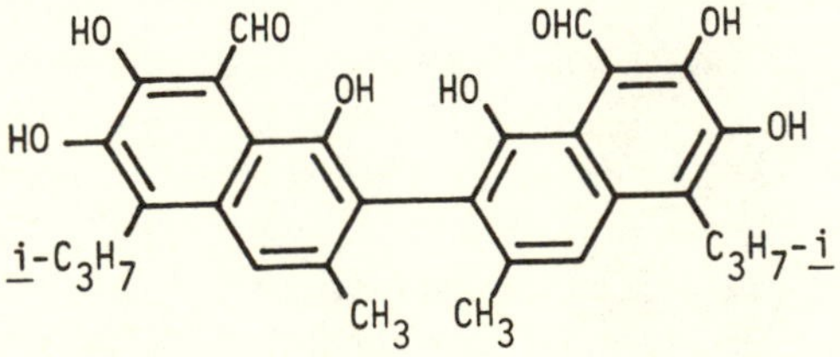

Gossypol

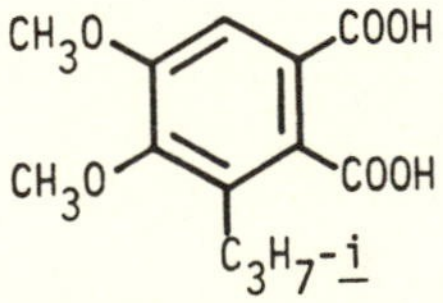

Apogossypolic acid

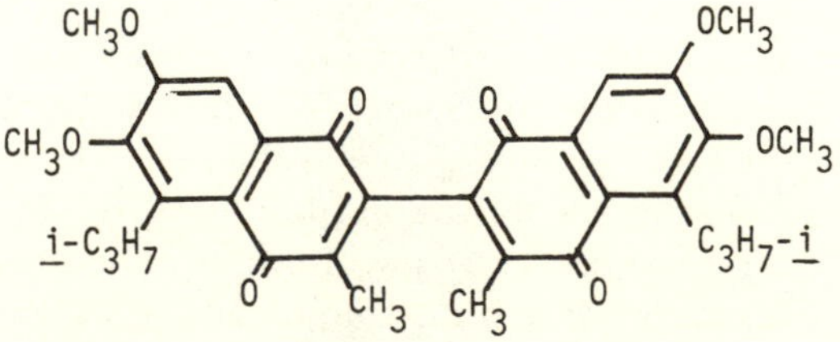

Desapogossypolone tetramethyl ether

C. S. Hudson's laboratory continued its work on sugar chemistry during the 1930s, dealing particularly with seven- and eight-carbon sugars, disaccharides, and the periodate oxidation of sugars.[69] A striking example of the value of Hudson's fundamental studies is his isolation and degradation of the seven-carbon sugar, sedoheptulose.[70] In 1954, M. Calvin found this sugar to be a key intermediate in photosynthesis,[71] and he was able to identify his material readily by using Hudson's methods, developed many years earlier.

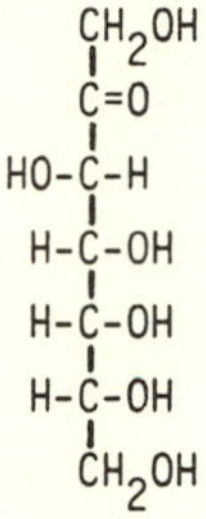

$$
\begin{array}{l}
CH_2OH \\
| \\
C=O \\
| \\
HO-C-H \\
| \\
H-C-OH \\
| \\
H-C-OH \\
| \\
H-C-OH \\
| \\
CH_2OH
\end{array}
\qquad \text{Sedoheptulose}
$$

Hudson's application of the specific oxidizing properties of periodic acid, HIO_4, or congeners, to sugars was not only the most important development in sugar chemistry in this period, but it also made available a reagent of the greatest usefulness in nucleic acid chemistry for structure determination of many antibiotics and synthetic products.

L. Malaprade and others[72] had shown that periodic acid not only cleaved 1,2-glycols to the corresponding aldehydes or ketones but also could be used as a quantitative reagent. This allowed Hudson to provide independent and conclusive evidence about two of the perennial problems in sugar chemistry: the configuration of C-1 in glycosides, and the ring size of the oxygen-containing ring, either a five-ring (furanose form) or a six-ring (pyranose form). The fundamental equations are illustrated.

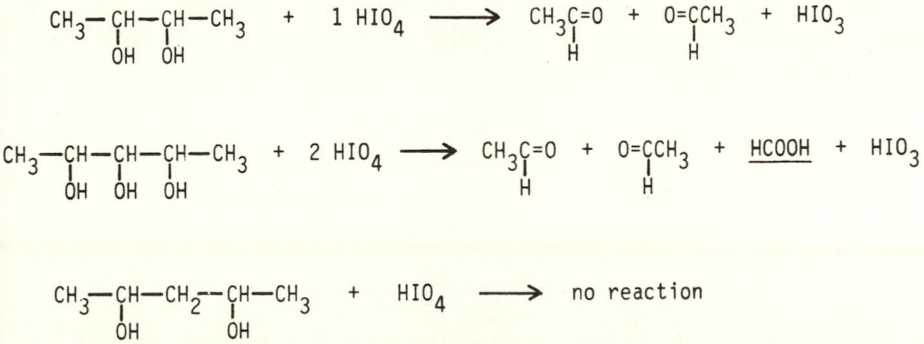

$$
CH_3-\underset{OH}{\underset{|}{CH}}-\underset{OH}{\underset{|}{CH}}-CH_3 \ + \ 1 \ HIO_4 \ \longrightarrow \ CH_3\underset{H}{\overset{|}{C}}=O \ + \ O=\underset{H}{\overset{|}{C}}CH_3 \ + \ HIO_3
$$

$$
CH_3-\underset{OH}{\underset{|}{CH}}-\underset{OH}{\underset{|}{CH}}-\underset{OH}{\underset{|}{CH}}-CH_3 \ + \ 2 \ HIO_4 \ \longrightarrow \ CH_3\underset{H}{\overset{|}{C}}=O \ + \ O=\underset{H}{\overset{|}{C}}CH_3 \ + \ \underline{HCOOH} \ + \ HIO_3
$$

$$
CH_3-\underset{OH}{\underset{|}{CH}}-CH_2-\underset{OH}{\underset{|}{CH}}-CH_3 \ + \ HIO_4 \ \longrightarrow \ \text{no reaction}
$$

Thus the methyl glycoside of a pyranose sugar would react with two moles of periodate, giving one mole of formic acid and a dialdehyde containing only two (C-1 and C-5) of the original five asymmetric carbons of the starting materials. Thus the configuration at C-1 could be correlated for a series of glycosides having the same ring size and configuration at C-5. The dialdehyde could be readily converted by oxidation and hydrolysis to D- or L-glyceric acid, thus showing the configuration at C-5. A furanose glycoside would consume only one mole of periodate and produce no formic acid.

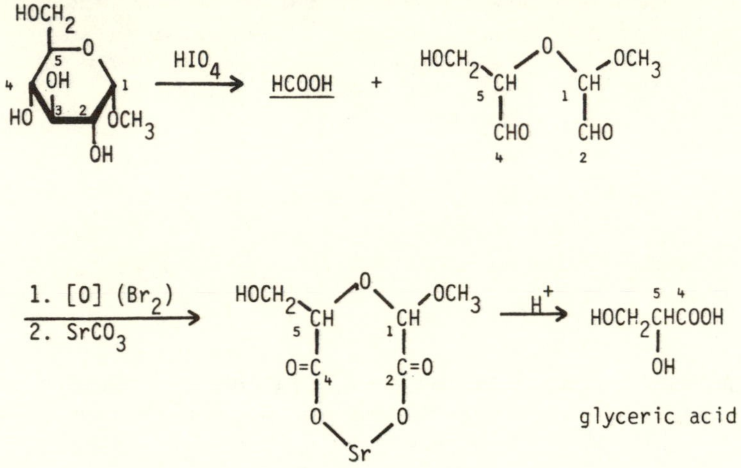

Hudson's masterly use of this procedure made it one of the most valuable tools for investigating sugar structures and configurations[73]–a true diagnostic test for 1,2-glycols or 1,2,3-triols. In general, NH_2 groups on carbon chains carrying OH groups also react with periodate in a similar fashion, a reaction useful in working with amino sugars and with many antibiotics such as streptomycin or actinospectacin. Periodate oxidation also yielded useful information about polysaccharides like starch and cellulose.

The periodate oxidation procedure was used to show the presence of furanose rather than pyranose rings in ribonucleosides. Most naturally occurring nucleosides and nucleotides in coenzymes and nucleic acids appear to have the furanose structure.

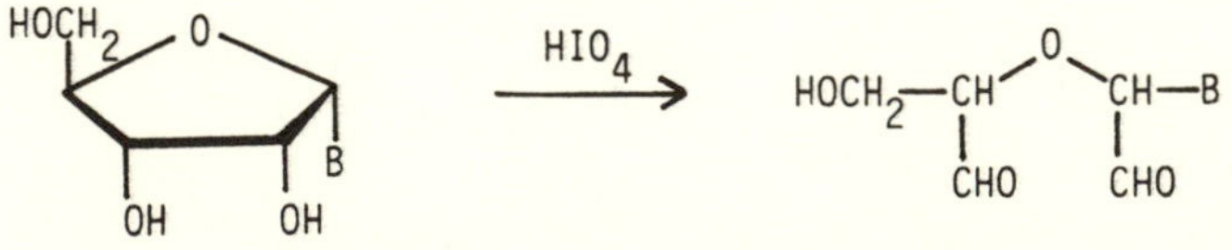

Ribonucleoside, B = adenine or similar base

M. L. Wolfrom at Ohio State continued work on the open-chain forms of sugars and started studies on cellulose and starch, using ethyl mercaptan and sulfur analysis to follow the decrease in molecular weight of the polysaccharide.[74]

During the 1930s, Karl Paul Link, in addition to his dicoumarol problem, studied the preparation of sugar acids relevant to the fruit pectins. These papers are of interest not only for the results obtained, but also for the training of two outstanding biochemists,[75] Carl Niemann (1908-1964) and Stanford Moore (1913-1982).

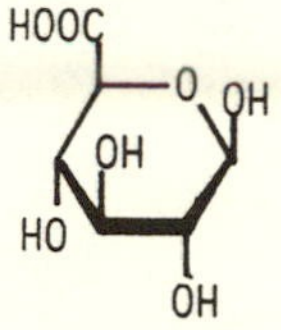

D-Glucuronic acid

The preceding section, which represents only part of the good research on natural products, amply documents the earlier statement that structural organic chemistry in this country was much better in the 1930s than in the 1920s, compared to the best work in the field. The American work was not only more polished technically than in the preceding decade but the problems investigated were of greater scientific value. What reasons may be given for this improvement?

One factor was undoubtedly the increase in job opportunities for organic chemists, in industrial research, in government laboratories, and in college and university teaching. This led to a marked increase in the number of strong graduate schools and graduate students, and probably a proportionate increase in the number of very able people in the field. After the long apprenticeship to German chemistry in the 19th century and the growth of an American tradition at Hopkins and other centers, outstanding leaders and institutions emerged–Jacobs, Roger Adams, W. E. Bachmann, Hudson, Folkers, Craig, biochemists like Doisy, du Vigneaud, Levene, Kendall, Link, and many more. The pressure from industrial research laboratories, especially pharmaceutical ones, for raising the standard of work on natural products became a major factor; first-class people, excellent supporting facilities, and the opportunity for intensive research until a problem was completed gave the university laboratories a compelling model. In reading JACS and JBC, one notes also the influence of strong groups in medical schools, agricultural colleges, and state experimental stations. Although the stimulus of outstanding physical chemists like G. N. Lewis, Pauling, A. A. Noyes, and C. A. Kraus was greater in physical organic chemistry than in natural product chemistry, nevertheless outstanding accomplishment in one area tended to raise the performance in other areas of chemistry. At the beginning of World War II, American structural chemistry had passed its colonial stage and could hold its own in world chemistry.

LITERATURE CITED

1. E. A. Doisy et al., JBC, **61**, 711 (1924).

2. E. A. Doisy, et al., Am. J. Physiol., 90, 329 (1929); JBC, 86, 499 (1930); 87, 357 (1930); 91, 791 (1931).

3. A. Butenandt, Naturwissenschaften, 17, 879 (1929); Nature, 130, 238 (1932); further discussion of structural work by Butenandt and others is given by L. F. Fieser, *Chemistry of Natural Products Related to Phenanthrene*, New York, 1936, pp. 193 ff.

4. E. A. Doisy et al., Proc. Soc. Exptl. Biol. Med., 32, 1182 (1935); estradiol had been prepared earlier by chemical reduction of the C=O in estrone.

5. W. M. Allen, JBC, 98, 591 (1932); H. E. Fevold, F. L. Hisaw and S. L. Leonard, JACS, 54, 254 (1932); E. Fels and K. H. Slotta, Proc. II. Inter. Nat. Cong. for Sex Research, p. 361, London, 1930.

6. K. H. Slotta et al., Ber., 67, 1624 (1934).

7. W. E. Bachmann, Wayne Cole and A. L. Wilds, JACS, 61, 974 (1939); 62, 824 (1940).

8. For Bachmann's career, see earlier essay.

9. Casimir Funk, J. Physiol., 45, 75 (1912); Funk was a versatile Polish biochemist; see Benjamin Harrow, *Casimir Funk*, New York, 1955.

10. P. Karrer et al., Helv. Chim. Acta, 14, 1036 (1931); 16, 557 (1933); synthesis: O. Isler et al., Experientia, 2, 31 (1946); Helv. Chim Acta, 30, 1911 (1947).

11. R. G Gould, Jr. and A. F. Thompson, JACS, 57, 340 (1935), (done in Conant's laboratory); R. C. Fuson and R. H. Christ, Science, 84, 294 (1936).

12. Crystalline vitamin A: J. G. Baxter and C. D. Robeson, JACS, 64, 2411 (1942); total synthesis of vitamin A: C. D. Robeson et al., ibid., 77, 4111 (1955).

13. Based on R. R. Williams' autobiographical account in *The National Cyclopaedia of American Biography*, vol. F, p. 204 (1934-1942).

14. Robert R. Williams and Tom D. Spies, M. D., *Vitamin B_1 and its Use in Medicine*, New York, 1938.

15. Structure: R. R. Williams, JACS, 57, 229 (1935); 58, 1063 (1936). Total synthesis: Williams and J. K. Cline, ibid., 58, 1504 (1936); Cline, Williams and J. Finkelstein, ibid., 59, 1052 (1937). As representative of the work of German groups, synthesis by R. Grewe, Z. Physiol. Chem., 242, 89 (1936).

16. A. E. Ruehle, JACS, 57, 1887 (1935).

17. R. Breslow, ibid., 79, 1762 (1957).

18. Max Tishler et al., 67, 2165 (1945); 69, 1487 (1947).

19. J. C. Keresztesy and J. R. Stevens, Proc. Soc. Exptl. Biol. Med., 38, 64 (1938); S. Lepkovsky, Science, 87, 169 (1938); P. György, JACS, 60, 983 (1938).

20. E. T. Stiller, et al., ibid., 61, 1237 (1939). Independent degradations were carried out by Richard Kuhn in Heidelberg.

21. S. A. Harris and K. Folkers, ibid., 61, 1245, 3307 (1939).

22. Summarizing paper: D. E. Metzler, M. Ikawa and E. E. Snell, ibid., 76, 648 (1954).

23. See articles on Lind and Anson, in *Dictionary of National Biography*. References to the ravages of scurvy could be multiplied endlessly from the historical writings of Francis Parkman, S. E. Morison, and others.

24. J. C. Beaglehole, *Life of Captain James Cook*, Stanford, 1974; Cook's success in preventing scurvy in his crews was due to his constant insistence on fresh food, fish, flesh, every variety of wild vegetables and fruits, "scurvy grass", fresh spruce beer, whenever opportunity allowed (p. 506). Cook received the Copley Medal of the Royal Society in 1776 for his report on his methods of protecting the health of his men (pp. 506, 704, and passim).

25. A. Szent-Györgyi, Biochem. J., 22, 1386 (1928) (from Cambridge University); JBC, 90, 385 (1931) (from E. C. Kendall's laboratory at the Mayo Clinic, Rochester, Minn.). An autobiographical account of his extraordinary career is in Ann. Rev. Biochem., 32, 1 (1963).

26. D. P. Grettie and C. G. King, JBC, **84**, 771 (1929); J. L. Svirbely and C. G. King, ibid., **94**, 483 (1931).

27. W. A. Waugh and C. G. King, Science, **76**, 630 (1932); JBC, **97**, 325 (1932).

28. J. L. Svirbely and A. Szent-Györgyi, Nature, **129**, 576 (1932); Biochem. J., **26**, 865 (1932).

29. E. L. Hirst, et al., Nature **131**, 617 (1933); synthesis: W. N. Haworth, E. L. Hirst et al., J. Soc. Chem. Ind., **52**, 645 (1933); JCS, 1419 (1933); T. Reichstein and A. Grüssner, Helv. Chim. Acta, **17**, 311 (1934); P. A. Wells et al., Ind. Eng. Chem., **29**, 1385 (1937).

30. Reviews: B. S. Gould, Vitamins and Hormones, **18**, 89 (1960); M. J. Barnes and E. Kodicek, ibid., **30**, 1 (1972).

31. On McCollum, see H. G. Day, *Biog. Mem. Nat. Acad. Scis.*, **45**, 263 (1974).

32. H. Steenbock, Science, **60**, 224 (1924) and many later papers.

33. For Steenbock, and work on the vitamin D complex and vitamins A, E, and K, see H. F. DeLuca and J. W. Suttie, Ed., *The Fat-Soluble Vitamins*, Madison, 1970.

34. L. F. Fieser and Mary Fieser, *Steroids*, New York, 1959, pp. 90 ff., structural elucidation of the vitamin D series.

35. H. F. DeLuca et al., Science, **180**, 190 (1973).

36. H. M. Evans and K. S. Bishop, Science, **56**, 650 (1922); biography of Evans by G. W. Corner, *Biog. Mem. Nat. Acad. Scis.*, **45**, 153 (1974).

37. H. M. Evans, O. H. Emerson and Gladys A. Emerson, JBC, **113**, 319 (1936).

38. Review on the chemistry of the tocopherols by L. I. Smith, Chem. Rev., **27**, 287 (1940).

39. E. Fernholz, JACS, **59**, 1154 (1937); **60**, 700 (1938). Fernholz, trained in Germany, was a gifted chemist who died very young.

40. P. Karrer et al., Helv. Chim. Acta, **21**, 520 (1938); L. I. Smith and H. E. Ungnade, Science, **88**, 37, 38, 40 (1938); JOC, **4**, 298 (1939); A. R. Todd et al., Nature, **142**, 36 (1938); JCS, 1382 (1938).

41. Reviews: E. A. Doisy, S. B. Binkley and S. A. Thayer, Chem. Rev., **28**, 477 (1941); H. Dam, Vitamins and Hormones, **24**, 295 (1966).

42. H. Dam, P. Karrer et al., Helv. Chim. Acta, **22**, 310, 1464 (1939); E. A. Doisy et al., JACS, **61**, 1612, 1928 (1939); JBC, **131**, 357 (1939).

43. E. A. Doisy et al., JACS, **61**, 2558 (1939); H. J. Almquist and A. A. Klose, ibid., p. 2557; L. F. Fieser, ibid., 2559, 2561.

44. For review, and a glimpse of a remarkable personality, see K. P. Link, Harvey Lectures, **39**, 162 (1943-44).

45. K. P. Link et al., JBC, **136**, 47 (1940); **138**, 21, 513 (1941).

46. C. A. Elvehjem et al., JACS, **59**, 1767 (1937); JBC, **123**, 137 (1938).

47. J. Goldberger, G. A. Wheeler and W. F. Tanner, Public Health Repts. (U.S.), **40**, 927 (1925).

48. W. W. Swingle and J. J. Pfiffner, Science, **71**, 321 (1930).

49. O. Wintersteiner and J. J. Pfiffner, JBC, **111**, 599 (1935); **116**, 291 (1936).

50. An excellent biography and character sketch of Kendall: D. J. Ingle, *Biog. Mem. Nat. Acad. Scis.*, **47**, 249 (1975).

51. H. L. Mason, C. S. Myers and E. C. Kendall, JBC, **114**, 613 (1936); **116**, 267 (1936); structural determinations: H. L. Mason, W. M. Hoehn and E. C. Kendall, ibid., **124**, 459, 475 (1938).

52. T. Reichstein, Helv. Chim. Acta, **19**, 29 (1936).

53. T. Reichstein and C. W. Shoppee, Vitamins and Hormones, **1**, 345 (1943). Structures had been established for all the twenty-eight compounds isolated at this time.

54. The first papers were JACS; **57**, 1755, 2358 (1935); by 1940, paper 111 in this series had been logged (Marker et al., **62**, 3009 (1940)) and the number climbed to 174 (Marker,

ibid., **71**, 4149 (1949)), the later papers being published from Mexico. Marker was honored for his role in progesterone preparation at the chemical meeting in Mexico City in 1975, Chem. Eng. News, December 15, 1975, p. 47.

55. R. E. Marker and E. Rohrmann, **61**, 3592 (1939); **62**, 518 (1940); Marker et al., ibid., **69**, 2167 (1947).

56. Details on Marker's varied career and his work on sapogenins and the progesterone process, in Fieser and Fieser, *Steroids*, pp. 817 ff, 548 ff; conversion of progesterone to estrone, an intermediate in the synthesis of antiprogestational compounds, pp. 479 ff.

57. Isolation: R. H. McCoy, C. E. Meyer and W. C. Rose, JBC, **112**, 283 (1935); configuration: ibid., **115**, 721 (1936).

58. H. E. Carter, ibid., **112**, 769 (1935); H. D. West and Carter, ibid., **119**, 103, 109 (1937) and later papers. Both Rose and Carter were distinguished investigators and teachers in biochemistry at Illinois, where biochemistry was a division of the chemistry department at Urbana.

59. R. Adams and R. S Long, JACS, **62**, 2289 (1940); Adams and N. J. Leonard, ibid., **66**, 257 (1944), one of a long series of papers; review: N. J. Leonard in *The Alkaloids*, R. H. F. Manske and H. L. Holmes, Eds., New York, 1950, vol. 1, pp. 107-164.

60. P. L. Julian and J. Pikl, JACS, **57**, 563, 755 (1935), and earlier papers. For Julian's career, see Louis Haber, *Black Pioneers of Science and Invention*, New York, 1970, pp. 87-102; and *National Cyclopedia of American Biography*. We have been told, by one who was there at the time, that Julian left Harvard because of the very hostile attitude of the laboratory stockroom staff, not because of any antagonism from faculty or students.

61. Biography of Craig by Stanford Moore, *Biog. Mem. Nat. Acad. Scis.*, 49, 49 (1978).

62. Reviews: A. Stoll, Chem. Rev., **47**, 197 (1950); A. L. Glenn, Quart. Rev., **8**, 192 (1954).

63. W. A. Jacobs and L. C. Craig, JBC, **104**, 547 (1934); **115**, 227 (1936); **122**, 419 (1938); **125**, 289 (1938).

64. By a group at Eli Lilly under E. C. Kornfeld, with collaboration of R. B. Woodward, JACS, **76**, 5256 (1954); **78**, 3087 (1956).

65. R. Adams et al., ibid., **62**, 196 (1940), and many later papers such as 64, 2087, 2653 (1942).

66. Summarizing paper (Gossypol XV): R. Adams, R. C. Morris, T. A. Geissman, D. J. Butterbaugh and E. C. Kirkpatrick, ibid., **60**, 2193 (1938).

67. R. Adams and B. R. Baker, ibid., **61**, 1138 (1939); **63**, 535 (1941).

68. Review of earlier work (Gossypol I): K. N. Campbell, R. C. Morris and R. Adams, ibid., **59**, 1723 (1937).

69. Claude S. Hudson, *Collected Papers*, vol. II, passim.

70. N. K. Richtmyer, R. M. Hann and C. S. Hudson, JACS, **61**, 343 (1939); the sugar was discovered in 1917 (H. L. LaForge and Hudson, JBC, **30**, 61 (1917)).

71. J. A. Bassham, M. Calvin, et al., JACS, **76**, 1760 (1954).

72. L. Malaprade, Bull. Soc. Chim., [4] **43**, 683 (1928); P. Fleury and J. Lange, Compt. Rend., **195**, 1395 (1932).

73. E. L. Jackson, *Organic Reactions*, **2**, 341 (1944); E. L. Jackson and C. S. Hudson, JACS, **58**, 378 (1936); **59**, 994 (1937); **61**, 959 (1939); application to starch and cellulose, ibid., **59**, 2049 (1937) and later papers.

74. M. L. Wolfrom et al., ibid., **58**, 1781 (1936) for example; for mercaptalation: ibid., **59**, 282 (1937); **61**, 2172 (1939).

75. Carl Niemann and K. P. Link, JBC, **106**, 773 (1934) and nine other papers with Link; Stanford Moore and Link, ibid., **133**, 293 (1940).

17

Molecular Rearrangements

As mentioned earlier the occurrence of rearrangements during organic reactions has fundamental implications for synthetic work and for the validity of structure proofs; the possibility of a rearrangement in the structure must be ruled out or the nature of the change must be established. The study of molecular rearrangements has contributed greatly to our knowledge of reaction mechanisms in general. Such rearrangements continued to be investigated after 1939 with newer experimental methods, particularly isotopic tracers, of increasing power and subtlety. The interpretations of some rearrangements aroused controversies which at times recall the arguments among medieval scholastics rather than scientific discussions.

C. W. Porter's monograph of 1928 on rearrangements is of particular interest for what it includes and omits.[1] It shows the general acceptance of G. N. Lewis' electronic theory of valence by chemists, including the proponents of the older dualistic theories like Julius Stieglitz. It gives some insight into the changes in Stieglitz's views (based on correspondence with Porter), which is useful because of Stieglitz's failure to publish his research and ideas after 1920. It is striking to the present-day reader that Porter does not mention Hans Meerwein's work on the camphene hydrochloride-isobornyl chloride rearrangement, which later led to a large body of important American work. Porter does, however, mention other examples of Meerwein's studies on rearrangements of carbon-carbon bonds. He has one reference each to C. K. Ingold and Robert Robinson.

The account of rearrangements by E. S. Wallis of Princeton in Gilman's *Treatise* of 1938 is organized around possible mechanisms for the reactions.[2] Wallis shows the usefulness of Whitmore's ideas of 1932 as a general explanation for many diverse rearrangements and also suggests that the transition state ideas of Eyring could be valuable.

The most significant American work in the 1914-1939 period was on rearrangements involving carbon-carbon bonds. An early example of this type is the pinacol rearrangement, reported in 1860 by R. Fittig[3] in Germany.

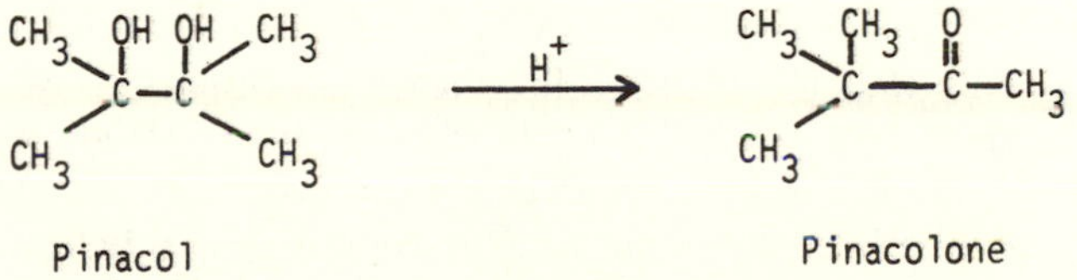

Pinacol   Pinacolone

Another class of rearrangements involves formation of an unsaturated hydrocarbon with loss of water illustrated in 1927 by Dorothy Bateman and C.

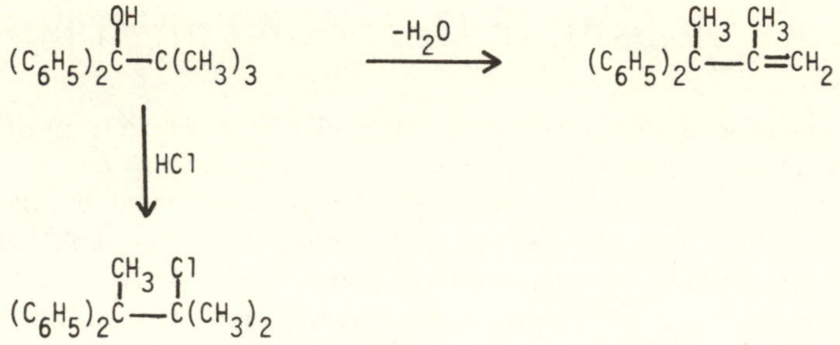

$$(C_6H_5)_2\overset{\overset{\displaystyle OH}{|}}{C}-C(CH_3)_3 \quad \xrightarrow{-H_2O} \quad (C_6H_5)_2\overset{\overset{\displaystyle CH_3}{|}}{C}-\overset{\overset{\displaystyle CH_3}{|}}{C}=CH_2$$

$$\downarrow HCl$$

$$(C_6H_5)_2\overset{\overset{\displaystyle CH_3}{|}}{C}-\overset{\overset{\displaystyle Cl}{|}}{C}(CH_3)_2$$

S. Marvel.[4] The reaction of HNO_2 and a primary amine represents another common case where rearrangement usually occurs.[5]

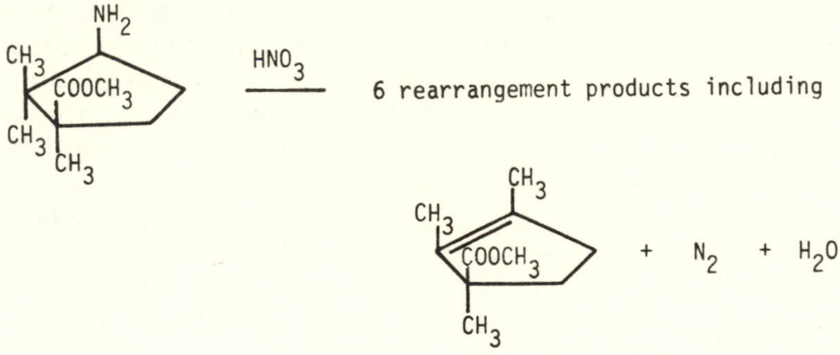

6 rearrangement products including

$+ \quad N_2 \quad + \quad H_2O$

The carbonium ion intermediate in carbon rearrangements was suggested by Meerwein, then at Bonn, Germany, for the camphene series in 1920-1922.[6] As mentioned in an earlier section, carbonium ions were first proposed by Stieglitz in 1899, and by J. F. Norris of MIT in 1907, to explain the behavior of *tert*-butyl alcohol and its chloride. Meerwein based his suggestion on the effect of ionizing solvents and acids, including HCl and Lewis acids ($SnCl_4$, $SbCl_5$, $FeCl_3$), in accelerating the rate. Ether as solvent inhibited the reaction, and the kinetics showed reasonably good first-order kinetics. Meerwein's interpretation of the rearrangement was modified by P. D. Bartlett[7] at Harvard by showing that it was not a first-order ionization, but a second-order or 3/2 order reaction, in which the chloride ion was pushed off with participation of acid at the same time that the migrating carbon attacked. The chloride ion entered the new structure on the opposite side from the departing carbon, thus forming isobornyl chloride with Walden inversion at the new chlorine-carrying carbon. Bartlett also showed a similar process to occur in the pinacol rearrangement.

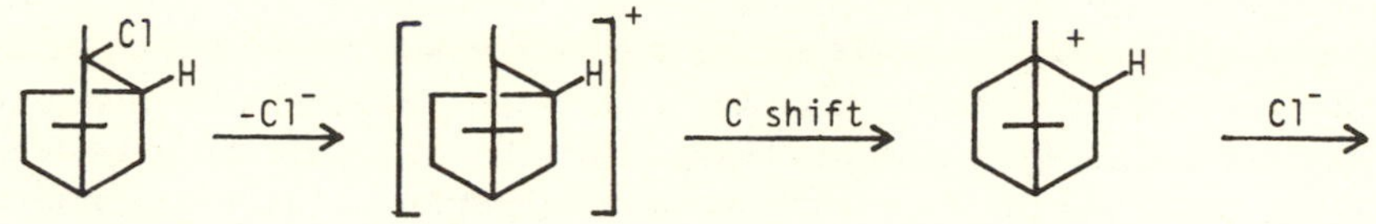

Camphene
hydrochloride

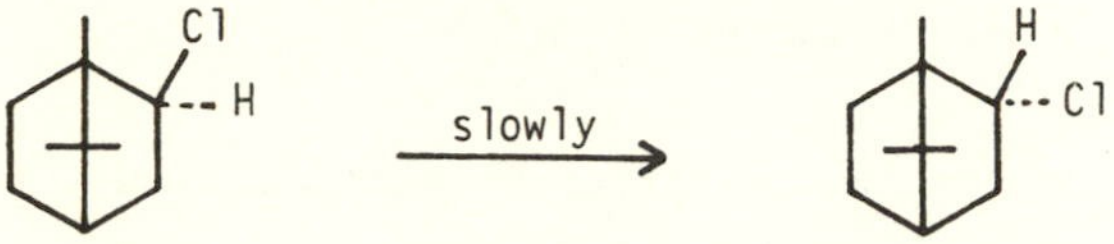

Isobornyl   Bornyl
chloride   chloride
                   (stable product)

These and many other rearrangements were correlated in 1932 in a very influential paper[8] by Frank C. Whitmore entitled "The Common Basis of Intramolecular Rearrangements". Whitmore (1887-1947, A.B. 1911, Ph.D. 1914, both at Harvard, with C. L. Jackson and E. P. Kohler) taught at Northwestern from 1920 to 1929, where he did research and wrote a monograph on organic mercury compounds, demonstrating the extraordinary energy which characterized his whole career. From 1929 until his death he was Dean of the School of Chemistry and Physics at Pennsylvania State College.[9] Here he started working on highly branched aliphatic compounds. It appears that he had been thinking about the mechanism of the pinacol rearrangement from his undergraduate days,[10] although his publications show no work on it. In 1924 he had tried unsuccessfully to make neopentyl chloride, $(CH_3)_3CCH_2Cl$, in order to get the corresponding mercury compounds.[11] His interest in highly branched aliphatic compounds was stimulated by the active group at Penn State under Dr. M. R. Fenske (1904-1971) studying petroleum products and distillation methods.

Whitmore's ideas on rearrangement mechanisms were not all new, as he clearly acknowledged in his paper, but his organization and application of these ideas was a very important contribution to organic chemistry. His extensive experimental studies on highly branched compounds illustrated the usefulness of his theory and developed important synthetic methods.

The key point in Whitmore's theory of rearrangements was the postulation of an electron-deficient intermediate which could stabilize itself by the

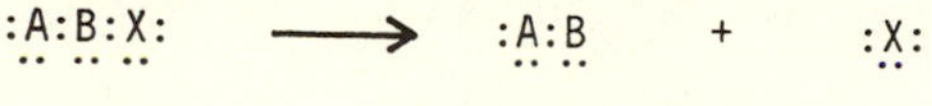

$$:A:B:X: \longrightarrow :A:B + :X:$$

migration of a group *with* its pair of bonding electrons. In the general case, structure ABX loses X, with its pair of electrons, where A and B are carbon- or nitrogen-containing groups, and X is a halogen or oxygen-containing group which can readily depart with its electron pair. The structure AB is electron-deficient, and it has three possibilities for stabilization: (1) it may pick up Y

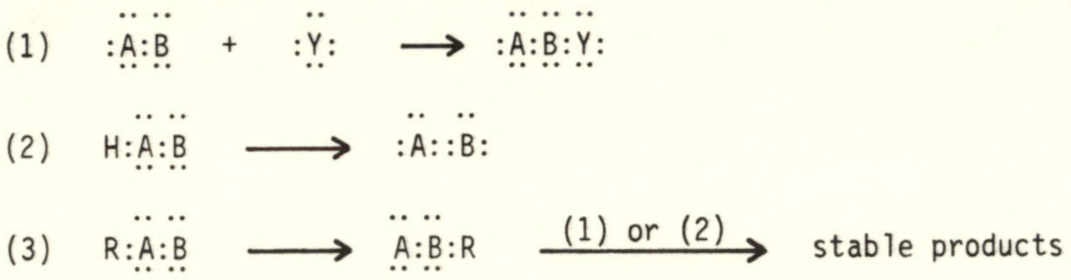

(1) $:A:B$ + $:Y:$ $\longrightarrow$ $:A:B:Y:$

(2) $H:A:B$ $\longrightarrow$ $:A::B:$

(3) $R:A:B$ $\longrightarrow$ $A:B:R$ $\xrightarrow{\text{(1) or (2)}}$ stable products

with eight electrons from solution; (2) it may lose a proton or other positive group from A, if one is attached, to form an unsaturated compound; (3) a group attached to A may migrate to B *with* its pair of electrons to form a new positive ion, which then stabilizes by processes (1) or (2). For example, the rearrangement of Marvel's compound[4] would take place following Whitmore's theory in the way shown.

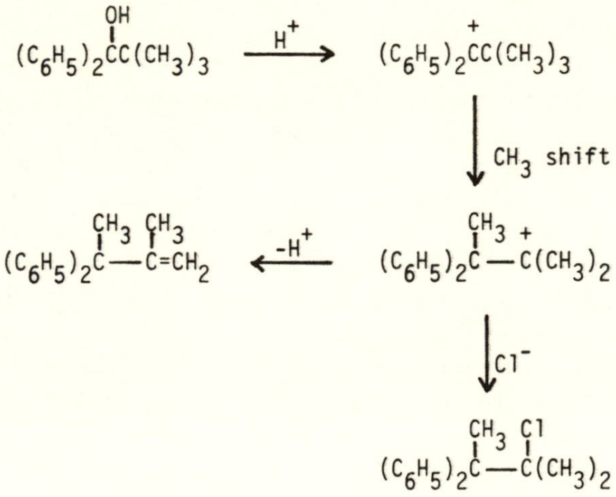

$$(C_6H_5)_2\overset{OH}{\underset{|}{C}}C(CH_3)_3 \xrightarrow{H^+} (C_6H_5)_2\overset{+}{C}C(CH_3)_3$$

$$\downarrow CH_3 \text{ shift}$$

$$(C_6H_5)_2\overset{CH_3}{\underset{|}{C}}-\overset{CH_3}{\underset{|}{C}}=CH_2 \xleftarrow{-H^+} (C_6H_5)_2\overset{CH_3}{\underset{|}{C}}-\overset{+}{C}(CH_3)_2$$

$$\downarrow Cl^-$$

$$(C_6H_5)_2\overset{CH_3}{\underset{|}{C}}-\overset{Cl}{\underset{|}{C}}(CH_3)_2$$

A more elaborate but important case is the dehydration of di-*tert*-butylcarbinol.[12] Here the carbonium ion stabilizes at 180° mainly by a methyl shift, followed by ejection of the *tert*-butylcarbonium ion, to yield seventy-seven percent of the pentene. The *tert*-butyl ion can lose a proton to give isobutene, or it can react with more isobutene to form di-, tri-, and te-traisobutenes. The rearranged 9-carbon ion can lose a proton to form two isomeric nonenes C_9H_{18}. At lower temperatures a larger proportion of nonenes is formed.

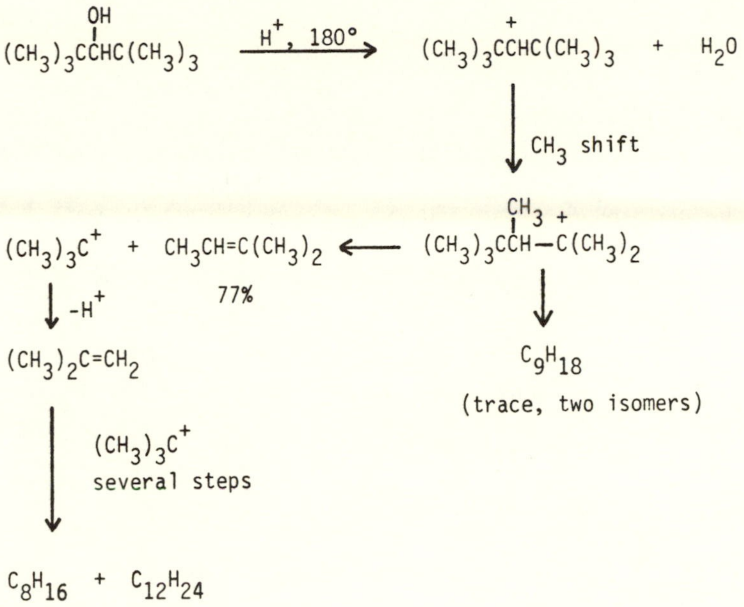

$$\underset{(CH_3)_3C\overset{\displaystyle OH}{\underset{|}{C}}HC(CH_3)_3}{} \xrightarrow{H^+, \ 180°} (CH_3)_3C\overset{+}{C}HC(CH_3)_3 \ + \ H_2O$$

$$\downarrow CH_3 \text{ shift}$$

$$(CH_3)_3C^+ \ + \ CH_3CH=C(CH_3)_2 \ \longleftarrow \ (CH_3)_3C\overset{\overset{\displaystyle CH_3}{|}}{C}H-\overset{+}{C}(CH_3)_2$$

$$\downarrow -H^+ \qquad\qquad 77\% \qquad\qquad\qquad \downarrow$$

$$(CH_3)_2C=CH_2 \qquad\qquad\qquad\qquad C_9H_{18}$$

(trace, two isomers)

$$\downarrow \begin{array}{c} (CH_3)_3C^+ \\ \text{several steps} \end{array}$$

$$C_8H_{16} \ + \ C_{12}H_{24}$$

The step indicated above, the attack of a carbonium ion on C=C, was an extremely important concept and explained the acid-catalyzed polymerization of isobutene as shown.[13] This reaction is involved in the polymerization of isobutene to butyl rubber and, with one important addition due to Bartlett and to Louis Schmerling of Universal Oil, explains the reaction of isobutane, *tert*-butyl chloride, isobutene, and a Lewis acid to form the branched octanes for engine fuel.[14] This is the *alkylation* of olefines by alkanes, an important process in petroleum technology.

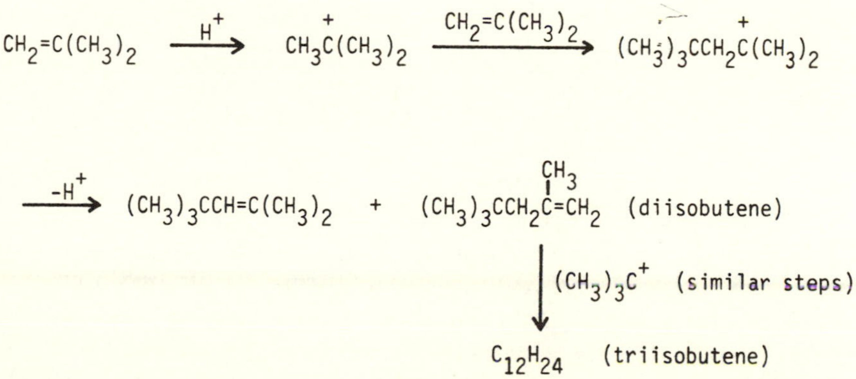

$$CH_2=C(CH_3)_2 \xrightarrow{H^+} CH_3\overset{+}{C}(CH_3)_2 \xrightarrow{CH_2=C(CH_3)_2} (CH_3)_3CCH_2\overset{+}{C}(CH_3)_2$$

$$\xrightarrow{-H^+} (CH_3)_3CCH=C(CH_3)_2 \ + \ (CH_3)_3CCH_2\overset{\overset{\displaystyle CH_3}{|}}{C}=CH_2 \quad \text{(diisobutene)}$$

$$\downarrow (CH_3)_3C^+ \quad \text{(similar steps)}$$

$$C_{12}H_{24} \quad \text{(triisobutene)}$$

The polymerization of isoprene units (isoprene is $CH_2=C(CH_3)CH=CH_2$) enzymatically in living systems to form polyisoprenoids, including products as diverse as camphor and the sex hormones, takes place by processes which follow the pattern of the scheme above for forming di- and tri-isobutene. Work on *in vitro* formation of the steroids from polyunsaturated compounds has been based on carbonium ion reactions. Whitmore's influence is evident in the accepted mechanism of the biological conversion of squalene, an open-chain polyunsaturated polyisoprene, to form the steroid lanosterol which is then converted to the many steroid types. These biochemical and synthetic pathways were elucidated after 1945, and they indicate the contribution of organic reaction mechanism studies to related fields like biochemistry.[15,16]

Whitmore's scheme also accommodates a series of rearrangements involving nitrogen compounds,[8] the Beckmann, Hofmann, Lossen, and Curtius reactions.

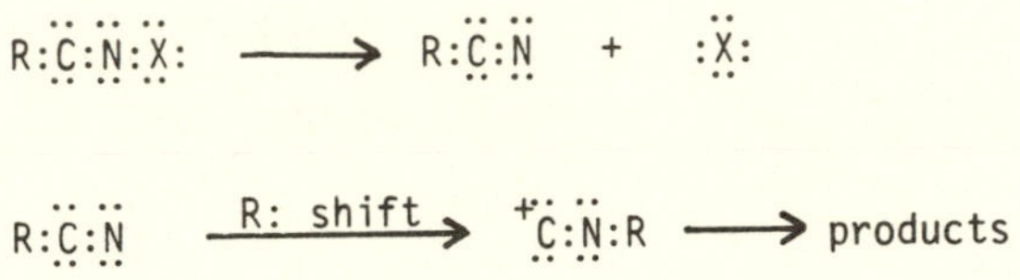

$$R:\ddot{C}:N:\ddot{X}: \longrightarrow R:\ddot{C}:\ddot{N} \ + \ :\ddot{X}:$$

$$R:\ddot{C}:\ddot{N} \quad \xrightarrow{\text{R: shift}} \quad {}^{+}\ddot{C}:\ddot{N}:R \longrightarrow \text{products}$$

The last three reactions follow the course below. In the Curtius rearrangement X is N_2, and there is no H on the N. These rearrangements had been studied particularly by Stieglitz, and by L. W. Jones and his students E. S. Wallis of Princeton and C. D. Hurd, later of Northwestern. Whitmore's theory of the migration of the R group with its electron pair is generally a suitable basis for explaining the facts.[17]

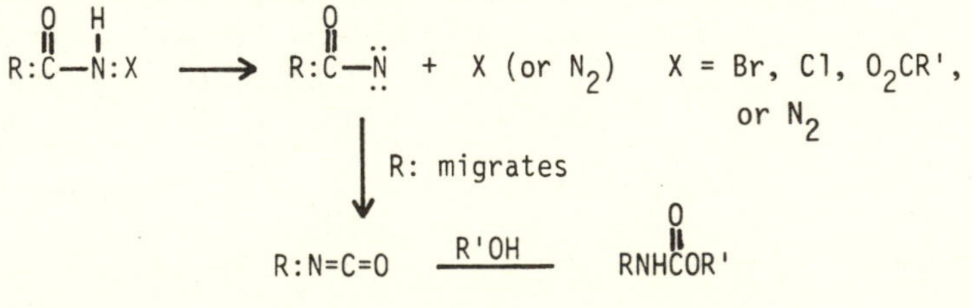

$$\underset{R:\overset{\overset{\text{O}}{\|}}{C}-\overset{\overset{\text{H}}{|}}{N}:X}{} \longrightarrow \underset{R:\overset{\overset{\text{O}}{\|}}{C}-\ddot{N}}{} + \text{X (or } N_2) \quad X = Br, \ Cl, \ O_2CR',$$

$$\text{or } N_2$$

$$\Big\downarrow \text{ R: migrates}$$

$$R:N=C=O \quad \xrightarrow{\text{R'OH}} \quad RNH\overset{\overset{\text{O}}{\|}}{C}OR'$$

Lauder W. Jones (A.B. 1892, Williams; Ph.D. 1897, Chicago with J. U. Nef) taught at Cincinnati, Minnesota, and then at Princeton from 1920 to 1937. He worked on rearrangements of nitrogen compounds and on applications of electron theories to reaction mechanisms.

Everett S. Wallis (1899-1961, B.S. 1921, Vermont; Ph.D. 1925, Princeton with L. W. Jones) taught at St. Johns and then at Princeton from 1930 to his death. He worked on free radicals and steroids as well as rearrangements.

The question of the existence of the carbonium ion as a discrete structure, or put in another way, the simultaneity of the loss of X, the shift of the R: group, and any later stages, requires comment. Bartlett's work quoted above on the Wagner-Meerwein rearrangement indicated a concerted process in which the carbonium ion is not free as such. This question of the "free" existence of carbonium ions recurs frequently, but it is not an essential point in the Whitmore scheme. Whitmore did not discuss the role of solvent in his scheme, but it was clear from Meerwein's work that solvent played a fundamental role, particularly in the carbonium ion rearrangements.

Some ingenious experiments by Jones and Wallis, using optically active R groups as migrating agents in the Curtius and similar rearrangements, show retention of optical activity.[18] In the Hofmann rearrangement of the optically active amide where the optical activity is due to hindered rotation around the phenyl-naphthalene bond, optical activity is retained in the product. This means that the migrating group is never separated completely from the carbon, or if it is, the new bond forms before rotation of the two aromatic groups which would destroy the optical activity.[19]

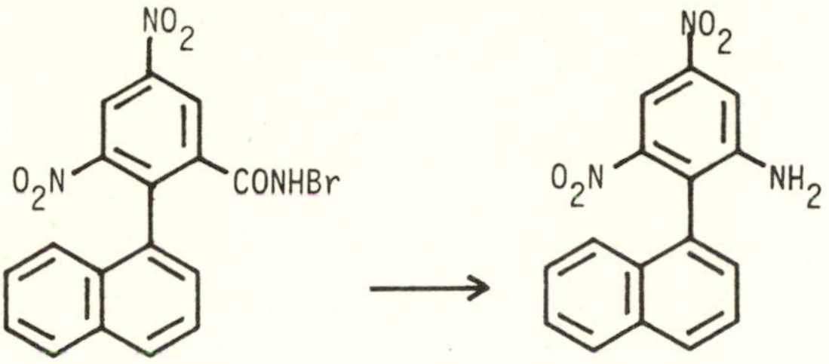

Several experiments showed that rearrangements carried out in the presence of triphenylmethyl radicals did not lead to incorporation of the triphenylmethyl residue, confirming the idea that the rearranging groups were not separated from each other.[20]

Later careful studies showed that the migration group in these nitrogen rearrangements retained its optical configuration. As Archer pointed out, however, this result was clear from work of W. A. Noyes in 1914 on camphor derivatives.[21]

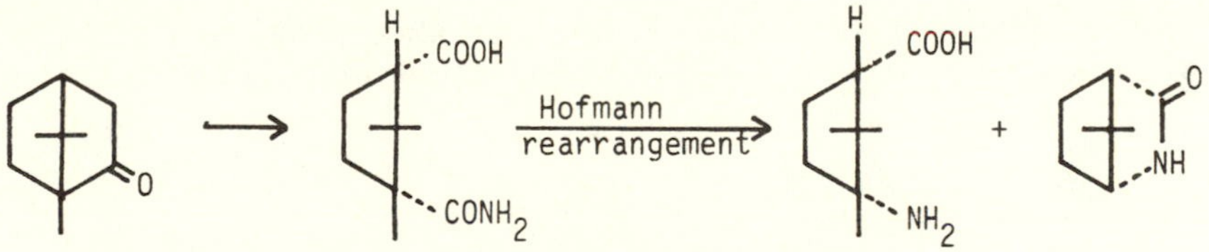

The benzilic acid rearrangement was shown by Frank H. Westheimer (then at Columbia) to take place in the negative ion[22] formed in a fast reversible step by addition of OH^-, followed by a slow shift of the phenyl group; this mechanism is compatible with ^{18}O tracer studies by Urey.[23] Although this rearrangement occurs in a negative ion, rather than in an electron-deficient one as in most of the Whitmore type, it was pointed out[22] that the benzilic acid rearrangement follows the Whitmore pattern, because the phenyl group shifts with its electrons to an electron-poor atom, the C of the C=O.

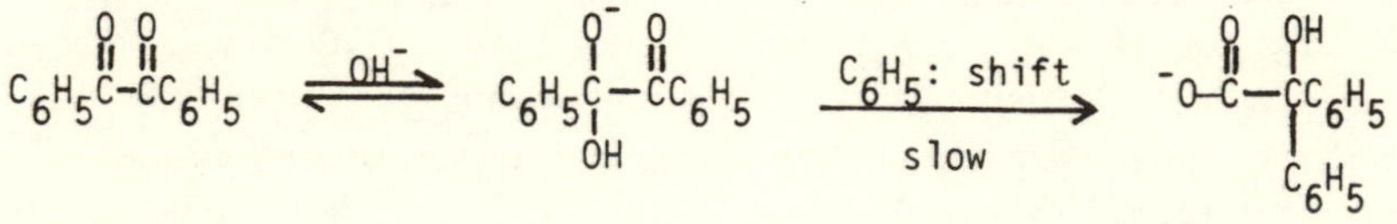

$$C_6H_5\overset{O}{\overset{\|}{C}}-\overset{O}{\overset{\|}{C}}C_6H_5 \;\rightleftharpoons\; \xrightarrow{OH^-}\; C_6H_5\overset{O^-}{\underset{OH}{\overset{|}{C}}}-\overset{O}{\overset{\|}{C}}C_6H_5 \;\xrightarrow[slow]{C_6H_5:\ shift}\; {}^-O-\overset{O}{\overset{\|}{C}}-\underset{\underset{C_6H_5}{|}}{\overset{OH}{\overset{|}{C}}}C_6H_5$$

Benzil   Benzilic acid

Whitmore's 1932 paper also included the allylic or 1,3-shift illustrated here, which he regarded as occurring in a positive ion.

$$CH_3CH=CHCH_2Br \;\rightleftharpoons\; CH_3\overset{Br}{\overset{|}{C}}HCH=CH_2$$

The allylic shift had been studied intensively in England, but William G. Young and Saul Winstein at the University of California at Los Angeles (UCLA) and their group are identified with it because of their outstanding work. They first recognized the rapid equilibration of the isomeric bromides below and devised methods for isolation of pure products.[24] Part of the driving force for this rearrangement is the resonance stabilization of the allyl carbonium ion, but other factors, such as the nature of the anion and the presence of acid, enter in.

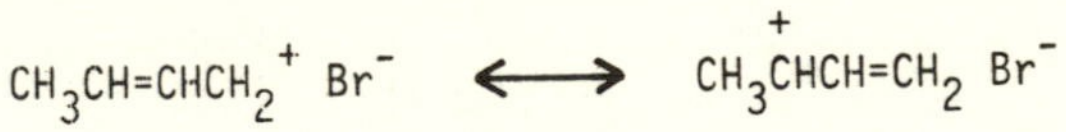

$$CH_3CH=CHCH_2{}^+\ Br^- \;\longleftrightarrow\; CH_3\overset{+}{C}HCH=CH_2\ Br^-$$

The usefulness of Whitmore's ideas in correlating so many diverse observations,[17] many of which we cannot discuss in detail, accustomed American organic chemists to thinking in terms of ionic intermediates, and therefore made the reaction schemes of the English school better known and more influential in this country.

There are numerous examples of rearrangements catalyzed by very strong bases; these may involve a divalent carbon intermediate (a carbene) of the

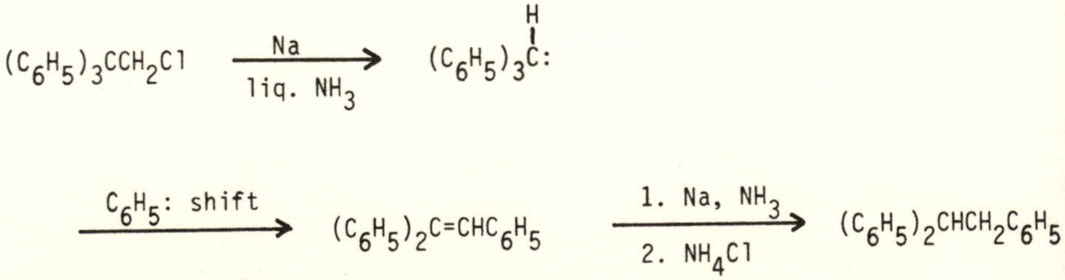

$$(C_6H_5)_3CCH_2Cl \;\xrightarrow[liq.\ NH_3]{Na}\; (C_6H_5)_3\overset{H}{\overset{|}{C}}:$$

$$\xrightarrow{C_6H_5:\ shift}\; (C_6H_5)_2C=CHC_6H_5 \;\xrightarrow[2.\ NH_4Cl]{1.\ Na,\ NH_3}\; (C_6H_5)_2CHCH_2C_6H_5$$

type postulated by Nef, followed by rearrangement. In the example shown, subsequent Na-NH$_3$ reduction occurs.[25] Formation of a diphenylacetylene from an unsaturated halide can occur.[26] This transformation might proceed

$$(C_6H_5)_2C\text{=}CHBr \xrightarrow[\text{liq. } NH_3]{KNH_2} C_6H_5C\text{≡}CC_6H_5 \quad + \quad KBr$$

via the vinyl carbene $(C_6H_5)_2C\text{=}C:$. Other rearrangements brought about by sodium amalgam and of uncertain mechanism are exemplified below.[27] The driving force here is probably the formation of the new aromatic ring and the acidity of the circled proton, because the sodium salt is the initial product of the rearrangement.

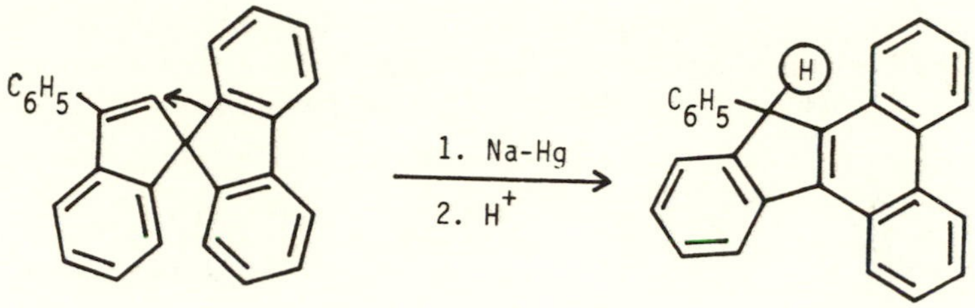

The action of acids on alkyl aryl ethers leads to alkylphenols, sometimes with rearrangement of the alkyl group. These reactions are similar to Friedel-Crafts reactions in being catalyzed by Lewis acids and showing, in some cases at least, intramolecular character.[28,29] Cases of purely thermal rearrangement of saturated ethers are known.

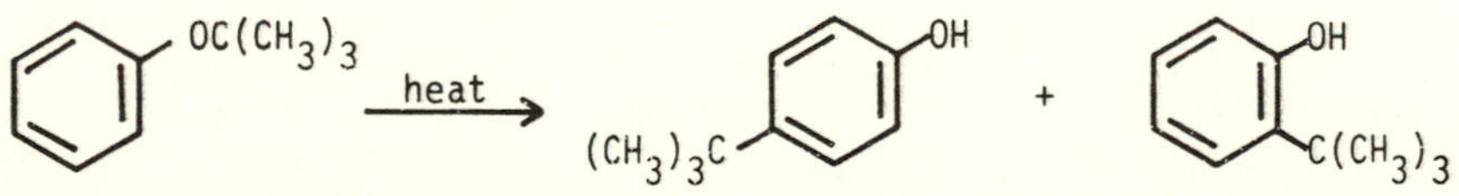

Other reactions involving a shift from a side chain to an aromatic nucleus, which were known from earlier European work, were extensively studied in this country. One was the acid-catalyzed reaction of N-chloroacetanilide to p-chloroacetanilide.[30] This was believed to involve the intermediate formation of free chlorine, but, in spite of experiments using radioactive chlorine, a satisfactory mechanism was not developed.[30]

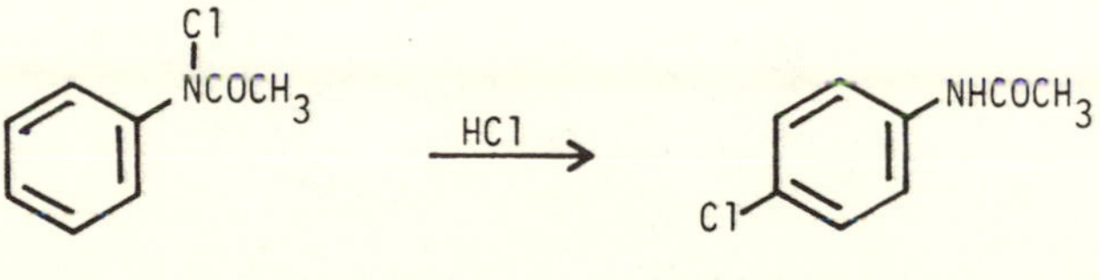

A reaction with a very modern look is the conversion of the ylide (doubly charged ion) into the aromatic substitution product.[31] Somewhat similar rearrangements were studied in detail years later by C. R. Hauser.

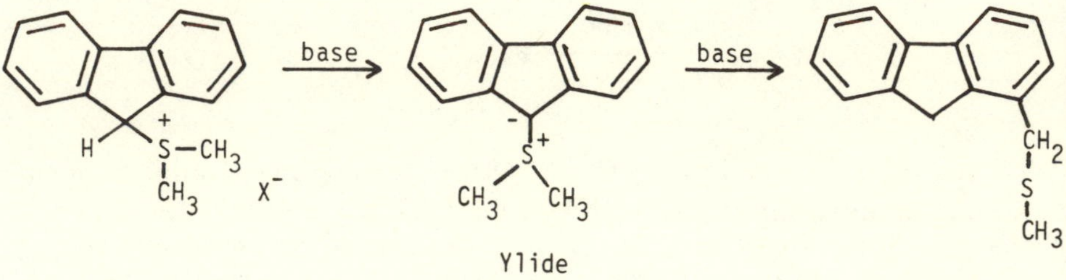

Ylide

The thermal rearrangement of allyl aryl ethers, discovered by L. Claisen in Germany in 1912, was shown by D. S. Tarbell and J. F. Kincaid at Rochester to be a first-order intramolecular reaction, for migration to both the ortho and the para positions.[32] These kinetic measurements were apparently the first in which the entropy of activation derived from Eyring's deve-

quantitative

lopment of the transition state theory was calculated and applied to an organic reaction. The entropy of activation was comparably negative for both o- and p-migrations indicating a highly ordered transition state probably similar for both reactions;[32] this was shown later by several lines of evidence to be correct. As the transition state implies, the allyl group after reaction is attached to the ortho position by the γ-carbon, not by the carbon attached to the ether oxygen.

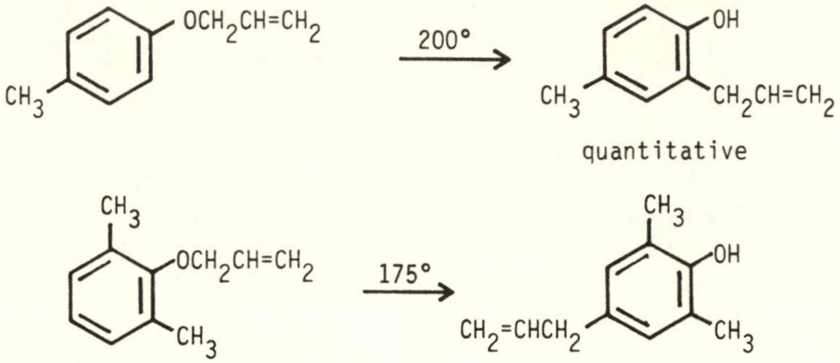

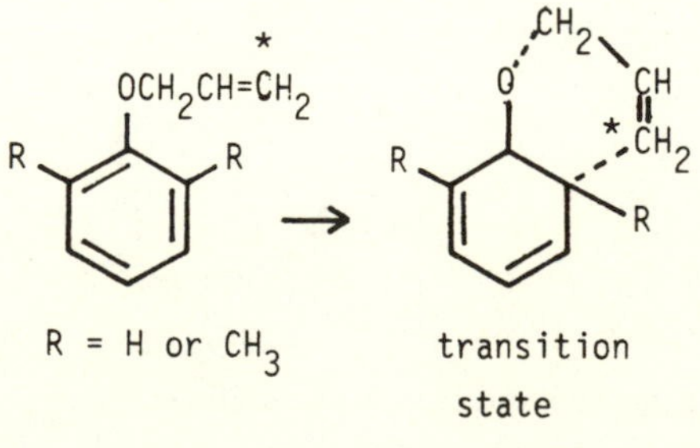

R = H or CH₃   transition
                        state

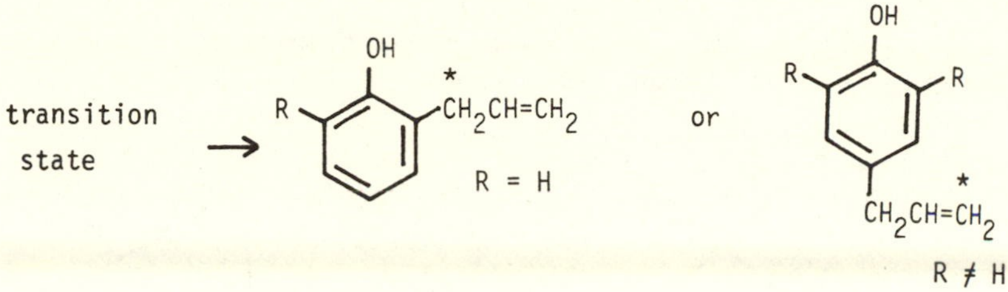

transition
state

$\rightarrow$

R—(ring with OH and $\overset{*}{C}H_2CH=CH_2$)

R = H

or

R—(ring with OH, two R groups, and $\overset{*}{C}H_2CH=CH_2$)

R ≠ H

Hurd and Pollack[33] showed that the reaction does not require an aromatic ring, allyl vinyl ethers rearranging smoothly to unsaturated aldehydes via a similar transition state.

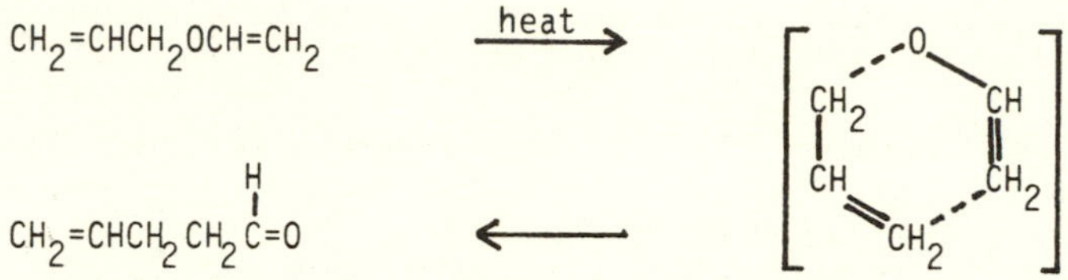

$$CH_2=CHCH_2OCH=CH_2 \quad \xrightarrow{\text{heat}}$$

$$\left[\begin{array}{c} CH_2 \cdots O \cdots CH \\ \\ CH \cdots CH_2 \\ CH_2 \end{array} \right]$$

$$CH_2=CHCH_2CH_2\overset{H}{\underset{}{C}}{=}O \quad \longleftarrow$$

After 1945, studies on the mechanism and synthetic application of the Claisen rearrangement, particularly the vinyl ether type, increased enormously. Numerous ingenious variants on the latter became of great synthetic value.[16] The Claisen rearrangement requires the grouping C=C-C-O-C=C, where the O-C=C may be part of an aromatic or heterocyclic ring or may be an isolated C=C.

An analogous reaction in an all-carbon system is the Cope rearrangement, mentioned briefly here because of its great later importance, both theoretical and synthetic;[34] it is shown below and it goes through a cyclic transition state similar to the Claisen.[35]

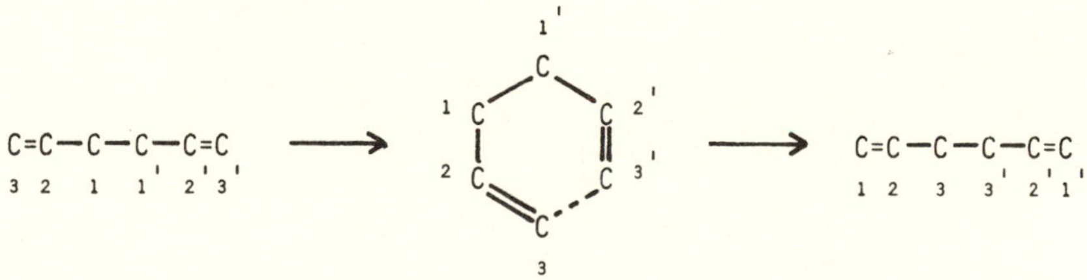

$$C=C-C-C-C=C \quad \longrightarrow \quad \text{(cyclic transition state)} \quad \longrightarrow \quad C=C-C-C-C=C$$
$$\begin{array}{cccccc} 3 & 2 & 1 & 1' & 2' & 3' \end{array} \qquad\qquad\qquad\qquad \begin{array}{cccccc} 1 & 2 & 3 & 3' & 2' & 1' \end{array}$$

A. C. Cope's career included his discovery of the Cope rearrangement, an object lesson in careful experimental work and intelligent observation. During the preparation of some unsaturated esters to be used in synthesizing new barbituric acids for pharmacological testing, he noted during distillation that there was an increase in refractive index as the distillation proceeded. He traced this increase to the new 1'-2' carbon-carbon double bond conjugated with the ester group.[34] Before the Instrumental Revolution made measurements of UV spectra simple, conjugation was most easily recognized by an increase in the refractive index. Cope's careful study of the reaction was followed by research by many other workers.

LITERATURE CITED

1. C. W. Porter, *Molecular Rearrangements*, New York, 1928. By 1960, it required a 2-volume collaborative work of nearly 1200 pages to review the field, dealing mainly with work after 1929: Paul De Mayo, Ed., *Molecular Rearrangements*, 2 vol., New York, 1963, 1964.
2. E. S. Wallis, in Henry Gilman, Ed., *Organic Chemistry, An Advanced Treatise*, New York, 1938, Vol. 1, pp. 720-801.
3. R. Fittig, Ann., **114**, 56 (1860).
4. Dorothy E. Bateman and C. S. Marvel, JACS, **49**, 2914 (1927).
5. W. A. Noyes and G. S. Skinner, **39**, 2692 (1917).
6. H. Meerwein and K. van Emster, Ber., **55**, 2500 (1922).
7. P. D. Bartlett et al., JACS, **59**, 820 (1937); **60**, 1585 (1938); **63**, 1273 (1941).
8. F. C. Whitmore, ibid., **54**, 3274 (1932).
9. C. S. Marvel in *Biog. Mem. Nat. Acad. Scis.*, **28**, 289 (1954). We are indebted to Dr. K. T. Finley of Brockport State College for the use of extensive unpublished information about Whitmore, and for Finley's paper on carbonium ions in Eastman Organic Chem. Bull., **41**, No. 2 (1969).
10. J. G. Aston, Chem. Eng. News, **23**, 1834 (1945).
11. F. C. Whitmore and E. Rohrmann, JACS, **61**, 1591 (1939).
12. F. C. Whitmore and E. E. Stahly, ibid., **55**, 4153, 5079 (1933).
13. F. C. Whitmore, Ind. Eng. Chem., **26**, 94 (1934).
14. F. C. Whitmore, Chem. Eng. News, **26**, 668 (1948); P. D. Bartlett et al., JACS, **66**, 1531 (1944); L. Schmerling, ibid., p. 1422.
15. Review, R. B. Clayton, Quart. Rev., **19**, 168 (1965).
16. W. S. Johnson, Bioorg. Chem., **5**, 51 (1976); R. B. Woodward and K. Bloch, JACS, **75**, 2023 (1953); W. G. Dauben et al., ibid., p. 3038.
17. An admirable account of the usefulness of Whitmore's theory, from the standpoint of 1950, is by G. W. Wheland, *Advanced Organic Chemistry*, New York, 1957, pp. 475-535. Insights into Whitmore's ideas, more detailed than in his 1932 paper, are given in letters between Whitmore and Wheland in 1944. Part of one of these has been published by M. D. Saltzman, JCE, **54**, 25 (1977). We have read this correspondence by the courtesy of Mr. Harry E. Whitmore and Dr. K. T. Finley.
18. L. W. Jones and E. S. Wallis, JACS, **48**, 169 (1926); Wallis and R. D. Dripps, ibid., **55**, 1701 (1933); review on Beckmann rearrangement, A. H. Blatt, Chem. Rev., **12**, 215 (1933).
19. Wallis and W. M. Moyer, JACS, **55**, 2598 (1933).
20. G. Powell, ibid., **51**, 2436 (1929); E. S. Wallis, ibid., p. 2982.
21. S. Archer, ibid., **62**, 1872 (1940); W. A. Noyes and L. F. Nickell, ibid., **36**, 118 (1914).
22. F. H. Westheimer, ibid., **58**, 2209 (1936); Westheimer credits P. D. Bartlett with pointing out the similarity to Whitmore's rearrangement theory.

23. I. Roberts and H. C. Urey, ibid., **60**, 880 (1938).
24. W. G. Young and S. Winstein, ibid., **57**, 2013 (1935); Young et al., **60**, 900 (1938); **61**, 2564 (1939); Young and J. F. Lane, **60**, 847 (1938); Young has many later papers.
25. C. B. Wooster and N. W. Mitchell, ibid., **52**, 1042 (1930) and later papers; E. E. Harris, ibid., p. 3633; L. Hellerman and R. L. Garner, ibid., **57**, 139 (1935).
26. G. H. Coleman and R. D. Maxwell, ibid., **56**, 132 (1934).
27. C. F. Koelsch, ibid., **56**, 480 (1934).
28. J. B. Niederl and S. Natelson, ibid., **53**, 1928 (1931); R. A. Smith, ibid., **55**, 3718 (1933); the Lewis acid BF_3 is an excellent catalyst (F. J. Sowa, S. D. Hinton and J. A. Nieuwland, ibid., p. 3402, and later papers).
29. S. Natelson, ibid., **56**, 1583 (1934); J. B. Niederl and C. H. Riley, ibid., p. 2412.
30. A. R. Olson et al., ibid., **59**, 1613 (1937); H. S. Harned and H. Seltz, ibid., **44**, 1475 (1922); S. F. Acree had worked on this reaction much earlier.
31. The ylide was prepared by C. K. Ingold and J. A. Jessop, JCS, 713 (1930). The rearrangement was recognized by G. E. Hilbert and L. A. Pinck, JACS, **60**, 494 (1938) of the U. S. Department of Agriculture.
32. Kinetics of Claisen rearrangement: D. S. Tarbell and J. F. Kincaid, ibid., **61**, 3085 (1939); **62**, 728 (1940); reviews, D. S. Tarbell, Chem. Rev., **27**, 495 (1940); *Organic Reactions*, **2**, 1 (1944). The intramolecular nature of the reaction was shown by C. D. Hurd and L. Schmerling, JACS, **59**, 107 (1937).
33. C. D. Hurd and M. A. Pollack, ibid., **60**, 1905 (1938); earlier observations by Claisen, and by W. M. Lauer and E. I. Kilburn, ibid., **59**, 2586 (1937), were on enol allyl ethers. For mechanism, C. D. Hurd and M. A. Pollack, JOC, **3**, 550 (1939).
34. A. C. Cope and Elizabeth M. Hardy, JACS, **62**, 441 (1940) and later papers.
35. Sara Jane Rhoads and N. Rebecca Raulins, *Organic Reactions*, **22**, 1-252 (1975) for recent work on Claisen and Cope rearrangements.

LOUIS P. HAMMETT

18.

Physical Organic Chemistry
A. Acids and Bases, 1914-1940

Physical organic chemistry includes reaction mechanisms, valence theory, structure and physical properties, and stereochemistry. We shall not follow the 1940 boundary slavishly, but the war years 1941-1945 do form a real division. Research after 1945, although it showed continuity in problems, was distinctly different in scale and approach to problems, instrumentation, the importance of American science on the world scene, and sources of research support. In reading papers prior to 1940, one is conscious of a simpler outlook in chemistry, of a field less complex, less highly specialized, and less fragmented. It was not necessarily a better world, but it was a different one, in terms of attitude and mood of the research workers, as well as in facilities.

In earlier sections we have documented the thesis that American physical organic chemistry had a long and productive history before 1914[1] and have discussed the development of the valence theory by G. N. Lewis, Pauling, and many others which provided a satisfactory theoretical basis for considering organic structures and reactions. In this and subsequent essays we plan to show the development of a number of fundamental topics, primarily but not exclusively from the American scene, and to focus on telling contributions in their development.

Although some organic reactions take place in the gas phase, we shall focus in this section on processes in solutions. Most biochemical processes take place in an aqueous medium and involve more complicated structures (enzymes, cell membranes) than the simple organic compounds. The effect of acids and bases on many organic reactions and on most biological processes is of fundamental importance.

The theory of acids and bases expressed by J. N. Brönsted in 1923 defined an acid as a source of protons, H^+, and a base as a proton acceptor.[2,3] This generalized concept of acids and bases is valid for all solvents, both aqueous and non-aqueous: HA and HB are acids, and A and B are bases.

$$HA + B \rightleftharpoons HB + A$$

This dissociation of a strong acid in aqueous solution, such as hydrochloric acid, HCl, or perchloric acid, $HClO_4$, illustrates this idea. Thus reaction proceeds completely to the right, H_2O and A^- are the bases, and HA and H_3O^+ are the acids. It is apparent that in water solution H_3O^+ is the strongest acid that can be formed, and by a similar argument, the strongest base that can exist in H_2O is the hydroxyl ion, OH^-. This "leveling" effect of a solvent was pointed out first by A. Hantzsch[4] in Germany and has become a key idea in acid-base chemistry.

$$HA + H_2O \rightleftharpoons H_3O^+ + A^-$$
$$A = Cl^-, ClO_4^-$$

Liquid ammonia, NH_3, is a stronger base (better proton acceptor) than water, and by the reciprocal relation which exists between acidic and basic strength, the amide ion, NH_2^-, derived from NH_3 by removal of a proton, is a much stronger base than the OH^- derived from water. The notion of liquid ammonia as a solvent comparable to water is present implicitly in the work of C. A. Kraus and E. C. Franklin[5] at Kansas in 1900.

Before we pursue experiments based on the Brönsted definition, the more general definition of acids and bases by Lewis should be mentioned.[6] A Lewis acid is an electron pair acceptor (H^+, $AlCl_3$, BF_3 and many others), and a Lewis base is an electron pair donor (OH^-, NH_3, R_2O, $R_2C=O$, $RCOOH$, $RCH = CHR$, and many more). This more inclusive idea has been of great value in mechanistic and synthetic thought.

A series of inorganic proton acids, including HCl and $HClO_4$, in glacial acetic acid, CH_3COOH, gives the equilibrium shown. Because CH_3COOH is a

$$HA + CH_3\overset{O}{\overset{\|}{C}}\text{-OH} \rightleftharpoons CH_3\overset{+OH}{\overset{\|}{C}}\text{-OH} + A^-$$
$$A = Cl^- \text{ or } ClO_4^-$$

weaker base (poorer proton acceptor) than H_2O, the reaction may not go to completion with all proton acids which form H_3O^+ completely in H_2O. It was found by Conant that the reaction is complete for $HClO_4$, but is incomplete for HCl.[7,8] The former is thus a stronger acid (better proton donor) than the latter. In aqueous solution, both HCl and $HClO_4$ appear to be equally good proton donors to H_2O, and both give H_3O^+ essentially completely. Thus $HClO_4$ can be used in glacial acetic acid to neutralize completely even weak bases, which are unaffected by H_3O^+ in H_2O. Solutions of proton acids in acetic acid were named *super acid solutions*. A very useful technique for titration of acids and bases in nonaqueous solutions has developed from work of this kind.[9]

James B. Conant (1893-1978, A.B. 1913, Ph.D. 1916, both at Harvard with E. P. Kohler and T. W. Richards) taught at Harvard from 1919 to 1933, when he became president of the university. His brief research career embraced a wide range of important problems in physical organic chemistry and reaction mechanisms as well as work on the structure of chlorophyll and an increasing interest in the processes occurring in living systems. He was high commissioner and then ambassador to West Germany from 1953 to 1957 and played a leading role in the scientific research of World War II, including the atomic bomb project. He conducted a ten-year study on American secondary school education, resulting in a number of books.[10]

By using mixtures of sulfuric acid, H_2SO_4, and water up to pure H_2SO_4, very weak bases can be protonated, and basic strengths can be determined for bases as weak as 2,4,6-trinitroaniline. L. P. Hammett devised an H_o function, which is to a first approximation a measure of the ability of a given acid solution to transfer a proton to a weak base.

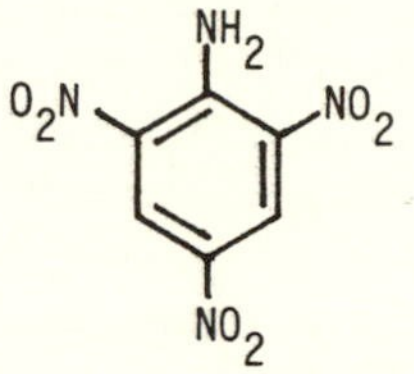

2,4,6-trinitroaniline

protonated methane

He found that benzoic acid was protonated by 100 percent H_2SO_4 as ex-

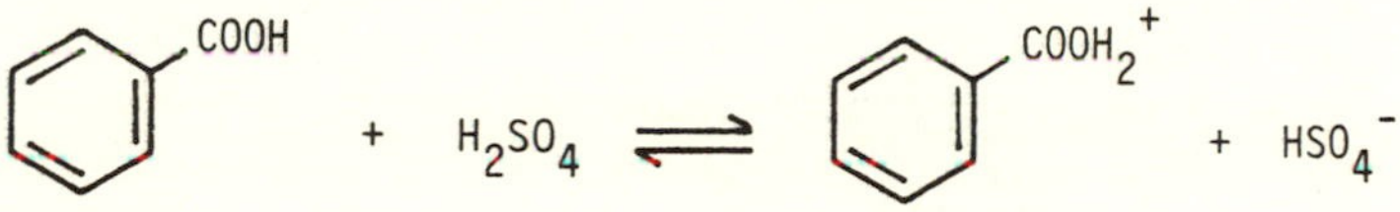

pected. However, 2,4,6-trimethylbenzoic acid and related compounds with two methyl groups next to the COOH group were protonated to yield the acylium carbonium ion, $ArCO^+$. The two ortho methyl groups thus caused the initial protonation product to lose an H_2O molecule. Electron-attracting groups like Br or NO_2 in the ortho positions did not promote acylium ion formation, so that it was not a mechanical crowding effect.[12]

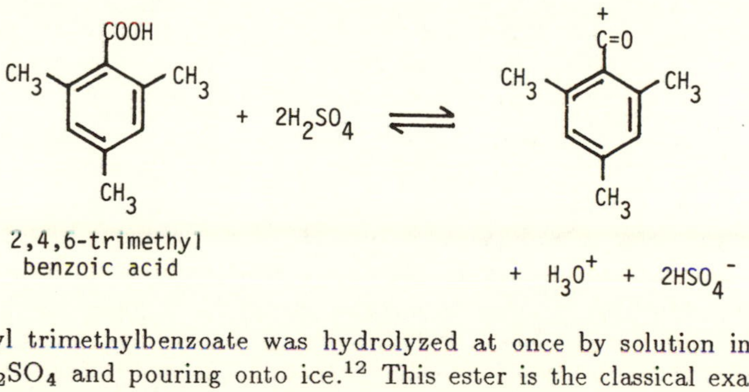

2,4,6-trimethyl benzoic acid

Methyl trimethylbenzoate was hydrolyzed at once by solution in 100 percent H_2SO_4 and pouring onto ice.[12] This ester is the classical example of steric hindrance in which reactions involving the C=O are normally very slow. M. S. Newman[13] found that the reverse of Hammett's reaction gave an excellent way of preparing the hindered esters by solution of the acid in 100 percent H_2SO_4, then addition of an alcohol followed by quenching in water.

$$\xrightarrow{H_2SO_4} \quad ArC=O \quad \xrightarrow{H_2O} \quad ArCOOH \; + \; H^+$$

Louis P. Hammett (1894-   ,A.B. 1916, Harvard; Ph.D. 1923, Columbia with H. T. Beans) taught at Columbia from 1920 to 1961. His classic book, *Physical Organic Chemistry* (1940), was a leading influence on the development of the field. In addition to his work on acid solutions, he pioneered in discovering quantitative relationships between rate and equilibria in organic reactions. His equation connecting the effects of substituents in aromatic rings on the acidity of substituted benzoic acids and their effects on rates of reactions was a milestone. Although his list of publications is not long, a high proportion of them is significant, and many are authentic classics.

Melvin S. Newman (1908-   , B.S. 1929, Ph.D. 1932, both at Yale with R. J. Anderson) in his doctoral research isolated and identified phthiocol (2-hydroxy-3-methyl-1, 4-naphthoquinone) from the wax of tubercle bacilli, presumably formed by the breakdown of vitamin K-like compounds. After postdoctoral work with Fieser at Harvard on carcinogenic aromatic compounds, he spent a productive career at Ohio State, working on aromatic chemistry, including the elegant synthesis and resolution of hexahelicene, steric effects, and varied synthetic and mechanistic problems. He edited a valuable book on steric effects.[15]

In recent work, FSO_3H-SbF_5 (*magic acid*) has been shown to be an extraordinarily powerful protonating agent, causing protonation and various subsequent reactions with such weak bases as saturated hydrocarbons like methane.[16] This mixture contains a very strong proton acid and a very strong Lewis acid and has enlarged the range of acidities and chemical horizons as much as the work of Hantzsch, Conant, and Hammett in the 1920s and 1930s.

Conant published an excellent paper on the relative strengths of very weak organic acids, most of them hydrocarbons.[17] This work is clearly a development of Kraus' studies on liquid NH_3 solutions, although Conant made his measurements semiquantitative.[18] The measurements were based on the fact that a stronger acid will liberate a weaker acid from its salts. By observing the behavior of a number of systems at equilibrium, using several experimental methods and making some reasonable assumptions, Conant arranged the acids in order of strengths or tendency to ionize, where the acid strength is given by the equilibrium constant for the process. Table 1 shows some of the relative acidities obtained. The absolute pKa values are approximate, but much later work has shown that they are reasonably correct. (The larger the pKa value, the weaker the acid).

$$RH \rightleftharpoons R^- + H^+ \qquad K = acid\ strength = \frac{[R^-]\,[H^+]}{[RH]}$$

$$pKa = -log\ K$$

These relative values are reasonable; alcohols, ROH, are acidic because the electronegative oxygen attains a negative charge with an octet of electrons around the oxygen. In *tert*-butyl alcohol, $(CH_3)_3COH$, the electron-donating effect of the CH_3 groups makes it more difficult for the proton to leave the oxygen and also cuts down on solvation of the negative ion. The reason for the acidity of the hydrogen attached to the *sp* hybridized carbon in acetylenes, $RC\equiv CH$, is not entirely clear, but it is an exceedingly useful synthetic property. Acetophenone, $C_6H_5COCH_3$, like other ketones with α-hydrogens, is acidic because the negative charge is delocalized over the oxygen as well as the carbon. In fluorene and triphenylmethane, $(C_6H_5)_3CH$, the negative charge remaining after removal of the proton is stabilized by delocalization in the aromatic rings.

Because potassium amide, $K^+NH_2^-$, will take a proton from $(C_6H_5)_2CH_2$, but not from $C_6H_5CH(CH_3)_2$, the pKa of NH_3 must be about 36. Thus the

Table 1. Relative acidities (Conant)

Compound	pKa in ether
CH_3OH	16
$(CH_3)_3OH$	19
$C_6H_5COCH_3$	19
$C_6H_5C\equiv CH$	21
fluorene[a]	25
$(C_6H_5)_3CH$	32.5
$(C_6H_5)_2CH_2$	35
$C_6H_5CH(CH_3)_2$	37

a

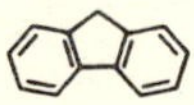

easily made bases sodamide, $NaNH_2$, and KNH_2 are very strong, capable of removing protons from very weak acids.[19] The great basic strength of the NH_2^- must be due to the two unshared pairs of electrons on the ion and to the additional fact that nitrogen is nearer the center of the periodic table than oxygen and hence less able than oxygen to tolerate a negative charge.

These experiments on weak acids, and many more recent ones which are in general agreement, are not only of theoretical interest but also of considerable synthetic importance. The reason is that a large family of reactions requires as the first step the removal of a proton by a base from carbon to form a carbanion (a carbon with eight electrons but only three groups attached, hence a negative charge). For example, a general synthetic reaction, used in the synthesis of vitamin A, involves the formation of a carbanion from an acetylene, $RC \equiv CH$, followed by reaction of this with the positive carbon of a carbonyl group or with a compound containing a group readily removed as a negative ion.

$$C_6H_5C \equiv CH \ + \ C_4H_9Li \ \longrightarrow \ C_6H_5C \equiv C^-Li^+ \ + \ C_4H_{10}$$

$$\downarrow CH_3I$$

$$C_6H_5C \equiv CCH_3 \ + \ LiI$$

The large basic strength of $NaNH_2$ has been emphasized. The alkyl lithium compounds, RLi, being the salts of saturated hydrocarbons (even weaker acids than NH_3) are even stronger bases than salts of NH_3, such as $NaNH_2$. However, $NaNH_2$ and the RLi compounds may react in ways other than by removing a proton to form a carbanion, such as addition to a carboxyl group. One way to avoid this is by converting a secondary amine into an amide ion which removes protons readily but attacks carbonyl groups only slowly. This idea was first developed in 1932 by the German chemist Karl Ziegler[21] into a method for making tri-α-substituted nitriles and carboxylic acids, such as $(C_2H_5)_2C(C_4H_9)COOH$.

$$RLi \ + \ R_2'NH \ \longrightarrow \ R_2'N^-Li^+ \ \xrightarrow{(C_2H_5)_2CHCN}$$

$$R_2'NH \ + \ (C_2H_5)_2\overset{-}{C}C \equiv N \ \ Li^+$$

$$\updownarrow$$

$$(C_2H_5)_2C = C = N^- \ Li^+ \ \xrightarrow{C_4H_9Br} \ (C_2H_5)_2\underset{\underset{C_4H_9}{|}}{C}CN$$

$$R' \ = \ C_2H_5 \ or \ \underline{i}\text{-}C_3H_7$$

Because carbanions are very important synthetic intermediates, it is of great synthetic utility to have strong bases available whose action is limited to carbanion formation. Since Ziegler's work on lithium dialkylamides, other workers have shown that basic strength can be greatly increased by proper choice of solvents or use of complexing agents which expose the base for more effective action in proton removal.

The broad Lewis concept of acids as electron acceptors and bases as electron donors[6,22] correlates a wide variety of phenomena, gives a sound basis for predicting chemical behavior, and is valuable in explaining important reactions and suggesting new ones. A general review of this topic would require consideration of large areas of organic and inorganic chemistry, and a few salient points may illustrate the value of the idea.

The compound or "salt" formed by a Lewis acid, trialkylboron, and a Lewis base, NH_3, is a typical example,[23] explored later in great detail by H. C. Brown. The most important Lewis acids are H^+, R_3B, BF_3, BCl_3, $AlCl_3$, $AlBr_3$, $FeCl_3$, $ZnCl_2$, $SnCl_4$, SbF_3, $HgCl_2$, SbF_5, and some organic compounds. Lewis bases with unshared electrons include N, P, and S derivatives in their lower valence state, and organic compounds with multiple linkages, such as $C=C$, $C=O$, $C=N$, $C \equiv N$, and others.

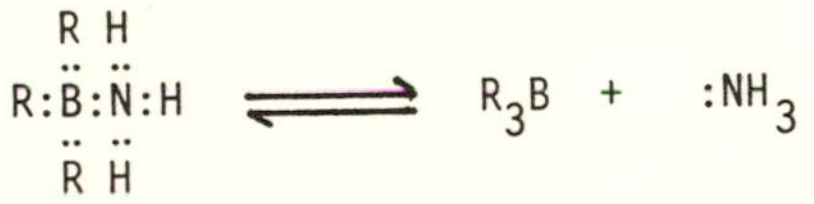

$$\begin{array}{c} R\ H \\ R\!:\!\overset{..}{B}\!:\!\overset{..}{N}\!:\!H \\ \overset{..}{R}\ \overset{..}{H} \end{array} \quad \rightleftharpoons \quad R_3B \ + \ :NH_3$$

Hans Meerwein[24] gave detailed experimental evidence for the importance of Lewis' acids[25] and pointed out that BF_3 is a very strong Lewis acid; this is because of the electron-attracting effect of the F atoms and the small size of the boron nucleus with its concentrated positive charge. Kraus[26] described stable compounds from BF_3, NH_3, and several amines. The use of BF_3 for catalyzing a variety of organic reactions was demonstrated by F. J. Sowa and others at Notre Dame;[27] one is shown here.

$$RCOOR' \ + \ C_6H_6 \ \xrightarrow{BF_3} \ RCOOH \ + \ C_6H_5R'$$

The very widely used Lewis acid $AlCl_3$ traditionally catalyzes the Friedel-Crafts reaction and its numerous variants.[28] The reaction was discovered by Charles Friedel (1832-1899), professor at the University of Paris, and the American, John Mason Crafts (1839-1917). Crafts had a varied career after graduating from Harvard in 1858, studying mining in Freiburg in 1859 and chemistry with Bunsen at Heidelberg and Wurtz and Friedel at Paris. He returned to this country in 1865, traveling in Mexico and the West inspecting mines. He became the first professor of chemistry at Cornell, and in 1870 professor at MIT. In 1874 he returned to Paris, worked with Wurtz, and then collaborated with Friedel, publishing over 100 papers with him and staying in Paris until 1891. He served MIT again as professor of chemistry, as president of the Institute, and then as a research worker.

The general mechanism for the Friedel-Crafts reaction[29] involves donation and acceptance of an electron pair. In some cases there is actual ionization of the Lewis salt, and in others probably only a polarization or partial electron shift.

$$C_6H_5\overset{O}{\overset{||}{C}}Cl \;+\; AlCl_3 \;\rightleftharpoons\; C_6H_5\overset{O}{\overset{||}{C}}{}^+ \, AlCl_4{}^- \overset{C_6H_6}{\longrightarrow} C_6H_5\overset{O}{\overset{||}{C}}C_6H_5 \;+\; HCl \;+\; AlCl_3$$

$$C_6H_5CH_2Cl \;+\; AlCl_3 \;\rightleftharpoons\; C_6H_5CH_2{}^+ \, AlCl_4{}^-$$

$$\downarrow C_6H_6$$

$$C_6H_5CH_2C_6H_5 \;+\; HCl \;+\; AlCl_3$$

A careful study of mercuric salts as Lewis acids shows the carbonium ion as the intermediate.[30]

$$2C_6H_5CH_2Cl \;+\; Hg(NO_3)_2 \longrightarrow 2C_6H_5CH_2NO_3 \;+\; HgCl_2$$

Mechanism:

$$C_6H_5CH_2Cl \;+\; Hg^{++} \xrightarrow{\text{slow}} C_6H_5CH_2{}^+ \;+\; HgCl^+$$

$$\text{fast} \downarrow NO_3{}^- \text{ or } H_2O$$

$$C_6H_5CH_2NO_3 \;+\; C_6H_5CH_2OH$$

Another example of Lewis acids generating a carbonium ion is of great practical importance in forming branched-chain hydrocarbons for motor fuel, as well as of fundamental chemical significance.[31] The complicated further reactions and rearrangements that lead to highly branched larger molecules can be reasonably explained. It is not certain that $AlBr_3$ is intrinsically a stronger Lewis acid than $AlCl_3$, but for mechanistic studies the former has the advantage that it is much more soluble in organic solvents.

$$(CH_3)_3CCl \ + \ AlBr_3 \ \rightleftharpoons \ (CH_3)_3C^+ \ AlBr_3Cl^-$$

$$\downarrow C_2H_5CH(CH_3)_2$$

$$(CH_3)_3CH \ + \ C_2H_5\overset{+}{C}(CH_3)_2$$

$$\downarrow AlBr_3Cl$$

$$\overset{Br}{\underset{|}{C_2H_5CHC(CH_3)_2}} \ + \ AlBr_2Cl$$

Addition of halogens to C=C under ionic conditions where free radicals are not formed is regarded on the basis of much evidence as attack of a Lewis acid derived from the halogen on the C=C, which acts as a Lewis base. The reaction is completed by interaction with solvent or with a negative ion from solution;[32] this is discussed in another essay.

Another representative of the Lewis acid class is the trialkyloxonium salts. Diethyl ether is a weak base toward protons in water solution, the equilibrium being to the left.

$$C_2H_5\overset{..}{\underset{..}{O}}C_2H_5 \ + \ H_3O^+ \ \rightleftharpoons \ C_2H_5\overset{..+}{\underset{\underset{H}{..}}{O}}C_2H_5 \ + \ H_2O$$

Boron trifluoride forms a stable "salt" with ether, and indeed this is the usual form in which BF_3 is handled, because it is a relatively stable liquid whereas BF_3 itself is a corrosive gas. By procedures worked out by Meerwein and others, it is possible to prepare trialkyloxonium ions as the fluoroborate or other salts.[23]

$$C_2H_5\overset{..}{\underset{..}{O}}C_2H_5 \qquad\qquad C_2H_5\overset{..}{\underset{..}{O}}C_2H_5 \ BF_4^-$$
$$BF_3 \qquad\qquad\qquad C_2H_5$$

boron trifluoride   triethyloxonium
etherate   fluoroborate

These salts are strong Lewis acids and transfer an ethyl group to any base stronger than diethyl ether, a reaction of considerable synthetic value.

$$(C_2H_5)_3O^+X^- \ + \ \overset{O}{\underset{\|}{HCN(CH_3)_2}} \ \longrightarrow \ \overset{OC_2H_5}{\underset{|}{HC=N^+(CH_3)_2X^-}} \ + \ C_2H_5OC_2H_5$$

It may be noted that oxonium salts with three carbon-oxygen bonds are stable if the accompanying base, such as BF_4^-, is very weak. They can also be stabilized if the oxygen is part of a benzenoid-like ring where the positive charge is stabilized by resonance, i.e., by being delocalized over a conjugated system, as in the pyrylium salts. The coloring materials of flowers are closely related to similar oxonium salts of somewhat more complex structure, such as pelargonidin chloride.

dimethylpyrylium
chloride

pelargonidin chloride

LITERATURE CITED

1. L. P. Hammett, *Physical Organic Chemistry*, 2nd ed., New York, 1970, pp. 97 ff. expresses a different view.
2. J. N. Brönsted, Rec. Trav. Chim., **42**, 718(1923); G. N. Lewis, *Valence*, p. 142.
3. J. N. Brönsted, Chem. Rev., 5, 231(1928).
4. A. Hantzsch, Z. Elektrochem., **29**, 224, 234 (1923).
5. E. C. Franklin and C. A. Kraus, ACJ, **23**, 304(1900) and later papers.
6. Lewis, *Valence*, p. 142; J. Franklin Inst., **226**, 293 (1938).
7. N. F. Hall and J. B. Conant, JACS, **49**, 3047, 3062 (1927); N. F. Hall and T. H. Werner, ibid., **50**, 2367 (1928).
8. Hantzsch (loc. cit., p. 228) recognized the unique (at that time) acid strength of $HClO_4$.
9. J. S. Fritz, *Acid-Base Titrations in Non-Aqueous Solvents*, Columbus, Ohio, 1952.
10. For Conant's career, see his autobiography, *My Several Lives*, New York, 1970; P. D. Bartlett, *Biog. Mem. Nat. Acad. Scis.*, **54**, 91 (1983).
11. L. P. Hammett and M. Paul, JACS, **56**, 827 (1934); Hammett, Chem. Rev., **13**, 161 (1933).
12. H. P. Treffers and L. P. Hammett, JACS, **59**, 1708 (1937).
13. M. S. Newman, ibid., **63**, 2431 (1941), and later papers.
14. See Hammett's autobiographical paper, JCE, **43**, 464 (1966).
15. *Steric Effects in Organic Chemistry*, New York, 1956.
16. G. A. Olah et al., JACS, **91**, 3261 (1969), and many other papers.
17. J. B. Conant and G. W. Wheland, JACS, **54**, 1212 (1932); W. K. McEwen, ibid., **58**, 1124 (1936). McEwen was a student of Conant who completed his research after Conant went into administrative work.
18. C. A. Kraus and R. Rosen, ibid., **47**, 2739 (1925); C. B. Wooster and N. W. Mitchell, **52**, 688 (1930). For preparation and synthetic reactions of metal acetylides in liquid ammonia. J. A. Nieuwland et al., JOC, **2**, 1 (1937).
19. N. S. Wooding and W. C. E. Higginson, JCS, 778 (1952).
20. The alkyl lithium compounds are now known to be associated to dimers (double molecules) or higher polymers in solution, but this does not remove their basic strength.
21. K. Ziegler and H. Ohlinger, Ann., **495**, 84 (1932). Ziegler, in addition to much elegant work in physical organic chemistry, developed as an outgrowth of this the method of making polyethylene now universally used.

22. A discussion of acids and bases from the various definitions: W. F. Luder and Saverio Zuffanti, *The Electronic Theory of Acids and Bases*, Wiley, 1946; W. F. Luder, Chem. Rev., **27**, 547 (1940).

23. Lewis, *Valence*, p. 98.

24. Professor at Marburg in South Germany for many years, a highly original chemist, nearly all of whose papers are classics.

25. H. Meerwein, Ann., **455**, 250 (1927).

26. Kraus and E. H. Brown, JACS, **51**, 2690 (1929).

27. J. F. McKenna and F. J. Sowa, ibid., **59**, 1204 (1937), for example; the use of BCl_3 as a strong Lewis acid catalyst is reviewed in W. Gerrard and M. F. Lappert, Chem. Rev., **58**, 1081 (1958).

28. C. Friedel and J. M. Crafts, Compt. Rend., **84**, 1392, 1450 (1877) and many later papers. The monograph by C. A. Thomas, *Anhydrous Aluminum Chloride in Organic Chemistry*, New York, 1941, describes the varied reactions of this Lewis acid as known at the time we are discussing. For Crafts' life, see C. R. Cross, *Biog. Mem. Nat. Acad. Scis.*, **9**, 159 (1920).

29. cf. K. H. Klipstein and G. Dougherty, Ind. Eng. Chem., **18**, 1328 (1926).

30. I. Roberts and L. P. Hammett, JACS, **59**, 1063 (1937).

31. P. D. Bartlett, F. E. Condon and A. Schneider, ibid., **66**, 1531 (1944); L. Schmerling, ibid., **67**, 1778 (1945).

32. E. M. Terry and L. Eichelberger, ibid., **47**, 1067 (1925); P. D. Bartlett and D. S. Tarbell, ibid., **58**, 466 (1936); **59**, 407 (1937).

33. H. Meerwein et al., JPr, **147**, 257 (1937); H. Meerwein, *Organic Synth.*, Coll. Vol. V, p. 1080, 1096; G. K. Helmkamp and D. J. Pettit, ibid., p. 1099.

HOWARD J. LUCAS

B. Reactions of Carbonyl Compounds
1914-1939

Carbonyl compounds comprise a wide variety of organic substances, including many types that are fundamental in the chemistry of living systems. Because reactions of carbonyl groups are usually catalyzed by both acids and bases, they offer ample opportunity for studying such catalysis, especially but not exclusively, in water solution. The increased knowledge of reaction mechanisms during the period under review was due in large part to research on carbonyl reactions.

The importance of J. N. Brönstead's ideas on acids and bases has been discussed above, as well as G. N. Lewis' more general definition of acids and bases. Brönsted advanced other useful concepts which workers in this country developed in many directions.

Reactions subject to acid-base catalysis *could* show general acid-base catalysis, in which every acid and base present in a solution, including those present as buffer components, could make its individual contribution to the overall catalytic effect. Although this idea might be regarded as a corollary of Brönsted's definition of acids and bases, general acid-base catalysis is not universally observed.

Brönsted found experimental support for the first explicit linear free energy relationship which connected the rate of an acid-catalyzed reaction (a kinetic or rate effect) with the dissociation constants of acids (a thermodynamic or equilibrium quantity). Brönsted and many others studied these relationships, which held quantitatively only for series of closely related acids and reactions. This research strongly influenced Hammett's work on linear free energy relationships and through him the work of generations of American chemists.

Brönsted recognized the effects of added salts on the rates of reactions, distinguishing two opposite effects. These ideas stemmed from Lewis' much earlier use of activity coefficients and ionic strength. Much research on salt effects was done in this country, but its relevance to organic chemistry was not great until the time of Saul Winstein (after 1940), and we shall not discuss it for the period ending in 1939. As a practical experimental matter, kinetic measurements on organic reactions are best carried out in solutions of constant ionic strength, which allows salt effects to be neglected.

Brönsted postulated that reactions went through a "critical complex" in which reactants came together in an unstable structure that could revert to starting materials or go on to products. His interest in this complex was partly due to the possible effects of added salts on "complexes" of different charge types, and he did not speculate about the energy or entropy changes involved in reaching the critical complex. Nevertheless, Brönsted's postulation was an important stage in the development of reaction rate theory. The later "transition state" theory of Polanyi, Wigner, Eyring, and others in the 1930s emphasized the energetics of the formation of the transition state. The transition state is similar to Brönsted's critical complex and both were elaborations on the earlier ideas of "polymolecules" as reaction intermediates postulated by Michael, Kekulé, and even earlier workers. Chemists have not always realized that the three-dimensional potential energy surfaces for reactions popularized by the later transition state theory embody very drastic simplications and assumptions, and that the presumed greater precision in definition over Michael and Kekulé may actually be very small.

The carbonyl group was early recognized to have a positive end, the carbon, which is the point of attack by bases (*nucleophiles*), and a negative end, the oxygen, which is the point of attack by proton or Lewis acids (*electrophiles*).

$$\begin{array}{c} R \\ \diagdown \\ \diagup \\ R \end{array} C{=}O \quad \longleftrightarrow \quad \begin{array}{c} R \\ \diagdown \\ \diagup \\ R \end{array} \overset{+}{C}{-}\overset{-}{O}$$

J. B. Conant in 1921 recognized in an important paper that the dipolar character of the carbonyl group explained the course of several carbonyl addition reactions.[1] The dipolar nature of the carbonyl group was supported by dipole moment measurements by C. P. Smyth at Princeton.[2] Conant's paper was a direct outgrowth of Lewis' ideas on electronic structure published in 1916; Conant acknowledged discussion of Lewis' ideas with Elliott Q. Adams from Berkeley in 1917, when both E. Q. Adams and Conant were in Washington working on war research.[3] It will be recalled that Lewis' monograph on valence was not published until 1923. This is an example of the informal transfer of central ideas from one person to another usually not shown in the printed literature.

Many characteristics of acid-base catalysis and addition reactions of carbonyl compounds are illustrated in the classical study of semicarbazone formation by carbonyls by Conant and P. D. Bartlett.[4] This work showed strikingly the role of rate and equilibria in determining products, and unlike earlier studies on the reactivity of carbonyl groups such as Michael's,[5] it was thoroughly modern in viewpoint. It utilized buffer solutions of constant ionic strength, measured pH accurately with the hydrogen electrode (the glass electrode was not yet generally available), gave a pH rate profile which showed a rate maximum, and demonstrated general acid-base catalysis. The papers as a whole show the very marked influence of Brönsted on American physical organic chemistry. They also show a notable advance in understanding over the earlier work of Acree and of Lapworth.[6] In turn, an advance in technology, the development of a practical ultraviolet spectrophotometer, allowed W. P.

Jencks more than twenty-five years later to give a more detailed mechanism of semicarbazone formation and to modify the picture somewhat.[7]

$$R'-\overset{\overset{O}{\|}}{C}-R \;+\; H_2NNHCONH_2 \;\rightleftharpoons\; R'-\overset{\overset{NNHCONH_2}{\|}}{C}-R \;+\; H_2O$$

The rate of the reaction to form the semicarbazone is sensitive to changes in R and R′; when R′ is aromatic or is $C(CH_3)_3$, the rate is much slower, and it is also slower when R is CH_3 than when R is H. There is no relation between the rate of formation of the semicarbazone and the equilibrium constant for the reaction. Thus a mixture of two carbonyl compounds, cyclohexanone and furfural, with semicarbazide forms the derivative from cyclohexanone very fast, but on standing some hours the sole product is the semicarbazone of furfural.

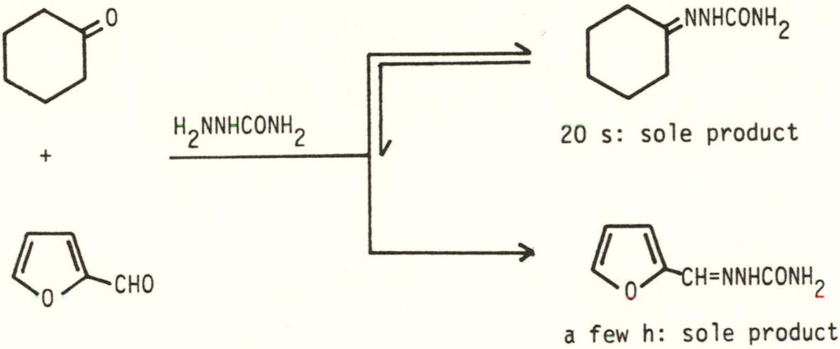

20 s: sole product

a few h: sole product

Similarly, furfural semicarbazone results from long reaction times when acetone, furfural, and semicarbazide are mixed, because the equlibrium constant favors the furfural derivative. Thus in these systems there is a *lack* of correlation between rate and equilibrium. As will be shown later, numerous examples of positive correlations between rate and equilibrium are now known.

The effect of acid on the rate of semicarbazone formation is a resultant of two opposing effects.[4,5,6] Acceleration is due to protonation of the carbonyl oxygen; the positively charged carbonyl carbon is then attacked rapidly by the nucleophilic semicarbazide.

$$R_2C=O \;+\; H^+ \;\rightleftharpoons\; R_2\overset{+}{C}OH$$

The decelerating effect of increasing acidity is due to the conversion of the basic semicarbazide into its salt, which is no longer nucleophilic and cannot attack the carbonyl carbon. The pH of maximum reaction rate clearly depends on the relative equilibrium constants for the two reactions shown.

$$H_2NNHCONH_2 \ + \ H^+ \ \rightleftharpoons \ H_3\overset{+}{N}NHCONH_2$$

Paul D. Bartlett (1907-   ), A.B. 1928, Amherst; Ph.D. 1931, Harvard with Conant) taught at Minnesota, at Harvard from 1934 to 1975, and as Welch Professor at Texas Christian University from 1975. His research covered a broad range of problems in reaction mechanisms, including both ionic and free-radical reactions. He published key work on bridgehead halogens, carbon rearrangements, vinyl polymerization, alkyation of unsaturated hydrocarbons, peroxides, reactions of elemental sulfur, mechanism of diazonium coupling, and many other topics.[8]

The hydrolysis of acetanilide was shown by D. R. Merrill and E. Q. Adams at Berkeley in 1917 to be catalyzed by protons,[9] the rate being proportional to both the acetanilide and proton concentrations. Correction was made for the protons furnished by $C_6H_5NH_3^+$ and CH_3COOH. This paper and the earlier excellent work by Adams at Berkeley on absorption spectra and structure make one wonder why Lewis did not keep Adams there to develop the promising field of physical organic chemistry.

$$C_6H_5NHCOCH_3 \ + \ H_2O \ \xrightarrow[\text{(0.1 N HCl)}]{H^+} \ C_6H_5\overset{+}{N}H_3 \ + \ CH_3COOH$$

Acid-catalyzed hydrolysis of oximes was reported by Stieglitz, the results being similar to his earlier work on hydrolysis of imido esters. In these reactions the proton acts as catalyst by adding to the oxygen.[10]

The esterification reaction and its reverse are accompanied by loss of OH from the carboxyl and cleavage of the H–OR bond in the alcohol (H⫫OR), in the most common type of acid-catalyzed esterification. This is shown by Reid's work reviewed earlier, and by Urey's ^{18}O tracer work.[11]

$$C_6H_5COOH \ + \ CH_3^{18}OH \ \overset{H^+}{\rightleftharpoons} \ C_6H_5CO^{18}OCH_3 \ + \ H_2O$$

Tracer studies and the occurrence of acid catalysis indicate a mechanism of the following general type; the reverse reaction is acid-catalyzed hydrolysis.

$$C_6H_5\overset{O}{\overset{\|}{C}}-OH \ + \ HCl \ + \ CH_3OH \ \rightleftharpoons \ \left[C_6H_5\overset{OH}{\underset{HOCH_3}{\overset{|}{\underset{|}{C}}}}-OH\right]^+ \ + \ Cl^-$$

$$C_6H_5\overset{O}{\overset{\|}{C}}-OCH_3 \ + \ H_2O \ \rightleftharpoons \ C_6H_5\overset{OH}{\underset{OH}{\overset{|}{\underset{|}{C}}}}-OCH_3 \ + \ HCl$$

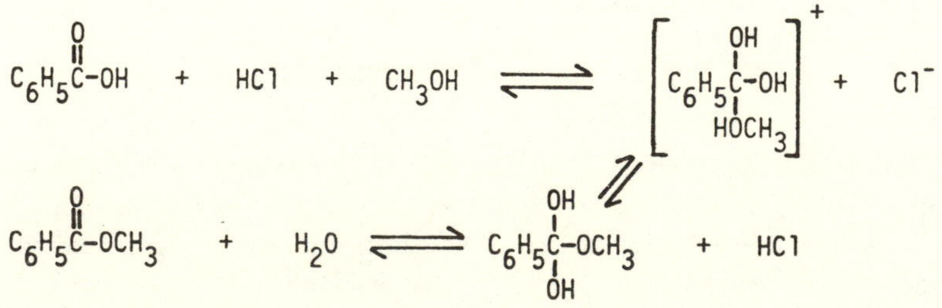

Acid-catalyzed oxygen exchange of acetone, shown by ^{18}O experiments by Mildred Cohn and Urey, occurs by a similar path.[12] In liquid ammonia NH_4Cl is the proton acid of the ammonia system and catalyzes the ammonolysis of the ester.[13]

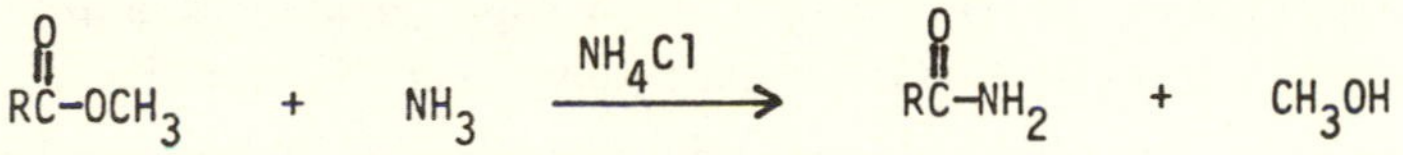

$$RC\overset{\overset{O}{\parallel}}{-}OCH_3 \quad + \quad NH_3 \quad \xrightarrow{NH_4Cl} \quad RC\overset{\overset{O}{\parallel}}{-}NH_2 \quad + \quad CH_3OH$$

This scheme of acid-catalyzed esterification and hydrolysis agrees in general with observations on the lowering of rate by ortho substitution (*steric hindrance*).[14] Rates of lactonization of hydroxy acids show acid catalysis similar to that in esterification.[15]

Hydrolysis and dimerization reactions of cyanamide and related compounds were studied at the American Cyanamid Company, probably because of the projected use of cyanamide for producing ammonia from atmospheric nitrogen during World War I. The reactions revealed several pathways depending on pH, and results fitted into a consistent scheme with quantitative agreement between theory and experiment.[16]

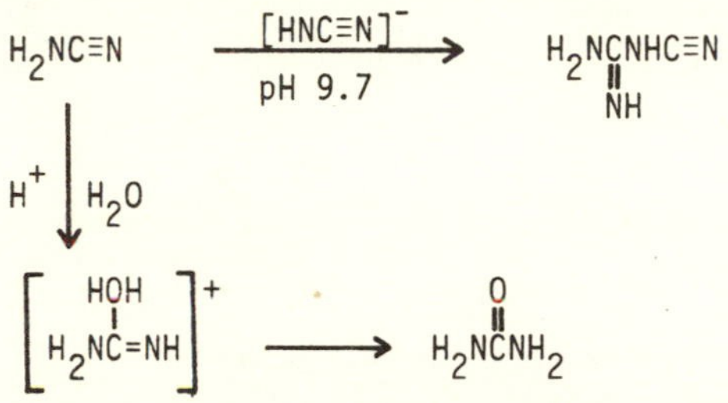

$$H_2NC\equiv N \quad \xrightarrow[pH\ 9.7]{[HNC\equiv N]^-} \quad H_2NCNHC\equiv N \ \overset{\parallel}{\underset{NH}{}}$$

$$H^+ \downarrow H_2O$$

$$\left[H_2N\overset{\overset{HOH}{|}}{C}=NH \right]^+ \quad \longrightarrow \quad H_2N\overset{\overset{O}{\parallel}}{C}NH_2$$

The acid-catalyzed hydrolysis (*inversion*) of sucrose to glucose and fructose was long a favorite reaction for demonstrating the particulars of acid catalysis. H. A. Fales at Columbia, G. Scatchard at Amherst, and others carried out detailed studies.[17] The mutarotation of glucose involves a change in rotation because the configuration of the 1-carbon is inverted, apparently by opening to the aldehyde form then cyclizing to yield the opposite configuration at C-1. It requires catalysis by both acids and bases[18] and shows general acid-base catalysis in the Brönsted sense.

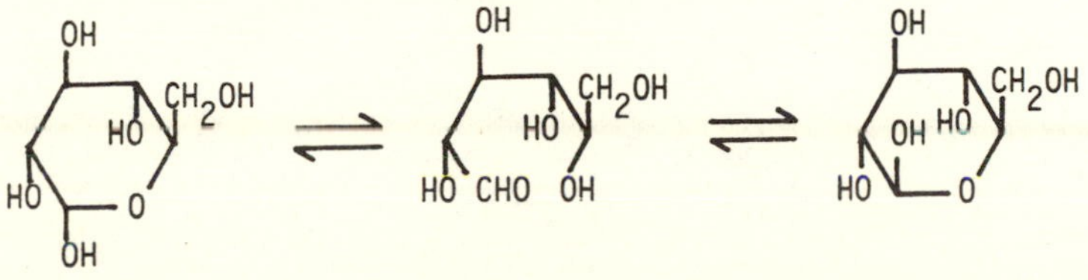

The role of both acids and bases in carbonyl additions is clearly shown in some aldol reactions, which form a new C-C bond by addition of a carbon nucleophile to a carbonyl group. A. C. O. Hann and A. Lapworth in 1904, at the Goldsmiths Institute in London, wrote a mechanism for the base-catalyzed condensation of acetaldehyde and ethyl acetoacetate,[19] with the nucleophilic carbanion attacking the carbonyl carbon and subsequent protonation and dehydration. H. D. Dakin in 1909 demonstrated that amino acids such as glycine, H_2NCH_2COOH, could catalyze a similar reaction.[20] E. P. Kohler and B. B. Corson supported the Lapworth course for an aldol condensation.[21] The aldol was in part cleaved to starting materials by base and in part dehydrated. This supported Lapworth's ideas and also disproved a suggestion which Ingold made without good evidence,[22] that an aldol condensation required the enol form of the carbonyl compound; $C_6H_5COCOOH$ cannot form an enol but readily gives the aldol.

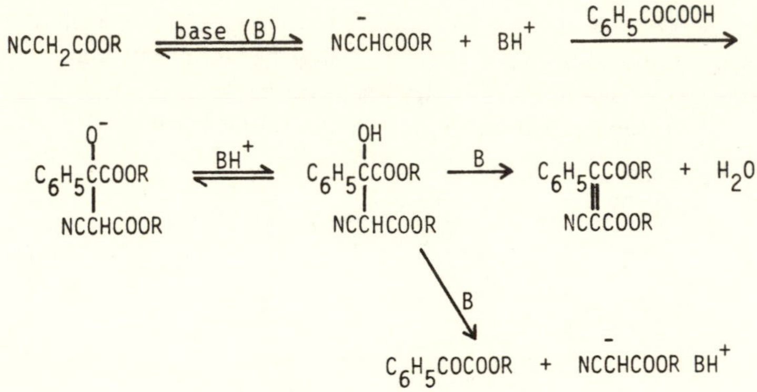

$$NCCH_2COOR \xrightleftharpoons{\text{base (B)}} NC\overset{-}{C}HCOOR + BH^+ \xrightarrow{C_6H_5COCOOH}$$

$$C_6H_5\underset{\underset{NCCHCOOR}{|}}{\overset{\overset{O^-}{|}}{C}}COOR \xrightleftharpoons{BH^+} C_6H_5\underset{\underset{NCCHCOOR}{|}}{\overset{\overset{OH}{|}}{C}}COOR \xrightarrow{B} C_6H_5\underset{\underset{NCCCOOR}{||}}{C}COOR + H_2O$$

$$\searrow{\scriptstyle B}$$

$$C_6H_5COCOOR + NC\overset{-}{C}HCOOR\ BH^+$$

A significant observation by K. C. Blanchard at New York University pointed out that urea, H_2NCONH_2, a weak base in water, was an excellent catalyst for an aldol reaction in acetic acid.[23] R. Kuhn in Heidelberg found that organic bases by themselves did not catalyze some aldol condensations, but he noted that a salt from an organic base and an organic acid was an excellent catalyst.

These various observations were combined, with characteristic economy and elegance, into a satisfying mechanistic scheme and a very useful synthetic procedure[24] by Arthur C. Cope, then of Bryn Mawr.[25] Cope confirmed the points above and obtained excellent yields in the following type of reaction, removing the water formed by distilling off the acetic acid solvent. He suggested a mechanism for the process.

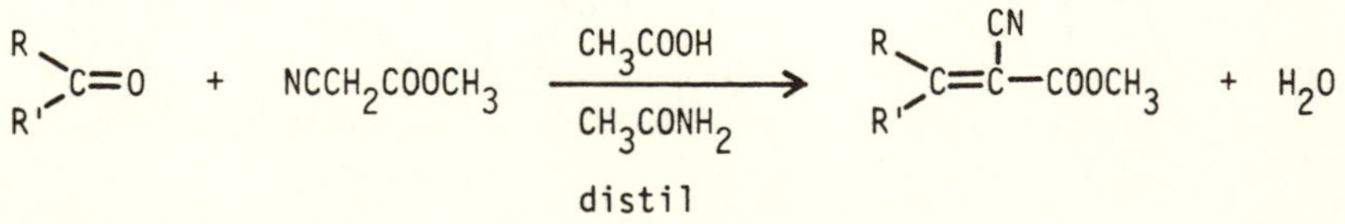

$$\overset{R}{\underset{R'}{>}}C=O + NCCH_2COOCH_3 \xrightarrow[\substack{CH_3CONH_2 \\ \text{distil}}]{CH_3COOH} \overset{R}{\underset{R'}{>}}C=\overset{\overset{CN}{|}}{C}-COOCH_3 + H_2O$$

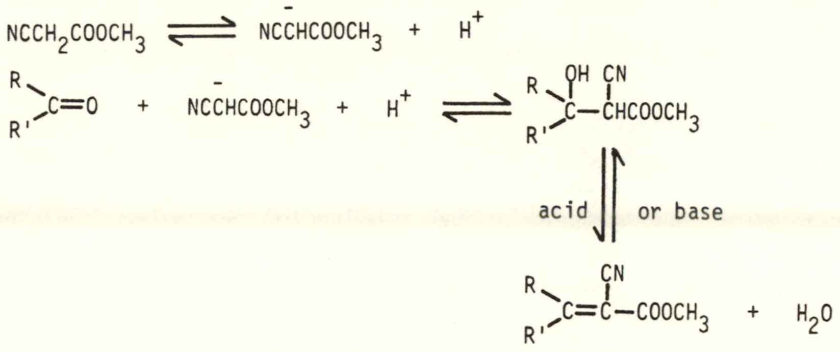

$$NCCH_2COOCH_3 \rightleftharpoons NC\overset{-}{C}HCOOCH_3 + H^+$$

$$\underset{R'}{\overset{R}{>}}C=O + NC\overset{-}{C}HCOOCH_3 + H^+ \rightleftharpoons \underset{R'}{\overset{R}{>}}\overset{OH\ CN}{\underset{}{C}}-CHCOOCH_3$$

$$\text{acid} \Updownarrow \text{or base}$$

$$\underset{R'}{\overset{R}{>}}C=\overset{CN}{\underset{}{C}}-COOCH_3 + H_2O$$

Acid can catalyze this aldol reaction, presumably via processes illustrated below, although base catalysis as above to remove a proton is more effective. The acetamide-acetic acid medium provides both Brönsted acids and bases, so that protons can be furnished or removed as needed. Removal of water shifts the final equilibrium toward product.

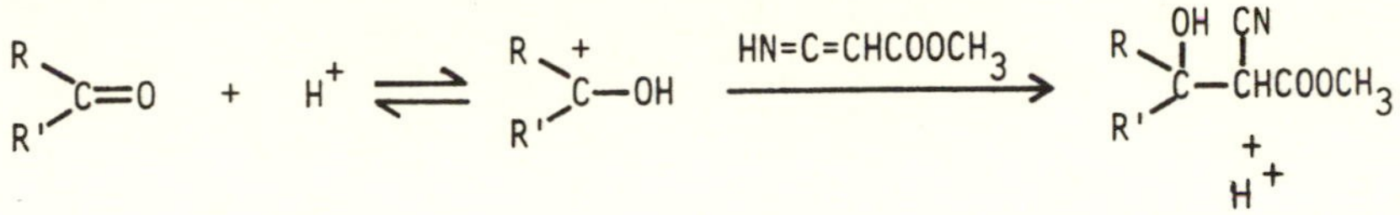

$$\underset{R'}{\overset{R}{>}}C=O + H^+ \rightleftharpoons \underset{R'}{\overset{R}{>}}\overset{+}{C}-OH \xrightarrow{HN=C=CHCOOCH_3} \underset{R'}{\overset{R}{>}}\overset{OH\ CN}{\underset{\underset{H^+}{+}}{C}}-CHCOOCH_3$$

A modified procedure by Cope used ammonium acetate in acetic acid and benzene; distillation removed benzene and water, a Dean-Stark water separator allowing benzene and any accompanying ketone to be returned to the reaction mixture.[26]

Although the removal of water by distillation with some water-immiscible solvent had been used for many years to shift an equilibrium reaction to completion, as in esterification,[27] Cope's procedure popularized the use of the Dean-Stark apparatus. This was developed originally for the quantitative determination of water in oils and emulsions, not as a reaction aid.[28] Simultaneously, I. N. Hultman from Eastman Kodak described a simple water separator designed expressly for removing water from organic reactions by distillation to ensure complete reaction.[29] Hultman is clearly entitled to the credit for this extremely useful device.

The formation of a carbanion by removal of a proton by base obviously requires some acidic character in the proton. This is due to the stabilization of the negative charge of the carbanion by delocalization over several atoms. In the cyanoacetate reaction above, such resonance stabilization is contributed by both electron-attracting groups, and the protons in $CH_2(CN)COOCH_3$ are therefore fairly acidic.

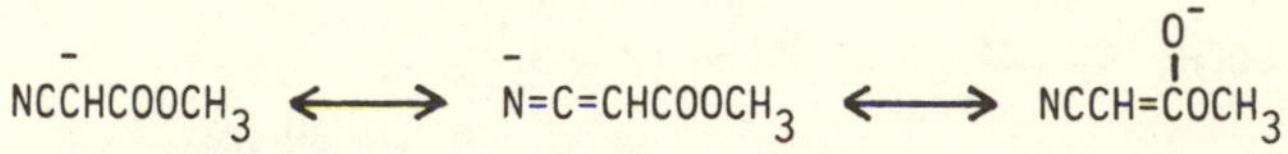

$$NCCHCOOCH_3 \longleftrightarrow N=C=CHCOOCH_3 \longleftrightarrow NCCH=\overset{O^-}{\underset{|}{C}}OCH_3$$

However, in the base-catalyzed aldol condensation of CH_3CHO with itself, a carbanion is undoubtedly involved in which the negative charge is stabilized by only the carbonyl group. It is quite probable that the carbanion derived from the weaker acid, such as CH_3CHO, reacts faster as a nucleophile with $C=O$ than the carbanion derived from a stronger acid, such as $CH_2(CN)COOCH_3$. This amounts to saying that the rate and/or the equilibrium of the reaction of a carbanion with a proton runs parallel to its reaction with the positive carbon of a $C=O$. This idea seems reasonable.

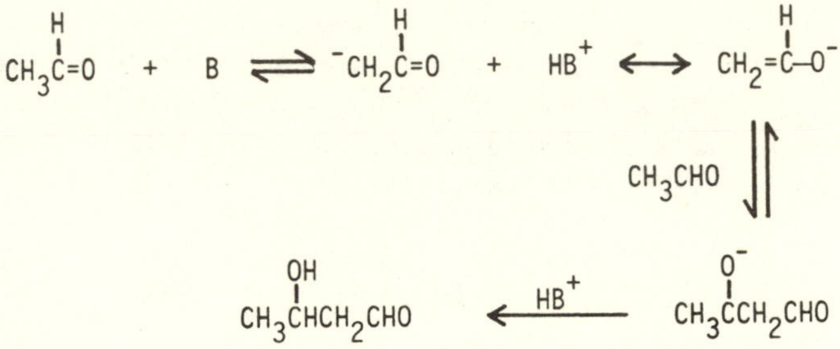

$$CH_3\overset{H}{\underset{|}{C}}=O \ + \ B \ \rightleftharpoons \ ^-CH_2\overset{H}{\underset{|}{C}}=O \ + \ HB^+ \longleftrightarrow CH_2=\overset{H}{\underset{|}{C}}-O^-$$

$$CH_3CHO \ \updownarrow$$

$$CH_3\overset{OH}{\underset{|}{C}}HCH_2CHO \ \xleftarrow{\ HB^+\ } \ CH_3\overset{O^-}{\underset{|}{C}}CH_2CHO$$

If the carbonyl group is made less reactive by electronic influences which make its carbon less positive, or by a high degree of substitution which hinders approach of a carbanion, the reactivity of the $C=O$ may be sharply diminished, even toward strong nucleophiles. The effect of substitution in decreasing addition to the $C=O$ is shown in the poor reaction of cyanoacetate under Cope's conditions[26] with diisobutyl ketone, $[(CH_3)_2CHCH_2]_2CO$, forty percent yield, and camphor, no yield. It is also shown by many studies on diortho-substituted aromatic ketones by Fuson and others.[30]

It is quite possible that carbanions or other nucleophiles present in very low concentration may initiate a series of equilibria, which may be unfavorable to product formation. However, if the series of equilibrium reactions can be capped by an irreversible process, the whole sequence can be pulled through to completion, an idea explicitly utilized by Hauser and by Spielman.

Fuson's principle of vinylogy,[31] which correlated a large number of known reactions under one rubric, stated that functional groups written this way, $A-(CH=CH)_n-D_1=E_2$, could and frequently did behave very similarly to those lacking the vinyl groups, $A-D_1=E_2$. Many of the reactions which were similar in the two groups involved carbanion intermediates, as in CH_3COOR compared to $CH_3CH=CHCOOR$, and the additional vinyl group(s) stabilized the negative charge by delocalization. Many illustrations of these and similar reactions were given by Richard Kuhn at Heidelberg and by Fuson's own work.[32]

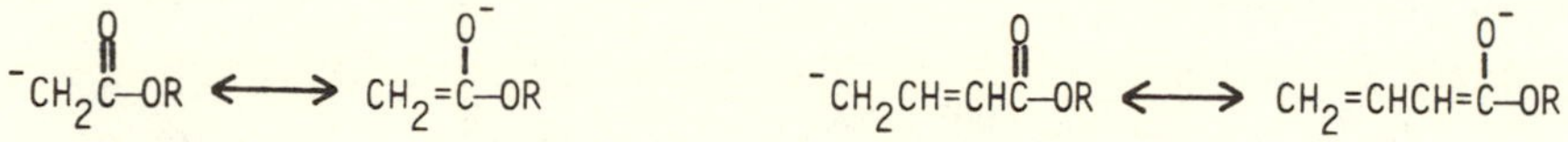

$$\overset{-}{C}H_2\overset{\overset{O}{\|}}{C}-OR \longleftrightarrow CH_2=\overset{\overset{O^-}{|}}{C}-OR \qquad\qquad \overset{-}{C}H_2CH=\overset{\overset{O}{\|}}{C}-OR \longleftrightarrow CH_2=CHCH=\overset{\overset{O^-}{|}}{C}-OR$$

The point made above, the high nucleophilic reactivity of carbanions at low concentration, is shown in the base-catalyzed reaction of aldehydes with malonic acid.[33] Here the most acidic hydrogens are on the carboxyl groups, but the reactive carbanion must be the one derived from the *weakest* acid, the CH_2 group. High rate of reaction and final irreversible steps must be the reason. A tertiary base (pyridine) and a secondary base are particularly effective catalysts here, and the reaction may involve intermediates from the C=O and the amine.[33]

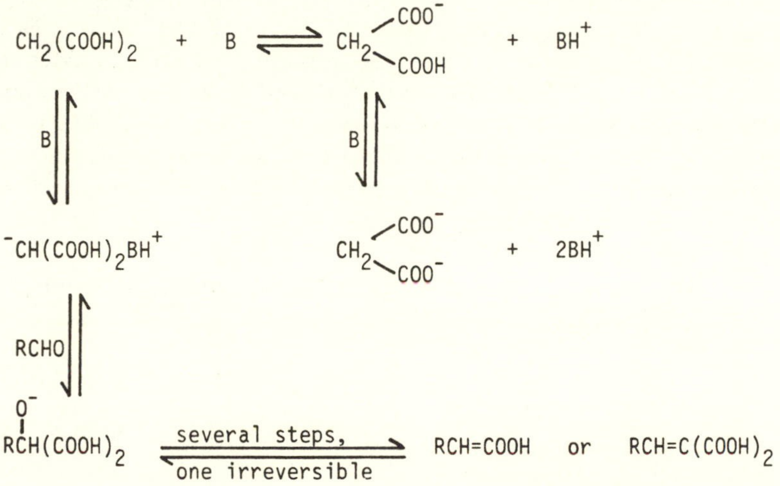

$$CH_2(COOH)_2 \;+\; B \;\rightleftharpoons\; CH_2\!\!\begin{array}{l}\diagup COO^-\\ \diagdown COOH\end{array} \;+\; BH^+$$

$$^-CH(COOH)_2BH^+ \qquad\qquad CH_2\!\!\begin{array}{l}\diagup COO^-\\ \diagdown COO^-\end{array} \;+\; 2BH^+$$

$$\overset{\overset{O^-}{|}}{R\overset{|}{C}H(COOH)_2} \;\;\xrightarrow[\text{one irreversible}]{\text{several steps,}}\;\; RCH=COOH \quad or \quad RCH=C(COOH)_2$$

The reversal of aldol condensations by bases has been known for a long time; a simple case is that of the condensation of acetone by base to the aldol ("diacetone alcohol"). The equilibrium is so unfavorable that a clever trick with the basic catalyst must be used to obtain the aldol.[34]

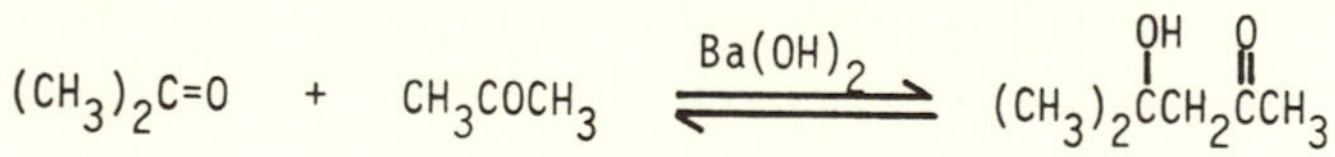

$$(CH_3)_2C=O \;+\; CH_3COCH_3 \;\xrightleftharpoons{Ba(OH)_2}\; (CH_3)_2\overset{\overset{OH}{|}}{C}CH_2\overset{\overset{O}{\|}}{C}CH_3$$

G. Akerlöf studied in detail the reversal of the condensation, "dealdolization", particularly the salt effect.[35] Westheimer showed that the reversal was catalyzed by buffers, amines, and amine salts, but not by tertiary amines and hence was not subject to general acid-base catalysis.[36] The reaction was catalyzed by OH^- *and* by molecular ammonia, primary amines, and secondary amines. It was suggested that a ketone-amine intermediate was involved, which could not form with tertiary amines.[37] The mechanism proposed by

Westheimer is shown.

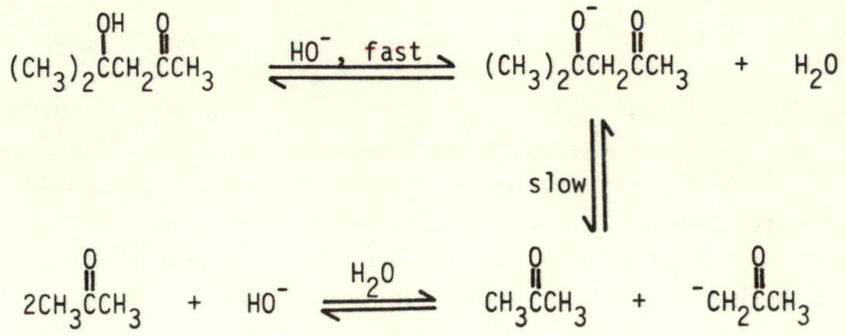

$$(CH_3)_2\overset{OH}{\underset{}{C}}CH_2\overset{O}{\underset{}{C}}CH_3 \quad \underset{\rightleftharpoons}{\xrightarrow{HO^-,\ fast}} \quad (CH_3)_2\overset{O^-}{\underset{}{C}}CH_2\overset{O}{\underset{}{C}}CH_3 \quad + \quad H_2O$$

$$slow \Updownarrow$$

$$2CH_3\overset{O}{\underset{}{C}}CH_3 \quad + \quad HO^- \quad \underset{\rightleftharpoons}{\xrightarrow{H_2O}} \quad CH_3\overset{O}{\underset{}{C}}CH_3 \quad + \quad {}^-CH_2\overset{O}{\underset{}{C}}CH_3$$

A family of reactions similar to the reversed aldol case was found by T. D. Stewart at Berkeley.[38] The rate of appearance of the final products, $(CH_3)_3N$, SO_3^-, and CN^- is in each case proportional to the OH^- concentration. The charge type of the ion from which the proton is abstracted is different in each case. These varied types of eliminations leading to the C=O compound in each case are parallel to elimination reactions to form C=C, which were to be extensively studied in the future.[39] At least some of these reverse carbonyl addition reactions do not show general acid-base catalysis, and hence the addition reactions themselves do not.

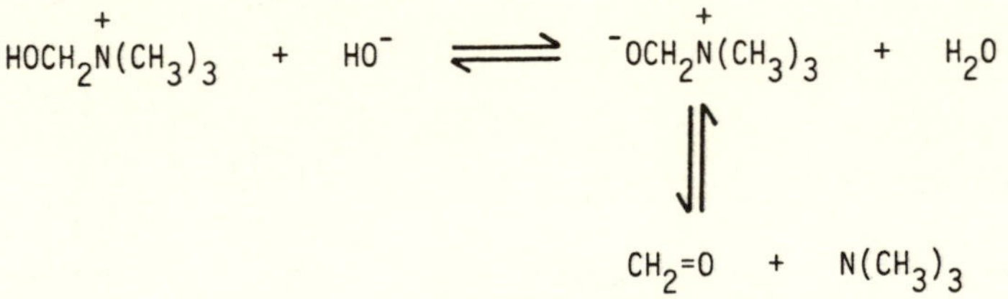

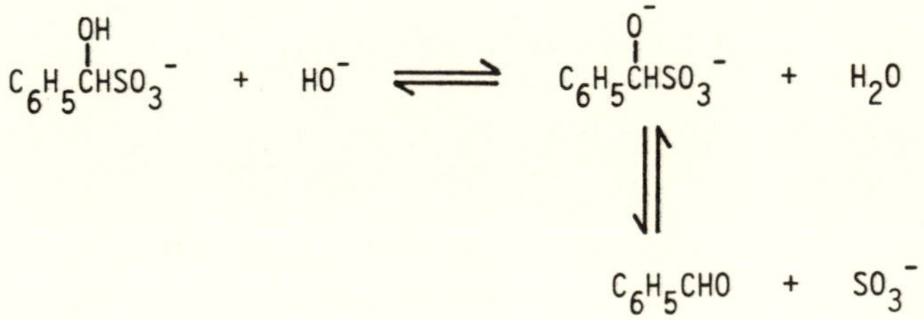

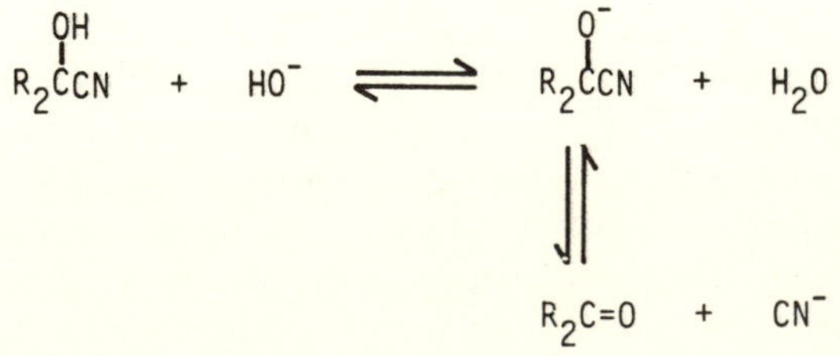

$$HOCH_2\overset{+}{N}(CH_3)_3 \quad + \quad HO^- \quad \rightleftharpoons \quad {}^-OCH_2\overset{+}{N}(CH_3)_3 \quad + \quad H_2O$$

$$\Updownarrow$$

$$CH_2=O \quad + \quad N(CH_3)_3$$

$$C_6H_5\overset{OH}{\underset{}{C}}HSO_3^- \quad + \quad HO^- \quad \rightleftharpoons \quad C_6H_5\overset{O^-}{\underset{}{C}}HSO_3^- \quad + \quad H_2O$$

$$\Updownarrow$$

$$C_6H_5CHO \quad + \quad SO_3^-$$

$$R_2\overset{OH}{\underset{}{C}}CN \quad + \quad HO^- \quad \rightleftharpoons \quad R_2\overset{O^-}{\underset{}{C}}CN \quad + \quad H_2O$$

$$\Updownarrow$$

$$R_2C=O \quad + \quad CN^-$$

The relative rate of HSO_3^- elimination at 20° C in alcohol solution was shown by T. D. Stewart and W. E. Bradley[40] in Table 1.

Table 1. Elimination of HSO_3^-

$$R_2\overset{+}{N}HCH_2SO_3^- \; \rightleftharpoons \; R_2\overset{+}{N}=CH_2 \; + \; HSO_3^-$$

R_2	Relative Rate
$-(CH_2)_5-$	1
$(CH_3)_2$	1.6
$(\underline{n}-C_5H_{11})_2$	3.0
$(C_2H_5)_2$	4.4
$(\underline{n}-C_4H_9)_2$	10
$(CH_2=CHCH_2)_2$	14
$(\underline{n}-C_3H_7)_2$	15
$(\underline{i}-C_3H_7)_2$	120
$(C_6H_5CH_2)_2$	280
$(HOCH_2CH_2)_2$	290
$(\underline{i}-C_4H_9)_2$	570

This order of reactivity is in the general direction of increasing electron donation toward the NCH_2 group by the alkyl groups, although the high value for $(HOCH_2CH_2)_2$ does not fit this pattern. Perhaps this group hydrogen bonds with solvent to promote the loss of HSO_3^-.

A. Lapworth at Manchester, England, measured the equilibrium with a variety of carbonyl compounds and HCN in alcohol.[41] His results show in general that substituents which increase the electron density on the carbonyl group make the equilibrium less favorable to product formation. This reaction is catalyzed by bases and hence involves the attack of the cyanide ion on the carbonyl carbon, followed by pickup of a proton from solvent. It was not determined whether the reaction was subject to general acid-base catalysis. The most striking examples of electron-donating groups cutting down on cyanide addition were aromatic ketones, like $C_6H_5COC_6H_5$, fluorenone, xanthone, and highly substituted carbonyl groups such as the one in camphor. Increased substitution in aliphatic ketones increased the stability of the cyanohydrin, perhaps by diminishing the acidity of the proton and thus its removal by the base to initiate the dissociation.

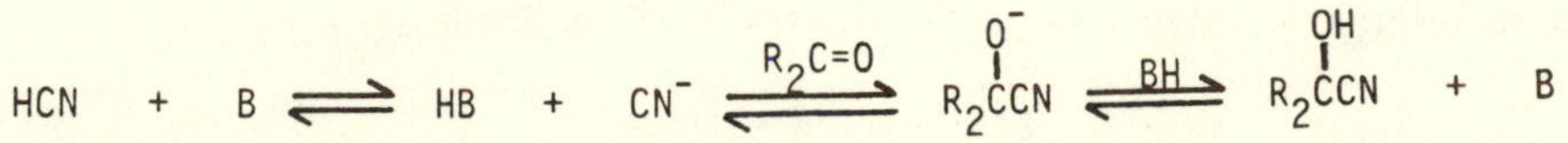

$$HCN + B \rightleftharpoons HB + CN^- \xrightarrow{R_2C=O} R_2\overset{O^-}{\underset{}{C}}CN \xrightarrow{BH} R_2\overset{OH}{\underset{}{C}}CN + B$$

Lapworth also suggested that α-aminonitriles, as below, could hydrolyze via the imonium salt,[42] an idea adopted by Stewart.[40]

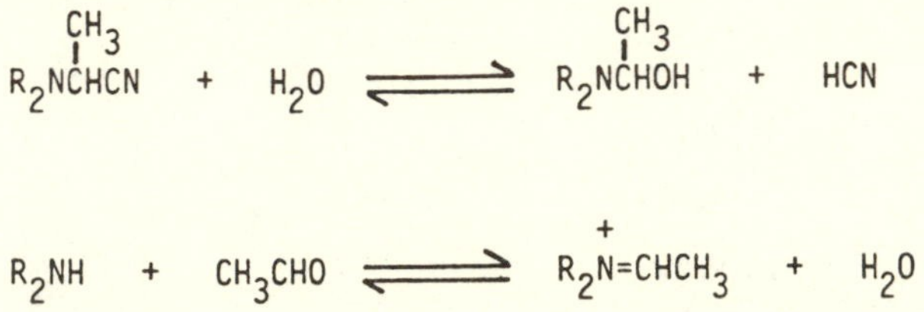

$$R_2N\overset{CH_3}{\underset{}{C}}HCN + H_2O \rightleftharpoons R_2N\overset{CH_3}{\underset{}{C}}HOH + HCN$$

$$R_2NH + CH_3CHO \rightleftharpoons R_2\overset{+}{N}=CHCH_3 + H_2O$$

G. E. K. Branch at Berkeley showed the rate of hydrolysis of the aminoacetal analog to be smaller with increasing acidity.[43] He attributed this behavior to a shift of the equilibrium by acid to the cyclic structure, whereas the hydrolysis takes place on the open-chain structure.

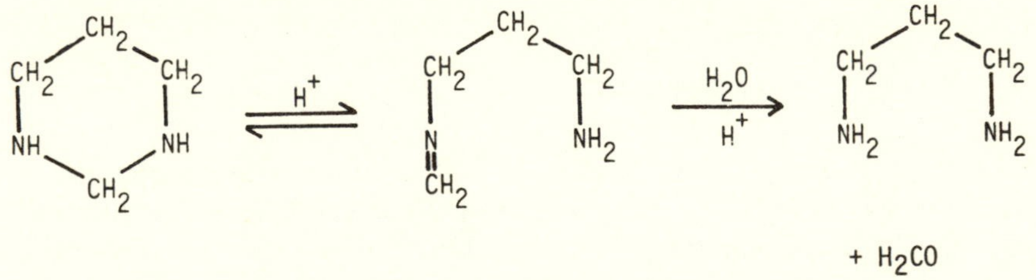

$$+ H_2CO$$

Branch (1886-1954, B.Sc. 1912, Liverpool; Ph.D. 1915, Berkeley) taught at Berkeley from 1915 until his death. He did research on acid catalysis, strength of acids, and reaction mechanisms. With Melvin Calvin, he wrote *The Theory of Organic Chemistry*, 1940, an important book, although not as influential as Hammett's.

The base-catalyzed self-condensation of esters to form β-ketoesters had been known for many years and many mechanisms had been suggested.[44] Carbanions clearly were involved; with ethyl acetate, the reaction could be written as shown. The β-ketoesters are valuable synthetic intermediates and, as described earlier, were the first carefully studied examples of tautomerism.

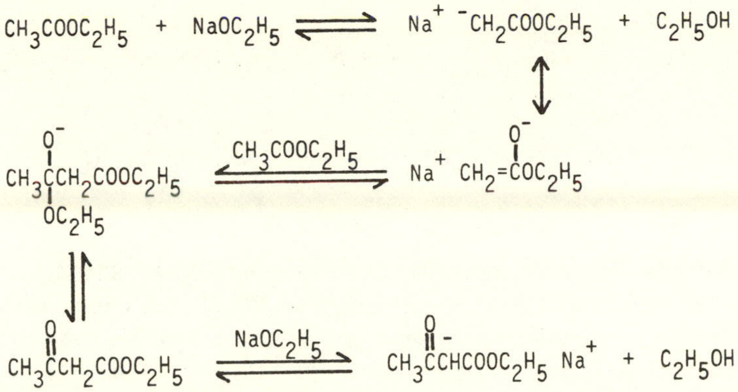

$$CH_3COOC_2H_5 \; + \; NaOC_2H_5 \; \rightleftharpoons \; Na^+ \; {}^-CH_2COOC_2H_5 \; + \; C_2H_5OH$$

$$\underset{\underset{OC_2H_5}{|}}{\overset{\overset{O^-}{|}}{CH_3CCH_2COOC_2H_5}} \quad \overset{CH_3COOC_2H_5}{\rightleftharpoons} \quad Na^+ \; \overset{O^-}{\underset{|}{CH_2=COC_2H_5}}$$

$$\underset{\overset{||}{O}}{CH_3CCH_2COOC_2H_5} \quad \overset{NaOC_2H_5}{\rightleftharpoons} \quad \overset{O}{\underset{||}{CH_3CCHCOOC_2H_5}} \; Na^+ \; + \; C_2H_5OH$$

Ethyl isobutyrate, $(CH_3)_2CHCOOC_2H_5$, did not condense with $NaOC_2H_5$, but it was found by M. A. Spielman at Wisconsin and C. R. Hauser at Duke that with stronger bases, mesitylmagnesium bromide or sodium triphenylmethyl, the reaction did occur.[45,46] In this case, the completion of the reactions can be ascribed to shifting the equilibria by forming the enolate irreversibly in the final step.[46]

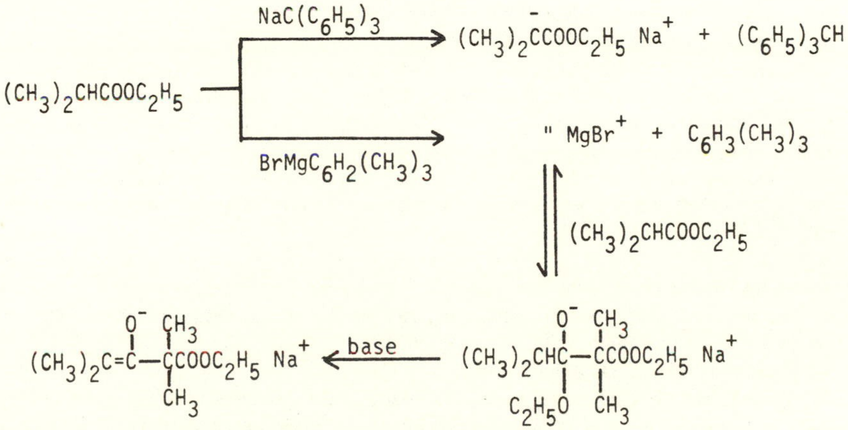

$$(CH_3)_2CHCOOC_2H_5 \quad\Bigg\langle \quad \overset{NaC(C_6H_5)_3}{\longrightarrow} \; (CH_3)_2\overset{-}{C}COOC_2H_5 \; Na^+ \; + \; (C_6H_5)_3CH$$

$$\overset{BrMgC_6H_2(CH_3)_3}{\longrightarrow} \quad ''\; MgBr^+ \; + \; C_6H_3(CH_3)_3$$

$$(CH_3)_2CHCOOC_2H_5$$

$$(CH_3)_2C=\overset{\overset{O^-}{|}}{C}-\underset{\underset{CH_3}{|}}{\overset{\overset{CH_3}{|}}{C}COOC_2H_5} \; Na^+ \quad \overset{base}{\longleftarrow} \quad (CH_3)_2CH\overset{\overset{O^-}{|}}{C}-\underset{\underset{CH_3}{|}}{\overset{\overset{CH_3}{|}}{C}COOC_2H_5} \; Na^+$$
$$\qquad\qquad\quad C_2H_5O$$

In other cases, condensations occurred in which enolization was impossible as the last step. These reactions must have occurred because the base used, sodium triphenylmethyl, was strong enough to convert the ethyl isobutyrate to its anion nearly completely and hence allow further reactions to occur.[47] These papers of Hauser were the first of an important series from his laboratory that provided significant theoretical and experimental advances on reactions catalyzed by strong bases. Hauser gave a general classification and discussion of mechanisms of reactions between a carbanion and a carbonyl group.

$$(CH_3)_2CHCOOR + NaC(C_6H_5)_3 \longrightarrow (CH_3)_2\overset{-}{C}COOR\ Na^+ + HC(C_6H_5)_3$$

$$(CH_3)_2\overset{-}{C}COOC_2H_5\ Na^+ + C_6H_5COCl \longrightarrow C_6H_5\overset{O}{\overset{\|}{C}}-\overset{CH_3}{\underset{CH_3}{\overset{|}{\underset{|}{C}}}}COOC_2H_5$$

Charles R. Hauser (1900-1970, B. S. Florida; Ph.D. 1928, Iowa) taught at Duke from 1929 to 1970. He published prolifically on condensations, cyclizations, and elimination and rearrangement reactions. He developed novel and useful synthetic methods and knowledge about reaction mechanisms, for which he had a real instinct. His work on dianions has been developed into an outstanding synthetic method by his student T. R. Harris at Vanderbilt.

Marvin A. Spielman (1906-1957, B.S. North Dakota; Ph.D. 1933, Minnesota with W. M. Lauer) was National Research Fellow at Yale, taught at Wisconsin from 1935 to 1940, and was a research chemist at Abbott Laboratories from 1940 to 1957. He worked on rearrangement reactions, long-chain fatty acids, steroids, anticonvulsant drugs, antibiotics, and the ester condensation reaction.

LITERATURE CITED

1. J. B. Conant, JACS, **43**, 1705, 2125 (1921).

2. C. P. Smyth, ibid., **46**, 2151 (1924).

3. J. B. Conant, *My Several Lives*, New York, 1970, p. 48; Conant worked for part of 1917 for the Bureau of Chemistry of the U. S. Department of Agriculture in Washington and later served in the Chemical Warfare Service of the U. S. Army.

4. J. B. Conant and P. D. Bartlett, JACS, **54**, 2881 (1932); F. H. Westheimer, ibid., **56**, 1962 (1934); the last paper deals with the reaction in water-60% methyl cellosolve mixtures.

5. A. Michael, ibid., **41**, 393 (1919); it seems likely that the experimental work reported was done several years before publication.

6. S. F. Acree and J. M. Johnson, ACJ, **38**, 308 (1907); Acree, ibid., **39**, 300 (1908); E. Barrett and A. Lapworth, JCS, **93**, 85 (1908).

7. W. P. Jencks, JACS, **81**, 475 (1959).

8. An informative (and informal) account of his research and his teaching methods, by his collaborators, is in *P. D. and the Bartlett Group at Harvard 1934-1974*, Fort Worth, 1975.

9. D. R. Merrill and E. Q. Adams, JACS, **39**, 1588 (1917); also W. A. Noyes and W. F. Goebel, ibid., **44**, 2286 (1922). Elliott Q. Adams (B.S. 1909, M.I.T.; Ph.D. 1914, Berkeley), physical chemist in Color Laboratory, Bureau of Chemistry, U. S. Department of Agriculture, 1917-1921; research laboratory, General Electric Company, Cleveland, from 1921. Probably Conant's contact with Adams occurred when they were both attached to the Bureau of Chemistry.

10. R. W. Johnson and J. Stieglitz, JACS, **56**, 1904 (1934); this was one of Stieglitz' few experimental papers after 1920.

11. I. Roberts and H. C. Urey, ibid., **60**, 2391 (1938); **61**, 2584 (1939); H. A. Smith, ibid., **61**, 254 (1939); review, J. N. E. Day and C. K. Ingold, Trans. Farad. Soc., **37**, 686

(1941). C. K. Ingold, *Structure and Mechnanism in Organic Chemistry*, Ithaca, 1953, 1969, gives a detailed discussion of carbonyl and many other organic reactions.

12. Mildred Cohn and H. C. Urey, JACS, **60**, 679 (1938). Harold C. Urey (1893-1913, B.S. 1917, Montana; Ph.D. 1923, Berkeley) taught at Hopkins, Columbia, Chicago, and California. In the 1930s he isolated deuterium and received the Nobel Prize in 1934; he also concentrated ^{13}C, ^{18}O and ^{15}N, making them available for tracer studies. After a prominent role in the atomic bomb project, he became an active worker on cosmic chemistry, the origin of the solar system, and early earth chemistry.

13. L. L. Fellinger and L. F. Audrieth, JACS, **60**, 579 (1938); Audrieth and J. Kleinberg, JOC, **3**, 312 (1938). Audrieth (1901-1966, B.S. 1922, Colgate; Ph.D. 1924, Cornell) taught at Illinois from 1928-1961. He carried out extensive studies on liquid ammonia solutions, on phosphorus and hydrazine compounds, and with M. Sveda, discovered in 1936 the sweet taste of the salt of cyclohexylaminesulfamic acid, $C_6H_{11}NHSO_3^-Na^+$ (cyclamate).

14. R. W. Hufferd and W. A. Noyes, JACS, **43**, 925 (1921); R. J. Hartman and A. M. Borders, ibid., **59**, 2107 (1937) and later papers; the classical work on steric hindrance is by Victor Meyer, Ber., **27**, 510 (1894) and later papers. However, the observation that ortho substituents decrease rate of reaction of functional groups is due to F. Kehrmann, Ber., **21**, 3315 (1888), and later papers on reactivity of ortho-substituted quinones.

15. H. S. Taylor and H. W. Close, JACS, **39**, 422 (1917).

16. G. Buchanan and G. Barsky, ibid., **52**, 195 (1930); **53**, 1270 (1931); Hammett, *Physical Organic Chemistry*, New York, 1940, pp. 339 ff.

17. G. Scatchard, JACS, **43**, 2387, 2406, (1921); H. A. Fales and J. C. Morrell, ibid., **44**, 2071 (1922).

18. J. N. Brönsted and E. A. Guggenheim, JACS, **49**, 2554 (1927); a key paper, published from Brönsted's laboratory in Copenhagen; C. S. Hudson and J. K. Dale, ibid., **39**, 320 (1917); F. H. Westheimer, JOC, **2**, 431 (1938); Hammett, op. cit., p. 337.

19. A. C. O. Hann and Arthur Lapworth, JCS, **85**, 46 (1904).

20. H. D. Dakin, JBC, **7**, 49 (1909), in the private laboratory of Dr. C. A. Herter in New York City.

21. E. P. Kohler and B. B. Corson, JACS, **45**, 1975 (1923); C. R. Hauser and D. S. Breslow, ibid., **61**, 793 (1939).

22. C. K. Ingold, JCS, **119**, 329 (1921).

23. K. C. Blanchard, et al., JACS, **53**, 2809 (1931); this was a preliminary communication, apparently never reported in full. Kenneth C. Blanchard (1900-   , A.B. 1914, Clark; Ph.D. 1929, MIT) was a biochemist teaching at the Washington Square campus of New York University from 1925 to 1942. From 1942 to 1965 he was on the faculty of pharmacology at Johns Hopkins. During World War II, he was active in the pharmacology of antimalarial drugs for the Office of Scientific Research and Development. He did not publish further on reaction mechanisms.

24. A. C. Cope, ibid., **59**, 2327 (1937). Cope does not quote the Blanchard paper, but his reaction conditions may have been suggested by their observation on urea-acetic acid as a favorable reaction medium.

25. For a tribute to Cope by Roger Adams, see D. S. and A. T. Tarbell, *Roger Adams*, Washington, D.C., 1981, pp. 193-194.

26. A. C. Cope, Corris M. Hofmann, Cornelia Wyckoff and Esther Hardenbergh, ibid., **63**, 3452 (1941); Cope and Evelyn H. Hancock, *Org. Synth.*, Coll. Vol. 3, 399.

27. P. S. Pinkney, *Org. Synth.*, **17**, 32 (1937).

28. E. W. Dean and D. D. Stark, Ind. Eng. Chem., **12**, 486 (1920).

29. I. N. Hultman, Anne W. Davis and H. T. Clarke, JACS, **43**, 366 (1921).

30. R. C. Fuson and J. T. Walker, JACS, **52**, 3269 (1930) and later papers.

31. R. C. Fuson, Chem. Rev., **16**, 1 (1935).

32. R. C. Fuson, et al., JACS, **58**, 2450 (1936); **59**, 893 (1937); R. Kuhn et al., Ber. **69**, 98(1936) and other papers.

33. Reviews; G. Jones, *Organic Reactions*, **15**, 204 (1967); J. R. Johnson, ibid., **1**, 226 (1942). Studies on condensations with malonic acids: E. J. Corey, JACS, **74**, 5897 (1952).

34. L. P. Kyriakides, JACS, **36**, 534 (1914); J. B. Conant and N. Tuttle, *Org. Synth.*, Coll. Vol. 1, p. 199.

35. G. Akerlöf, JACS, **48**, 3046 (1926).

36. F. H. Westheimer and H. Cohen, ibid., **60**, 90 (1938); J. Miller and M. Kilpatrick, ibid., **53**, 3217 (1931).

37. A detailed scheme for the ketimine function was proposed much later by W. P. Jencks, *Catalysis in Chemistry and Enzymology*, New York, 1969, p. 121.

38. T. D. Stewart and L. H. Donnally, JACS, **54**, 2333 (1932); Stewart and H. P. Kung, ibid., **55**, 4813 (1933); Stewart and C. H. Li, ibid., **60**, 2782 (1938). C. H. Li (b. 1913) became a leader in research at California on polypeptide hormones.

39. E. D. Hughes and C. K. Ingold, Trans. Farad. Soc., **37**, 657 (1931); W. H. Saunders and R. A. Williams, JACS, **79**, 3712 (1957) and later papers; W. H. Saunders and A. F. Cockerill, *Mechanisms of Elimination Reactions*, New York, 1973.

40. T. D. Stewart and W. E. Bradley, JACS, **54**, 4172, 4183 (1932). Stewart (1890-1958, B.S., Ph.D. 1916, both at Berkeley) taught at Berkeley from 1917. He worked on kinetics and mechanisms of organic reactions and was author with his colleague C. W. Porter of a successful textbook in organic chemistry.

41. A. Lapworth and R. H. F. Manske, JCS, 2533 (1928); 1976 (1930).

42. W. Cocker, A. Lapworth and A. Walton, ibid., 449 (1930).

43. G. E. K. Branch, ibid., **38**, 2466 (1916). E. Q. Adams is credited with help in interpreting the results of this paper.

44. J. M. Snell and S. M. McElvain, ibid., **53**, 2310 (1931); D. C. Roberts and McElvain, ibid., **59**, 2007 (1937), and refs. therein.

45. M. A. Spielman and M. T. Schmidt, ibid., **59**, 2009 (1937); C. R. Hauser and W. B. Renfrow, ibid., **59**, 1823 (1937), and refs. therein.

46. C. R. Hauser, ibid., **60**, 1957 (1938).

47. W. B. Renfrow and C. F. Hauser, ibid., **60**, 463 (1938); Hauser et al., ibid., **60**, 1960 (1938).

C. Addition to the Carbon-Carbon Double Bond

As described in an earlier essay hydroxylation of the double bond in maleic and fumaric acid by permanganate gives *cis* hydroxylation. Addition of bromine was shown by several groups of workers around 1911 to add *trans* to maleic and fumaric acids, a discovery usually claimed by and for Arthur Michael, although he never gave a definitive proof.[1] It was thus demonstrated by 1912 that either *cis* or *trans* addition could occur, depending on the reagent and groups adding.

Ethel M. Terry and Lillian Eichelberger at Chicago[2] showed that bromination of sodium maleate in water with excess sodium bromide (to prevent addition of HOBr to the double bond) gave eighty percent *cis* addition, whereas sodium fumarate gave ninety percent *trans* addition.[3] This meant that the

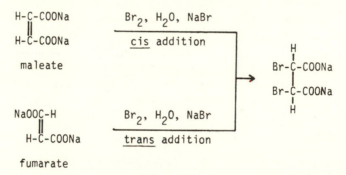

process shown here was taking place. Ethel M. Terry (Mrs. Herbert McCoy) (A.B. 1907, Ph.D. 1913, both at Chicago) was associated with the Chicago department for many years. Lillian Eichelberger (Mrs. Ralph Cannon, B.S. 1915, Mississippi State College for Women; Ph.D. 1921, Chicago) made her career in Chicago as research instructor at the university 1921-1924, as biochemist at the Municipal Tuberculosis Sanitarium 1924-1928, and at the university's department of medicine from 1928. The University of Chicago was a pioneer in advanced training for women, and it is difficult to find another significant paper from a predominantly male university with two women as the only authors.

The maleic acid bromination in water was thus unusual in giving mainly the product of *cis* addition, *meso*-dibromosuccinic acid. These results agreed with, but did not prove, the idea that bromination of both maleate and fumarate went through the same intermediate, perhaps an ion formed by electrophilic attack of bromine on the double bond. This implies a two-step addition process with isomerization at some stage to a common intermediate.

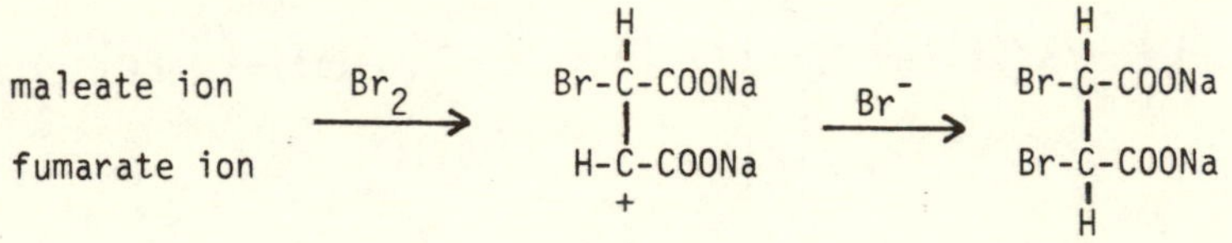

$$
\begin{array}{ccccc}
\text{maleate ion} & \xrightarrow{\ \text{Br}_2\ } &
\begin{array}{c} H \\ | \\ Br-C-COONa \\ | \\ H-C-COONa \\ + \end{array}
& \xrightarrow{\ \text{Br}^-\ } &
\begin{array}{c} H \\ | \\ Br-C-COONa \\ | \\ Br-C-COONa \\ | \\ H \end{array}
\\
\text{fumarate ion} & & & &
\end{array}
$$

A. W. Francis suggested in 1925 that bromine first ionized to form positive and negative bromine ions, the positive ion attacking the double bond and the resulting carbonium ion picking up a negative bromide ion.[4] The

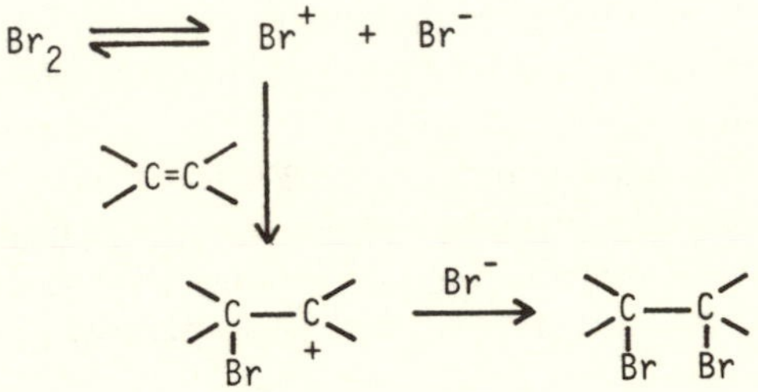

notion of positive halogens was proposed much earlier,[5] but it seems improbable that a symmetrical compound like a bromine molecule would ionize appreciably in the manner proposed by Francis prior to its reaction with the double bond. Francis, however, found evidence for the formation of his positively charged carbon intermediate in the formation of $ClCH_2CH_2Br$ from ethylene and bromine in concentrated sodium chloride solution. An analogous compound, $BrCH_2CH_2ONO_2$, was found under similar conditions in sodium nitrate solution.

$$
CH_2{=}CH_2 \;+\; [Br^+] \;\longrightarrow\; [\overset{+}{CH_2}{-}CH_2Br] \;\xrightarrow{\ Cl^-\ }\; ClCH_2CH_2Br
$$

or equivalent

Electrophilic attack on the double bond in a two-stage process had been suggested by Stieglitz[6] and Lucas[7]. Ingold[8] represented the concurrent formation of dibromides and bromohydrins where the curved arrow indicates the potential drift of electrons (*polarization*) within the double bond. Carothers' ideas of addition to a multiple linkage envisioned a nucleophilic attack, in this case by bromide ion or its equivalent.[9] However, it seems more likely that attack on a double bond, with its extra pair of electrons, would be made by an electron-deficient group, an electrophile, rather than by an electron-rich entity, a nucleophile, such as bromide ion, Br^-.

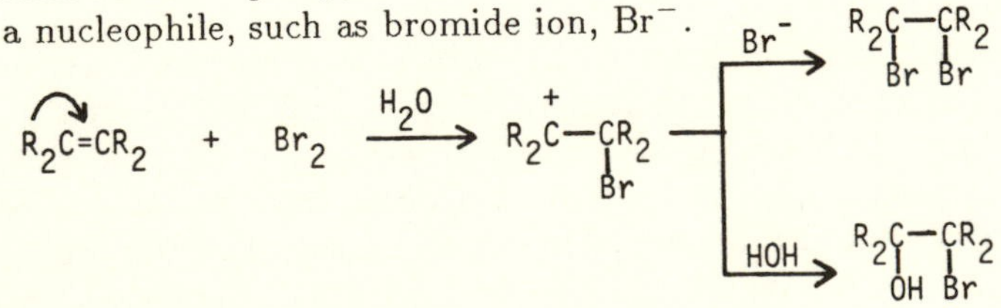

Kinetic and synthetic evidence supporting the two-stage scheme of addition to a double bond initiated by an electrophile was supplied by P. D. Bartlett and D. S. Tarbell in 1936-1937. A study of the addition of bromine in methanol to stilbene showed that the effects of added bromide ion and acid

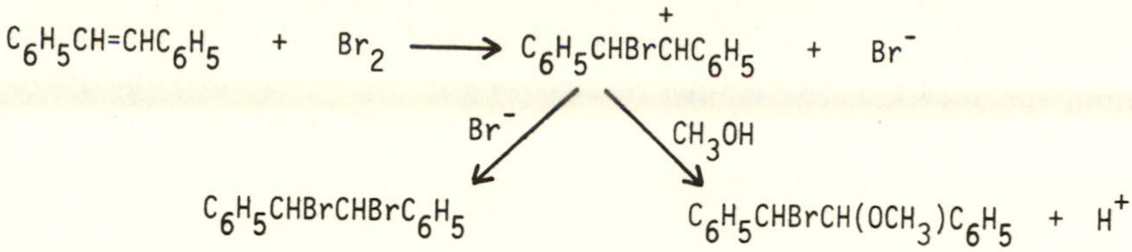

$$C_6H_5CH=CHC_6H_5 \quad + \quad Br_2 \quad \longrightarrow \quad C_6H_5CHBr\overset{+}{C}HC_6H_5 \quad + \quad Br^-$$

$$Br^- \swarrow \qquad \searrow CH_3OH$$

$$C_6H_5CHBrCHBrC_6H_5 \qquad\qquad C_6H_5CHBrCH(OCH_3)C_6H_5 \quad + \quad \overset{+}{H}$$

on the product ratio and the kinetics could only be explained by the scheme shown.[10] This paper contained a general theory of addition to double bonds under polar conditions, of which bromination was a special case. The paper supported the general electronic scheme of the English school by giving a specific case where it was a necessary explanation; thus it aided in the dissemination of the English ideas in this country. However, American physical organic chemistry was not a transplant from England.

The action of bromine water and chlorine water on sodium dimethylmaleate and dimethylfumarate led to two pairs of stereoisomeric β-lactones.[11]

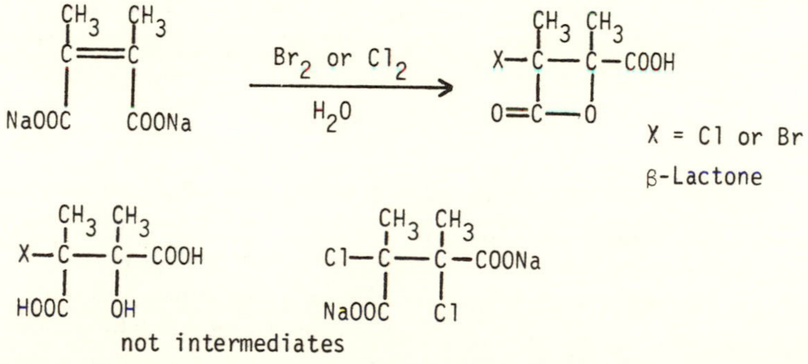

$$\begin{array}{cc} CH_3 & CH_3 \\ C\!=\!\!C \\ NaOOC & COONa \end{array} \xrightarrow[H_2O]{Br_2 \text{ or } Cl_2} \begin{array}{cc} CH_3 & CH_3 \\ X\!-\!C\!-\!\!C\!-\!COOH \\ O\!=\!C\!-\!O \end{array}$$

X = Cl or Br

β-Lactone

$$\begin{array}{cc} CH_3 & CH_3 \\ X\!-\!C\!-\!C\!-\!COOH \\ HOOC & OH \end{array} \qquad \begin{array}{cc} CH_3 & CH_3 \\ Cl\!-\!C\!-\!C\!-\!COONa \\ NaOOC & Cl \end{array}$$

not intermediates

Experiments showed that the corresponding halohydrin or the dichloro acid could not be intermediates, and hence a two-step process, completed by cyclization to the β-lactone, was indicated. I. Roberts and G. E. Kimball at Columbia at once suggested that the intermediate ion postulated here with a halogen and a positive charge on adjacent carbons, might better be written

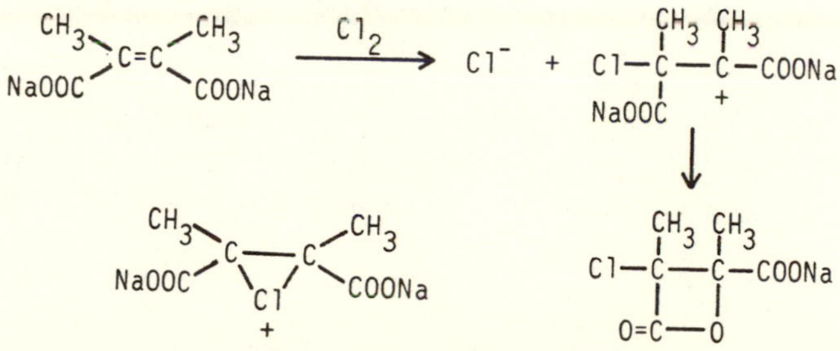

$$\begin{array}{cc} CH_3 & CH_3 \\ C\!=\!C \\ NaOOC & COONa \end{array} \xrightarrow{Cl_2} Cl^- + \begin{array}{cc} CH_3 & CH_3 \\ Cl\!-\!C\!-\!C\!-\!COONa \\ NaOOC & + \end{array}$$

$$\downarrow$$

$$\begin{array}{cc} CH_3 & CH_3 \\ C\!-\!\!\!-\!C \\ NaOOC \diagdown Cl \diagup COONa \\ + \end{array} \qquad \begin{array}{cc} CH_3 & CH_3 \\ Cl\!-\!C\!-\!C\!-\!COONa \\ O\!=\!C\!-\!O \end{array}$$

as a cyclic chloronium ion.[12] This ring would be opened by intramolecular attack of the carboxylate ion, in this case, to form the chloro-β-lactone. This idea, which seemed obvious once it was proposed, has had a long and useful life. The intermediate formation of halonium ions explains the stereochemistry of neighboring group reactions, as observed by H. J. Lucas and his collaborators.[13] The example shown illustrates the application of the halonium ion idea. This example is also important as the starting point for Winstein's neighboring group work, which he developed in his independent career later with great rigor and originality.

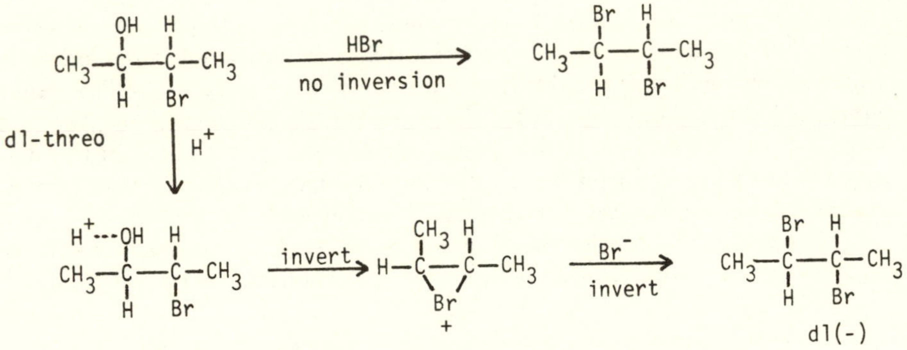

bromonium ion
(neighboring group effect)

Howard J. Lucas (1885-1963, B.S. 1907, M.A. 1908, both at Ohio State) taught at Throop Institute (later Caltech) from 1913 to 1955. A self-taught scientist, he was an accomplished synthetic chemist and an outstanding physical organic chemist. He trained students like W. G. Young and Saul Winstein who made California a center for research on reaction mechanisms. His *Organic Chemistry*, New York, 1935, proved to be an outstanding textbook, the first to utilize the newer theoretical ideas. His work involved mechanism of addition reactions to carbon-carbon bonds (both double and triple) and the formation of complexes between heavy metal ions (like silver) and unsaturated compounds.[14]

Hypochlorous acid, HOCl, was generally considered to add *trans* to the carbon-carbon double bond, although the evidence for this was mainly by analogy to the *trans* bromination of maleate and fumarate esters. R. Kuhn reported that HOCl added 100 percent *cis* to maleic acid, and 80 percent *trans* - 20 percent *cis* to fumaric acid.[3] The resulting stereoisomeric acids HOOCCH(OH)CH(Cl)COOH were assigned configurations by Kuhn from the ratio of the first and second acid dissociation constants, a method which gave correct results with the *meso-* and *dl*-disubstituted succinic acids, whose configurations were known unambiguously from the resolution of the *dl*-form into optically active forms.

The *cis* addition of halogen and HOCl to maleic acid and its ions made the maleic-fumaric pair undesirable as general examples. It was felt, again on intuitive rather than very rigorous grounds, that HOCl added to cyclohexene

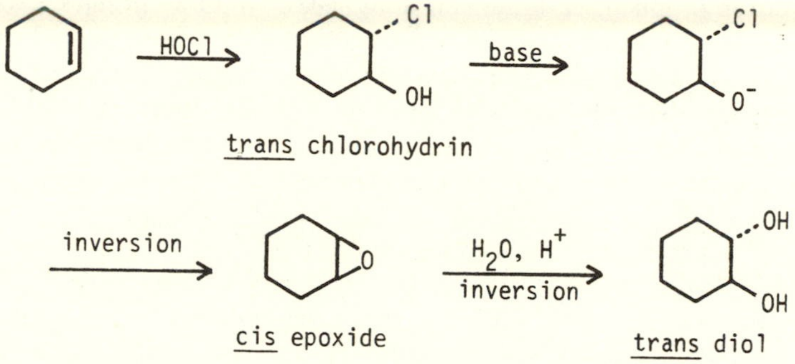

trans chlorohydrin

cis epoxide   trans diol

to give the *trans* chlorohydrin, and that this formed the epoxide with base with inversion. The epoxide was considered to be a *cis* ring, because a *trans* epoxide would be too highly strained as a three-ring structure to be stable. This view was supported by Bartlett's preparation of a stereoisomeric chlorohydrin,[15] presumably *cis*, by reduction of chlorocyclohexanone with

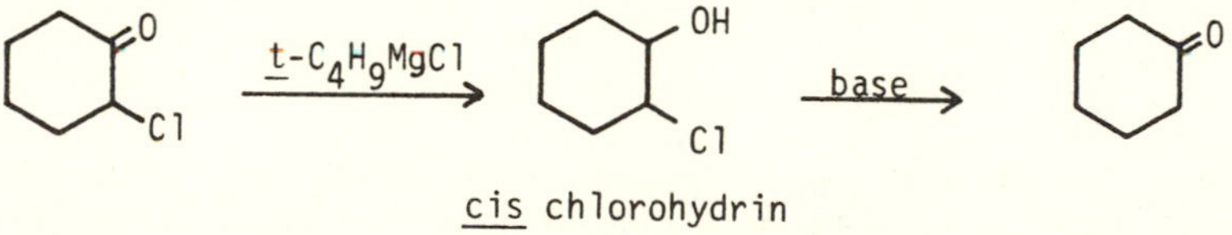

cis chlorohydrin

tert-C_4H_9MgCl. It had been known since 1920 that acid-catalyzed hydration of the (presumed) *cis*-epoxide of cyclohexene gave the *trans*-diol, which meant that an inversion took place in opening the epoxide ring.[16] It was therefore natural to consider that inversion occurred in the closure of the chlorohydrin to form the epoxide. This assumption had support from conclusive evidence that displacement of a group or atom from a saturated secondary carbon atom by external attack occurred with inversion by the displacing group (see next essay). The assumptions in these deductions have been stressed to show the steps in the chain of argument. Since the Instrumental Revolution various spectroscopic methods can be used to support these assigned configurations, but it must be remembered that spectroscopic methods of structure determination are empirical and require compounds of known configuration for comparison. X-ray crystallography is, however, free of most of this ambiguity.

It was clearly desirable to have further studies on the stereochemistry of addition to double bonds where the possible disturbing effects of adjacent carboxyl groups or ions, as with maleic and fumaric acids, would be absent. A series of elegant papers by Geza Braun on the *cis* and *trans* hydroxylation by known methods of the double bond in crotonic acid, $CH_3CH=CHCOOH$, and related compounds furnished some useful correlations of configurations, especially of some four-carbon sugars.[17] Braun's work, however, did not answer the questions above.

Lucas supplied an important demonstration of addition, cyclization, and elimination reactions in which some configurations were proved and the rest inferred. All subsequent work has supported his inferences, so that the complete series, derived from both *cis* and *trans* 2-butene, is established. The preparation and analysis of the 2-butenes was a laborious task involving assumptions about *trans* addition and elimination.[18] The elimination of bromine

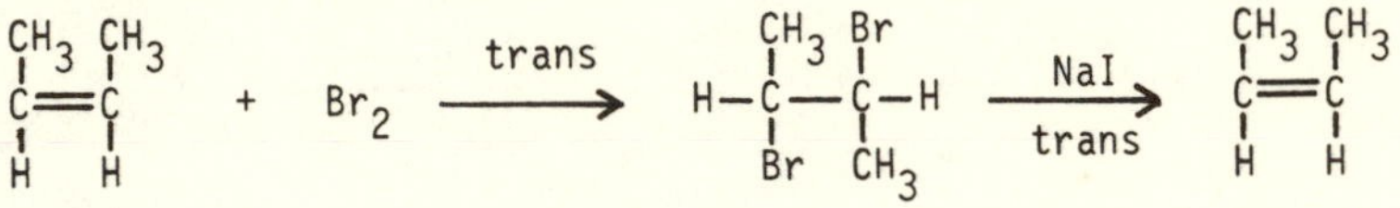

from 1,2-dibromo compounds by sodium iodide was shown to be *trans* by Winstein, Pressman, and Young.[18] The compounds 1,2-dibromobutane, *meso*-2,3- and *dl*-2,3-dibromobutane all react with sodium iodide at different rates.

The following sequence was carried out with both *cis* and *trans* 2-butenes. Addition of bromine to *cis*-2-butene was proved to be *trans* by showing that the resulting pure dibromide was the *dl* form, because it could be obtained optically active. It was shown also that addition to *trans*-2-butene was *trans*, because the *meso* dibromo compound was obtained.

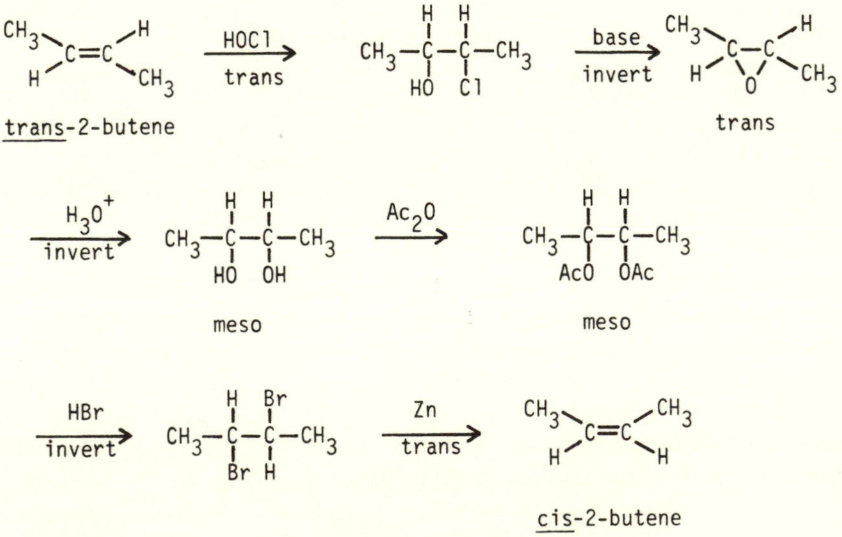

Configurations proved

Addition of HOCl must be *trans* in order to fit the sequence.[19] The configurations assigned to the 2-butenes by G. B. Kistiakowsky[18] from the heats of hydrogenation fit Lucas' scheme. It is worth pointing out, as an example of the caution with which physical methods should be used, that the first assignment of configurations to the 2,3-epoxybutanes by electron diffraction was opposite to Lucas' assignment from his reaction sequences.[20] The electron diffraction results when recalculated on a different model gave results agreeing with the chemical assignment.[21]

The idea that "isomers" existed, differing only in charge distribution, goes back at least to J. J. Thomson in 1907. Such "electron isomers" were called *electromers* by H. S. Fry in 1911, and were discussed in detail in connection with the dualistic theories of molecular structure.[22] Electromers derived from different distributions of electrons in double bonds were proposed in electronic terms by Carothers,[9] Lucas,[7] and others, according to Lewis' electronic theory, and the impression is given that electromers in some cases are stable enough to be separated. Dipole moment measurements by Smyth at Princeton in 1925 on several unsaturated hydrocarbons showed that electromers, if present, were in amounts too small to give a measurable dipole moment.[23] The existence of electromers was espoused by M. S. Kharasch, who reported the interconversion of electromers of 2-pentene.[24]

There is obviously room for a wide gradation of types here, from electromers with independent existence to a single structure in which the electron distribution is unsymmetrical due to the substituents. It is not always clear exactly what is meant by the term. The idea of electromers, although indefensible on quantum mechanical grounds, was neverless useful, because it focused attention on the addition of unsymmetrical reagents to unsymmetrical double bonds. The Markovnikov rule empirically predicted the mode of addition illustrated, and this was in agreement with the electron-donating effect of the CH_3 group making the terminal carbon more negative and the middle carbon more positive, following Lucas, Ingold, and Michael.

$$CH_3CH{=}CH_2 \;+\; H^+Br^- \longrightarrow CH_3\underset{\underset{Br}{|}}{C}HCH_3$$

Much of the experimental work had been done using HBr addition to unsaturated compounds, and the results were confusing and contradictory. Kharasch therefore undertook a careful study of the addition of HBr to allyl bromide to get experimental support for his ideas on electromers and electronegativity of groups, which differed somewhat from those of Lewis, Lucas, and Ingold. This is a notable example of results of the first importance arising from the investigation of a questionable theory.

Morris S. Kharasch (1895-1957) was born in the Ukraine but came to America and received his undergraduate and Ph.D. degrees at Chicago in 1917 and 1919. After a brief stay at the University of Maryland, he returned to the Chicago faculty in 1928, where he remained the rest of his life.[25] His demonstration of the peroxide effect in HBr addition led him to the discovery of whole series of free-radical reactions, a field which he largely created and of which he became the acknowledged master. Kharasch was brilliant, sometimes caustic, highly original, one of the gifted sceptics who are needed periodically to reinvigorate a science by questioning the current dogmas. He trained many outstanding graduate students and postdoctoral fellows.

Kharasch and his student Frank R. Mayo showed in a 1933 paper, which was a masterpiece of deductive reasoning and experimental skill,[26] that the

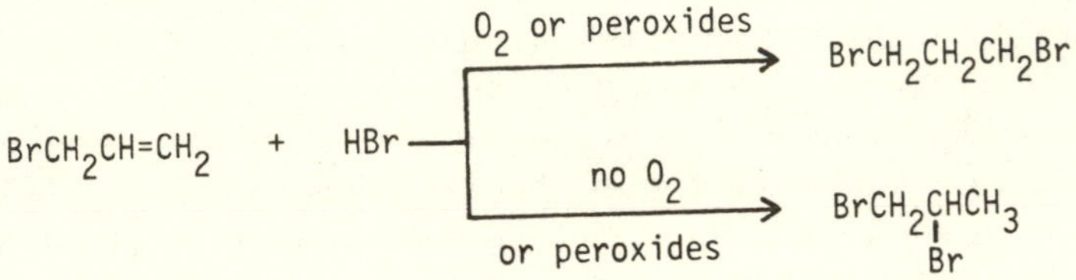

$$BrCH_2CH=CH_2 \quad + \quad HBr \quad
\begin{cases}
\xrightarrow{O_2 \text{ or peroxides}} BrCH_2CH_2CH_2Br \\
\xrightarrow[\text{or peroxides}]{\text{no } O_2} BrCH_2\underset{Br}{CH}CH_3
\end{cases}$$

direction of addition of HBr to allyl bromide depended on the presence or absence of peroxides (*peroxide effect*). Thus peroxides caused anti-Markovnikov addition, and absence of peroxides (or more simple experimentally, presence of antioxidants) gave "normal" (Markovnikov) addition. A peroxide effect was shown in the addition of HBr to series of about a dozen unsymmetrical ethylenes, and in 1937 Kharasch published a mechanism for a free-radical chain reaction.[27] For allyl bromide, the suggested path is shown.

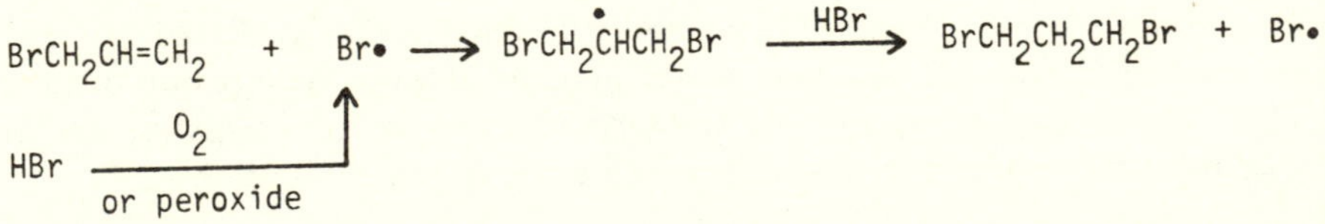

$$BrCH_2CH=CH_2 \quad + \quad Br\bullet \longrightarrow BrCH_2\overset{\bullet}{C}HCH_2Br \xrightarrow{HBr} BrCH_2CH_2CH_2Br \quad + \quad Br\bullet$$

$$HBr \xrightarrow{O_2 \text{ or peroxide}}$$

A free-radical mechanism was also suggested in less detailed terms by two English chemists, D. H. Hey and W. A. Waters, in a notable review article[28] in 1937.

The free-radical mechanism for the peroxide effect was another manifestation of the active interest of the 1930s in free radicals as intermediates in addition polymerization, in high temperature reactions, and in photochemistry.[29] Kharasch's originality lay not so much in suggesting a free-radical process as in the scope and variety of new free-radical reactions which he discovered.[25,30]

The above mechanism can be reasonably explained in terms of bond strengths and stabilities of free radicals[30] just as the ionic Markovnikov addition has a reasonable explanation. The peroxide effect work emphasizes some fundamental points in chemical research and in scientific work in general. The first is the importance of careful experimental work in which all variables are controlled. A corollary of this is that in cases of extensive experimental work with discordant results, the existence of an unknown variable should be suspected. Another point is that extended speculation about a reaction should be based on a knowledge of the mechanism of the reaction. In absence of peroxides the addition of HBr to multiple linkages is similar to that of HCl and H_2SO_4, which do not show a peroxide effect. The presence of peroxides changes completely the mechanism of reaction with HBr.

It should be emphasized that the addition of chlorine or bromine to double bonds can be accelerated by light; this is not an ionic reaction, as discussed in the first part of this section, but is a free-radical chain reaction involving Cl· or Br·, formed by dissociation of the halogen by light to the atoms, which start the chain. The same is true of substitution of hydrogen on carbon by halogen in presence of light, frequently done in the gas phase instead of the liquid phase.

In an earlier section we gave F. C. Whitmore's mechanism of polymerization of isobutene by acid; this is regarded as proton addition to form a tertiary carbonium ion which adds to the double bond of another molecule. The BF_3-catalyzed polymerization must involve a similar process.

Detailed studies of both rate and equilibria for hydration of unsaturated compounds show that the rate is proportional to the concentration of catalyzing acid and that the free energy change is small, that is, the equilibrium point is not predominantly toward either product.[31]

$$CH_3CH{=}C(CH_3)_2 \quad + \quad H^+ \quad \underset{HNO_3}{\overset{}{\rightleftharpoons}} \quad CH_3CH_2\overset{+}{C}(CH_3)_2 \quad \underset{H_2O}{\overset{}{\rightleftharpoons}} \quad CH_3CH_2\underset{OH}{C}(CH_3)_2$$

The attack of electrophilic ions on double bonds, analogous to the first step in ionic halogenation, is shown in the formation of 1:1 complexes between silver ion and ethylene and other unsaturated compounds, including some dienes.[32] Lucas measured equilibrium constants and documented the complexing shown by silver, cuprous, and mercury salts.[33] He found that the metal complexes did not hydrate and explained the stability of the silver ion

$$\mathrm{C{=}C} \quad + \quad Ag^+ \quad \rightarrow \quad \overset{Ag}{\underset{}{C{-}C}} \quad \longleftrightarrow \quad \overset{Ag}{\underset{}{C{-}C}} \quad \longleftrightarrow \quad \underset{Ag^+}{C{-}C}$$

complex by the contribution of resonance forms depicted.[34] A similar explanation was given for the stability of mercury (Hg^{++}) complexes with cyclohexene;[34] such complexes may be intermediates in hydration or formation of covalent-bonded addition mercuration products, such as C(OH)CHgOH.

It had been known for many years that mercury salts add to double bonds; mercuric acetate in methanol adds with incorporation of solvent.[35] Marvel had shown that such compounds contain a covalent C-Hg bond, because they could be obtained in stereoisomeric forms due to the asymmetry of the carbon carrying mercury,[36] as in $C_6H_5CH(OCH_3)CH(HgOAc)COOR$. Lucas regarded his "mercurinium" ions as intermediates in mercuration, just as bromonium ions are intermediates in bromine addition.[34] Whether mercuration gave *trans* addition to the double bond was not settled.

$$>C=C< \quad + \quad Hg(OAc)_2 \quad + \quad CH_3OH \quad \longrightarrow \quad >\underset{CH_3O}{C}-\underset{HgOAc}{C}<$$

Lucas thus utilized the ideas of resonance in his research and extensively in his textbook. His colleague Linus Pauling was the principal proponent of resonance, but Lucas was an independent thinker who would have been abreast of modern developments in theory whatever his location.[37]

Mercuric sulfate and sulfuric acid catalyze the once commercially important hydration of acetylene to acetaldehyde.[38] The Notre Dame group under J. A. Nieuwland had proposed organomercury compounds as intermediates in this reaction.[39] A kinetic study showed that the initial rate was proportional to the acetylene concentration and to the square of the $HgSO_4$ (or $Hg(HSO_4)_2$) concentration.[40] This suggested that the acid-catalyzed hydration of a complex $C_2H_2Hg_2(HSO_4)_4$ might be involved.[40]

$$HC\equiv CH \quad + \quad H_2O \quad \xrightarrow[H_2SO_4]{HgSO_4} \quad [CH_2=CHOH] \quad \longrightarrow \quad CH_3CHO$$

In summary, addition to C=C in polar solvents is electrophilic and usually *trans*, via a two-stage process which may involve cyclic ions as intermediates. Metal ions such as silver and mercury form (presumably) cyclic ions by addition to C=C. The addition of HBr in presence of peroxides is a free-radical chain reaction and gives anti-Markovnikov products with unsymmetrically substituted C=C. Photochemical addition reactions are free-radical processes. Ionic elimination reactions producing C=C are usually *trans*.

LITERATURE CITED

1. In addition to refs. above, a review by E. C. Frankland, JCS, **101**, 678 (1912); A. Michael, JACS, **40**, 704, 1674 (1918).
2. E. M. Terry and Lillian Eichelberger, ibid., **47**, 1067 (1925).
3. Their figures on chlorination were corrected by R. Kuhn and T. Wagner-Jauregg, Ber., **61**, 511 (1928), who got results similar to the bromination reactions.
4. A. W. Francis, JACS, **47**, 2340 (1925).
5. In 1901 by W. A. Noyes and by Stieglitz; cf. W. A. Noyes, Chem. Rev., **17**, 17 (1935); Noyes, JACS, **45**, 2959 (1923) suggested that *during reaction* Cl_2 could ionize to Cl^+ and Cl^-.

6. J. Stieglitz, ibid., **44**, 1304 (1922).

7. H. J. Lucas et al., ibid., **46**, 2475 (1924); **47**, 1459, 1462 (1925). Lucas used the electronic ideas of G. N. Lewis, but Stieglitz did not.

8. C. K. Ingold, Annual Repts. of the Chem. Soc., 131 (1928); R. Robinson, *Outline of an Electrochemical (Electronic) Theory of the Course of Organic Reactions*, Institute of Chemistry of Great Britain and Ireland, London, 1932.

9. W. H. Carothers, JACS, **46**, 2226 (1924). Although this paper is impressive in its generalizing power, it scarcely justifies Roger Adams' praise that Carothers "in essence includes in his discussion (of the double bond) everything that has since been written on this particular subject", *Biog. Mem. Nat. Acad. Scis.*,**20**, 296 (1938).

10. P. D. Bartlett and D. S. Tarbell, JACS, **58**, 466 (1936). The addition of the elements of ROX (X = Cl or Br, R = CH_3 or alkyl) had been studied preparatively by several groups, including J. B. Conant and E. L. Jackson, ibid., **46**, 1727 (1924); K. Meinel, Ann., **510**, 129 (1934); **516**, 231 (1935).

11. D. S. Tarbell and P. D. Bartlett, JACS, **59**, 407 (1937).

12. I. Roberts and G. E. Kimball, ibid., **59**, 947 (1937).

13. Bromonium ions, Lucas and S. Winstein, ibid., **61**, 1576, 2845 (1939); iodonium ions, Lucas and H. K. Garner, **72**, 2145 (1950); chloronium ions, Lucas and C. W. Gould, Jr., ibid., **63**, 2541 (1941).

14. W. G. Young and Saul Winstein, *Biog. Mem. Nat. Acad. Scis.* **43**, 162 (1973). Lucas' contributions to both teaching and research have too frequently been overlooked by writers on Caltech and American chemistry.

15. P. D. Bartlett, ibid., **57**, 224 (1935).

16. J. Böeseken and J. van Giffen, Rec. trav. chim., **39**, 184 (1920); also R. Kuhn and F. Ebel, Ber., **58**, 919 (1925).

17. Geza Braun, JACS, **51**, 228 (1929); **52**, 3167 (1930) and later papers. Braun was an International Research Fellow from Hungary, who worked at both Chicago and Harvard.

18. Synthesis and analysis of the 2-butenes: W. G. Young, R. T. Dillon and H. J. Lucas, ibid., **51**, 2528 (1929); analysis, **52**, 1953 (1930); W. G. Young and S. Winstein, ibid., **58**, 102 (1936). *Trans* elimination of bromine by NaI: S. Winstein, D. Pressman and W. G. Young, ibid., **61**, 1645 (1939). G. B. Kistiakowsky at al., ibid., **57**, 879 (1935), separated the pure *cis* and *trans* 2-butenes by a laborious distillation procedure. Lucas obtained pure materials via the epoxides or the diols.

19. C. E. Wilson and H. J. Lucas, ibid., **58**, 2396 (1936).

20. L. O. Brockway and P. C. Cross, ibid., **58**, 2407 (1936).

21. Idem, ibid., **59**, 1147 (1937).

22. H. S. Fry, Z. physikal. Chem., **76**, 387 (1911); L. W. Jones, JACS, **36**, 1268 (1914); J. Stieglitz and B. A. Stagner, ibid., **38**, 2052 (1916).

23. C. P. Smyth and C. T. Zahn, ibid., **47**, 2501 (1925).

24. M. S. Kharasch and F. R. Darkis, Chem. Rev., **5**, 590 (1928); Kharasch and O. Reinmuth, JCE, **8**, 1732 (1931). Workers at Mount Holyoke College in collaboration with Kharasch reported isolation of electromers of 2-pentene with slightly different physical properties; JACS, **51**, 3034, 3041 (1929).

25. See F. H. Westheimer, *Biog. Mem. Nat. Acad. Scis.*, **34**, 123 (1960); D. S. Tarbell, Chem. Eng. News, **54**, 115 (1976), issue of April 6.

26. M. S. Kharasch and F. R. Mayo, JACS, **55**, 2468 (1933).

27. Kharasch, H. Engelmann and F. R. Mayo, JOC, **2**, 288 (1937).

28. D. H. Hey and W. A. Waters, Chem. Rev., **21**, 169 (1937).

29. See sections on polymerization and free radicals. For photochemistry up to this period, see W. A. Noyes, Jr. and P. A. Leighton, *The Photochemistry of Gases*, New York, 1941.

30. The authoritative contemporary review of the peroxide effect in addition of many reagents to unsaturated compounds is F. R. Mayo and C. Walling, Chem. Rev., **26**, 351 (1940). C. Walling, *Free Radicals in Solution*, New York, 1957 is an excellent survey.

31. H. J. Lucas and W. F. Eberz, JACS, **56**, 460, 1230 (1934); Lucas and S. Winstein, ibid., **59**, 1461 (1937).

32. S. Winstein and Lucas, ibid., **60**, 836 (1938).

33. Lucas at al., ibid., **59**, 45 (1937).

34. Lucas, F. R. Hepner and S. Winstein, ibid., **61**, 3102 (1939).

35. G. F. Wright, ibid., **57**, 1994 (1935), and refs therein.

36. L. T. Sandborn and C. S. Marvel, ibid., **48**, 1409 (1926).

37. Correspondence of Linus Pauling and W. G. Young with DST.

38. A review of the reaction: J. A. Nieuwland and R. R. Vogt, *The Chemistry of Acetylene*, New York, 1945, pp. 115-123.

39. R. R. Vogt and J. A. Nieuwland, JACS, **43**, 2071 (1921); G. F. Hennion, Vogt and Nieuwland, JOC, **1**, 159 (1936).

40. Lucas, at al., JACS, **59**, 722 (1937). Later, H. Lemaire and Lucas, ibid., **77**, 939 (1955) suggested a mercurinium salt as the reaction intermediate in the mercury-catalyzed addition of acetic acid to 3-hexyne $(C_2H_5C{\equiv}CC_2H_5)$.

D. The Walden Inversion

In replacement of a group at an asymmetric carbon, the replacing group might occupy the same position in space that the replaced group left; the configuration would thus be unaltered or retained. Conversely, the entering group might occupy the opposite position in space from the leaving group, giving the opposite or inverted configuration.

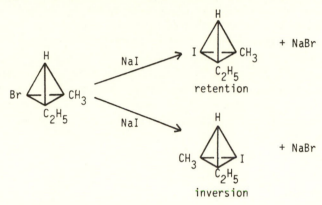

Paul Walden (1863-1957), a Baltic German, found that (−)chlorosuccinic acid, $HOOCCH_2CHClCOOH$, could be converted by one sequence of reactions into its mirror image, the (+) acid. Hence in this sequence there had been an overall *change of configuration* (or *inversion*), frequently called the "Walden inversion". He found other sequences in which an optically active compound, after several transformations, was regenerated with the same sign of rotation, thus with overall *retention of configuration*.[1] Known for this and other significant researches, Walden also published much on the history of chemistry, including a good history of classical organic chemistry from 1880 to 1935.[2]

The phenomenon of inversion, or retention of configuration in reactions at a saturated carbon atom, was clearly of fundamental importance to the understanding of the mechanism of reactions in which bonds attached to an asymmetric carbon were broken. From Walden's sequences of reactions, one could only say that those resulting in overall inversion involved either one or some odd number of inversions. Those sequences resulting in no inversion involved either no inversion at all or some even number of inversions. It was impossible in Walden's time to identify the reactions in the sequence at which inversion or retention occurred because there was no way of determining the relative configurations of the compounds involved.

This ambiguity about configurations of reaction products disturbed organic chemists because it represented a serious gap in the structural theory of organic compounds. Explanations for inversion were given by leading chemists like Alfred Werner for inorganic complexes, and by Emil Fischer,[3] but the clearest picture was proposed by G. N. Lewis in his book in 1923.[4]

Lewis suggested that the entering group attached itself temporarily to the face of the asymmetric carbon opposite the leaving group. The latter then detached, there was a shift of the valences around the central carbon, and substitution was thus accompanied by inversion.

With the development of quantum mechanics and the knowledge of energy requirements for reactions, A. R. Olson, also of Berkeley, was able to give a theoretical rationalization of Lewis' mechanism for inversion. This was based on the idea that the most economical path energetically was for a displacing group to start forming a bond at the same time that the leaving group was breaking its bond.[5] In the transition state the asymmetric carbon was planar, and the reaction was finished by complete bond formation for the entering group and departure of the leaving group. The analogy commonly used to describe inversion by this mechanism was that the bonds around the asymmetric carbon had turned inside out like an umbrella in a high wind.

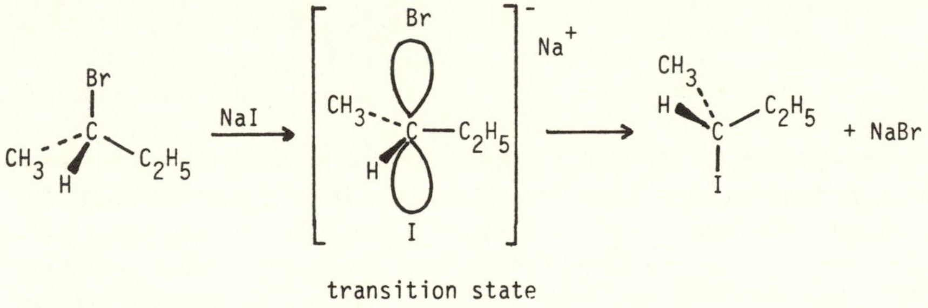

transition state

This scheme suggested that every replacement (or displacement) was accompanied by inversion *if* the bond from the asymmetric carbon was broken. Olson supported this conclusion by kinetic studies of the complicated series of reactions when optically active α-bromo acids were treated with chloride ion in aqueous solution.[6] Similar conclusions were reached in Germany by M. Polanyi on other experimental grounds.[5]

J. Kenyon and H. Phillips in England had actually provided evidence, starting in 1923, of inversion of each displacement in systems more clear-cut and simpler than those of Olson. The English workers carried out sequences similar to Walden's, but in which two steps did not disturb the bond from the asymmetric carbon and hence did not cause inversion. The first example is shown below, with the optical rotations (α) and occurrence of inversion indicated.[7] The retention of configurations during step I (acetylation) and step IV (hydrolysis of acetate) was demonstrated by later tracer experiments on ester hydrolysis. Retention during step II was assumed because the asymmetric carbon bond C-O was not disturbed during preparation of the tosylate.

This means that the inversion which occurred took place in step III, the displacement of tosylate ion by acetate ion. Displacement of tosylate by other nucleophiles ($RO^- Na^+$, $R'COO^- K^+$) with inversion was also demonstrated.[7] The significance of these results was emphasized by F. Richter[8] and later by L. P. Hammett.[9]

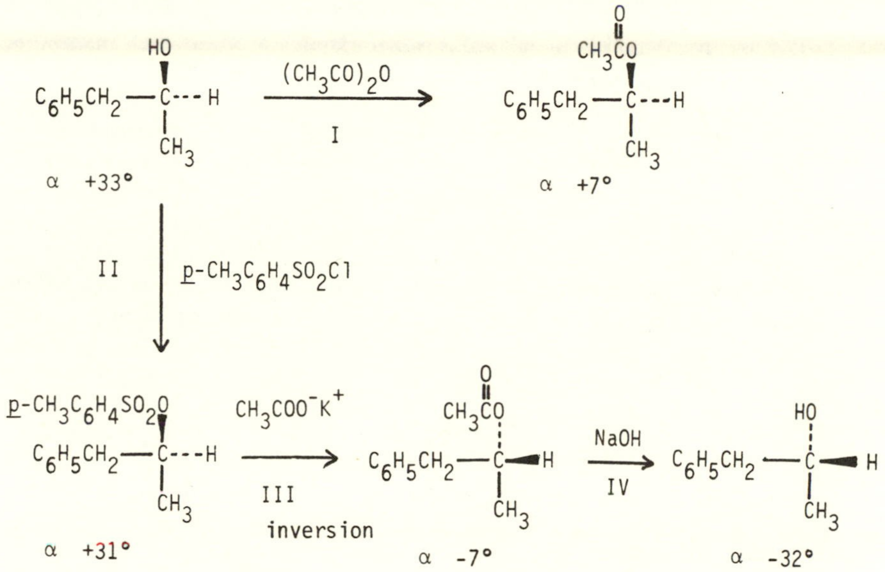

The advent of radioactive iodide, coupled with earlier observations on the racemization of optically active alkyl iodides by iodide ion, allowed E. D. Hughes in C. K. Ingold's laboratory at University College, London, to demonstrate, with a finality only rarely attainable in scientific experiments, that each displacement by iodide ion is accompanied by inversion.[10] The reaction rate is second order, depending on the concentrations of both the alkyl iodide and inorganic iodide, thus being an S_N2 reaction (substitution - nucleophilic - second order) in the Ingold terminology. A nucleophilic reagent is one that seeks out a relatively positive center.

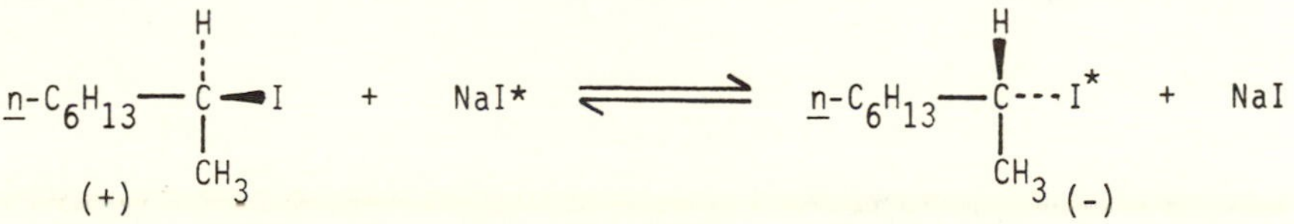

I* = radioactive I

If each displacement goes with inversion, each passage will cause the loss of optical activity from *two* molecules of alkyl iodide. The formation of each molecule of (−) iodide cancels the optical activity of one molecule of (+) iodide, because together they form a (±) pair. If the sodium iodide is radioactive, the rate of radioactive exchange with the alkyl iodide is readily measured and should equal one half the rate of racemization. This is actually observed, and therefore the inversion on displacement is proved in this case. Extension of this idea to all second-order nucleophilic displacements, whether by ions or by neutral molecules such as NH_3, (nucleophilic because of its unshared pair of electrons), has been justified by much further work.

Displacement reactions within a molecule also proceed with inversion, as in the formation of an epoxide from a *trans* chlorohydrin.[11]

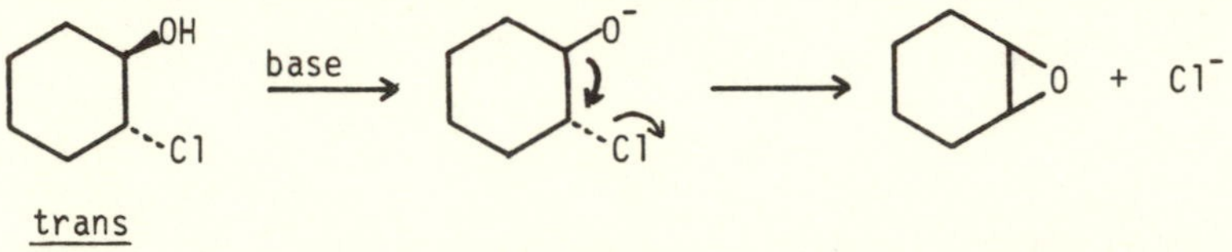

trans

The *trans* ring opening of an epoxide by aqueous acid may be pictured with H_2O as a nucleophile attacking the protonated epoxide from the rear.

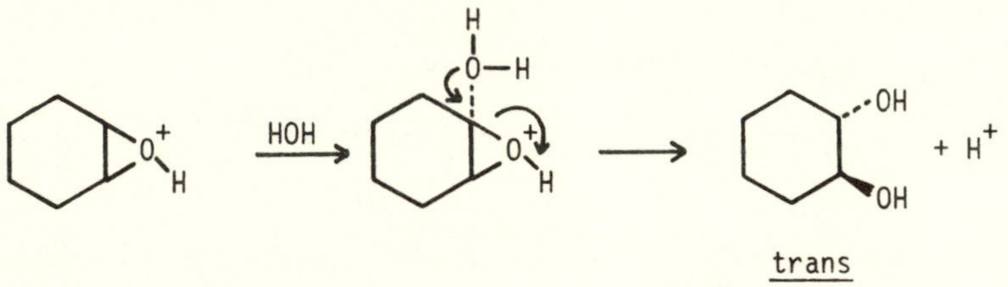

trans

If this picture of the displacement reaction is correct, an organic halide so constructed that the planar transition state is impossible should not give the displacement. The demonstration of this by P. D. Bartlett and L. H. Knox in 1939 was a signal contribution to the development of physical organic chemistry.[12] The bridgehead halogen compound below, which they synthesized, cannot have a planar transition state because the saturated bridged ring makes a rigid structure, and the chlorine is extremely stable to displacement conditions. The lack of reactivity of the chlorine was not due to the degree of substitution, because the open-chain chlorine compound is extremely reactive. The latter has no structural bar to the formation of a planar transition state.

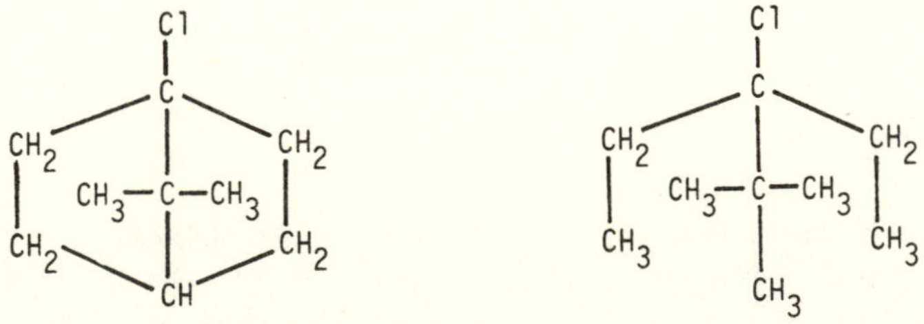

The explanation for displacements with apparent lack of inversion is probably that two inversions occur. Some such cases probably involve a cyclic intermediate. The action of phosgene on the optically active octyl alcohol shown forms the chlorocarbonate with no inversion. This can lose CO_2, and the Cl enters the same side of the asymmetric carbon from which the CO_2 is departing, thus with no inversion.[13] However, in the presence of pyridine, the chlorocarbonate gives the inverted chloride, because although this process involves two stages, only one occurs with inversion.[13]

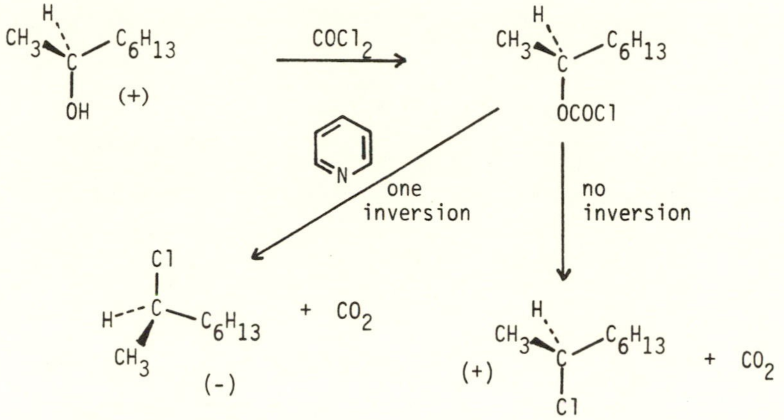

The stereochemistry of the S_N1 reaction of Ingold in which a halide is attacked by solvent was not worked out until after 1940. This reaction, whose rate depends only on the concentration of the halide and is not increased by added base, was called by Hammett in this country *solvolysis*, or *polymolecular reaction*.[14] Both these names imply that the reaction involves solvent removing the leaving group, stabilizing the carbonium ion, and reacting with it. Solvolysis of tertiary compounds can lead to partial inversion, retention of configuration, or structural rearrangements. The unraveling of these possibilities was to be done by many workers after 1940.

LITERATURE CITED

1. P. Walden, Ber., **30**, 3146 (1897).
2. P. Walden, *Geschichte der organischen chemie seit 1880*, Berlin, 1941, reprinted 1972; D. S. Tarbell, JCE, **51**, 7 (1974).
3. E. Fischer, Ann., **381**, 123 (1911); A. Werner, Ber., **44**, 873 (1911).
4. G. N. Lewis, *Valence and the Structure of Atoms and Molecules*, New York, 1923, p. 113.
5. A. R. Olson, J. Chem. Phys., **1**, 418 (1933); E. Bergmann, M. Polanyi and A. Szabo, Z. physik. Chem., **B20**, 161 (1933).
6. A. R. Olson and F. A. Long, JACS, **56**, 1294 (1934); M. J. Young and A. R. Olson, ibid., **58**, 1157 (1936).
7. H. Phillips, JCS, **123**, 44 (1923), and later papers; review in J. Kenyon and Phillips, Trans. Faraday Soc., **26**, 451 (1930).
8. F. Richter, Chem. Rev., **10**, 365 (1932).
9. L. P. Hammett, *Physical Organic Chemistry*, New York, 1940, pp. 160 ff.

10. E. D. Hughes et al., JCS, 1525 (1935) and later papers.

11. P. D. Bartlett, JACS, **57**, 224 (1935) and refs. therein.

12. P. D. Bartlett and L. H. Knox, ibid., **61**, 3184 (1939); for a different type of bridgehead halogen, see Bartlett and S. G. Cohen, ibid., **62**, 1183 (1940).

13. H. Phillips, J. Kenyon et al., JCS, 108, 1232 (1932); ibid., 179 (1933).

14. J. Steigman and L. P. Hammett, JACS, **59**, 2536 (1937) and later papers; A. M. Ward, JCS, 2285 (1927).

E. Studies on Relative Reaction Rates and Linear Free Energy Relationships 1914-1939

The effect of changes in structure on the rates of reaction of organic compounds is of fundamental importance in planning synthesis of organic compounds and understanding the structure and the mechanism of organic reactions. Changes in structure may include the effect of lengthening the carbon chain, as in comparing CH_3Br with $n\text{-}C_6H_{13}Br$; effects due to isomerism, as in the behavior of the primary alcohol $n\text{-}C_4H_9OH$ compared with the isomeric tertiary alcohol $(CH_3)_3COH$; and effects due to other functional groups, such as in the pair CH_3COCH_2Cl and $CH_3COCH_2CH_2CH_2Cl$, on the reactivity of the halogen. The relative reactivity of the C=O groups in CH_3COCH_3 and $CH_3COC(CH_3)_3$ represents the effects of other important types of structural change on reaction rates. In the benzene series phenol, C_6H_5OH, reacts with electrophilic reagents, such as bromine, at a rate several powers of ten faster than nitrobenzene, $C_6H_5NO_2$, reacts with bromine.

In comparing reactivities the same reaction must be used throughout a series, and if competing reactions occur with some members of a series, that must be recognized. Furthermore, the effects of rate vs. equilibrium must be distinguished; does a product form rapidly but change slowly to the thermodynamically more stable product? The possibility of linear free energy relationships, a quantitative connection between rates and equilibria in a given series, must be considered; this topic is considered below. It is obviously necessary, in measuring relative reactivities, to make sure that the reaction conditions are comparable. It is of no significance to compare the reactivity of two compounds toward aqueous sodium hydroxide, for example, if one compound is completely soluble in this medium while the other is virtually insoluble. This seems like an elementary point, but it has sometimes been overlooked.

In the present section we shall consider some of the many studies on relative reactivity for the 1914-1939 period that are of particular significance for understanding reaction mechanisms, theories of structure, and synthetic work.

The dualistic theory of valence, in vogue before G. N. Lewis' postulation of the shared electron bond and not entirely given up until the 1920s, was extended by some to the alternating polarity idea.[1] This proposed that the carbons in a saturated chain were alternately more positive and less positive. If this were true, the rate of reaction of an alkyl halide, for example, should alternate as the length of the chain increased, CH_3CH_2Br and $CH_3CH_2CH_2Br$.

An important series of papers by Conant[2,3] investigating halogen exchange showed that the alternating polarity theory was invalid. The reactions were measured at different temperatures and the rates extrapolated to
a common temperature for comparison; this allowed a comparison of rates
differing by a large factor. All reactions gave good rate constants, calculated
for a bimolecular reaction: Rate = k [RCl][NaI]. A few significant values are
shown in Table 1; clearly there is no support for the alternating polarity
theory.

Table 1

Relative Reactivity of RCl + NaI $\longrightarrow$ RI + NaCl, 50°, Acetone Solution

Compound	Relative Reactivity
$\underline{n}\text{-}C_4H_9Cl$	1.00 (standard)
C_2H_5Cl	1.94
$\underline{n}\text{-}C_3H_7Cl$	1.03
$\underline{n}\text{-}C_{12}H_{25}Cl$	1.00
$(CH_3)_2CHCl$	0.015
$(CH_3)_3CCl$	0.018
$cy\text{-}C_6H_{11}Cl$	<0.0001
$C_6H_5CH_2Cl$	195
$C_6H_5CH_2CH_2Cl$	0.91
$CH_2=CHCH_2Cl$	79
$\underline{p}\text{-}O_2NC_6H_4CH_2Cl$	1370
$\underline{p}\text{-}ClC_6H_4CH_2Cl$	553
$2,4\text{-}(O_2N)_2C_6H_3CH_2Cl$	50800

Primary halides react faster than secondary, and these faster than tertiary, a result in agreement with Lewis' explanation of Walden inversion. If

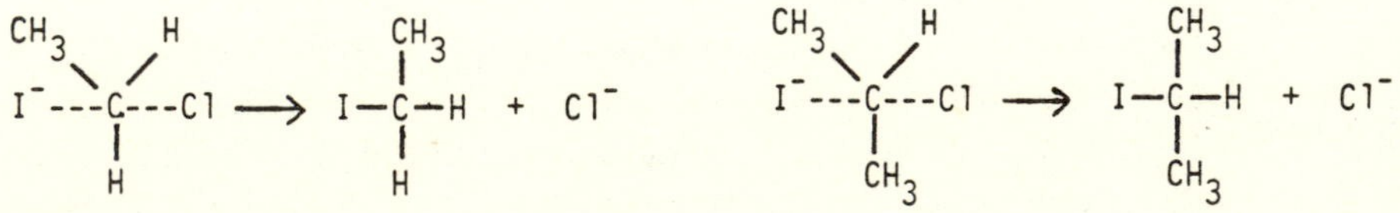

the I^- approaches from the rear of the C–Cl bond, the reactions would be slower if approach was impeded by two or more alkyl groups, compared to hydrogen.

The effect of unsaturation or an aromatic ring on the reactivity of a chlorine on an adjacent carbon is striking and agrees with many qualitative observations. The effect of substitutents in the aromatic ring of $C_6H_5CH_2Cl$ in increasing the rate follows the same order as their effect in increasing the acid dissociation constant of the benzoic acids, ArCOOH. There is thus a qualitative relation between kinetics and thermodynamics here. Conant's work antedates most of the Ingold work on displacement and solvolysis reactions.

The reactions of alkyl halides with amines as a function of structure of both components were studied by S. M. McElvain at Wisconsin[4] and C. R. Noller at Stanford.[5] They found that the displacement reaction to form quaternary ammonium salts became slower in the order primary > secondary > tertiary. The elimination reaction was predominant with tertiary halides, displacement being very slow. Increasing substitution around the nitrogen atom slowed down displacement greatly. The very low reactivity of cyclohexyl halides observed by Conant was also seen in the reaction of cyclohexyl bromide with amines. All of these effects were explored in great detail in later work.

$$RBr \ + \ R'_3N \ \longrightarrow \ RNR'_3{}^+Br^-$$

$$(CH_3)_3CBr \ + \ R_3N \ \longrightarrow \ R_3NH{}^+ \ Br^- \ + \ CH_2{=}C(CH_3)_2$$

Carl R. Noller (1900-1980, B.S. 1922, Washington University at St. Louis; Ph.D. 1926, Illinois with Roger Adams) was a research chemist at Eastman Kodak Company for two years and then taught at Stanford from 1929 to 1970. He did research on the nature of the Grignard reagent, the structure of natural products, and stereochemistry. He was the author of a very successful textbook of organic chemistry.

The use of p-toluenesulfonate (tosylate) esters instead of bromides for displacement reactions of branched compounds was shown by D. L. Tabern and E. W. Volwiler[6] to be preferable for two reasons. There is no rearrangement in the preparation of the tosylate from the alcohol, in contrast to the bromide, and displacement of the tosylate is faster (tosylate is a better "leaving group" than bromide). A similar rearrangment by treating $(C_2H_5)_2CHOH$ with HI had been observed in 1875 by Wagner and Saytzev.[7] Substituted malonic esters were used to synthesize barbituric acids, and tosylates were a very useful synthetic innovation.

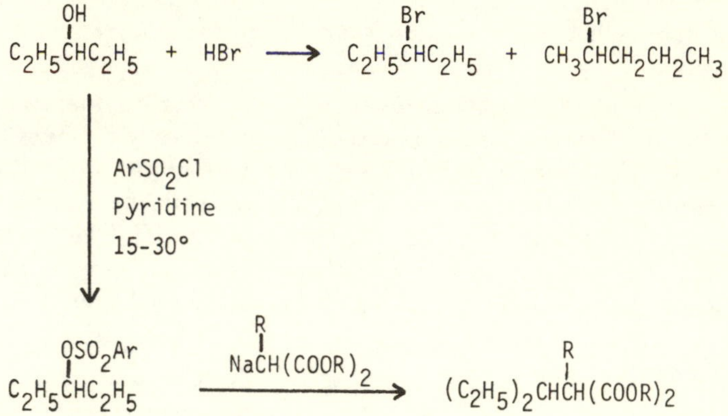

$$
\underset{C_2H_5\overset{\overset{\displaystyle OH}{|}}{C}HC_2H_5}{} \;+\; HBr \;\longrightarrow\; \underset{C_2H_5\overset{\overset{\displaystyle Br}{|}}{C}HC_2H_5}{} \;+\; \underset{CH_3\overset{\overset{\displaystyle Br}{|}}{C}HCH_2CH_2CH_3}{}
$$

$$
\downarrow \quad \begin{array}{l} ArSO_2Cl \\ Pyridine \\ 15\text{-}30° \end{array}
$$

$$
\underset{C_2H_5\overset{\overset{\displaystyle OSO_2Ar}{|}}{C}HC_2H_5}{} \quad \xrightarrow{\;\overset{\overset{\displaystyle R}{|}}{NaCH(COOR)_2}\;} \quad (C_2H_5)_2\overset{\overset{\displaystyle R}{|}}{C}H CH(COOR)_2
$$

Mounting evidence demonstrated a qualitative and quantitative difference between reactions of tertiary halides or alcohols and the corresponding primary and secondary compounds. J. F. Norris[8] showed that *tert*-butyl alcohol reacts with HCl to form the chloride 200,000 times as fast as *n*-butyl alcohol, and that the Lewis acid $ZnCl_2$ increases the rate of reaction of alcohols with HCl. This forms the basis of the Lucas test for distinguishing primary, secondary, and tertiary alcohols.[9] An interesting complexity was introduced by the observation that it is virtually impossible to replace the OH by Cl in the primary alcohol $(CH_3)_3CCH_2OH$ without rearranging the carbon skeleton (see essay 17).

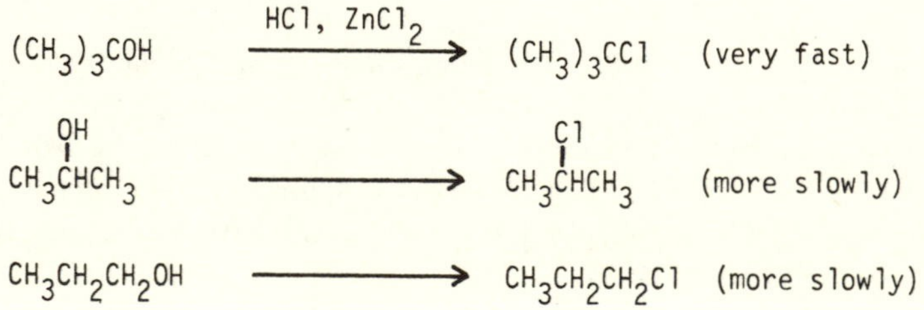

$$
(CH_3)_3COH \quad \xrightarrow{\;HCl,\ ZnCl_2\;} \quad (CH_3)_3CCl \quad \text{(very fast)}
$$

$$
\underset{CH_3\overset{\overset{\displaystyle OH}{|}}{C}HCH_3}{} \quad \longrightarrow \quad \underset{CH_3\overset{\overset{\displaystyle Cl}{|}}{C}HCH_3}{} \quad \text{(more slowly)}
$$

$$
CH_3CH_2CH_2OH \quad \longrightarrow \quad CH_3CH_2CH_2Cl \quad \text{(more slowly)}
$$

HCl cleaves *tert*-butyl ethers very much faster than ethers of primary or secondary alcohols,[8,10] and *tert*-butyl bromide is converted by aqueous bases almost completely to the alcohol with little butene formation, whereas isopropyl bromide gives both alcohol and propylene.[11]

Oxalyl chloride yields esters with primary alcohols, but chlorides are formed from tertiary alcohols.[12]

$$C_2H_5OH \ + \ (COCl)_2 \ \longrightarrow \ (COOC_2H_5)_2$$

Diethyl oxalate

$$(CH_3)_3COH \ + \ (COCl)_2 \ \longrightarrow \ (CH_3)_3CCl$$

Norris showed that acylation of alcohols with p-nitrobenzoyl chloride grows much slower as branching in the alcohol increases, and also that tertiary alcohols reacted more slowly than secondary, and secondary more slowly than primary alcohols. This is the reverse of the order of reactivity with HCl to form the chlorides.[13]

Table 2

Second-Order Rate Constants at 25° in Ether, 1-Molar Solutions

$$\underline{p}\text{-}O_2NC_6H_4COCl \ + \ HOR \longrightarrow \underline{p}\text{-}O_2NC_6H_4COOR \ + \ HCl$$

R	Relative Rate Constant
C_2H_5	(1.00)
$\underline{n}\text{-}C_4H_9$	0.87
$(CH_3)_2CHCH_2$	0.27
$(CH_3)_2CHCH_2CH_2$	0.86
$(CH_3)_2CH$	0.12
$C_2H_5(CH_3)CH$	0.09
$(CH_3)_3C$	0.03
$(C_2H_5)_2CCH_3$	0.016
$C_6H_5CH_2$	0.20
$C_6H_5CHCH_3$	0.006

The rates of reaction are greater for tertiary alcohols and halides than for primary, where the indicated bond is being broken. However, for the reactions involving the cleavage of the bond between O and H as in Table 2 and numerous other examples, the tertiary alcohols are the slowest.

$$(CH_3)_3C \!-\! OH \quad \xrightarrow{\text{HCl}} \quad (CH_3)_3C\!-\!Cl \ + \ H_2O$$

$$(CH_3)_3C \!-\! Cl \quad \xrightarrow[\text{or base}]{\text{H}_2\text{O}} \quad (CH_3)_3C\!-\!OH \ + \ (CH_3)_2C\!=\!CH_2$$

$$(CH_3)_3C\!-\!O\!-\!H \ \text{slower than} \ (CH_3)_2CH\!-\!O\!-\!H \ \text{slower than} \ CH_3CH_2\!-\!O\!-\!H$$

The explanation for these observations is that the CH_3 group repels electrons more than H; hence in $(CH_3)_3COH$, the OH group leaves with its electrons more readily than in CH_3CH_2OH. The H of the OH group leaves as H^+ without its electrons more slowly from the tertiary alcohol than from the primary alcohol. These simple ideas of electron-donating effects of alkyl groups are very useful in predicting relative reactivities, although they need to be supplemented by consideration of solvation of ions and of steric (spacial) effects of groups. Although Norris, as noted earlier, recognized the special character of tertiary alcohols and halides, and indeed had suggested in 1907 the tertiary butyl carbonium ion as an intermediate in the reactions of tertiary butyl compounds,[14] he never attempted to develop a systematic rationalization of his extensive observations. In 1926 W. H. Carothers found some anomalies in the reactivities, emphasizing that relative reactivities are reversed by changing the reagent used, and that reaction mechanisms must be known before reasonable explanations are possible.[15]

In 1935, E. D. Hughes of C. K. Ingold's laboratory at University College, London, gave a condensed account of the Ingold scheme of classifying displacement reactions as $S_N 2$ (nucleophilic substitution, second order) and $S_N 1$ (nucleophilic substitution, first order). The latter is illustrated by the hydrolysis of tertiary butyl chloride where the rate is independent of added base; the first-order elimination reaction, usually prominent with tertiary chlorides, was then indicated by E1. This general scheme had been published in 1933 in England by the Ingold group,[16] but Hughes' paper seems to be the first appearance in the American literature of the $S_N 1$-$S_N 2$ notation.[17]

The role of solvent in the $S_N 1$ ionization was demonstrated by Bartlett, Winstein, and Hammett, the latter coining the terms "solvolysis" for the first-order reaction of tertiary halides and benzyl halides with solvent.[18] The $S_N 1$-solvolysis reaction was to be most intensively investigated in this country, probably more than its intrinsic importance in the overall panorama of organic chemistry justified. It is discussed in other sections of this account.

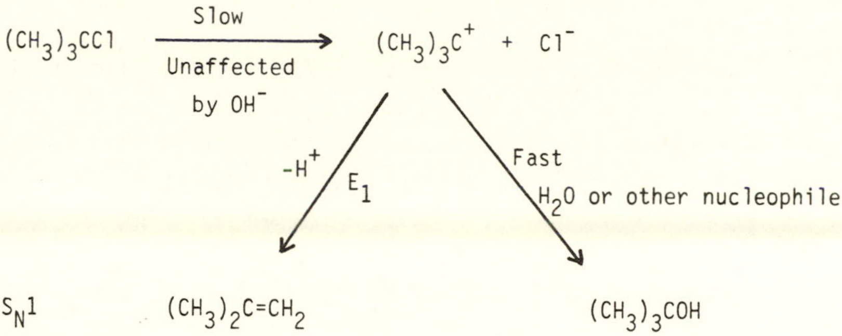

$$(CH_3)_3CCl \xrightarrow[\substack{\text{Unaffected} \\ \text{by } OH^-}]{\text{Slow}} (CH_3)_3C^+ + Cl^-$$

$S_N 1$ $(CH_3)_2C=CH_2$ $(CH_3)_3COH$

$-H^+$ E_1 Fast H_2O or other nucleophile

$S_N 2$ $CH_3\underset{Br}{CH}C_2H_5 + NaI \longrightarrow CH_3\overset{I}{CH}C_2H_5 + NaBr$

with inversion

The esterification of an alcohol by a carboxylic acid was investigated both in its equilibrium and its acid-catalyzed rate, Table 3.

$$RCOOH + CH_3OH \xrightarrow[CH_3OH]{HCl} RCOOCH_3 + H_2O$$

Of many studies on the relation between structure of the acid, the alcohol, and rate, those of Hilton A. Smith are the most comprehensive and reliable.[19] Smith measured the rates at several temperatures and obtained energies and entropies of activation, although these showed little variation in the whole series, except for benzoic acid. The small relative rates for benzoic acid and for very highly branched acids, such as diisopropylacetic acid, are notable. Newman later proposed a general empirical "rule of six", which correlated the low rate of reaction of very highly substituted compounds.[20]

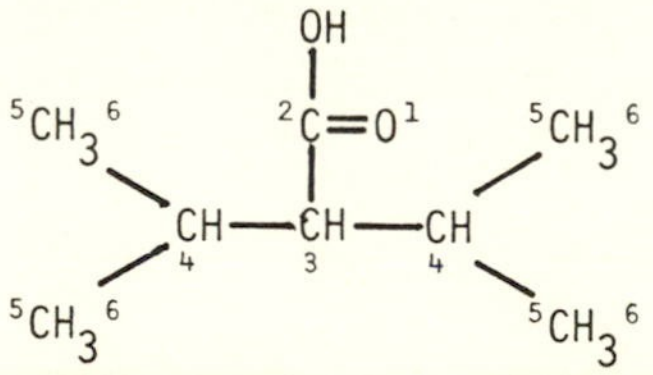

"rule of six"
12 H in the 6 position—
highly hindered

Table 3 shows *rates*, not equilibria. The equilibrium reached in esterification of various acids and alcohols was studied by many,[21] but the results gave no particular theoretical insight. Tertiary alcohols were found to dehydrate rather than esterify, depending on the reaction conditions.

Table 3

Relative Rates of Esterification of RCOOH with Dry HCl in CH_3OH

0.005 M HCl in CH_3OH

R	Relative Rate Constant
CH_3	1.00
$CH_3CH_2CH_2CH_2$	0.49
$(CH_3)_2CH$	0.33
$(CH_3)_3C$	0.033
$(C_2H_5)_2CH$	0.0084
$C_6H_5CH_2$	0.44
$(C_6H_5)_2CH$	0.031
C_6H_5	0.00033
$[(CH_3)_2CH]_2CH$	0.0

H. A. Smith (1908- , A.B. 1930, Oberlin; Ph.D. 1934, Harvard with G. B. Kistiakowsky) taught at Lehigh and from 1941 on at the University of Tennessee, where he became vice-chancellor for graduate study and research.

There were numerous studies on reactivity and equilibria by several workers at Berkeley, some of them significant; they were marked, however, by choice of very complicated reaction systems, so that the results were not as widely recognized as they deserved.[22] We have discussed the study of rate and equilibrium of semicarbazone formation by Conant and Bartlett elsewhere.

Marguerite Wilcox and R. F. Brunel at Bryn Mawr in 1916 measured equilibria and rates in the system below, following the reaction by the change in optical activity. Tertiary alcohols reacted more slowly than primary and secondary, and the equilibrium was less favorable for tertiary alcohols.[23]

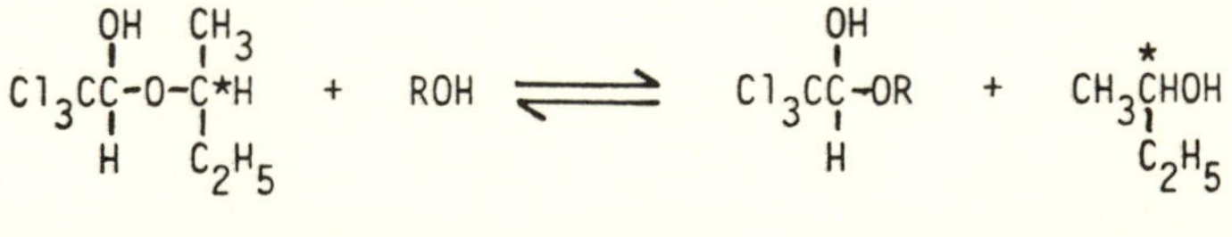

optically active   optically active

Roger F. Brunel (1881-1924, A.B. 1903, Colby; Ph.D. 1906, Hopkins with Remsen) worked as a research assistant to Arthur Michael, taught at Syracuse and then at Bryn Mawr from 1912 to 1924, succeeding E. P. Kohler and being succeeded by L. F. Fieser. He disagreed with the formidable Michael in some independent papers. A lecture room is named in his memory at Bryn Mawr.

H. Adkins reported extensive measurements on rates and equilibria of acetal formation from many aldehydes and alcohols. The results, as far as they follow a pattern, were analogous to the semicarbazone work of Conant and Bartlett.[24]

$$RCHO + 2 R'OH \rightleftharpoons RCH(OR')_2 + H_2O$$

M. S. Kharasch and his workers examined the cleavage of unsymmetrical mercury compounds. Their assumption that the group appropriating the proton was more "electronegative" than the one retained by the mercury can be regarded only as valid for the reaction studied.[25] Lively polemics with several chemists ensued, but fortunately Kharasch was soon involved in his work on the peroxide effect and the resulting free-radical studies which were of far greater fundamental importance.

$$RHgR' + HCl \longrightarrow \begin{cases} RH + R'HgCl \\ R'H + RHgCl \end{cases}$$

Another series of papers on "relative reactivities" based solely on equilibrium measurements may be discussed here. The oxidation-reduction equilibrium of quinones and hydroquinones is one of the few organic oxidation reactions which is strictly and rapidly reversible and yields a reversible potential E, just as a hydrogen or zinc electrode does. These can be treated by the methods of thermodynamics and resolved into the "normal potential" E_o (the potential under standard conditions) and increments due to the hydrogen ion concentration and the ratio of oxidized and reduced forms. In the early 1920s several groups, including those of V. K. La Mer at Columbia and Conant at Harvard, studied the oxidation-reduction potentials of quinones to see how the potentials varied with structure.[26]

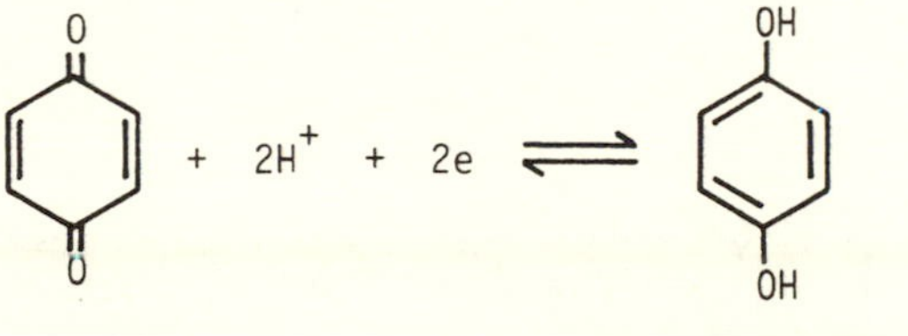

$$+ 2H^+ + 2e \rightleftharpoons$$

p-Benzoquinone   Hydroquinone

Representative quinone potentials[26,27] are shown in Table 4. A lower E_o value means that the quinone is a weaker electron acceptor (or oxidizing agent), hence the equilibrium above is moved toward the quinone as E_o decreases. Thus methyl groups make the quinone a weaker oxidizing agent. The polycyclic aromatic compounds also have lower potentials, indicating that the aromatic (π electron) systems become less stable compared to the oxidized nonaromatic quinone system. This decreased aromatic stabilization as the number of rings increases is in agreement with a mass of qualitative observations.

Table 4

E_o of Quinone-Hydroquinone Systems in Ethanol, 25°

Quinone	E_o (volts)
p-Benzoquinone	+ 0.711
o-Benzoquinone	(+ 0.806)[a]
Chloro-p-benzoquinone	+ 0.736
Tetrachloro-p-benzoquinone	+ 0.703
Tetramethyl-p-benzoquinone	+ 0.466
1,4-Naphthoquinone	+ 0.484
9,10-Phenanthrenequinone	+ 0.460
9,10-Anthraquinone	+ 0.156

[a] Estimated from water solution

1,4-Naphthoquinone　　　　　　1,2-Naphthoquinone

9,10-Anthraquinone　　　　　　9,10-Anthrahydroquinone
Favored

The occurrence of intermediates (free radicals or semiquinones) in the hydroquinone oxidation, in which only one electron has been removed, was demonstrated by L. Michaelis, as discussed in the section on free radicals. Analogous odd electron (free-radical) intermediates appear to be formed in biochemical oxidation-reduction processes. Photographic developers are hydroquinones or aminophenols readily oxidized by the activated silver halide grains in an exposed film, the silver ion being reduced to black metallic silver. The aminophenols form unstable quinone imines on oxidation, and here also one-electron transfer leads to free-radical intermediates.

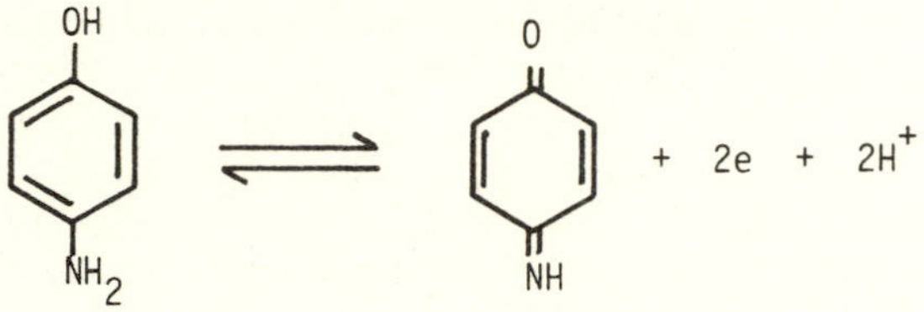

$\underline{p}$-Aminophenol $\underline{p}$-Benzoquinonimine

One of the important and influential developments in these years was the recognition of quantitative relationships between rates of reaction and the equilibria of closely related processes. The study of such *linear free energy relationships* has been a major activity of physical organic chemists since that time.[28] As the Conant-Bartlett semicarbazone study showed, there is no *universal* relation between rate and equilibrium. Another frequently quoted example is gaseous H_2 and O_2, they do not form H_2O in the absence of a catalyst or a spark at a measurable rate at room temperature, although the equilibrium favors H_2O formation overwhelmingly.

Brönsted and Pedersen in Copenhagen first noted a simple relation between the rate of an acid- or base-catalyzed reaction and the dissociation (equilibrium) constant of the catalyzing acid or base.[29] Here k_A is the rate constant of a reaction catalyzed by the acid HA, of dissociation constant K_A; g_A and B_A are constants for the series of reactions under study. B is greater than zero and less than unity.

$$\log k_A = B_A \log K_A + \log g_A$$

Hammett's immediate interest in the Brönsted equation has been noted earlier; his discovery of linear free energy relationships for reactions other than acid-catalyzed ones was in the series of reactions shown here; this paper of 1933 is one of his best and most clear-cut.[30] For the reaction below, log k (the rate constant) gives a linear plot against log K_A, the dissociation constant for RCOOH, for a series of different R groups, where RCOOH is a substituted benzoic acid.

$$RCOOCH_3 + N(CH_3)_3 \longrightarrow RCOO^- N(CH_3)_4^+$$

$$K_A = \frac{[RCOO^-][H^+]}{[RCOOH]}$$

However, there is *no* linear free energy relation between the rate constant for hydrolysis of these esters and K_A. This was attributed by Hammett, no doubt correctly, to the fact that in hydrolysis a different bond in the ester is being broken. This would make the reaction subject to more steric effects from changes in R than the reaction with $N(CH_3)_3$.

$$RC{\overset{O}{\overset{\|}{-}}}OCH_3 + H_2O \longrightarrow RC{\overset{O}{\overset{\|}{-}}}OH + HOCH_3$$

Hammett, and also G. N. Burkhardt, in England, extended the linear free energy relationships to many reactions of aromatic compounds with substituents in the *m*- and *p*- positions.[28] Hammett tabulated fifty-two reactions for which the "Hammett equation" held reasonably well.[31]

$$\log k - \log k^\circ = \rho\sigma$$

From these numerous rate studies, Hammett showed that this equation was followed for *m*- and *p*- substituents. Here k is the measured rate or equilibrum constant; k° is the rate or equilibrium constant for the unsubstituted compound; σ (sigma, for substituent) is determined by the ratio (K/K_{subst}) where K and K_{subst} are the dissociation constants for the unsubstituted and the substituted benzoic acids. Rho (ρ for the reaction) is characteristic of the reaction studies and is determined from the slope of a plot of log k against σ. A few examples are given.

C. R. Hauser and W. B. Renfrow at Duke had shown a linear relation between $\log k_{rate}$ and $\log K_A$ for the Hofmann rearrangement of bromoamides.[32]

$$ArCONHBr \xrightarrow{OH^-} ArCONBr^- + H_2O \longrightarrow ArN=C=O + Br^-$$

Norris and W. H. Strain[33] found a qualitative parallel between rate and K_A for the reaction of esterification with a diazomethane.

$$ArCOOH + (\underline{p}-CH_3C_6H_4)_2CN_2 \longrightarrow N_2 + ArCOOCH(C_6H_4CH_3-\underline{p})_2$$

Renfrow and Hauser[34] saw a similar relation and determined the Hammett rho and sigma "constants" for the Lossen hydroxamic acid rearrangement for substituents in the Ar group; the agreement with the Hammett equation was less satisfactory for substituents in the C_6H_5 group.

$$(C_6H_5\overset{\overset{O}{\|}}{C}NOCOAr)^- K^+ \xrightarrow{NH_3} C_6H_5NHCONH_2 + ArCOO^- K^+$$

The Hammett equation thus allows reasonably accurate predictions of many reaction rates and shows that the electronic effect of *m*- and *p*- substituents is relatively independent of the reaction. Ortho substituents do not usually follow the equation, because steric effects in the *o*-position, as well as hydrogen bonding, solvation, and other factors vary unpredictably. A negative σ value for a substituent is shown by groups that donate electrons to the aromatic ring, and a positive sigma is shown by groups known to be electron-attracting. For one of the most powerful of these electron-attracting groups, the nitro group, the sigma value was shown by Hammett to be variable, depending on the reaction.

Hammett's generalization, although empirical in origin, afforded important quantitative and predictive relationships, and thus markedly diminished the empirical nature of some domains of organic chemistry. Like many successful generalizations, further research extended, qualified, and altered them;[35] several different sets of σ ρ values were developed which gave better agreement with experiment in certain types of reaction. The determination of σ ρ "constants" became a routine research procedure, and the significance of the linear free energy relationships for giving information about the nature of the transition state was a common subject for speculation. The more subtle developments along these lines came after 1940.

LITERATURE CITED

1. E. J. Cuy, JACS. **42**, 503 (1920) gives discussion and leading references to the alternating polarity theory.
2. J. B. Conant and W. R. Kirner, ibid., **46**, 232 (1924) and later papers. Other less clear cut evidence against alternating polarity, with useful bibliography is in H. J. Lucas and A. Y. Jameson, ibid., **46**, 2475 (1975).
3. A rather uncritical collection of many comparisons of relative rates of reaction is given by H. Adkins, in *Organic Chemistry*, edited by Henry Gilman, 1st ed., **1**, 802-857 (1938).
4. S. M. McElvain et al., JACS, **53**, 690 (1931) and later papers.
5. C. R. Noller and R. Dinsmore, ibid., **54**, 1025 (1932).
6. D. L. Tabern and E. W. Volwiler, ibid., **56**, 1139 (1934).
7. G. Wagner and A. Saytzev, Ann., **179**, 313 (1875).
8. J. F. Norris and Hazel B. Taylor, JACS, **46**, 753 (1924); Norris and G. W. Rigby, ibid., **54**, 2088 (1932).
9. H. J. Lucas, ibid., **52**, 802 (1930).
10. E. L. Gustus and P. G. Stevens, ibid., **54**, 3461 (1932) for cleavage by acyl iodides.
11. H. E. French et al., ibid., **56**, 1346 (1934).
12. Roger Adams and L. F. Weeks, ibid., **38**, 2514 (1916).
13. J. F. Norris, A. A. Ashdown and F. Cortese, ibid., **47**, 837 (1925); **49**, 2640 (1927). Similar relative reactivities were found for the isocyanate-alcohol reaction to form

urethans, T. L. Davis and J. M. Farnum, ibid., **56**, 883 (1934). This reaction after World War II became industrially important in the preparation of polyurethans.

14. J. F. Norris, ACJ, **38**, 627 (1907); J. Stieglitz, ibid., **21**, 101 (1899), had first suggested carbonium ions. These early American ideas are usually overlooked, in spite of the vast amount of later discussion on carbonium ions and displacement reactions.

15. W. H. Carothers, JACS, **48**, 3192 (1926).

16. E. D. Hughes, ibid., **57**, 708 (1935), and refs. therein.

17. It is not discussed in C. K. Ingold's review of his system, Chem. Rev., **15**, 225 (1934).

18. J. Steigmann, N. T. Farinacci and L. P. Hammett, JACS, **59**, 2536, 2542 (1937); **60**, 3097 (1938).

19. H. A. Smith, ibid., **61**, 254, 1176 (1939); **62**, 1136 (1940); **63**, 605 (1941); **66**, 1494 (1944), this contains a summary, including a few measurements from other laboratories.

20. M. S. Newman et al., ibid., **74**, 3929 (1952); also in *Steric Effects in Organic Chemistry*, M. S. Newman, Ed., New York; 1956, pp. 203 ff.

21. H. Adkins et al., JACS, **57**, 193 (1935); **59**, 1694 (1937).

22. T. D. Stewart et al., JACS, **55**, 4813 (1933); **54**, 4172, 4183 (1932); H. C. Biddle and C. W. Porter, ibid., **36**, 84 (1914); **37**, 1571 (1915).

23. Marguerite Wilcox and R. F. Brunel, ibid., **38**, 1821 (1916).

24. H. Adkins et al., ibid. **56**, 442 (1934); **53**, 1043 (1931); Adkins, ref. 3.

25. M. S. Karasch et al., ibid., **54**, 674 (1932); JCE, **11**, 82 (1934).

26. V. K. La Mer and Lillian E. Baker, JACS, **44**, 1954 (1922); J. B. Conant et al., ibid., p. 1382 and refs. therein; Conant and L. F. Fieser, ibid., **44**, 2480 (1922); **46**, 1858 (1924) and many later papers by Fieser and Mary Fieser.

27. Hammett, op. cit., p. 392; L. F. Fieser, in Gilman's *Organic Chemistry*, 1938, Vol. 1, p. 96.

28. The classic summary is by the chief discoverer, L. P. Hammett, *Physical Organic Chemistry*, 1940; Chem. Rev., **17**, 225 (1935). Similar observations were made by G. N. Burkhardt, Nature, **136**, 684 (1935), and later papers. A detailed monograph is J. E. Leffler and E. Grunwald, *Rates and Equilibria of Organic Reactions*, New York, 1963.

29. J. N. Brönsted and K. J. Pedersen, Z. physik. Chem., **108**, 185 (1924).

30. Hammett and H. L. Pfluger, JACS, **55**, 4079 (1933).

31. Hammett, ibid., **59**, 96 (1937); *Physical Organic Chemistry*, pp. 184 ff.

32. C. R. Hauser and W. B. Renfrow, Jr., JACS, **59**, 121 (1937); Hauser, et al., ibid., **57**, 1056 (1935).

33. J. F. Norris and W. H. Strain, ibid., **57**, 187 (1935); this reaction was later studied in detail by J. D. Roberts et al., ibid., **73**, 760(1951) and later papers.

34. Renfrow and Hauser, ibid., **59**, 2308 (1937).

35. D. H. McDaniel and H. C. Brown, JOC, **23**, 420 (1958); H. H. Jaffe, Chem. Rev., **53**, 191 (1953).

F. Color and Chemical Structure

The use of dyes from natural sources, usually plants, has been common since prehistoric times, and as a result of the progress in organic chemistry in the nineteenth century the plant dyes were found to be organic compounds. The most widely used vegetable dyes were alizarin (or madder) and indigo. Carl Graebe and Carl Liebermann in Germany in 1869 proved the structure and synthesized alizarin very readily. After many years of research Adolph Baeyer at Munich established the structure of indigo in 1883 and developed several syntheses for it.

Royal or Tyrian purple was the most highly prized dye of antiquity; obtained from a Mediterranean mollusc, it was so costly that its use was allowed only by royalty, hence the expression "born to the purple". It required 12,000 molluscs to furnish a gram of pure dye that was then readily shown to be a dibromoindigo.

The commercial syntheses of alizarin and indigo brought about important economic consequences. The German dye industry acquired a firm base because the manufacture of synthetic alizarin was relatively easy technically and very profitable. The development of a technically feasible indigo synthesis was far more difficult and expensive, but it was also highly profitable once in operation. The German dye and chemical industry soon far surpassed the English and French in scope and research activity.[1,2]

The production of synthetic alizarin and indigo also had far-reaching effects in several geographical areas. The indigo plant was widely grown in India, and in the colonial period it was an important crop in South Carolina; the madder plant, the source of alizarin, was extensively cultivated in Europe and the Middle East. The synthetic dyes rapidly displaced the material from plant sources, causing widespread economic distress.

The synthesis of many other types of colored compounds, azo compounds, triphenylmethanes, aniline black, and indanthrenes, for example, and of pH indicators, which changed color with change in acidity, accompanied the growth of organic chemistry in the 19th century. Many other colored products from natural sources were isolated and their structures determined. Most of these synthetic and natural colored compounds were not useful as dyes, but there was a lively interest in the relation between chemical structure and color. The early literature on this topic is voluminous, but a detailed presentation of most of it would be unprofitable.[3]

In 1876 O. N. Witt in Germany recognized that color was accompanied by the presence of various unsaturated groups called *chromophores* (color bearers). These included NO_2, $C=O$, $N=N$, $C=C$, $C=S$, $N=O$, and the *para* quinonoid group. *Auxochromic groups*, also recognized, which shifted the color toward longer wavelengths, i.e., from yellow toward purple, were OH, OCH_3, NH_2, or NR_2. These are effective only in the position para to the chromophore, and are groups with unshared pairs of electrons.

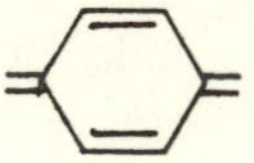

para quinonoid group

Color is due to selective absorption of light in the visible region of the spectrum; the unabsorbed light is reflected or transmitted, causing the color. As soon as absorption spectra could be measured in the ultraviolet (UV) it was found that unsaturated compounds generally absorbed in the UV and that, in some cases, absorption bands reached into the visible, causing color. The measurement of absorption spectra in the UV and the visible was initiated by W. N. Hartley in England in the 1870s, and the experimental procedures were well worked out, if laborious. Many research papers, such as those by Emma Carr from Mount Holyoke College, describe the general procedure with a spectroscope and photographic plates.[4]

Until 1930 the theories of selective absorption of radiation in the UV and visible were based on the classical wave theory of radiation instead of the quantum theory. One reads that "sympathetic vibrations of charged particles" cause absorption,[5] or that damping of vibration of aromatic rings by substituents leads to shifts of absorption in the visible,[6] or (in 1910) that absorbers of light are negative electrons with free periods corresponding to absorption bands.[7]

Baeyer[6] thought that the color of Döbner's violet, a triphenylmethane dye, was due to oscillation of a chlorine from one amino group to another which he pictured as shown. He wrote a similar formulation for the color of benzaurin.

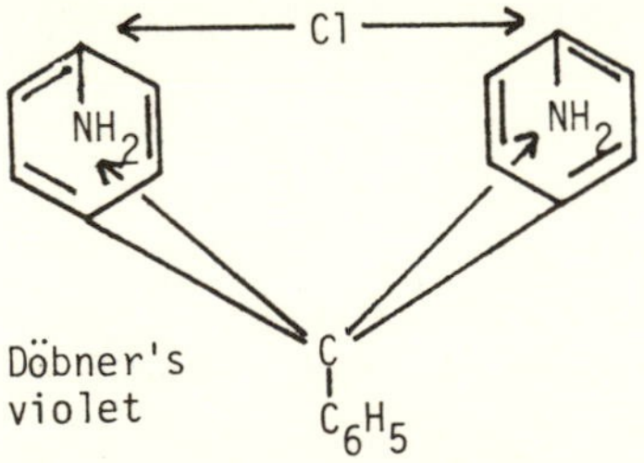

Döbner's violet

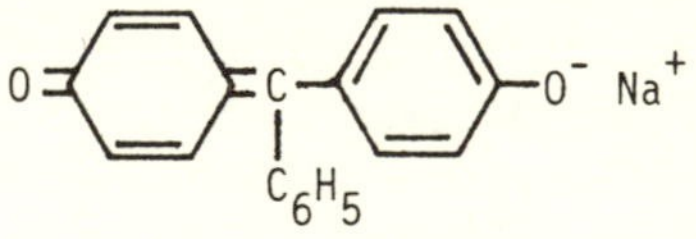

benzaurin

In an elegant instrumental study of crystal violet, E. Q. Adams and L. Rosenstein[5] at Berkeley in 1914 decided that the colored form was the cation, in which oscillations of charge could occur from one nitrogen atom to another.

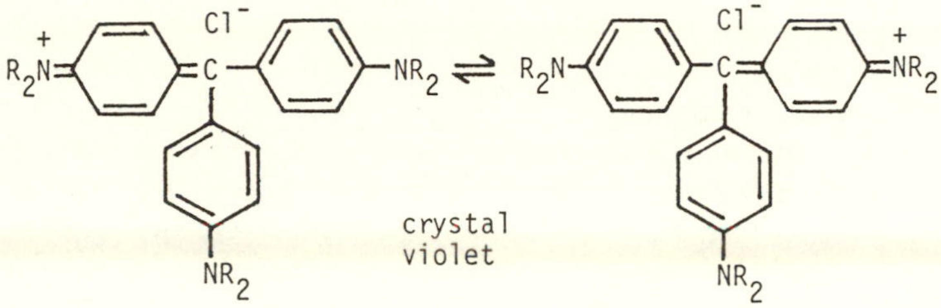

crystal
violet

Acree[8] showed that the colored form of indicators like phenolphthalein and sulfonphthalein was a phenoxide ion in which the negative charge could oscillate, although he did not emphasize the oscillatory possibilities.

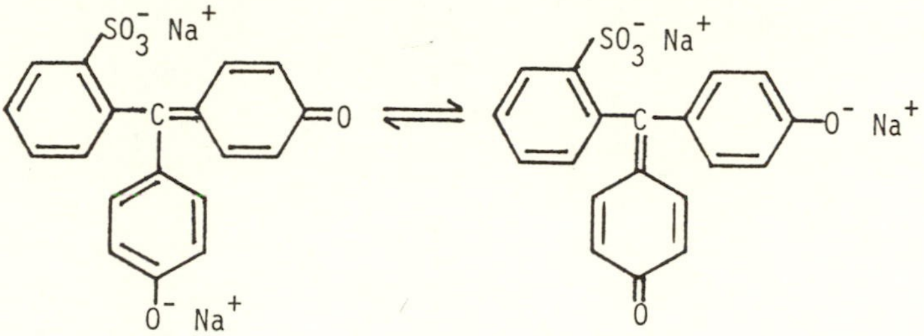

H. A. Lubs and W. M. Clark in the U. S. Department of Agriculture's laboratories in Washington, using an improved form of the hydrogen electrode to measure pH, showed that the sulfonphthaleins formed a very useful set of pH indicators.[9,10] Clark further measured oxidation-reduction potentials of dyes, and some of these proved suitable as indicators in oxidation-reduction titrations. Important as Clark's work was in physical chemistry and the life sciences, he did not do much speculating about the causes of color.

With the advent of ideas about resonance and electron delocalization around 1930, a fundamentally new definition of color was possible, which, nevertheless, included the ideas of Adams, Acree, and others. C. R. Bury in a key paper suggested that instead of actual oscillation of charges the ions were resonance hybrids with the charge delocalized.[11] He did not, however, explain why a charged resonance hybrid ion should absorb in the visible and be colored.

By 1930 the quantum theory of electronic transition in the UV and visible regions was well established, and it was known that absorption of radiation of a given wavelength is due to a transition from the ground electronic state to an excited electronic state. A resonance hybrid requires less energy to raise it to an excited electronic state than is needed for a saturated or less highly conjugated molecule. Thus, following Planck's fundamental equation, a smaller energy *difference* between the ground state and the excited state means absorption at longer wavelength and therefore color, if the energy difference corresponds to a wave length in the visible region. A saturated molecule requires much more energy to raise it to an excited electronic state and therefore absorbs only high-energy quanta in the UV region.

It still was not explained why delocalization or resonance did lower the energy of the excited electronic states, and indeed G. N. Lewis and Melvin Calvin in 1939 provided a reasonably successful correlation of electronic spectra and structure using a combination of classical oscillation ideas and the quantum theory.[12] L. Pauling[13] gave the most detailed explanation for the effect of resonance on electronic transitions. For the compounds of the cyanine dye series, of which many thousands were synthesized by L. G. S. Brooker and others at the Eastman Kodak Company for testing as sensitizers for photographic emulsions, the resonance idea gives a reasonable explanation for the absorption spectrum.[14] The Lewis-Calvin treatment also fits the observed absorption data well.

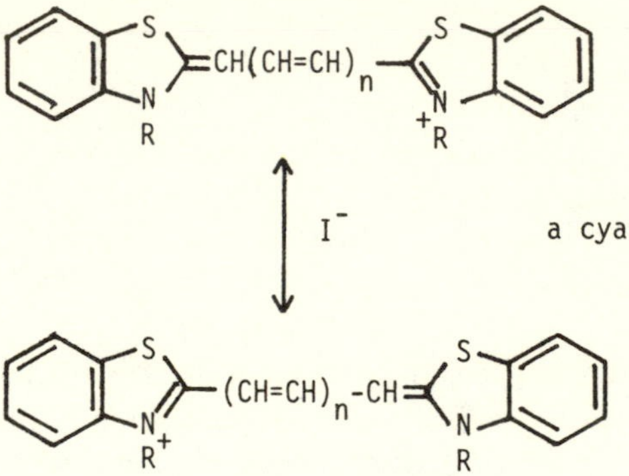

a cyanine dye

Leslie G. S. Brooker (B.Sc. 1923, Ph.D. 1926, University of London with Samuel Smiles) was a research associate with Eastman Kodak in Rochester, N.Y., from 1926 to 1970. His work on cyanine dyes as photographic sensitizers was fundamental for color photography and for ultraviolet-sensitive photographic plates.

For compounds such as indigo which do not exist as ions, Pauling suggested resonance between various dipolar ionic structures, as shown for indigo.[15]

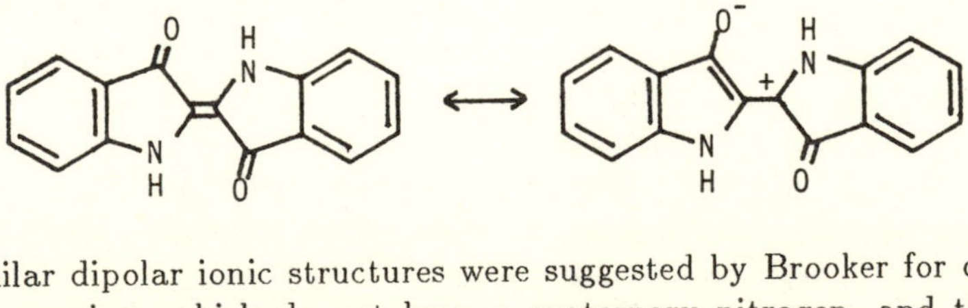

Similar dipolar ionic structures were suggested by Brooker for dyes related to cyanines which do not have a quaternary nitrogen, and this was supported by dipole moment measurements.[14]

Carr observed[4] that the UV spectra of cyclopropyl ketones **2** showed absorption at longer wavelengths than saturated ketones **1**, but at shorter wavelengths than unsaturated ketones **3**. This meant that the cyclopropane ring had resonance interaction with the C=O similar to but weaker than that of the C=C, and much stronger than the saturated C-C. This was an early indication of some π bond character in the cyclopropane ring. Carr also found,[16] by measurement in the far (vacuum) ultraviolet at 160-240 nm, that the first absorption band in simple C=C is at longer wavelengths the more carbons are attached to the C=C; the nature of the saturated carbon is of secondary importance. This regularity was later used by Barton as diagnostic for a highly substituted C=C.

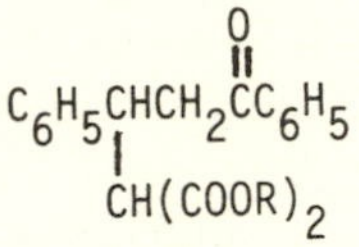

$$C_6H_5CHCH_2\overset{\overset{O}{\|}}{C}C_6H_5 \qquad C_6H_5CH\!-\!\overset{\overset{O}{\|}}{CH}CC_6H_5 \qquad C_6H_5CH\!=\!\overset{\overset{O}{\|}}{C}CC_6H_5$$
$$\underset{}{CH(COOR)_2} \qquad \underset{(COOR)_2}{\overset{C}{\diagdown\diagup}} \qquad CH(COOR)_2$$

$$\underline{1} \qquad\qquad\qquad \underline{2} \qquad\qquad\qquad \underline{3}$$

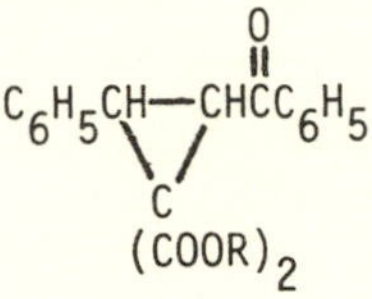

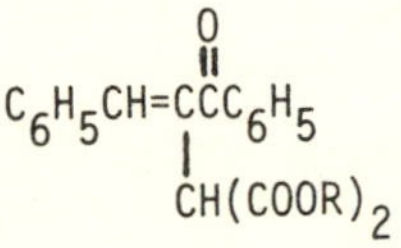

Emma P. Carr (1880-1972), B.S. 1905, Ph.D. 1910, Chicago with Stieglitz) taught at Mount Holyoke College from 1910 to 1946. She built up an active research group in electronic absorption spectroscopy, including the experimentally difficult vacuum region at short wavelengths where the atmosphere absorbs.

The measurement of electronic spectra was reported in detail by R. C. Gibbs at Cornell for phenolphthalein and related compounds.[17] L. C. Anderson at Ann Arbor showed that the absorption spectra of ketones were different in acid, corresponding to addition of protons or of Lewis acids.[18] Wallace R. Brode showed that pH could be determined by measuring the change in absorption spectra of indicators.[19]

$$R_2C\!=\!O \quad + \quad H^+ \quad \rightleftharpoons \quad R_2C\!=\!\overset{+}{O}H$$

W. R. Brode (1900-1974, B.S. 1921 Whitman; Ph.D. 1925, Illinois with Roger Adams) taught at Ohio State from 1928 to 1948 and directed the National Bureau of Standards from 1947 to 1958. He was scientific adviser to the Secretary of State from 1958 to 1960. His research was on dyes, color and structure, and spectroscopy.

The commercial synthesis of dyes in America did not gain the heights reached in England and Germany in this period. Nevertheless, the brilliant work by Eastman Kodak Co. in color photography and the development of man-made fibers requiring new types of dyes stimulated innovations in research and synthesis of coloring chemicals.[20] American chemists made contributions to practical and theoretical knowledge.

LITERATURE CITED

1. The commercially important dyes are reviewed by R. E. Rose, JCE, **3**, 973 (1926), which has swatches of cloth dyed with the characteristic color of the dye. L. F. Fieser, ibid., **7**, 2609 (1930), gives an interesting account of the chemistry, synthesis, and technology of alizarin.
2. L. F. Haber, *The Chemical Industry in the Nineteenth Century*, Oxford, 1951, 1968. The best account of the rise of the German dye industry and its relation to research in the German universities is J. J. Beer, "The Emergence of The German Dye Industry", Urbana, 1959, Illinois Studies in the Social Sciences, Vol. 44, pp. 57-93.
3. The older speculations may be found in J. B. Cohen, *Organic Chemistry for Advanced Students*, Vol. II, New York, 1913, pp. 346 ff.; F. Henrich, *Theories of Organic Chemistry*, New York, 1922, trans by T. B. Johnson and Dorothy A. Hahn.
4. Emma P. Carr and C. Pauline Burt, JACS, **40**, 1590 (1918).
5. E. Q. Adams and L. Rosenstein, ibid., **35**, 1883 (1913); **36**, 1452 (1914).
6. A. Baeyer, Ann., **354**, 163 (1907); Baeyer is quoting Hartley here.
7. W. W. Strong, ACJ, **44**, 85 (1910); a review article.
8. S. F. Acree et al., JACS, **39**, 648 (1917) and later papers.
9. H. A. Lubs and W. M. Clark, J. Wash. Acad. Sci., **5**, 609 (1915); **6**, 481 (1916); biography of Clark by H. B. Vickery, *Biog. Mem. Nat. Acad. Scis.*, **39**, 1 (1967).
10. W. M. Clark, Chem. Rev., **2**, 127 (1925) and many later papers.
11. C. R. Bury, JACS, **57**, 2115 (1935); Bury was at the University College of Wales.
12. G. N. Lewis and M. Calvin, Chem. Rev., **25**, 273 (1939), with many refs. to earlier works.
13. L. Pauling, in Gilman's *Organic Chemistry*, 1938, Vol. II, pp. 1888-1890; Proc. Nat. Acad. Sci., **25**, 577 (1939); *The Nature of the Chemical Bond*, Ithaca, 1940, pp. 281, 431. An approximate quantum mechanical calculation of excited electronic levels for aromatic compounds is given by A. L. Sklar, J. Chem. Phys., **7**, 984 (1939).
14. L. G. S. Brooker, et al., **62**, 1116 (1940), and later papers. L. F. Werner, ibid., **42**, 2309 (1920), made observations on cyanine dyes which agree with the resonance scheme.
15. Pauling, Proc. Nat. Acad. Sci., ref. 14. Pauling used the *cis* structure for indigo, but it is now known to be *trans*.
16. Emma P. Carr and Hildegard Stücklein, J. Chem. Phys., **4**, 762 (1936) and accompanying papers.
17. W. R. Orndorff, R. C. Gibbs, S. Alice McNulty et al., JACS, **48**, 1327, 1994, (1926), and later papers: also E. R. Riegel and M. C. Reinhard, ibid., **48**, 1334 (1926), for detailed experimental procedure.
18. L. C. Anderson and C. M. Gooding, ibid., **57**, 999 (1935).
19. W. R. Brode, ibid., **46**, 581 (1924); W. C. Holmes (in W. M. Clark's laboratory), ibid., **46**, 627, 631 (1924) and later papers.
20. *Chemistry in the Economy*, American Chemical Society, Washington, 1973, pp. 102 ff.

G. Free-Radical Chemistry, 1914-1939

We have described Moses Gomberg's important discovery of the first free radicals containing trivalent carbon, the triphenylmethyl free radicals. During the 1914-1939 period there were significant advances in the field. M. S. Kharasch and others found that free radicals, much more reactive than the triphenylmethyl type, were intermediates in many reactions. G. N. Lewis' prediction that structures with an odd number of electrons (free radicals or "odd molecules") should be detectible by their paramagnetic properties was confirmed by N. W. Taylor of Berkeley[1], and this served as a quantitative analytical procedure for free radicals, although it was subject to corrections of uncertain magnitude.

Gomberg and his colleagues at Michigan, particularly W. E. Bachmann and L. C. Anderson, studied the triphenylmethyl type of free radical in detail. Bachmann and F. Y. Wiselogle in 1936, for example, showed that pentaarylethanes dissociated reversibly into free radicals and combined with oxygen, like triphenylmethyl.[2] L. C. Anderson reported measurements of electronic spectra of triphenylmethyl derivatives.[3] Other physical properties of free radicals of this type, such as heats of dissociation, were studied by J. B. Conant, H. E. Bent (both at Harvard), C. B. Wooster (Brown), and Karl Ziegler in Germany.[4]

L. Pauling and G. W. Wheland attributed the formation of triphenylmethyl radicals to resonance stabilization of the radical compared to the undissociated product.[5] Clearly there were other factors to be considered,[4] but the resonance idea was of fundamental importance.

The ability of highly unsaturated groups, such as $C \equiv C$, to promote dissociation was examined by C. S. Marvel, who found compounds such as $[C_6H_5C{\equiv}CC(C_6H_5)_2]_2$ to give evidence of dissociation to free radicals; here the $C_6H_5C{\equiv}C$ group takes the place of C_6H_5 in the free radical.[6] Marvel found some novel and complicated rearrangement reactions in the acetylenic derivatives. Marvel and Roy showed the effect of various substituents on the stability of triphenylmethyl radicals, using magnetic susceptibility to determine the amount of free radical.[7]

Conant introduced a different way of preparing triarylmethyl radicals by the reduction of the corresponding carbonium ion, adding one electron from vanadous ion as reducing agent.[8]

$$(C_6H_5)_3C^+ \underset{-e}{\overset{+e}{\rightleftarrows}} (C_6H_5)_3C\cdot$$

An instructive example of excessive dogmatism and the unwillingness to recognize new discoveries was the controversy over the radical prepared by C.

F. Koelsch in 1932 from the carbinol shown. The referee of the manuscript on the radical in 1932 did not consider it a free radical and recommended rejection of the paper. Twenty-five years later, examination of the original sample of the radical by electron spin resonance (ESR) showed that it was completely dissociated and was, in fact, a highly stabilized radical,[9] as Koelsch had thought in 1932.

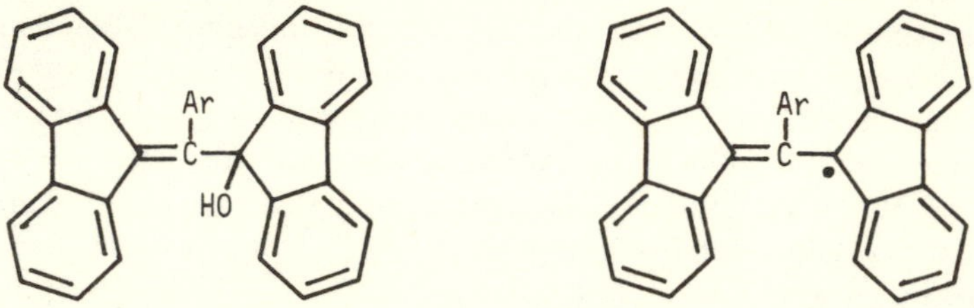

Michaelis of the Rockefeller Institute demonstrated in the 1930s that a quite different type of free radical was an intermediate in the oxidation of hydroquinones and aromatic diamines. The oxidation takes place in two one-electron steps; the intermediate radical ion with an unpaired electron is stable in favorable cases, as shown by its paramagnetic properties.[10] Michaelis called these free radicals *semiquinones*, and it is probable that many biological oxidations go through a similar odd-electron intermediate.[11]

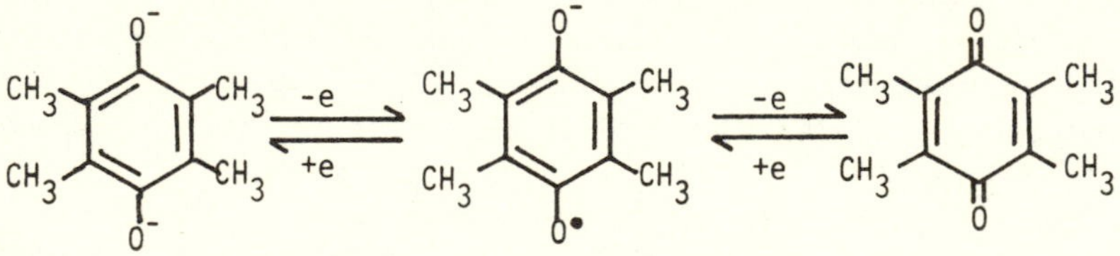

Leonor Michaelis (1875-1949, 1896 medical degree from Berlin) did research and teaching on medical and biochemical problems in Germany until 1922. After three years each in Japan and at Johns Hopkins, he worked at the Rockefeller Institute from 1929 to 1941. His research covered a wide range of problems, including buffer solutions, pH, isoelectric points of proteins, oxidation-reduction processes, and iron metabolism. The method of permanent hair waving with ammonium thioglycolate is based on his observations on the reduction of disulfide groups in keratin (a protein present in hair).

Arnold Weissberger's work indicated that air oxidation of α-hydroxyketones as well as hydroquinones goes through a free-radical intermediate, similar to a semiquinone.[12] Weissberger (1898-1984, Ph.D. 1924, Leipzig with A. Hantzsch) began his distinguished career at Leipzig from 1923 to 1933. Following a fellowship at Oxford from 1933 to 1936, he worked in the Eastman

Kodak Research Laboratories from 1936 to 1963. He studied dipole moments, stereochemistry, autoxidation reactions, and synthetic organic chemistry. In addition to contributing to the chemistry of color photography, he edited two extensive series of books on heterocyclic compounds and techniques of organic chemistry.

The stability of the semiquinone radical type as well as the triphenyl-methyl series is explained by Pauling and others as due to delocalization of the unpaired electron over the extensive aromatic π electron system.[5] Koelsch's radical above offers a large number of delocalization possibilities.

It is traditional knowledge that some organic compounds take up oxygen from the air (autooxidation); fats become rancid by autoxidation, for example. A. M. Clover of Parke Davis in 1922 showed in an excellent study[13] that ethers form peroxides from air, a process catalyzed by acetaldehyde, CH_3CHO. Although Clover did not suggest a free-radical chain reaction for the autoxidation reaction, evidence was gradually accumulated in the next decade that many oxidation processes did in fact involve this type of reaction[14]. H. L. S. Bäckstrom and H. S. Taylor of Princeton, and F. Haber and R.Willstätter in Germany led in this research.

$$C_2H_5OCH_2CH_3 \xrightarrow{O_2} C_2H_5OCHCH_3 \xrightarrow[H^+]{H_2O} CH_3CHO + C_2H_5OH + H_2O_2$$
$$\quad\quad\quad\quad\quad\quad\quad\quad\overset{OOH}{|}$$

Our account is not concerned in detail with free-radical reactions in the gas phase or with photochemical gas phase reactions. The demonstration of the formation of the very reactive methyl radical, CH_3, in the gas phase by F. Paneth and W. Hofeditz in Germany, and the extensive studies of F. O. Rice at Johns Hopkins on thermal free-radical reactions in the gas phase were noteworthy.[15] The high temperature decomposition of hydrocarbons (*cracking*) developed by W. M. Burton,[16] a Remsen Ph.D., at Standard Oil of Indiana, was a process of great economic importance, but although it undoubtedly involved a free-radical chain route, it was too complicated to be studied as such in detail.

Kharasch's demonstration of the free-radical chain reaction in the addition of HBr in presence of peroxides to carbon-carbon double bonds has been already reviewed. He also discovered numerous other peroxide-catalyzed reactions involving free radicals,[17,18] and in later years he unmasked whole new families of free-radical chain reactions,[19] many of practical or synthetic importance. Kharasch's work is marked by careful experiments and a high degree of originality in conception and interpretation. It was one of the important contributions of American organic chemistry and opened up whole fields of new investigation and new ways of looking at reaction mechanisms. It fostered the increasing research on photochemistry of organic compounds, because many free-radical reactions can be initiated by light. The important free-radical chain reactions of vinyl polymerization are discussed elsewhere.

One type of aromatic reaction perceived as a free-radical process is the Gomberg-Bachmann reaction of diazonium salts in base with other aromatic nuclei.[20] The yields are not high in this method, probably due to further reactions and the (unknown) way in which the extra H is lost in the second step.

$$ArN_2{}^+OH^- \longrightarrow Ar\bullet \;+\; N_2 \;+\; HO\bullet$$

$$Ar\bullet \;+\; Ar'H \xrightarrow{-H\bullet} Ar{-}Ar'$$

Another free-radical reaction is addition of sodium in dimethyl ether, $(CH_3)_2O$, or glycol dimethyl ether, $(CH_3OCH_2)_2$, but not in diethyl ether $(C_2H_5)_2O$, to naphthalene and other aromatic compounds.[21] This product, colored and strongly electrically conducting, was shown later to be a radical anion; addition of proton sources (H_2O, C_2H_5OH) produces mixtures of 1,2- and 1,4-dihydronaphthalenes by secondary process. Free-radical ions of this type occur as intermediates in several processes[22] discovered since 1939.

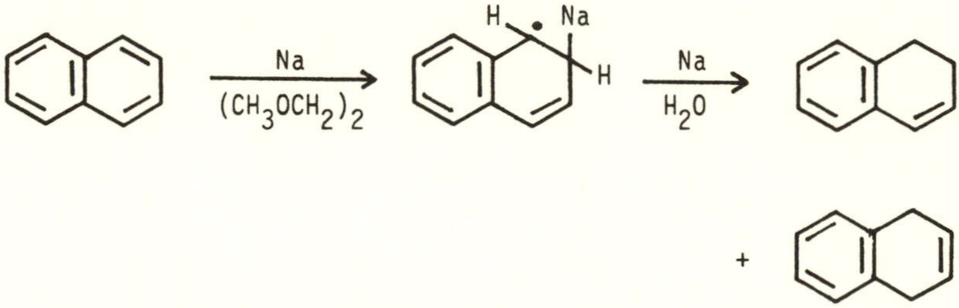

The metal ketyls, formed by reduction of ketones, such as benzophenone, by metallic sodium or Mg-MgI$_2$ in ether, are generally regarded as free radicals in equilibrium with the dimer.[23]

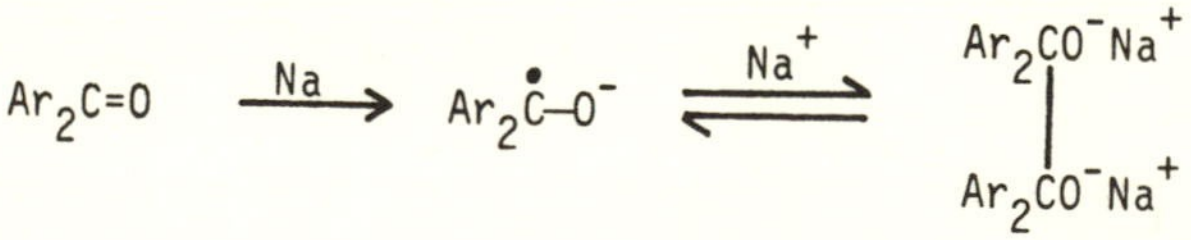

The discovery of deuterium (heavy hydrogen) by H. C. Urey provided a convenient way of demonstrating exchange of hydrogen atoms by adsorbed compounds, most logically explained by dissociation on the catalytic surface to atoms and radicals. For example, H. S. Taylor at Princeton showed exchange reactions of these and similar types.[24] Deuterium tracer techniques became very valuable in many reaction mechanism studies.

$$CH_4 + D_2 \xrightarrow[Ni]{184°} CH_3D + HD$$

$$CH_4 + CD_4 \xrightarrow{Ni} CH_3D + CD_3H$$

$$D_2 + CH_2=CH_2 \xrightarrow[Ni]{-80° \text{ to } 65°} CHD=CH_2 + CH_2DCH_2D$$

LITERATURE CITED

1. N. W. Taylor, JACS, **48**, 854 (1926).
2. W. E. Bachmann and F. Y. Wiselogle, JOC, **1**, 354 (1936).
3. L. C. Anderson, JACS, **57**, 1673 (1935) and earlier papers.
4. J. B. Conant, J. Chem. Phys., **1**, 427 (1933), H. E. Bent et al., JACS, **58**, 165 (1936) and other papers; K. Ziegler and Lisa Ewold, , Ann., **473**, 163 (1929); C. B. Wooster, JACS, **58**, 2156 (1936).
5. L. Pauling and G. W. Wheland, J. Chem. Phys., **1**, 362 (1933).
6. H. E. Munro and C. S. Marvel, JACS, **54**, 4445 (1932).
7. M. F. Roy and C. S. Marvel, ibid., **59**, 2622 (1937).
8. J. B. Conant and A. W. Sloan, ibid., **45**, 2466 (1923) and later papers; M. Gomberg and W. E. Bachmann, ibid., **52**, 2455 (1930), using $Mg\text{-}MgI_2$ as reducing agent.
9. C. F. Koelsch, ibid., **54**, 4744 (1932); **79**, 4439 (1957).
10. L. Michaelis, Chem. Rev., **16**, 243 (1935); L. Michaelis et al., JACS, **53**, 2953 (1931); **60**, 202, 1678 (1938); J. B. Conant and N. M. Bigelow, JACS, **53**, 676 (1931).
11. H. R. Mahler and E. H. Cordes, *Biological Chemistry*, New York, 1966, pp. 361, 573-575.
12. T. H. James and A. Weissberger, JACS, **60**, 98 (1938) and many earlier papers by Weissberger.
13. A. M. Clover, ibid., **44**, 1107 (1922); **46**, 419 (1924).
14. H. L. S. Bäckstrom, ibid., **49**, 1460 (1927); Z. Phys. Chem., **25B**, 99 (1934); H. S. Taylor and A. J. Gould, JACS, **55**, 859 (1933); G. E. K. Branch and M. A. Joslyn, ibid., **57**, 2388 (1935); F. Haber and R. Willstätter, Ber., **64**, 2844 (1931).
15. F. Paneth and W. Hofeditz, ibid., **62**, 1335 (1929); F. O. Rice et al., JACS, **54**, 3529 (1932); F. O. Rice and K. K. Rice, *The Aliphatic Free Radicals*, Baltimore, 1935.
16. W. M. Burton, Ind. Eng. Chem., **10**, 484 (1918).
17. F. R. Mayo and C. Walling, Chem. Rev., **27**, 351 (1940).
18. M. S. Kharasch et al., JACS, **59**, 1655 (1937), and later papers.
19. C. Walling, *Free Radicals in Solution*, New York, 1957.
20. M. Gomberg and W. E. Bachmann, JACS, **46**, 2339 (1924); Gomberg and J. C. Pernet, ibid., **48**, 1372 (1926). The free radical mechanism was suggested by D. H. Hey and W. A. Waters, Chem. Rev., **21**, 169 (1937).
21. N. D. Scott, J. F. Walker and V. I. Hansley, JACS, **58**, 2442 (1936); **60**, 951 (1938).
22. N. Kornblum et al., ibid., **88**, 5662 (1966); J. F. Kim and J. F. Bunnett, ibid., **92**, 7463 (1970); G. A. Russell and W. C. Danen, ibid., **88**, 5663 (1966) and later papers.
23. W. E. Bachmann, ibid., **55**, 1179 (1933) and earlier papers; C. B. Wooster and J. G. Dean, ibid., **57**, 112 (1935); $Mg\text{-}MgI_2$ reaction, Gomberg and Bachmann, ibid., **49**, 236 (1927).
24. H. S. Taylor et al., ibid., **58**, 1445 (1936) and later papers. On adsorption on catalytic surfaces, R. W. Harkness and P. H. Emmett, JACS, **56**, 490 (1934).

ERNEST H. VOLWILER

19

Research in the
Pharmaceutical Industry
1914-1939

The development of research programs in pharmaceutical firms after 1914 provided employment for many organic chemists. Although these research laboratories were not as large as those of other parts of the chemical industry, they formed a significant segment of organic industrial research. Furthermore, they published organic research fairly steadily in the JACS and in other journals as they became available. The larger pharmaceutical research laboratories became major centers for synthetic organic chemistry and for work on natural products, a trend which was accentuated by the emergence of the vitamins and sulfa drugs in the late 1930s, and particularly by the rise of antibiotics in later years. The pharmaceutical laboratories also contributed some excellent work in physical organic chemistry; some of these contributions are reviewed elsewhere in this work.

It is impractical to describe in detail the evolution of each of the major pharmaceutical research enterprises.[1] When fully developed, these contained organic chemistry and research programs in the biomedical sciences, particularly pharmacy which included the formulation of drugs, pharmacology, biochemistry, and arrangements for clinical testing.

A number of the American firms which later developed important research laboratories began in the 19th century, namely Merck (1818), Sharp and Dohme (1845), Squibb (1858), Lilly (1876), Parke Davis (1866), Upjohn (1885), and Abbott (1888).[1] Some of these started as analytical laboratories or as distributors for imported products. Before 1914 American drug firms mainly prepared galenicals and proprietary remedies. The "patent medicines", most of which had dubious efficacy or were positively harmful,[2] were usually prepared and marketed by small operators, who did not develop into major pharmaceutical firms.

Research in Germany prior to 1914 had produced a number of widely used synthetic medicinal compounds, such as aspirin (analgesic), sulfonal and barbital (sedative, hypnotic), and Salvarsan (antisyphilitic).[3] These and other synthetic medicinals were available in this country, but they were tightly controlled by German firms and were manufactured either in Germany or in this country by German-controlled firms.

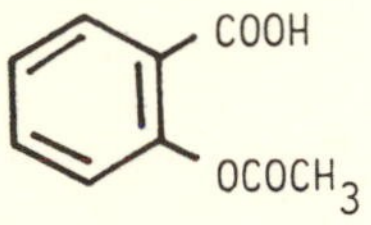

Aspirin

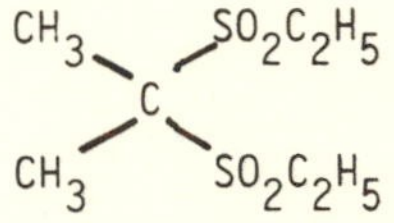

Sulfonal

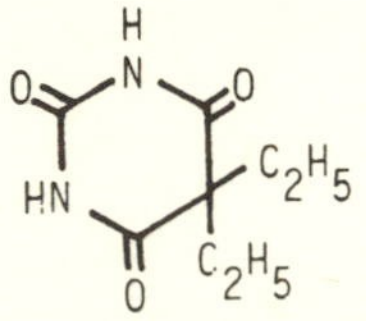

Barbital (Veronal)

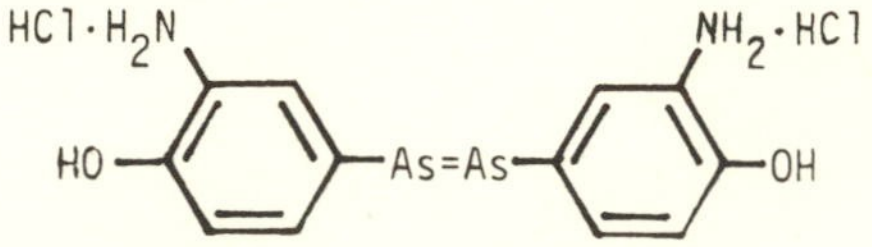

Salvarsan (arsphenamine)

The extent to which German research dominated the organic chemistry of drugs is strikingly shown by a tabulation of the number of patents on drugs or intermediates issued to representative German and American firms taken from the Decennial Indices of *Chemical Abstracts*, Table 1.

A few papers on organic chemistry or biochemistry from pharmaceutical research laboratories appeared in JACS, ACJ, and JBC before 1914. The number of papers on organic chemistry published by pharmaceutical laboratories in JACS from 1914 to 1939 is shown in Table 2.

The number of publications from industrial laboratories is not necessarily proportional to research activity, because much good research may appear only as patents or may not be published at all. Publications from some pharmaceutical laboratories showed a sharp increase after 1932. It is reported[4] that Lilly hired its first chemist, a graduate of the new School of Pharmacy at Purdue, in 1886, but research did not become a major activity at Lilly until after 1920. The figures in Table 2 do not include publications in JBC or papers dealing with procedures for concentrating materials from natural sources unless pure compounds were obtained. Nevertheless, Table 2 does give a rough idea of research activity in organic chemistry in pharmaceutical firms through 1939. It does not, of course, take into account firms which became active research centers after 1939.

We have chosen to survey the growth of research at Abbott Laboratories as a representative, major pharmaceutical firm. A high proportion of Abbott sales over the period 1914-1939 was based on products developed by organic synthesis, to a greater degree than that of comparable firms.[5]

Abbott was founded by Dr. Wallace C. Abbott, a native of Vermont, who combined a medical practice with the operation of a drug store in Ravenswood, Illinois, starting in 1888. He soon began producing and marketing plant extracts as solutions or pills, naming his enterprise the Abbott

Table 1

TOTAL PATENTS ISSUED ON DRUGS AND INTERMEDIATES TO GERMAN AND AMERICAN FIRMS, 1907-1936

A. German Firms

Firm	1907 — 1916	1917 — 1926	1927 — 1936
E. Merck & Co.	52	25	65
C. F. Boehringer & Söhne	53	10	18
Farbwerke, vorm. Meister Lucius and Brünings	9 columns of titles, dyes and many drugs	2 columns, partly drugs	1(a)
Farbenfabriken, vorm. F. Bayer & Co.	12 columns, dyes, many on drugs especially barbiturates	1.5 columns, partly drugs	4(a)
I. G. Farben	—	—	78 columns, partly drugs

B. American Firms

Firm	1907 — 1916	1917 — 1926	1927 — 1936
Abbott Laboratories	0	3	51
Merck & Co.	0	1	22
Eli Lilly	1	3	51
W. S. Merrell & Co.	0	0	7
Hynson, Westcott & Dunning	0	0	9
Parke Davis	0	4	19
Sharp & Dohme	0	0	19
Upjohn	0	2(b)	0
Burroughs, Wellcome	0	0	1
Squibb	0	1	60

(a) The low figures here are due to mergers which made these firms part of the I. G. Farben in 1926. For the transformations of German chemical firms, see L. F. Haber, *The Chemical Industry: 1900-1930*, Oxford, 1971.

(b) Under the name of the Upjohn Chief Chemist, Dr. Frederick W. Heyl. It is clear from Haynes, *American Chemical Industry*, 6, p. 455ff., that Upjohn owned or licensed numerous patents issued to individuals.

Table 2

Papers in JACS from Pharmaceutical

Companies, 1914-1939 Inclusive

Firm	First Publication	Papers 1914 – 1931	Papers 1933 – 1939	Total 1914 – 1939
Parke Davis	1915	34	4(b)	38
Burroughs, Wellcome	1930	13	20	33
Merck	1931	2	27	29
Upjohn	1915	12	10	22
Abbott	1921	11(a)	7	18
Lilly	1924	4	9	13
Sharpe and Dohme	1926	9	6	15
Hynson, Westcott and Dunning	1927	3	4	7
Merrell	1932	1	6	7
Squibb	1938	–	5	5

(a) This includes six papers from the Dermatological Research Laboratory, purchased by Abbott in 1922. (b) An additional group of over 20 papers might be added, published jointly from Penn State and Parke Davis, with R. E. Marker and Oliver Kamm as coauthors; these are part of the large number of Marker's publications on steroid chemistry during these years.

Alkaloidal Company in 1894. His business grew and expanded its scope, stimulated by Dr. Abbott's energy, editing, writing, and the organization of a sales force to visit physicians. By 1909 annual sales were $445,000 and several able associates had joined the firm, which changed its name to Abbott Laboratories in 1914. Dr. A. S. Burdick, a physician who had joined Abbott in 1904, realized that the company needed synthetic medicinals and a research program to produce them to replace the declining alkaloidal extract business. In 1912 Burdick hired a young Ph.D. from Harvard, Franklin Summers, to assist Fred Young, the company production manager, in developing more synthetic products. Summers seems to have made little impact on Abbott and is not listed in any edition of *American Men of Science*. With the cutting off of German drugs and chemical intermediates in 1914, Abbott, like all other American chemical enterprises, faced a serious crisis.

As part of Abbott's contribution to a cooperative effort to produce the necessary chemicals and intermediates, Young manufactured hydroquinone for a year or two. This was not a successful operation, and Abbott turned to the production of three drugs, formerly obtained from Germany and urgently requested by the Army and Navy. These were the sedative barbital, the local anaesthetic procaine, and the analgesic and antipyretic (fever-reducing) cinchophen. Abbott also undertook the American production of the new and effective antiseptic developed for military use by the British-American biochemist, H. D. Dakin. This was called Chloramine-T or Chlorazene; the aqueous solution was "Dakin's solution", and its antiseptic activity was due to slow release of chlorine.

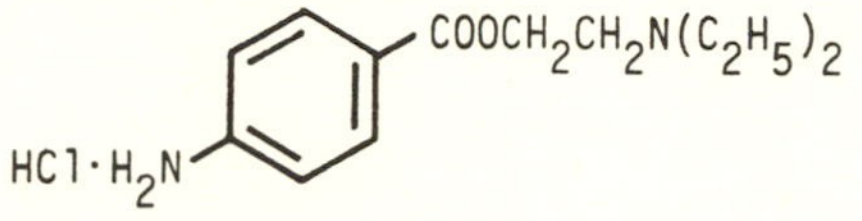

Procaine (Novocaine)

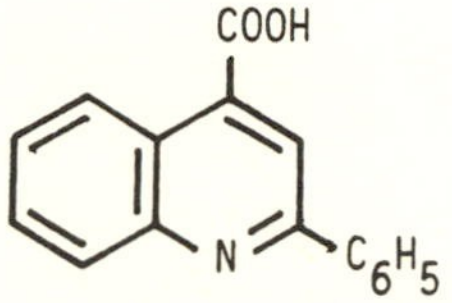

Cinchophen (Atophan)

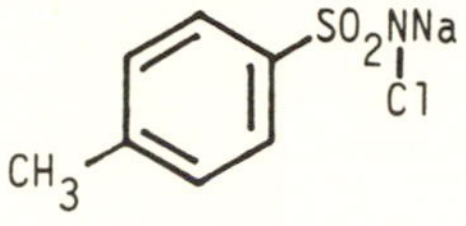

Chloramine-T

Although Young was an able production man, the manufacture of these materials ran into difficulties due to lack of experience and, doubtless also, to variable quality of the available wartime intermediates. Abbott at this time had a laboratory which did chemical and biolgical quality control work on products, but not research. The company obviously needed expert chemical assistance, and this soon became available from two unusual men.

One was Roger Adams, of the University of Illinois, whose career we have described earlier. Adams was engaged by Dr. Abbott in 1917 as a chemical consultant for Abbott Laboratories, a connection which lasted many years, and Adams eventually served on the Abbott Board of Directors. The valuable chemical advice which Abbott gained from Adams proved useful in the immediate problems of production. Adams had started research at Illinois in the field of local anaesthetics in 1917, synthesizing procaine and numerous analogs.[6] A valuable long-range benefit to Abbott from the Adams connection was his recommendation to Abbott of his first Ph.D. student at Illinois, Ernest H. Volwiler.

Volwiler graduated from Miami University in Ohio in 1914, finishing his undergraduate course in three years. He went to Illinois for graduate work in 1914 and received a Master's degree with C. G. Derick. During his graduate work, Volwiler was in charge of the "preps" lab in 1916 and 1917 as senior graduate student. Derick left Illinois for an industrial position (Roger Adams was his replacement) and Volwiler was offered a position in Cincinnati at $1800 per year, the going rate for Ph.D.s at that time. He was uncertain, after various conflicting pieces of advice, whether to accept the offer or to stay at Illinois for his Ph.D. He decided to stay at Illinois when the Illinois glass blower, Paul Anders, argued with him that it would soon be the same in this country as in Germany: he would be foolish not to take his doctorate which would be important to his future career.

After completing his Ph.D. with Adams in 1918, Volwiler was offered a commission in the Chemical Warfare Service, in which Adams was then a Major. Because Adams told him that the war would soon be over and little more would be done on chemical warfare, Volwiler followed Adams' advice to work for Abbott. During his first year at the plant, Volwiler worked as a production chemist, and by the end of that period the production of procaine, barbital, and cinchophen was moving well. Turning then to research on local anaesthetics, he hired several Illinois men, including Elmer B. Vliet, who was to have a long record of service at Abbott, and Floyd Thayer, who eventually became vice-president in charge of chemical sales.

One of the first successful new drugs introduced by Abbott after Volwiler's arrival was Butyn (butacaine),[7] superior to procaine for some uses as a local anaesthetic. This utilized the current availability of n-butyl alcohol, produced along with acetone by a special fermentation of starches. This process was developed in England by Dr. Chaim Weizmann, later the first President of Israel, to produce the acetone urgently needed for the manufacture of explosives.

$$H_2N-\underset{}{\underset{}{\bigcirc}}-COO(CH_2)_3N(C_4H_9)_2$$

Butacaine (Butyn)

E. B. Vliet reported an improved method of making dibutylamine,[8] a component of Butyn, from butyl bromide and calcium cyanamid ("lime nitrogen"). The latter material became available in this country during World War I for the fixation of atmospheric nitrogen for fertilizer and explosives; it was to have been manufactured on a large scale at a plant at Muscle Shoals, Alabama. Thus, in addition to its intrinsic medical usefulness, Butyn is emblematic of the developing American pharmaceutical industry based on research, of the advances in chemical technology, of the way in which results from one area of research might be important in an entirely different one, and also of the relations between research in universities and industrial laboratories in this country.

The Abbott research group increased gradually during the 1920s. Volwiler became chief chemist in 1921, director of research in 1930, and vice-president in charge of research and development in 1933. He directed the research activities in all fields soon after his arrival, and the total research effort was set up as a unit with its own budget of the order of $100,000 per year in the mid 1920s. Abbott sales in 1923 were $2,154,000, a new high and an increase of $500,000 over 1922.

The drug Salvarsan was an object of intensive research when German supplies were cut off.[9] Several laboratories essayed its manufacture on a commercial scale, but the only group to produce a good product was the Dermatological Research Laboratory (DRL) in Philadelphia, founded by a successful dermatologist, Dr. J. F. Schamberg, as a non-profit research institute. Dr. G. Raiziss, author of a monograph on organic arsenic compounds,[10] directed the chemical work. Abbott acquired DRL in 1922, increasing Abbott's research staff with able people and broadening its line of products.

The manufacture of barbiturates like barbital in this country led to much research on synthesis and testing of analogs.[11] Volwiler and D. L. Tabern (a Ph.D. with Moses Gomberg at Michigan and instructor at Cornell who joined Abbott in 1926) developed Nembutal (pentobarbital) as a sedative and hypnotic; the sulfur analog, Pentothal, became a very important intravenous anaesthetic.[12] Like most barbiturates, these are used as the more soluble sodium salts. Although there were numerous barbiturates in addition to Nembutal for use as sedatives, Pentothal was suitable as an intravenous anaesthetic comparable to inhalation anaesthetics like ether; it became very widely used and was valuable in surgery in World War II.[13] Nembutal and Pentothal were important contributors to Abbott sales and to the company's reputation.

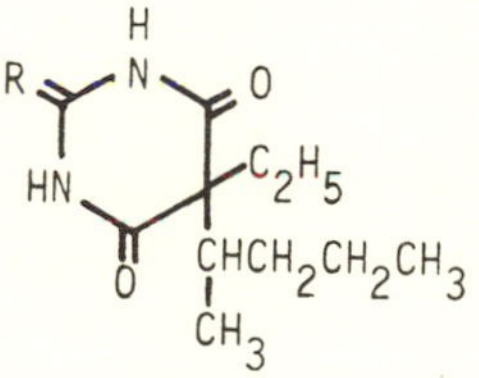

R = O, Nembutal (Pentobarbital)
R = S, Pentothal (Thiopental)

The merger with the Swan-Myers Company of Indianapolis in 1930 brought Abbott new research people and new products, particularly ephedrine, of which Abbott became the major producer. This compound, originally isolated from a Chinese plant, is a potent pressor agent (increasing blood pressure) and is useful in treating asthma, hay fever, and other conditions. It is somewhat similar in physiological action to Adrenalin (epinephrine), a component of the natural regulatory mechanism of the nervous system.

Ephedrine   Adrenalin (epinephrine)

Although Abbott was not active in chemical research on the structure and synthesis of the vitamins, a striking feature of the 1930s, particularly at Merck, the company did develop vitamin A and D concentrates from halibut liver oil (Haliver oil). It had been licensed in 1929 under the University of Wisconsin's Steenbock patents to market vitamin D products obtained by irradiating ergosterol solutions with ultraviolet light; the product was marketed as Viosterol. Other vitamin products were bought from the manufacturers, formulated, and sold; they became important sources of income.

Abbott weathered the depression, as did other pharmaceutical firms of the 1930s, without decreasing its research staff and hired several additional Ph.D.s in organic chemistry in the middle of the decade. A development laboratory was created to solve the problems of large scale synthesis of compounds prepared on a small scale in the research laboratories, if large amounts were needed for further testing and, with the clinically successful compounds, for eventual factory production. The difficulties involved in scaling up a laboratory procedure from gram to kilogram quantities of product are always troublesome and frequently formidable. The development group under Norman Hansen (another Illinois chemist) began in 1926 and became an indispensable part of the chemical research activity.

In 1939, Abbott total sales were \$11,850,000. A strong research group of about eight Ph.D.s in organic chemistry worked with other teams necessary to test and develop new medicinal agents. The new \$500,000 research building and the extensive chemical manufacturing facilities occupied a site in North Chicago, to which Abbott had moved in 1922. The firm's chemical research was oriented primarily toward synthesis, and its leadership was excellent. As a whole it became one of the country's major pharmaceutical houses with a solid reputation for stable growth, high quality research, and reliable products. Abbott also expanded in many other ways, such as in developing international sales. The Abbott research organization successfully met the serious problems posed to the pharmaceutical industry by World War II, even when they were in fields in which the company had no previous experience, such as large scale fermentation to produce penicillin.

It was a long way from Dr. Abbott's alkaloidal extracts of 1888 to the Abbott Laboratories of 1939. A similar story could be told of the development of the other major pharmaceutical firms of 1939.

LITERATURE CITED

1. Accounts of most American pharmaceutical firms, with some information about their research activities, are in Williams Haynes, *American Chemical Industry, Vol. 6,* New

York, 1949. The dates of founding are taken from Haynes' accounts of the individual firms and represent the earliest date of the firm as a recognizable unit. Many changes of ownership and partial changes of name occurred in some cases.

2. Patent medicines received widespread publicity around the turn of the century, and the Pure Food and Drug Act of 1906 brought some control over them.

3. See *Merck Index* (several editions, 5th, 1940); G. L. Jenkins, W. H. Hartung, K. E. Hamlin, Jr., and J. B. Data, *The Chemistry of Organic Medicinal Products*, 4th ed., New York, 1957, and earlier editions.

4. E. J. Kahn, Jr., *All in a Century, The First Hundred Years of Eli Lilly and Company*, n.p., 1976, p. 25.

5. This account is based on Herman Kogan, *The Long White Line, the Story of Abbott Laboratories*, New York, 1963; on the chemical publications from Abbott, and on an interview with E. H. Volwiler, Aug. 2, 1976. Haynes, op. cit., pp. 2-5, has a brief account of the history of Abbott. For more detail on Adams, see D. S. and A. T. Tarbell, *Roger Adams: Scientist and Statesman*, Washington, 1981.

6. R. Adams et al., JACS, 59, 2248 (1937).

7. O. Kamm, R. Adams and E. H. Volwiler, U. S. Patent, 1,358,751 (1920); Chem. Abstr., 15, 412 (1921); O. Kamm, JACS, 42, 1030 (1920) reported the synthesis of a number of Novocaine analogs, but not of Butyn itself, which was being patented. This paper was published jointly from Illinois and Abbott. Oliver Kamm (1888-1974, Ph.D. 1915, Illinois) was on the staff at Illinois from 1917 to 1920 and worked with the preps lab. He wrote a widely used book on qualitative organic analysis and joined Parke Davis in 1920 as director of research. Among other projects he directed research on the separation of two hormones from the pituitary gland, later called oxytocin and vasopressin, shown to be polypeptides and synthesized by V. du Vigneaud after World War II. Kamm also supported Marker's work on steroids at Penn State in the 1930s. An alternative synthesis of Butyn was reported by Adams and Volwiler in U. S. Patent 1,676,470 (1928); Chem. Abstr., 22, 3265 (1928).

8. E. B. Vliet, JACS, 46, 1305 (1924).

9. Many papers appeared in JACS and JBC in the 1915-1925 period on preparation of intermediates, analogs, and analytical procedures related to Salvarsan (called arsphenamine in this country).

10. G. W. Raiziss and J. L. Gavron, *Organic Arsenical Compounds*, New York, 1923.

11. Many papers from pharmaceutical and university laboratories could be cited; H. A. Shonle and A. Moment, JACS, 45, 243 (1923), is representative.

12. E. H. Volwiler and D. L. Tabern, ibid., 52, 1676 (1930); 57, 1961 (1935); D. L. Tabern and E. F. Shelberg, ibid., 55, 328 (1933).

13. Early discussion of Pentothal as an anaesthetic; L. Goodman and A. Gilman, *The Pharmacological Basis of Therapeutics*, New York, 1941, p. 136; the 4th edition of this book, 1970, p. 95, shows the wide acceptance of Pentothal as an anaesthetic.

WALLACE H. CAROTHERS

20

Studies on Polymerization and Synthetic Polymers, 1914-1939

Polymers are high molecular weight molecules derived from low molecular weight units; the combining of small molecules (monomers) to form large ones (polymers) is called *polymerization*. Many naturally occurring materials fundamental to the functioning and reproduction of living systems are polymers, sometimes called biopolymers. These include, for example, polymers derived from simple sugars, such as starch, glycogen, and cellulose; proteins, composed of α-amino acid residues; nucleic acids (DNA and RNA), whose molecular weights may be many million; and natural rubber. The nucleic acid building blocks are *nucleotides*, consisting of a combination of a heterocyclic base (e.g. adenine), a sugar, usually D-ribose or 2-deoxy-D-ribose, and a phosphate ester unit. The unraveling of nucleic acid structures and the explanation of their functions have been particularly rapid since 1945. The "double helix" of DNA and its function as the carrier of genetic information have been widely discussed in non-technical publications.

The synthetic polymers to be discussed in this section include well-known materials such as nylon, used for textile fibers and for other plastic products; rubber-like materials, such as neoprene and butyl rubber; and other plastics, such as polystyrene and Lucite (polymer of methyl methacrylate).

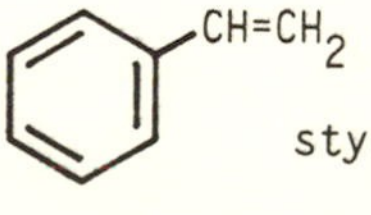

CH=CH$_2$ styrene

CH$_2$=CCOOCH$_3$ CH$_3$ methyl methacrylate

Several polymeric materials of industrial importance were developed in this country before 1914. One of these was Bakelite, made by base-catalyzed reaction of phenol and formaldehyde. Phenol has three reactive ring positions and formaldehyde can combine with two phenol molecules, so that very large molecules can be formed. Leo H. Baekeland in 1907 worked out the conditions under which this reaction yields the thermosetting plastic Bakelite, useful for molding and casting. Baekeland (1863-1944), born and educated in Belgium, invented the photographic paper Velox and moved to this country, where he worked on industrial electrochemistry before his development of Bakelite.[1]

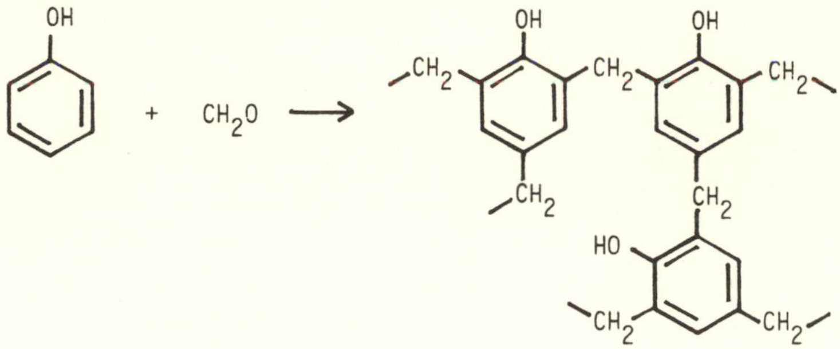

Polymers from glycerin and phthalic acid (glyptal resins) were investigated by the General Electric Company[2] in 1912. Within the next few decades, modification using different polyols and dibasic acids (isophthalic and maleic) brought these alkyd resins to top importance in the U.S. industrial coatings industry.

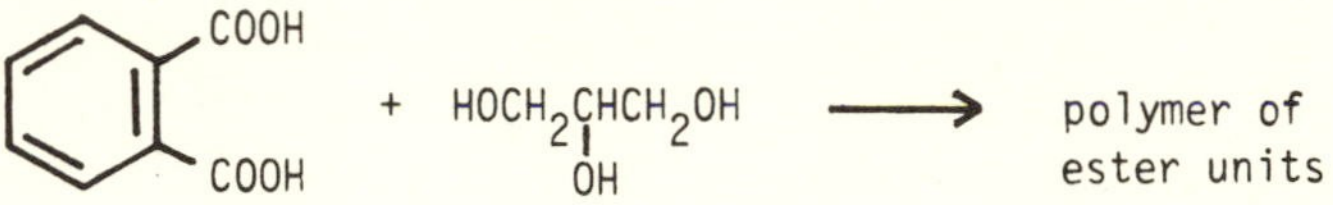

The extraordinary expansion of knowledge of both the fundamental and applied aspects of polymer chemistry was due to work in the 1914-1939 period, and the impetus of the work done in the 1930s, particularly in this country, continued long after World War II. Information about the structure and properties of the naturally occurring polymers soared, and the striking progress still continues.

The spectacular rise of polymer chemistry as a separate field in the 1930s was due in large part to the accomplishments of Wallace H. Carothers (1896-1937), an organic chemist of remarkable ability.[3] Carothers, after graduating from Tarkio College in Missouri, earned his Ph.D. at Illinois in 1924 with Roger Adams, working on the development of the platinum oxide catalyst for catalytic reduction. After brief stints as instructor at Illinois and at Harvard, during which he published enough independent work to show his ability and originality, he moved to Du Pont in 1928. There he was given several Ph.D. collaborators to work on problems of his choice. He had evidently been thinking about high molecular weight compounds for some time. In nine years at Du Pont he not only clarified the conceptual basis for polymer formation and structure but also did the research[4] from which nylon and neoprene rubber eventually developed. He suffered from periodic depressions and committed suicide before the practical importance of his work became generally known.

By 1928, work by H. Staudinger[5] in Germany had established that polystyrene was a high molecular weight material composed of styrene units, $(CH_2CHC_6H_5)_n$. Cellulose was known to be a high molecular weight material with six-membered glucose rings repeating, and natural rubber was thought correctly to be a linear polymer of isoprene units, $(CH_2C(CH_3)=CHCH_2)_n$, cross-linked by sulfur atoms during vulcanization.[6] There was nevertheless an aura of uncertainty about the structure of both synthetic and naturally occurring polymers in 1928. It was thought by many that polymers consisted of relatively small molecules held into large aggregates by "secondary valence forces" or other molecular forces of unspecified nature. These ideas persisted in the nucleic acid field until the high molecular weights of nucleic acids were demonstrated[7] around 1940.

Carothers[8] classified polymers into two main types, condensation polymers and addition polymers. The former are obtained by step-wise reactions to form ordinary valence linkages by the elimination of a small molecule such as water or an alcohol.[8] A linear polyester would thus be formed.[9] Dacron is a polyester condensation polymer of this type, and nylon 6,6 is a polyamide[10] with the units attached by CONH linkages, as they are in natural proteins. Condensation polymers have "end groups" at the ends of the chains, which sometimes can be chemically detected.

$$HOOCCH_2CH_2COOH \quad + \quad HOCH_2CH_2OH$$

$$\downarrow 250°$$

$$HOOCCH_2CH_2COO(CH_2CH_2OOCCH_2CH_2COO)_nCH_2CH_2OH \quad + \quad H_2O$$

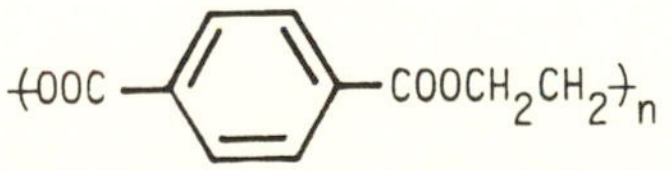

$$\{OOC\!-\!\!\langle\rangle\!\!-\!COOCH_2CH_2\}_n$$

$$\{(CH_2)_6NHCO(CH_2)_4CONH\}_n$$

Dacron, poly(ethylene terephthalate)   66 nylon

The other class, addition polymers, is formed from ethylenic compounds or ions. These, under the influence of light or catalysts that generate free radicals, add to each other to form large molecules. Addition polymers must also have end groups, because it is very unlikely that a large ring will be formed.

$$CH_2=CHR \quad \xrightarrow[\text{agent}]{\text{catalytic}} \quad \{CH_2CHR\}_n$$

After three years of experimental work and study, Carothers published his review on polymerization[8] in which he proposed the above classification and surveyed the whole field of polymer chemistry. This remarkable paper shows such insight and such a comprehensive knowledge of organic chemistry that some points will be quoted from it. He notes that the limits of the molecular weights of condensation polymers are set by the viscosity and rate of diffusion; further, the reactive groups decrease in concentration and are less likely to react on a kinetic basis.[8a] Silicone polymers arouse his interest.[8b] He adopts the Sachse-Mohr theory for carbocyclic rings and says that large rings exist probably as parallel chains.[8c] He suggests that large-ring formation might be brought about by adsorption on a suitable surface, and that large rings in natural products may be formed this way.[8d] This idea was more original in 1931 that it appears now, because at that time theories of enzyme structure and mechanism of their action were as vague and imprecise as some of the ideas about high polymer structure.

He concludes that the failure to detect functional end groups in a polymer is not proof that they are absent, because they may merely be unreactive.[8e] He remarks that six-membered rings must either be uniplanar and hence strained, or that the two forms must be readily interconvertible and hence pass through a planar phase, where they would be strained.[8f]

He suggests that the formation of high polymers by addition polymerization is a chain reaction.[8g] This proposal had been made elsewhere by Carothers and other workers, who had shown the catalytic effect of light and peroxides and the effect of inhibitors.[11] He enumerates the reasons for preferring the structural theory over the "association" theory for polymers. He emphasizes the necessary molecular weight distribution of polymers and comments shrewdly on the relation between physical properties and structure of high polymers.[8h]

Carothers showed an intuitive understanding of the physical chemistry involved in many problems in polymer chemistry, although he thought primarily in the structural terms of an organic chemist. The mathematical-statistical approach to polymer chemistry by Paul J. Flory found an experimental starting point in the work of Carothers' group.

Carothers' own experiments on condensation polymerization were more fundamental than his work on addition polymerization. His polyamide high polymers[10] (nylon type) were not produced on a commercial scale until after his death, but his work on the addition polymerization[11] of chloroprene, $CH_2=C(Cl)CH=CH$, led during his lifetime to the production of neoprene rubber.

The chloroprene story is an example of the interaction of university and industrial research, as well as of serendipity in research. J. A. Nieuwland of Notre Dame had prepared both the dimer of acetylene, monovinylacetylene (MVA), $CH_2=CHC\equiv CH$, and the trimer, divinylacetylene (DVA), by the action of cuprous chloride on acetylene. E. K. Bolton of Du Pont heard Nieuwland describe his experiments at a chemical meeting. After Bolton learned that he was abandoning his research on the project, the industrial chemist asked Nieuwland to consult for Du Pont where work would be continued on these materials. Bolton intended to study MVA as an intermediate for butadiene, a possible precursor of a synthetic rubber, and to examine MVA polymers themselves.

Bolton suggested to Carothers, when he arrived at Du Pont, that he prepare some really pure MVA for the study in polymerization. Arnold Collins designed and built a still to purify the mixture obtained from acetylene and cuprous chloride. Distillation gave MVA, DVA, and another higher boiling fraction. Examination of these portions after an intervening weekend revealed that the higher boiling fraction had changed to a rubbery material, obviously a polymer. Analysis showed the rubbery material to contain chlorine. This could have come only from Nieuwland's catalyst. Apparently it was Carothers who deduced that HCl had added to MVA to form chloroprene, which had polymerized.

Chloroprene thus polymerized to the useful rubber-like material named neoprene.[12] Neoprene is superior to natural rubber in some cases, being more resistant to benzene and other hydrocarbon solvents than natural rubber, for example.

$$CH_2=CHC\equiv CH \ + \ HCl \longrightarrow CH_2=CHC=CH_2 \xrightarrow{\text{polymerize}} (CH_2CH=CCH_2)_n$$

Vinylacetylene   Chloroprene   Neoprene

Carothers also studied many other reactions of chloroprene and of DVA, with results of scientific but not technological importance. It might be said that Carothers was fortunate in working in polymer chemistry at the time when it was ready for a dramatic expansion. His ability, however, was so outstanding that he would have made important contributions to organic chemistry in any era.

Julius A. Nieuwland (1878-1936, B.S. 1899, Notre Dame; Ph.D. 1904, Catholic University) was ordained in 1903 and taught at Notre Dame until his death. In addition to his work on the chemistry of acetylene and on boron trifluoride, he was a distinguished botanist and the founding editor of the *American Midland Naturalist*. He was professor of botany at Notre Dame from 1904 to 1918.

The mechanism of addition (vinyl) polymerization, particularly the free-radical nature of the process, was not clear in 1931, although it was realized that a chain process was probably involved.[11,13] The demonstration in the 1930s that free-radical chain reactions did occur in other systems made this increasingly probable. A classic paper[14] by Paul J. Flory in 1937, supported by kinetic measurements made by Carothers' group,[15] defined the conceptual framework for nearly all the later work on free-radical addition polymerization.

Flory (1910-1985) graduated from Manchester College in Indiana, and after taking his Ph.D. in physical chemistry at Ohio State in 1934 with H. L. Johnston, he joined Carother's group at Du Pont. Flory had extensive experience in both academic and industrial polymer research and has been professor at Stanford since 1961. His book, *Principles of Polymer Chemistry* (1951), at once became a classic, and he received the Nobel Prize in 1974 for his work on polymers.[16]

Flory's 1937 paper proposed a three-stage process for free-radical polymerization, which, while not entirely new, was more comprehensive and satisfactory than earlier ones. As supplemented and supported by later work by C. S. Marvel, C. C. Price, P. D. Bartlett, F. R. Mayo, Cheves Walling, M. S. Matheson, and others, the *initiation step* involved the formation of a free radical from the monomer. This took place usually by addition of a free radical from decomposition of a peroxide or other source of free radicals, although initiation might be a photochemical or purely thermal process.

Initiation was followed by addition of the monomer radical to another monomer molecule, the *propagation stage*. The addition of this dimer radical to another monomer was repeated many times until the polymer radical had a high molecular weight.

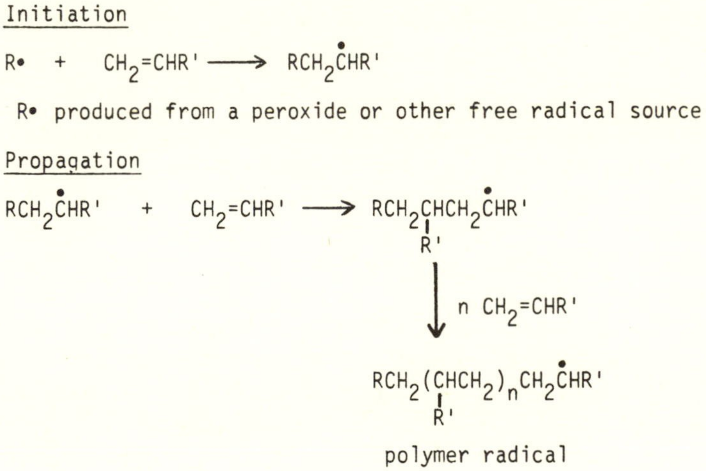

Initiation

$$R\bullet \ + \ CH_2=CHR' \longrightarrow RCH_2\overset{\bullet}{C}HR'$$

R• produced from a peroxide or other free radical source

Propagation

$$RCH_2\overset{\bullet}{C}HR' \ + \ CH_2=CHR' \longrightarrow RCH_2\underset{R'}{CH}CH_2\overset{\bullet}{C}HR'$$

$$\downarrow \ n \ CH_2=CHR'$$

$$RCH_2(\underset{R'}{CH}CH_2)_nCH_2\overset{\bullet}{C}HR'$$

polymer radical

Growth of the polymer was *terminated* by one of several possible events. Two radicals might combine or otherwise react with each other, an infrequent process, or the polymer radical might pick up an atom from some other component of the reaction mixture, generate a new free radical from it, and itself decay to a stable polymer molecule. The new radical formed by this *chain transfer* might initiate the formation of a new, growing polymer, or might be used up in some other way.[17] The polymerization of dienes such as chloroprene follows a similar course, and copolymerization of mixtures of monomers can be treated by extensions of Flory's scheme.

Termination

R"• = polymer radical

R"• + R"• $\longrightarrow$ R"—R"   (combination)

R"• + R"'X $\longrightarrow$ R"X + R"'•   (chain transfer)

R"'• + CH$_2$=CHR' $\longrightarrow$ R"'CH$_2$ĊHR'   (initiation)

$\downarrow$ further polymerization

The polymerization of isobutene (discussed earlier) was catalyzed by protons or Lewis acids such as BF$_3$ to form very high molecular weight polymers (butyl rubber).[18] Their formation can be explained by an analogous set of initiation, propagation, and termination reactions. In this acid-catalyzed polymerization, the reactive intermediates are not free radicals but carbonium ions; they may undergo rearrangement of the carbon skeleton of the Whitmore type. The butyl polymers produced by this process are of molecular weight up to 300,000. They are more stable than natural rubber because they contain almost no carbon-carbon double bonds.[18]

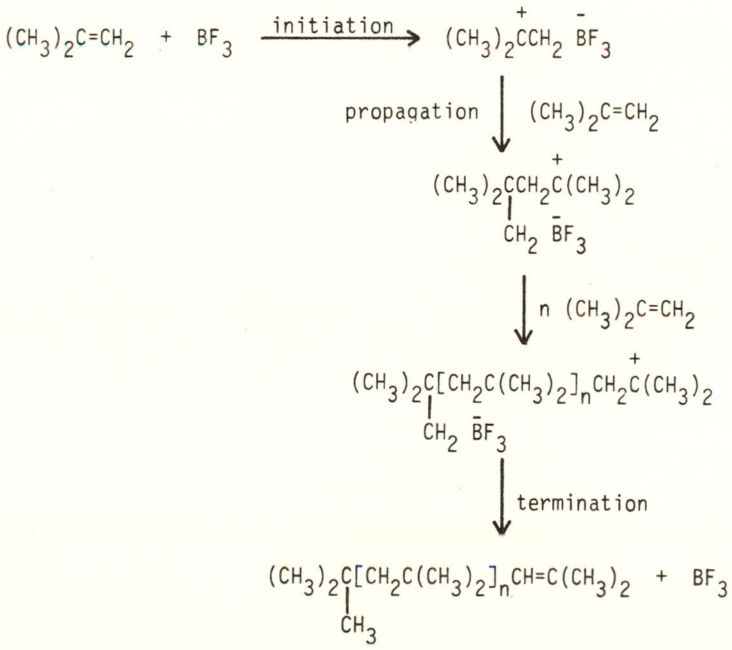

$$(CH_3)_2C=CH_2 \ + \ BF_3 \ \xrightarrow{\text{initiation}} \ (CH_3)_2\overset{+}{C}CH_2 \ \overset{-}{BF_3}$$

propagation $\downarrow$ (CH$_3$)$_2$C=CH$_2$

$$(CH_3)_2\overset{+}{C}CH_2C(CH_3)_2$$
$$\underset{CH_2 \ \overset{-}{B}F_3}{|}$$

$\downarrow$ n (CH$_3$)$_2$C=CH$_2$

$$(CH_3)_2\overset{+}{C}[CH_2C(CH_3)_2]_nCH_2C(CH_3)_2$$
$$\underset{CH_2 \ \overset{-}{B}F_3}{|}$$

$\downarrow$ termination

$$(CH_3)_2\overset{|}{C}[CH_2C(CH_3)_2]_nCH=C(CH_3)_2 \ + \ BF_3$$
$$\underset{CH_3}{|}$$

Although much work had been done in European universities on addition polymerization, particularly by Staudinger, the first American university chemist to carry out a major study on addition polymerization was C. S. Marvel at Illinois, starting in the middle 1930s and continuing for over fifty years at Illinois and Arizona. Marvel's work was important not only for his significant results, but also because he introduced generations of research students, postdoctorates, and academic colleagues to the chemistry of synthetic polymers. Marvel and his collaborators exercised an important influence on the increasing industrial research on polymers. His example made polymer chemistry a popular field in many university laboratories, and his experience was particularly valuable in the World War II research on synthetic rubber.

Marvel's work prior to World War II was mainly on the structure of addition polymers. Clearly a monomer, $CH_2=CHR$, can polymerize in three ways: to form head-to-tail, head-to-head structures, or a mixture of the two modes. The repeating units (neglecting end groups) of the first two types are as shown.

$$CH_2=CHR \longrightarrow \{CH_2CHCH_2CH\}_n \quad or \quad \{CH_2CHCHCH_2\}_n$$
$$\qquad\qquad\qquad\quad R \quad R \qquad\qquad\qquad R \; R$$

$$\text{Head-to-tail} \qquad\qquad\qquad \text{Head-to-head}$$

For a specific example, if vinyl acetate, $CH_2=CHOCOCH_3$, is polymerized and hydrolyzed, the head-to-tail form would be $(CH_2CHOHCH_2CHOH)_n$, the head-to-head would be $(CH_2CHOHCHOHCH_2)_n$, and a random polymer would have some units of both types. Marvel and C. E. Denoon found[19] that periodic acid, HIO_4, a reagent specific for the $CHOHCHOH$ group, showed no reaction with polyvinyl alcohol, indicating a head-to-tail structure. Marvel found that other addition polymers had a similar arrangement. Staudinger had concluded from pyrolysis products that polystyrene was also a head-to-tail structure.[20] These regularities of structure are undoubtedly caused by resonance stabilization of the unpaired electron on the carbon carrying the substituent group R', compared to the isomeric radical.

The simplest monomer, ethylene, was converted into a useful polymer, polyethylene, by British chemists, using high pressure and temperature. Later Karl Ziegler in Germany developed a low pressure method using organometallic compounds as catalysts. As an example of the way in which important advances frequently have a long history it is interesting to note that two American chemists had come upon indications of both these techniques by 1930.

Marvel and M.E.P. Friedrich, in a study with a very different objective,[21] had observed that ethylene was converted by ethyl or butyllithium to a saturated compound. A white solid was isolated in one run, but the reaction was not pursued in detail.

$$CH_2{=}CH_2 \quad \xrightarrow{\quad C_2H_5Li \quad} \quad {+}CH_2CH_2{+}_n$$

J. B. Conant, using the high pressure equipment of the physicist P. W. Bridgman, observed that several unsaturated hydrocarbons, including isoprene, styrene, and indene, were converted to solid materials at 12,000 atmospheres pressure. The product from the high pressure polymerization of isoprene was a transparent, tough, rubber-like material. Conant did not study ethylene itself; he notes ruefully in his autobiography that he might have obtained a commercially useful polyethylene if he had carried the experiments further using ethylene.[22]

LITERATURE CITED

1. On the development of Bakelite, see L. H. Baekeland, Ind. Eng. Chem., **1**, 149 (1909).
2. Williams Haynes, *American Chemical Industry*, New York, 1949, Vol. VI, p. 187. *Chemistry in the Economy*, American Chemical Society, Washington, 1973, pp. 153-160.
3. Biography by Roger Adams, *Biog. Mem. Nat. Acad. Scis.*, **20**, 293 (1939). An account of a hiking trip in Austria in 1936 by Adams, Carothers, and J. F. Norris is summarized in our biography of Adams. The Illinois Archives contain much Adams-Carothers correspondence.
4. H. Mark and B. S. Whitby, Eds., *Collected Papers of Wallace Hume Carothers on High Polymeric Substances*, New York, 1940. This volume contains an abbreviated version of Adams' biography cited above. J. K. Smith and D. A. Hounshell, "Wallace H. Carothers and Fundamental Research at Du Pont", *Science*, **229**, 436 (1985).
5. H. Staudinger et al., Ber., **62**, 241 (1929); polymerization of styrene as a chain reaction: Staudinger, ibid., **53**, 1031 (1920).
6. K. H. Meyer and H. Mark, Ber., **61**, 593 (1928) and later papers. On rubber, H. Staudinger and J. Fritschi, Helv. Chim. Acta, 5, 785 (1922) and many later papers. The enormous literature on polymers is summarized by K. H. Meyer, *Natural and Synthetic High Polymers*, New York, 1950.
7. J. M. Gulland et al., JCS, 30 (1944), and refs. therein to earlier studies by Caspersson and others.
8. W. H. Carothers, Chem. Rev. **8**, 353 (1931); in *Collected Papers*, it is numbered p. 81-140. Our quotations use the latter page numbers. (8a) p. 93; (8b) p. 99; (8c) pp. 101-102; (8d) p. 102, note; (8e) p. 103; (8f) p. 109; (8g) p. 119; (8h) pp. 129-136.
9. Carothers and G. H. Dorough, JACS, **52**, 711 (1930).
10. Some of Carothers' work on polyamides is described, with J. W. Hill, ibid., **54**, 1566 (1932).
11. Carothers et al., ibid., **53**, 4203 (1931); H. Staudinger, ref. 5; H. S. Taylor and W. H. Jones, ibid., **52**, 1111 (1930).
12. J. A. Nieuwland, W. S. Calcott et al., ibid., **53**, 4197 (1931); U. S. Patent 1,811,959 (1931); Chem. Abstr., **25**, 4892 (1931). We are indebted to Dr. R. M. Joyce for the account of the discovery of the rubbery material from the distillation fraction.
13. By Staudinger in 1920 (ref. 5), among others. Chain reactions in addition polymerization of ethylene were supported by evidence of H. S. Taylor and A. A. Vernon, JACS, **53**, 2527 (1931); O. K. Rice and D. V. Sickman, ibid., **57**, 1384 (1935).
14. P. J. Flory, ibid., **59**, 241 (1937); this paper discusses earlier attempts in the 1930s to formulate a satisfactory kinetic scheme for free-radical addition polymerization.
15. H. B. Starkweather and G. B. Taylor, ibid., **52**, 4708 (1930).

16. Interview with Flory by D. W. Ridgway, JCE, **54**, 341 (1977).

17. Further detailed discussion of the free-radical processes in polymerization is given in Flory's monograph and by C. Walling, *Free Radicals in Solution*, New York, 1957.

18. R. M. Thomas et al., JACS, **62**, 276 (1940), a joint paper summarizing many patents on butyl rubber from the Standard Oil Development Co. of N.J. and the I. G. Farbenindustrie laboratories in Germany. More details on the valuable industrial properties of butyl rubber are given in Esso papers by R. W. Thomas, W. J. Sparks, P. K. Frolich et al., Ind. Eng. Chem., **32**, 299, 731, 1283 (1940).

19. C. S. Marvel and C. E. Denoon, Jr., JACS, **60**, 1045 (1938); for general review of synthetic polymers, see C. S. Marvel and E. C. Horning in *Organic Chemistry*, Henry Gilman, Ed., 2nd ed., 1943, p. 701-773.

20. H. Staudinger and A. Steinhofer, Ann., **517**, 35 (1935).

21. M. E. P. Friedrich and C. S. Marvel, JACS, **52**, 376 (1930).

22. P. W. Bridgman and J. B. Conant, Proc. Nat. Acad. Scis. U.S., **15**, 680 (1929); Conant et al., JACS, **52**, 1659 (1930); **54**, 628 (1932); J. B. Conant, *My Several Lives*, pp. 62-63; Bridgman and Conant, U. S. Patent 1,952,116 (to E. I. Du Pont), Chem. Abstr., **28**, 3415 (1934). The preparation of technically useful polyethylene by polymerization of ethylene at 100°-400° and 500-3000 atm. is disclosed by E. W. Fawcett et al., U.S. Patent 2,153,533 (to Imperial Chemical Industries), Chem. Abstr., **33**, 5552 (1939).

21
The Instrumental Revolution, 1930-1955

As we have seen, organic chemistry correlated successfully an enormous number of facts on the extraordinarily simple structural theory of Kekulé, van't Hoff, and their contemporaries in the 19th century. However, the laboratory techniques and equipment available for research in organic chemistry did not change strikingly from 1875 to 1930. Laboratories were somewhat more convenient, with (usually) better ventilation, electricity, steam, and gas available by 1930. Purification was limited essentially to crystallization for solids, and to distillation through very inefficient columns (especially under reduced pressure) for liquids. Materials which could neither be obtained crystalline nor distilled were usually not investigated further. Organic chemists developed great facility in the art of obtaining crystals from unpromising mixtures and oils. Criteria of purity were limited in general to the sharpness of the melting point for solids, and the refractive index, density, and optical rotation (if the compound was optically active) for liquids. Determination of empirical formulas by elemental analysis and molecular weight measurements were necessary preliminaries to any structural determination. The analytical procedures of Liebig, Dumas, and other 19th century chemists had been improved in sensitivity and accuracy but remained fundamentally the same.

The advance of knowledge in any field of science is closely related to the experimental techniques and instrumentation available. The whole history of western science illustrates this, from the invention of the astronomical telescope by Galileo and of the microscope by Leewenhooek to the recent development of radio astronomy and of nuclear magnetic resonance. Verification, modification, and rejection of new theoretical ideas are dependent on having experimental methods for testing them. Furthermore, new theories are frequently suggested by observations made with more searching experimental equipment.

New instruments are the product of both research and the development of technology based on research. The paucity of experimental methods and instrumentation available to organic chemists began to change with increasing speed in the 1930s. The Instrumental Revolution, as we have called it,[1] gave the organic chemist by 1955 powerful methods for purification and structure proof; some of these derived from techniques known for many years, which now became routinely available with relatively trouble-free instruments. New discoveries in physics underlay the development of some new instruments such as nuclear magnetic resonance spectrometers (NMR). The new instrumentation allowed completely new ways of determining structures and of studying reactions of organic compounds. In days or weeks chemists could now solve problems which would have required months or years of careful work, or which might have been completely insoluble by classical methods. Many reaction rates and mechanisms, impossible to study before, now became accessible to attack.

In addition the Instrumental Revolution had significant repercussions in other fields, such as biochemistry and molecular biology. The new techniques were first applied to a wide variety of relatively simple organic compounds, and the knowledge thus gained illuminated the more formidable problems of structure and chemical behavior of materials of fundamental importance in the life sciences. Much of the spectacular development of detailed knowledge about the processes occurring in living systems, which marked the decades following World War II, was due to this general sequence of events. Thus structural organic chemistry, with its limited number of basic principles, acts as a bridge between quantum mechanics which explains chemical bonding, and the complexities of living systems. It is striking that these complexities can be represented in terms derived from the behavior of relatively simple organic molecules.[2]

Most of the techniques and instruments comprising the Instrumental Revolution are the subjects of individual monographs, large numbers of scientific publications, and scientific journals devoted solely to them. Our purpose will be to indicate the effects of the revolution in experimental organic chemistry and the kinds of new information made available, as well as the evolution of fresh ideas based on the new techniques.

A natural result of the Instrumental Revolution was that many of the new techniques were challenging and fascinating enough to attract some chemists to specialize exclusively on one of them. The research worker and teacher in organic chemistry who did not wish to devote all his energy to mass spectrometry, for instance, nevertheless had to know enough about the application and limitation of the method to use it intelligently and to instruct students about it.

We shall discuss some of the key elements in the Instrumental Revolution with illustrations of their experimental and theoretical impact.

Measurements of acidity (pH) were early done by indicators which changed color at a definite pH range, and also by electrical conductivity in aqueous solution. W. Mansfield Clark in U.S. government laboratories developed a reliable hydrogen electrode for pH determinations and a series of indicators and buffer solutions for setting up standards.[3] This was a fundamental contribution to the study of reactions catalyzed by acids and bases, including enzymatic processes. The hydrogen electrode could be used only in aqueous solutions which did not poison the platinized electrode. The principle of the glass electrode was known to Helmholtz in the 1880s, but successful application required a reproducible way of making the electrodes and a suitable sensitive amplifier for the minute current produced. MacInnes and Dole[4] at the Rockefeller Institute improved the preparation of glass electrodes around

1930. Arnold Beckman at Caltech designed an electronic amplifier in 1934 and established a company (later known as Beckman Instruments) to manufacture his pH meters in the late 1930s. This was an extremely successful venture.[5]

The Beckman pH meter became a standard and indispensable instrument in all laboratories where pH measurements were important for acid-base titrations, for kinetic measurements where acid was produced or consumed, and for measurements in buffer solutions. It revolutionized the measurement and control of pH.

The Beckman Laboratories played such a large role in the Instrumental Revolution that a discussion of their moving spirit, Arnold Beckman, is desirable. Beckman (1900- , B.S. 1922, Illinois) received his Ph.D. from Caltech in 1928 with R. D. Dickinson and was a faculty member there from 1926 to 1939. The extraordinary success of his firm and instruments (both financial and scientific) made him a scientific leader and a major benefactor of Caltech and of the University of Illinois. He was chairman of the board of trustees of Caltech for many years. He developed even more useful instruments after the success of the pH meter.[5]

Electronic spectra are caused by the absorption of light in the ultraviolet (UV) or in the visible region which is accompanied by raising an electron from the ground state to an excited state. As discussed earlier, W. N. Hartley (1846-1913) in England measured the electronic spectra of many organic compounds[6] starting in 1878. Such spectra appeared in the literature in increasing numbers from 1900. Even in the late 1930s, however, obtaining a UV or visible spectrum was a laborious and specialized process, involving photographing and developing spectroscopic plates, and wavelength and intensity calibrations.

The DC amplifier developed by Beckman for the pH meter could be used to measure the current generated by photocells. His group therefore designed a spectrophotometer, introduced in 1941, which made routine the determination of electronic spectra from about 220 nm through the visible region.[5,7] This was the Beckman DU, a sturdy instrument that could be used for quantitative analysis at selected wavelengths. It was therefore useful for measuring rates of chemical reactions where a change in absorption occurred in the UV or visible region. Because many enzymatic reactions could be set up so that there was a change in absorption intensity as the reactions progressed, the "Beckman" became indispensable in biochemical laboratories and was a workhorse instrument for several decades. It was a contributing factor in the rapid development of biochemistry after 1945, and when it was introduced was particularly useful for the quantitative analysis of vitamin A and D concentrates. The spectrophotometric analysis of these materials was far more rapid than the laborious animal assay methods previously required.

The Beckman was frequently called simply "the DU"; the letters came from the fact that it was the fourth (D) model built of the ultraviolet (U) instrument. It required hand setting for each wavelength and hand recording of the optical density reading. In a few years, automatic recording instruments were available which gave a tracing of the optical density of a complete spectrum. Despite the convenience of the automatic recording instruments, the Beckman DU continued to be sold by the thousands, its simplicity, relatively low price (about $1,000 for the first models), and versatility making it probably the commonest major laboratory instrument. It made possible extensive correlations of electronic absorption and structure for organic compounds, in addition to its analytical applications. The Beckman instrument allowed rapid recognition of conjugated systems and many aromatic systems. Thus the electronic spectrum could distinguish between $CH_3CH=CHCH_2COCH_3$ and $CH_3CH_2CH=CHCOCH_3$, or between a benzene and a naphthalene derivative, for example.

The infrared (IR) spectrum is at wavelengths beyond the visible region. The range useful for organic compounds (1000-15,000 nm) corresponds to energies involved in the stretching or bending of bonds in organic and many inorganic compounds. Because the energies are smaller than those of the UV or visible region, it is experimentally a more difficult spectrum to study, requiring very sensitive galvanometers.

Nevertheless, considerable work had been done before 1900 in parts of this spectral region, and in 1905 W. W. Coblentz (1873-1962) of the U.S. Bureau of Standards reported measurements for 130 organic compounds in the 1-15 micron region. In many cases he was able to recognize absorption bands characteristic of organic groups, which varied little from one compound to another.[8] This laborious and monumental study led to numerous publications of IR spectra of organic compounds in the following years,[9] and some serviceable data accumulated. In the late 1930s a key series of papers by O. R. Wulf, U. Liddel, and their collaborators, appropriately at the Bureau of Standards, demonstrated the significance of IR spectroscopy in organic chemistry,[10] particularly the detection of hydrogen bonding. Key studies were also published by W. H. Rodebush at Illinois.[11] The IR spectra were highly useful for determining the presence of functional groups, both saturated and unsaturated, and were normally conclusive for determining identity, or lack of it, of two samples. This was because there were bands characteristic of vibrations of the molecule as a whole (skeletal or "fingerprint" vibrations) in addition to the bands due to the vibration of specific groups. The IR spectra were particularly helpful in demonstrating hydrogen bonding, involving usually a bond between an OH or NH and some atom with unshared electrons, such as oxygen or nitrogen. It was possible from an IR spectrum to decide at once whether a $C\equiv N$, a $C=O$, a $COOCH_3$, or a $C=OOC=O$ (anhydride) group was present, to mention only a few examples.

The IR spectrometers before 1945 were specialized research instruments, difficult to build and laborious to operate; the development during World War II of sensitive and rugged detecting devices made possible greatly improved and simplified IR spectrometers. Thus after 1945 commercial spectrometers became available, which could be used more or less routinely for study and characterization of organic compounds. The first commercial models available were "single beam" instruments, meaning that two curves had to be run, one for the background absorption, of which water was the most troublesome, and the other for the sample; the ratio of absorption at each point had to be determined and the absorption curve plotted manually. This was particularly laborious in the region where most C=O groups showed very useful absorption, but where the water vapor in the air showed a series of sharp bands.

By 1955, however, commercial "double beam" instruments were available which compensated automatically for the background absorption and gave a direct tracing of the absorption due to the sample. These spectrometers could also be used quantitatively for analysis of mixtures and for following rates of reactions. The determination of the IR spectrum became more common, as well as usually much more informative, than the determination of the melting point. The IR spectrometer became an indispensable adjunct of research, so that its malfunction in a laboratory severely curtailed research work.

The Raman technique,[12] discovered by the Indian physicist C. V. Raman in 1928, gives in general the same sort of information as IR spectroscopy about atomic and molecular vibrations, although the physical basis of the Raman effect is quite different. The Raman radiation is weak, but it has the great advantage that it is applicable to aqueous solutions, and by 1939 a large amount of Raman frequency data had been accumulated.[13] The advent of the laser gave a very intense light source and the laser Raman later became a highly useful instrument.

A striking triumph of Raman spectroscopy was the demonstration[14] by J. T. Edsall in 1936 that α-amino acids in aqueous solution actually existed almost completely as the dipolar ions; he observed vibrations corresponding to the charged ions, but none corresponding to the uncharged structure.

$$\underset{\underset{NH_2}{|}}{RCHCOOH} \quad \rightleftharpoons \quad \underset{\underset{\overset{+}{N}H_3}{|}}{RCHCOO^-}$$

Nuclear magnetic resonance spectroscopy (NMR), although it had some similarity to earlier work on the behavior of atomic beams in magnetic fields, was discovered simultaneously in 1946 by K. Bloch at Stanford and E. M. Purcell at Harvard;[15] the Nobel Prize was awarded to the two groups.

John D. Roberts at MIT and later at Caltech was one of the first to demonstrate its extraordinary value to organic chemistry, both for structure determination and for measurements of rates of rapid organic reactions. The availability of commercial instruments from the Varian Company and others from about 1955 allowed the rapid exploration and utilization of the method in many laboratories.

The NMR effect is due to the fact that nuclei of atoms such as ^1H, ^{13}C, ^{15}N, ^{19}F, and ^{31}P (but not ^{12}C and ^{16}O) can exist in two (or more) magnetic energy spin levels. The transitions lie in the energy range of radio waves, i.e., very small energies, and can be observed in a strong magnetic field. The importance of this effect for organic compounds is that the transition energy (or the corresponding frequency of transition) depends on the environment of a given atom, ^1H (proton) for example, in an organic molecule. The nearby atoms shield a given nucleus from the imposed external magnetic field and hence affect its transition frequency to a degree that is characteristic of the environment of the atom in question. In addition, the possible energy levels of a given nucleus can interact and be split by the spins of adjacent nuclei. This means that an NMR spectrum can be extremely informative about structure, to a degree far greater than other spectroscopic methods. For example, one could distinguish instantly between isomers, such as $(CH_3)_3C(CH_2)_2COOH$, $(CH_3)_2CH(CH_2)_3COOH$, and $CH_3(CH_2)_5COOH$, by the NMR spectrum, where chemical or other spectroscopic methods would be either laborious or inconclusive. The other elements mentioned, ^{13}C, ^{15}N, ^{19}F, and ^{31}P, may be equally informative. The field has become a major area of research with its own journals, reviews, and constantly improving instrumentation. Relative configurations can be assigned conclusively in many cases by NMR.

Furthermore, NMR measurements at several temperatures give the potential energy barrier to rotation around single bonds, as in the case of the C-N bond of dimethylformamide, $HCON(CH_3)_2$. This is a great advance, because it allows measurements of interconversion of conformational isomers and of rates of reactions, including shifts of hydrogen atoms, which cannot be obtained by other methods. Altogether, NMR is probably the most powerful single instrumental technique so far applied to organic chemistry. The ^{13}C and ^{15}N NMR developed rapidly in the later years, and NMR spectroscopy continued to expand its usefulness in many directions, almost without limit. A proton (^1H) NMR spectrum and an IR spectrum are taken routinely on new compounds and frequently allow a complete assignment of structure. Reactions can be run in NMR tubes, and the formation of intermediates or of final products can be followed directly. Thus optimum conditions for a synthetic reaction can be determined using a small amount of material.

The development of magnets of very high field, depending on superconductivity at liquid helium temperatures, made the technique even more useful and yielded new kinds of information available from no other technique. Roberts' comment[15] has been verified in ever increasing measure.

Electron spin resonance (ESR) utilizes the behavior of an unpaired electron in a strong magnetic field;[16] it is thus applicable to the detection and measurement of free radicals, even in very low concentrations. The method is the best way of detecting unpaired electrons; it demonstrates, for example, that free radicals actually are intermediates in addition polymerization. Methods of "spin labeling", in which a free-radical label is put on a compound of biological interest, afford information about the structure of large molecules. The use of ESR in showing the free-radical nature of Koelsch's hydrocarbon has already been mentioned. Although not as widely applicable to organic chemistry as NMR spectroscopy, ESR measurements are useful in following free-radical reactions in solution.

Mass spectroscopy, introduced by F. W. Aston at Cambridge and by A. J. Dempster at Chicago about 1918-1919, allows the detection of positive ions differing in their mass to charge ratios.[17] Its first application demonstrated the existence and relative abundance of isotopes, and it was gradually applied to organic compounds for characterization and quantitative analysis of mixtures of hydrocarbons. An early paper from Berkeley in 1931 by H. R. Stewart and A. R. Olson describes the mass spectrometry of saturated hydrocarbons.[18]

By 1950 commercial spectrometers were available which were effective for relatively low mass numbers; as in other cases, the range and accuracy of the instruments were improved rapidly, with satisfactory resolving power for ions above 1000. High resolution instruments, with an accuracy of one part in a million, became available, and with a computer connection allowed determination of the empirical formulas of many fragments from a complex material.[19]

The mass spectrometer produces positive ions from organic compounds by bombarding them in the gas phase with electrons. The resulting ions generally include the molecular ion (the parent compound minus an electron) plus a varied assortment of smaller ions formed by decomposition through several paths of the parent compound. This "cracking pattern" is usually characteristic of the compound, and if the structure of the original is unknown frequently gives much structural information. Klaus Biemann at MIT and Carl Djerassi at Stanford applied mass spectroscopy to organic compounds with particular success. Mass spectroscopy requires only very small amounts of material $(10^{-6}$ g) and can be used to determine the amino acid sequence in polypeptides and in studying other natural products, such as steroids, alkaloids, and polyisoprenes, some of which are available in very minute amounts. Like NMR, mass spectroscopy developed satellite techniques which increased its usefulness enormously, and also like NMR, it became an independent field of research. Frequently NMR and mass spectral examination of an unknown sample allows complete assignment of structure.

The observation that the wavelength of X-rays was of the order of magnitude of interatomic distances suggested that the diffraction of X-rays by crystals should give information about the arrangement of molecules in the crystal and of atoms in the molecule.[20] This proved to be a very fruitful idea, and the method was developed particularly by the Braggs at Manchester and Cambridge,[21] by Kathleen Lonsdale, Dorothy Crowfoot Hodgkin, and Rosalind Franklin in England, and by American workers. The latter included David Harker and Isabella and Jerome Karle, of the Naval Research Laboratory in Washington. The number of women who made important contributions to this field is notable.

The first organic compound to have its structure established by X-ray diffraction was hexamethylenetetramine by R. G. Dickinson and A. L. Raymond at Caltech[22] in 1923. Linus Pauling at Caltech became an expert on X-ray crystallography of complex inorganic compounds, and the laboratory there was one of the leading American centers for this technique. The determination of the structure of hexabromocyclohexane and its all-equatorial bromines by Bilicke and Dickinson at Caltech was important in establishing the chair form for cyclohexane derivatives, as mentioned in an earlier essay.

David Harker (1906-   , B.S. 1928, Berkeley; Ph.D. 1931, Caltech) taught at Johns Hopkins, Brooklyn Polytechnic Institute, and from 1959 was director of biophysics and of the Center for Crystallography at the Roswell Park Center in Buffalo. He made fundamental contributions to the application of X-ray to structural problems.

Up to 1945, two facts limited the usefulness of X-ray crystallography of organic compounds. First, the compound must possess a high degree of symmetry, otherwise the interpretation of the complex diffraction patterns required extremely laborious calculations. Therefore complete structure analyses for only a few unsymmetrical compounds, such as cholesteryl iodide,[23] had been done. Second, an atom of high scattering power (high atomic number) was required to furnish a point of reference; such atoms as Br, I, Fe, and other metallic elements were the usual ones employed. Elements such as C, H, N, and O did not have enough scattering power to be effective by themselves.

Electronic computers in conjunction with X-ray diffraction cameras removed the first limitation; the necessary extended calculations could be done rapidly by computer and thus structures of compounds of low symmetry could be determined rapidly. Hodgkin's[24] assignment of structure to vitamin B-12 in 1955 was a landmark in structure determination by X-ray. The limited amount of material available precluded conventional structure determination by chemical degradation. Later improvements in technique made the heavy element unnecessary, so that X-ray crystallography became the method of choice for structure determination if single crystals of the proper type could be obtained, a condition by no means routinely met.

Pauling's work on protein structure was supported by X-ray work at Caltech by R. B. Corey.[25] The postulation of the double-helix structure for DNA by J. D. Watson and F. H. C. Crick relied in part on the X-ray work on DNA by Rosalind Franklin at Kings College, London.[26] Franklin's work receives proper attention in the book by Anne Sayre.[27] X-ray crystallography eventually worked out complete structures for some enzymes.[28]

A modification of the X-ray technique allowed the determination of the absolute configuration of tartaric acid; this showed that Emil Fischer's assignment was correct and allowed the assignment of absolute configurations to complex natural products and synthetic optically active compounds.[29]

In an earlier essay we have described the painstaking and successful determination of the structures of the cardiac drugs or "heart poisons" by Walter A. Jacobs at the Rockefeller Institute before the resources brought by the Instrumental Revolution were available.

How would a chemist after the Instrumental Revolution attack the structural problem of the cardiac drugs, strophanthidin and digitoxigenin, isolated from a plant source and of unknown structure?[1] The purity would be demonstrated by chromatographic techniques, perhaps aided by NMR and mass spectroscopy. The mass spectrometer would give the molecular weight, or information from which this might be derived, and probably the empirical formula. The IR spectrum would show the presence of OH groups, and of the CHO in strophanthidin, and of the lactone C=O in both strophanthidin and digitoxigenin. The size of the lactone ring would be shown by the C=O frequency in the IR, as compared to that of the saturated lactone. The ^{1}H NMR spectrum would show the protons on the lactone ring, confirm the structure of the lactone ring as shown, and show the presence of two angular CH_3 groups in digitoxigenin. In strophanthidin, one CH_3 group would appear, and the CHO group would be shown conclusively by the band for the CHO proton. The ^{13}C NMR probably would allow the complete carbon skeleton to be established. By this time, the modern chemist would suspect that he or she was dealing with steroid derivatives and could write very nearly correct structures. In conjunction with the foregoing information, an X-ray crystallographic determination would show the complete three-dimensional structure of the molecule.

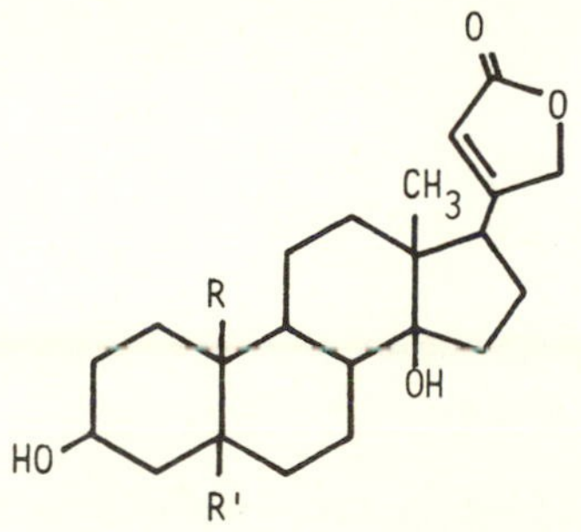

Strophanthidin, R = CHO; R' = OH

Digitoxigenin, R = CH_3; R' = H

This picture of the work of a modern chemist is the way the structure for batrachotoxin was determined by Bernhard Witkop of the National Institutes of Health, and the Karles of the Naval Research Laboratory. This compound, a steroid derivative of more complicated structure than strophanthidin or digitoxigenin obtained in minute quantities from Colombian frogs, is extraordinarily toxic and has very interesting neuropharmacological properties.[30]

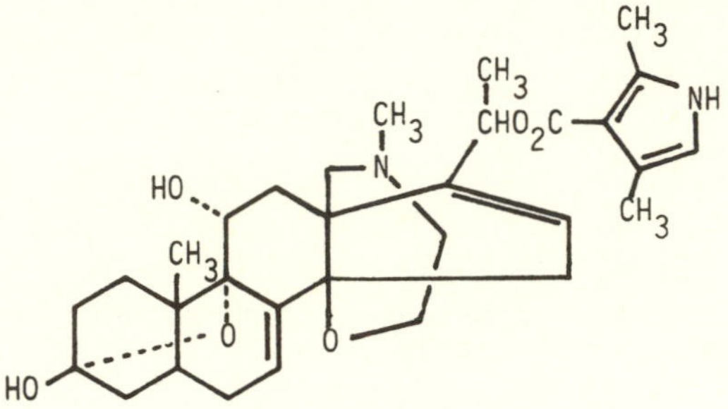

Batrachotoxin

Bernhard Witkop (1917- , Ph. D. 1940, Munich, with H. Wieland) was fellow and instructor at Harvard from 1947 to 1951. From 1951 he has been at the National Institutes of Health, since 1955 as chief of the chemical laboratory, Institute of Arthritis, Metabolic, and Digestive Diseases. His work has included a variety of problems, many on the borderline between chemistry and biochemistry, including natural products, non-enzymatic cleavage of proteins, carcinogenesis, and biochemical mechanisms.

Reasonable purity is usually necessary for spectroscopic or X-ray methods to be effective. Accompanying the Instrumental Revolution were equally important discoveries of new powerful methods of separation and purification, which differ as greatly in effectiveness from the previous crystallization and inefficient distillation procedures as an NMR spectrometer does from a melting point bath. Most of these involved partition processes, i.e., separation of components of a mixture by distribution between a solid and solvent, between two immiscible liquids, or between a gas stream and a liquid.

This involves in the simplest case a solution of a mixture running down a column of solid adsorbent, such as alumina or silica. Components of the mixture in a successful separation will be adsorbed on the solid with differing degrees of firmness and can be eluted from the column at differing rates, depending on the compounds and nature of the eluting solvent. This procedure is generally credited to the Russian botanist M. Tswett[31] about 1905, used by him for the separation of natural pigments from leaves, mainly the chlorophylls and accompanying compounds. The procedure was ignored for many years, was rediscovered in Europe by R. Kuhn and others in 1931, and was soon widely used for separating complex mixtures.[32] By proper choice of solvents and adsorbents based on extended experience, sharp separations of closely related compounds are possible. The coating of very small glass spheres with adsorbents and the use of high pressures to secure a reasonable rate of liquid flow greatly increases the effectiveness of column chromatography; high pressure liquid chromatography (HPLC) has become a very efficient separation method.

An important extension of column chromatography is the use of ion exchange resins as solid adsorbents. These are solid polymers carrying attached acid or basic groups, usually sulfonic acids or quaternary ammonium groups.[33] These columns are of great value in separating acidic or basic compounds, usually using an aqueous buffer to carry the mixture. Modern work on natural products, such as nucleotides or other phosphates, and on separation of polypeptides and amino acids would be impossible without the ion exchange columns. The Moore-Stein quantitative method for analysis of mixtures of amino acids, indispensable for advances in protein and polypeptide chemistry, uses ion exchange columns and buffers of varying pH for separation.[34] Moore-Stein amino acid analyzers are made commercially available by Beckman Instruments.

The ion exchange resins, essentially providing water-insoluble acids and bases, superseded older very involved methods of purifying water-soluble acids and bases, particularly those like amino acids and naturally occurring phosphates which are only slightly soluble in organic solvents. Ion exchange resins resulted from advances in the chemistry and technology of polymers and thus represent yet another example of progress in one area furthering research in a seemingly totally unrelated field.

Ion exchange columns developed during World War II revolutionized the chemistry of the rare earth elements because these elements, hitherto separated only partially by long series of recrystallizations, were now relatively readily available in pure form.[35] Some of them became very important as infrared sensitive materials, the basis for the IR sensitive optical instruments for seeing in the dark. They were also highly useful in expanding the scale of NMR spectra by forming complexes with functional groups in organic compounds.

Two English biochemists, A. J. P. Martin and R. L. M. Synge,[36] showed in 1941 that excellent qualitative and semiquantitative separations of mixtures of amino acids and of other classes of compounds were possible by spotting samples on filter paper and allowing solvents to ascend or descend the vertical sheets of filter paper by capillary action, like ink spreading on a blotter. The components of the mixture moved on the paper at different rates, depending on their solubility in the solvents used and the firmness of their adsorption to the cellulose of the paper. This relatively simple yet highly effective technique played a key role after 1945. It allowed, for example, separation of compounds labeled with radioactive atoms such as radiocarbon or tritium, and it was the technical basis for M. Calvin's classical studies on photosynthesis.[37] It was also indispensable in F. Sanger's equally classical work on the structure of insulin,[38] and in many other investigations, such as the separation of mixtures of steroids. Mixtures of polycyclic aromatic compounds are separable by paper chromatography.[39]

A very valuable modification of paper chromatography is thin-layer chromatography (TLC), in which a thin layer of silical gel or some other adsorbent on a glass plate or cellulose acetate film replaces filter paper. Application of solvents to the plates causes samples to migrate at rates depending on the compounds, solvents, and adsorbents used. This is a somewhat more versatile system than paper chromatography and has become exceedingly helpful for examining reaction mixtures in organic chemistry.[40] The equipment and procedures are simple and widely applicable.

The techniques of affinity chromatography, gel filtration, and electrophoresis (migration of material in an electric field at controlled pH) are used principally for purification of high molecular weight compounds from natural sources.

P. J. W. Debye (1884-1966), the great physicist and physical chemist,[41] developed light scattering into an important method of studying particle size and shape in solutions of polymers,[42] and hence of determining molecular weights. It initiated a new era in the study of synthetic polymers particularly.

The Svedberg (1884-1971) in Sweden developed a mechanical ultracentifuge which produced valuable information about the molecular weights of proteins. Svedberg's ultracentrifuge was a massive and extremely expensive instrument. The physicist Jesse W. Beams (1898-1977) of the University of Virginia devised a relatively simple, air-driven ultracentrifuge,[43] which was manufactured by Beckman Instruments. It became an indispensable tool in the characterization and separation of proteins and other natural high molecular weight materials. Beams' instruments were developed to rotate up to a million times per second at which speeds the rotor disintegrated. Laboratory models give a separation force of many thousand times the force of gravity.

Vapor phase chromatography or gas liquid chromatography (VPC or GLC) is a relatively simple yet extraordinarily useful procedure for qualitative and quantitative examination of mixtures which can be carried in the gas phase. It was proposed in 1941 by Martin and Synge and demonstrated experimentally by Martin and A. T. James in 1952.[44] The mixture to be examined is volatilized in a stream of inert gas (helium or argon) and passed over a finely divided solid with a surface coating of a very high boiling liquid or wax in a metal tube at elevated temperatures. The components of the mixture are partitioned between the gas stream and the packing, depending on the conditions, and the emergence of each component from the end of the column is detected by measuring the change in thermal conductivity of the gas stream or by other more sensitive methods.

This technique, with relatively simple instrumentation, revolutionized the analysis and purification of mixtures which can be put into the vapor phase. It is superior in efficiency to fractional distillation by many powers of ten, can be applied quantitatively to very small samples, and also allows collection of purified samples which can then be examined by NMR or IR. Combined with a mass spectrometer, this technique can separate complex mixtures and usually identify the components. The combination of VPC and mass spectroscopy allows the detection and quantitative determination of various toxic pollutants in water or soil. Compounds of very low volatility, such as steroids, can be separated and identified. It can be applied to non-volatile compounds, like sugars or amino acids, by converting them to more volatile derivatives. Many reaction mixtures of apparently straightforward organic reactions, of which only the main components had been identified (or misidentified) by earlier techniques, were found on VPC analysis to be far more complex than previously thought. Frequently all of the material can be accounted for by VPC and the structures established. Organic compounds for use as reagents or for synthesis can be checked for purity very readily.

Like other instruments, VPC units were constantly improved, made more efficient and sensitive, and made applicable to wider classes of compounds. However, even in its first relatively simple incarnation, VPC was adopted rapidly, almost universally, and used with significant results.

It had been known since the 19th century that the magnitude, and in some cases the sign, of the optical rotation of optically active compounds varied with the wave length of light used to measure it; this was called optical rotatory dispersion (ORD). This variation had been measured in many cases, using the readily available monochromatic light from sodium lamps, mercury arcs, and a few other sources. Empirical equations had been devised to fit these rotatory dispersion curves,[45] such as the Drude equations. It was known that if the optical rotation was measured at a wave length close to an absorption band of the sample, a very striking change in the magnitude of the rotation occurred, called the Cotton effect.[46]

With development of spectropolarimeters that could furnish monochromatic polarized light from the visible into the ultraviolet range, where electronic absorption of groups like aromatic rings and carbonyl groups occurred, the effects of change in configuration on the Cotton effect could be measured. Carl Djerassi at Stanford investigated this problem and developed a method for determining absolute configurations.[47] A theoretical basis for the assignment of absolute configurations by the sign of the Cotton effect was worked out.[47] The circular dichroism measurements were similar but required more elaborate instruments.

Carl Djerassi (1923- , A.B. 1942, Kenyon College; Ph.D. 1945, Wisconsin with A. L. Wilds) was a leader in the Syntex developments in steroids and oral contraceptives. He combined an extremely productive academic research career at Wayne State and Stanford with active direction of work on products of applied biological importance in the research laboratories of Syntex and Zoekon. His energy, versatility, imagination, and concern about the social aspects of science make him a unique figure in American chemistry.

The determination of Cotton effects was done better, as instrumentation became available, by circular dichroism (CD) measurements. These showed the difference in absorption of a chromophore, such as $C=O$, of one circular polarized beam compared to absorption of the other.[47]

The ORD and CD measurements became useful not only in correlating absolute configurations, but also in dealing with the interaction of biologically active compounds with their sites of action in living systems. The ORD-CD curves also showed the coiling of polypeptide or nucleic acid chains into helical arrangements and allowed the process of coiling or uncoiling to be followed.[47]

H. C. Urey followed his discovery of deuterium (^2H or D) in 1931 by concentration of ^{13}C, ^{15}N, and ^{18}O; these tracers allowed definitive experiments to be done on many problems of mechanism of organic reactions. Their application to the mechanism of biochemical processes was started by Rudolph Schoenheimer (1898-1941), David Rittenberg (1906-1970), and others at Columbia before World War II.[48] The development of cyclotrons in the 1930s produced some radioactive elements useful as tracers, such as ^{126}I used by Ingold in studying the Walden Inversion.

The discovery of nuclear fission and the work on the atomic bomb during World War II made new radioactive isotopes available for reaction mechanism studies. Radioactive carbon, phosphorus, tritium, sulfur, iodine, and a few others opened up new paths of investigation, which were rapidly developed by many workers to elucidate reaction pathways in organic reactions and in processes of living systems. The combination of isotopic labeling with the instrumental and separation procedures described above led to an increase in knowledge of chemistry and biochemistry which can only be described as revolutionary.

The details of enzymatic processes were probed, as in the classic experiments of F. H. Westheimer and B. Vennesland on NAD, a coenzyme involved in many biochemical oxidation-reduction processes.[49] The pathway of conversion of sugars in living systems to acetate and then to polyisoprenoids including the steroids, such as cholesterol, was almost completely worked out by Konrad Bloch in this country and by F. Lynen in Germany and J. W. Cornforth in England.[50] Equally impressive were the studies on photosynthesis by Calvin mentioned earlier, and the biogenesis of the nucleic acids by many workers. These problems required a broad knowledge of the organic chemistry involved and the development of methods of synthesis of labeled compounds, as well as the determination of the position of the label in the products formed by enzymatic processes. The results of these and similar studies in other fields not only explained some biochemical processes, but also allowed more rational attempts to synthesize useful medicinal agents for diseases such as cancer and many others.

The decades following 1945 witnessed a Biochemical Revolution which changed the whole approach to investigating the fundamental processes in living systems. No matter that each advance in knowledge opened up new and awe-inspiring complexities, the new knowledge already gained pointed the way to further significant experiments. Whether life processes would ultimately be completely understood was beside the point; the quest for new knowledge generated its own dynamics, and there was no question that broader understanding would result. The social implications of scientific discovery came to be an increasingly serious subject for discussions by scientists and concerned non-scientists.

In spite of the enormous effect of the Instrumental Revolution in increasing the rate of research, in broadening the scope of problems accessible to study, and in providing new theoretical insight, it had some disadvantages. One drawback was the expense of new instruments. Mass spectrometers, NMR spectrometers, X-ray equipment, computers, and most of the other new techniques required large sums to purchase and operate. Furthermore, constant improvements in performance (and cost) meant that instruments became obsolete rapidly, a drawback if one wished to do research that was competitive with the leading laboratories. The operation and maintenance of the later more elaborate instruments required highly skilled people, who normally could look after only one or two types of instruments on a full time basis. Such experts were difficult to obtain, train, and retain.

The large sums needed to purchase and operate such instruments were available mainly from government agencies. Faculty members, therefore, had to spend much time in writing proposals and reports for research grants, and also in reviewing research proposals from other laboratories. The accounting procedures for large grants required additional personnel and increased the complexities of laboratory operations. Some attempts made to set up "national centers" of instrumentation, to which samples could be sent for instrumental examination, were not very satisfactory. There was an inevitable delay in getting results back, and no distant center, however well-equipped and conscientiously run, is as satisfactory as having instrumentation close at hand. It may be necessary to make repeated measurements to obtain the best results, and this is impossible or very difficult with a distant laboratory.

The training of undergraduate or masters' students in research in liberal arts colleges became more difficult as instrumentation became more expensive. For example, Kohler at Bryn Mawr from 1895 to 1912 could do as good research as that going on at any university in the country, and his students were just as well trained. After the 1930s it became increasingly difficult to do competitive research in small colleges and small universities, although many of them did very well.

Another disadvantage, particularly for graduate students and some young Ph.D.s, was a more subtle one. The increased reliance on instrumental results in research led to a decreasing degree of immediacy of experience in research results. A student may receive an NMR spectrum of a new compound run on an instrument for which he or she knows little of the physical basis, and of whose actual operation he knows practically nothing, a common yet unsatisfactory occurrence. Is the instrument working properly, and has it been properly calibrated? The basis of practically all measurements or separations by the instrument is empirical; it depends on a comparison, either implicit or explicit, of the behavior of compounds of known structure and properties with the behavior of the sample in question.

Inexperienced workers tend to develop a "black box" mentality. What the instrument (or more likely the chart from it) seems to tell them, they accept uncritically. They do not inquire if the instrument has been shown to be operating properly. More seriously, they do not question the breadth or the relevance of the underlying information on which generalizations from the instrumental results are based. All instrumental results must be examined critically with a reasonable knowledge of the applicability and limits of the technique in question. Those responsible for training research workers must inculcate a certain scepticism of instrumental results. The presumption is that the obvious result from the instrument means what it purports to mean, but this presumption must be questioned critically before one accepts it as probably true. Regardless of the instrumental information available, it is still desirable to obtain an elementary analysis on all samples.

The sweeping changes already wrought in organic chemistry by the Instrumental Revolution presage continuing invention and modification of physical methods for investigating molecular structure. Their application to the growing and in some ways still unperceived problems of the amazing science of the compounds of carbon promises still greater revelations.

LITERATURE CITED

1. D. S. Tarbell, Chem. and Eng. News, Centennial Edition, April 6, 1976, p. 117. The name has been adopted by others.
2. For an impressive account of the relation between research in biology and the physical sciences, see Ernst Mayr, *The Growth of Biological Thought*, Cambridge, 1982.
3. W. Mansfield Clark, *The Determination of Hydrogen Ions*, Baltimore, 1920, later eds., 1923, 1928.
4. D. A. MacInnes and M. Dole, JACS, **52**, 29 (1930); Ind. Eng. Chem., Anal., Ed., **1**, 57 (1929); M. Dole, *The Glass Electrode*, New York, 1941.
5. D. S. Tarbell and A. T. Tarbell, JCE, **57**, 133 (1980).
6. W. N. Hartley, and A. K. Huntington, Proc. Roy. Soc., **28**, 233 (1879) and many later papers in JCS.
7. History of the Beckman DU instruments: A. O. Beckman et al., Anal. Chem., **49**, 280A (1977); introduction of the DU, H. H. Cary and A. O. Beckman, J. Opt. Soc. Am., **31**, 682 (1941).

8. W. W. Coblentz, *Investigations of Infra-Red Spectra*, Carnegie Publication 35, Carnegie Institution of Washington, 1905. This monograph gives ref. to earlier IR measurements. Biography of Coblentz by W. F. Meggers, *Biog. Mem. Nat. Acad. Scis.*, **39**, 55 (1967).

9. F. K. Bell, JACS, **47**, 2192, 2811, 3039 (1925) and later papers; R. H. Gillette and F. Daniels, ibid., **58**, 1139, 1143 (1936).

10. S. B. Hendricks, O. R. Wulf, G. E. Hilbert and U. Liddel, ibid., **58**, 1991 (1936) and numerous earlier papers, mainly showing H-bonding by IR.

11. A. M. Buswell, V. Deitz and W. H. Rodebush, J. Chem. Phys., **5**, 84 (1937) and later papers; M. D. Joesten and L. J. Schaad, *Hydrogen Bonding*, New York, 1974.

12. C. V. Raman and K. S. Krishnan, Nature, **121**, 501 (1928). Raman (1888-1970) was knighted and received the Nobel Prize in 1930 for his discovery.

13. James H. Hibben, *The Raman Effect and its Chemical Applications*, New York, 1939.

14. J. T. Edsall, J. Chem. Phys., **4**, 1 (1936); **5**, 225 (1937).

15. E. M. Purcell, H. C. Torrey and R. V. Pound, Phys. Rev., **69**, 37 (1946); F. Block, W. W. Hansen and M. E. Packard, ibid., **69**, 127 (1946); J. D. Roberts, *Nuclear Magnetic Resonance*, New York, 1959. Roberts says (p. 1) the development of NMR "is now recognized as one of the most important events of the last fifty years for the advancement of organic chemistry". See also L. M. Jackman and S. Sternhell, *Application of Nuclear Magnetic Resonance Spectroscopy in Organic Chemistry*, New York, 1969.

16. P. Ayscough, *Electron Spin Resonance in Chemistry*, London, 1967.

17. F. W. Aston, *Mass-Spectra and Isotopes*, London, 1933; A. J. Dempster, Phys. Rev., **11**, 316 (1918); F. W. Aston, Phil. Mag., **38**, 707 (1919). Aston (1877-1945) received the Nobel prize in 1922.

18. H. R. Stewart and A. R. Olson, JACS, **53**, 1236 (1931).

19. K. Biemann, *Mass Spectrometry*, New York, 1962; F. W. McLafferty, *Interpretation of Mass Spectra*, New York, 1966; J. H. Beynon, *Mass Spectroscopy and Its Applications to Organic Chemistry*, Amsterdam, 1960.

20. M. Laue, W. Friedrich and P. Knipping, Sitzungsber. Bayer. Akad. Wiss. (Math. phys. Klasse) 1912, p. 303; Ann. der Physik, **41**, 971 (1913).

21. W. H. Bragg and W. L. Bragg, *X Rays and Crystal Structure*, London, 1915, and later eds; R. W. G. Wyckoff, *The Structure of Crystals*, New York, 1924, and later eds.

22. R. G. Dickinson and A. L. Raymond, JACS, **45**, 22 (1923).

23. C. H. Carlisle and Dorothy Crowfoot Hodgkin, Proc. Roy. Soc., **184** A, 64 (1945).

24. D. C. Hodgkin et al., Nature, **176**, 325 (1955).

25. L. Pauling and R. B. Corey, JACS, **72**, 5349 (1950); L. Pauling, R. B. Corey and H. R. Branson, Proc. Nat. Acad. Scis., **37**, 205 (1951).

26. Rosalind E. Franklin and R. G. Gosling, Nature, **171**, 740 (1953) and later papers.

27. Anne Sayre, *Rosalind Franklin and DNA*, New York, 1975.

28. G. Kartha, Accts. Chem. Res., **1**, 374 (1968) and refs. therein.

29. B. M. Bijvoet et al., Nature, **168**, 271 (1951).

30. Based on ref. 1; for batrachotoxin, T. Tokuyama, J. Daly, B. Witkop, I. L. Karle and J. Karle, JACS, **90**, 1917 (1968); **91**, 3931 (1969).

31. M. Twsett, Ber.Deut. bot. Ges., **24**, 235 (1906); trans. by H. H. Strain and J. Sherma, JCE, **44**, 238 (1967); R. Kuhn and E. Lederer, Ber., **64**, 1349 (1931) and many later papers. Excellent discussion, with refs., for many of the partition procedures are in R. V. Dilts, *Analytical Chemistry*, New York, 1974.

32. L. Zechmeister and L. Cholnoky, *Principles and Practice of Chromatography*, tr. by A. L. Bacharach and F. A. Robinson, New York, 1943, and later publications by Zechmeister. *Modern Practice of Liquid Chromatography*, J. J. Kirkland, Ed., New York, 1971, gives summaries of various chromatographic methods.

33. B. A. Adams and E. L. Holmes, J. Soc. Chem. Ind., **54**, 1T (1935); F. Helfferich, *Ion Exchange*, New York, 1962.

34. W. H. Stein and S. Moore, JBC, **176**, 338 (1948); **178**, 79 (1949); D. H. Spackman, W. H. Stein and S. Moore, Anal. Chem., **30**, 1190 (1958); P. B. Hamilton, ibid., **35**, 2055 (1963).

35. F. H. Spedding et al., JACS, **76**, 2545 (1954).

36. A. J. P. Martin and R. L. M. Synge, Biochem. J., **35**, 1358 (1941) and many later papers; *Paper Chromatography*, I. M. Hais and K. Macek, Eds., New York, 1963.

37. J. A. Bassham, M. Calvin et al., JACS, **76**, 1760 (1954) and earlier papers; Bassham and Calvin, *The Path of Carbon in Photosynthesis*, Englewood Cliffs, 1957.

38. F. Sanger and E. O. P. Thompson, Biochem. J., **53**, 353, 366 (1953); F. Sanger et al., ibid., **60**, 535, 541, 556 (1955).

39. D. S. Tarbell et al., JACS, **77**, 767 (1955); **78**, 2228 (1956); JOC, **24**, 887 (1959).

40. J. E. Meinhard and N. F. Hall, Anal. Chem., **21**, 185 (1949); J. K. Kirchner, J. M. Miller and C. J. Keller, ibid., **23**, 420 (1951); E. V. Truter, *Thin Film Chromatography*, New York, 1963.

41. On Debye, J. W. Williams, *Biog. Mem. Nat. Acad. Scis.*, **46**, 23 (1975); W. O. Baker, Proc. R. A. Welch Foundation, XX, 154 (1976).

42. P. J. W. Debye, J. Phys. Colloid Chem., **51**, 18 (1947) and other papers.

43. On Beams see Walter Gordy, *Biog. Mem. Nat. Acad. Scis.*, **54**, 2 (1983).

44. A. J. P. Martin and A. T. James, Biochem. J., **50**, 679 (1952); S. Dal Nogare and R. S. Juvet, Jr., *Gas-Liquid Chromatography*, New York, 1962.

45. For review, P. A. Levene and A. Rothen, in H. Gilman's *Organic Chemistry*, 1st ed., 1938, Vol. 2, chapter 21.

46. A. Cotton, Compt. rend., **120**, 989 (1895).

47. Carl Djerassi, *Optical Rotatory Dispersion*, New York, 1960; the octant rule for establishing absolute configurations: W. Moffitt, R. B. Woodward, A. Moscowitz, W. Klyne and C. Djerassi, JACS, **83**, 4013 (1961); Pierre Crabbé, *Optical Rotatory Dispersion and Circular Dichroism in Organic Chemistry*, San Francisco, 1965.

48. See R. E. Kohler, Jr., in *Historical Studies in the Physical Sciences*, **8**, 257 (1977) for an account of Schoenheimer's tracer work in biochemistry. Schoenheimer's own account of his early experiments, *The Dynamic State of Body Constituents*, Cambridge, 1942, is a classic of biochemistry.

49. F. H. Westheimer, Birgit Vennesland et al., JACS **73**, 2403 (1953), **75**, 5018(1953).

50. R. B. Clayton, Quart. Rev., **19**, 169, 201 (1965).

22

Chemistry During and After
World War II: an Overview

World War II fundamentally altered chemistry in all branches. Research and teaching changed markedly from 1939 to 1946.

In June 1940 an executive order of President Roosevelt mobilized American scientists for war-related problems. Under this order Vannevar Bush, J. B. Conant, Roger Adams, and others formed the National Defense Research Committee which later expanded into the Office of Scientific Research and Development (OSRD). This office gradually drew most of the active research chemists of the country into some type of war-related investigations. Military problems, such as explosives and war gases, and medical problems, such as the synthesis of possible antimalarial compounds, fell within the OSRD. Important studies outside this office included those on penicillin, synthetic rubber, and the vast Manhattan (atomic bomb) project, which were also controlled by the federal government.

The result was a drastic curtailment of normal, peacetime work in both university and industrial laboratories. The military draft took most undergraduate and some graduate students; those left worked on war research. Faculty members turned almost universally to war work in their own institutions or were posted to other laboratories. Many, perhaps most, of the projects in chemistry had solid scientific content, but the results understandably were not published until after the war, if then. In addition, many colleges and universities ran large, time-consuming teaching programs for young service personnel. Roger Adams, always a strong proponent of fundamental chemical research, wrote a former student in 1943:[1] "After another year I feel our graduate schools will be so depleted that there will be very little opportunity for anything but more or less haphazard investigations. It has been my hope that enough students could be retained so that the continuity of scientific research in this country could be maintained throughout the war period but I doubt whether this will be the case."

In spite of the dislocating effects of the war on normal research activities, there were some favorable consequences. Many of the striking changes of the Instrumental Revolution resulted in part from technological advances based on war research. Adequate funds for projects allowed both the purchase of good technical equipment and the hiring of postdoctorate or very experienced research assistants. Furthermore, the success of the wartime research tended to emphasize the national benefits from scientific research and to suggest that the national interest would be served by government support of good scientific investigations in peacetime.

As a result good chemists could obtain outside support after 1945 from a variety of sources, both public and private. The National Institutes of Health, the National Science Foundation (established in 1950), the Petroleum Research Fund administered by the American Chemical Society, the Sloan Foundation, and the Research Corporation provided the principal financial aid for organic chemists. Chemical industry also backed university research substantially through fellowships and unrestricted grants. Not that this was without cost to the research supervisors; the disadvantages in time and effort expended in enlarging operations and maintaining highly specialized equipment have been detailed previously.

New reactions and reagents (such as the metal hydrides) enlarged the scope of synthetic work, just as the Instrumental Revolution completely altered the methods of determining structures of organic compounds. Organic chemists were able to tackle new problems involving the structures of natural products and to strive for their total synthesis. Stereospecific syntheses for molecules with numerous asymmetric centers multiplied, the leader in this field being R. B. Woodward. Conformational studies based on D. H. R. Barton's work were utilized in synthetic, mechanistic, and biochemical studies. Increased understanding of the mechanism of organic reactions and the availability of radioactive or isotopic tracer elements grew symbiotically. Many workers showed how living systems formed natural products, and research clarified the mechanism of enzymatic reactions. Molecular biology and the double helix were solidly based on the structural definition of the nucleic acids in England in the decade from 1945 to 1955. The discovery of penicillin's extraordinary clinical value in controlling bacterial infections led to a drive for analogs; chemical study revealed whole new ranges of antibiotics effective against pathogens unresponsive to the penicillins.

Biochemistry and medicinal chemistry thus benefited dramatically from these researches. The advent of antibiotic therapy was one of the great steps in medical history, and medicinal chemistry became less empirical and more amenable to rational planning, although it still retained a large empirical component. Organic chemistry was thus a fundamental, if largely unheralded, source of advancing knowledge in the life sciences.

Because of the interference of the wartime research, the important years in the 1940-1955 period were those from 1946 to 1955. Although we do not cover this period as thoroughly as earlier ones, these essays contain a comprehensive account of research on natural products and significant problems in physical organic chemistry.

The most notable characteristic of science after 1945 was its almost explosive growth. More support from public and private funds, more special instruments, and more research positions opening in industry, government laboratories, and academia primed the pump in organic chemistry as in other fields.

Membership in the American Chemical Society[2] grew from 24,000 in 1939 to 75,000 in 1955. The number[3] of Ph.D.s granted in chemistry increased from 464 in 1939 to 1010 in 1955. In the war years the number dropped, but rose rapidly to around 1000 by 1950. The fraction of these degrees in organic chemistry remained fairly constant over this period.[4] Graduate degree programs in chemistry increased rapidly, many schools finding it necessary to introduce them to attract and hold a strong faculty in chemistry. The volume of published research expanded much more than the larger number of Ph.D.s granted would indicate, mainly because of the many postdoctorates from this country and abroad paid by more generous funds available.

Unfortunately, specialization, or fragmentation as it was frequently put, accompanied growth. This not only meant greater numbers of specialized chemical research journals, research monographs, and annual reviews of various fields, but also floods of research papers. To stay abreast of the work in any field required the survey of a swelling amount of material published here and abroad. Since almost no one could command all of the current literature in organic chemistry, it was inevitable that reading became as specialized as research.

The *Journal of Organic Chemistry* was started in 1936, and after 1945 it became a highly respected publication. The *Journal of Polymer Science*, founded in 1946, not only became the standard periodical in the field, but was soon subdivided. After 1955 the number of specialized journals publishing organic chemistry multiplied rapidly. A few of the more important ones were *Tetrahedron* (1957), *Tetrahedron Letters* (1959), the *Journal of Medicinal Chemistry* (1959), and the *Journal of Heterocyclic Chemistry* (1964). The techniques brought in by the Instrumental Revolution soon produced their own journals; some of these were useful for organic chemistry, such as the *Journal of Chromatographic Science* (1962). The journals not only increased in size, but also in excellence. For example, the *Journal of Organic Chemistry* published 592 pages in 1936-1937 inclusive, 617 pages in 1945, 1816 pages in 1955, and 4459 pages in 1965. The *Journal of the American Chemical Society* was generally regarded after 1945 as the world's most important publication for chemistry.

Pervading scientific endeavor was the sense that research would pay off. In all research the pay off was in part the heightened stature of both researcher and laboratory, and larger, more productive, and better funded teams to direct. In industrial research the pay off was in marketable products and in meeting the competition of other companies with active research programs. In some industrial and some academic research, the pay off was in increased success in preventing or treating disease. If no immediate pay off was visible, the better understanding of organic chemistry was a good, long range investment, a firm basis for the advancement of science.

LITERATURE CITED

1. R. Adams to DST, March 19, 1943.

2. *A Century of Chemistry*, H. Skolnik and K. M. Reese, Eds., Washington, 1976, p. 456.

3. L. R. Harmon and H. Soldz, *Doctorate Production in United States Universities 1920-1962*, Washington, 1963, p. 10.

4. The figures for Illinois for percent of organic chemists in their total Ph.D. degrees are 47 (1930-40), 56 (1940-50), 47 (1950-60); *Roger Adams*, p. 176.

23

The Chemistry of
Natural Products, 1940-1955

The variety and extent of work on the structure and synthesis of natural products in 1939 in this country were shown in earlier essays. This field grew in significance and research activity over the following decades, aided by the Instrumental Revolution, the elaboration of new synthetic procedures and reagents, and the discovery of new classes of compounds, including antibiotics. Progress derived from, and in turn benefited, organic chemistry as a science, and knowledge of the mechanism of life processes increased. Contributions from government research laboratories, such as the National Institutes of Health, and from the laboratories of the pharmaceutical companies became increasingly prominent. The field of natural products continued to expand after our somewhat arbitrary time limit of 1955.

The types of chemical structures elucidated comprised a very wide range, particularly the increasing number of antibiotics (or mold metabolites), many of which had novel structural features and showed unusual reactions. The structures varied all the way from the relatively small, but chemically very confusing, penicillin series to large ring compounds (the macrolides), complex polysaccharides like the streptomycins, polypeptides like bacitracin, polyisoprenes like fumagillin, polycyclic compounds such as the tetracyclines, and many less easily classifiable structures.

Syntheses of most of the new natural products required control of the stereochemistry of new asymmetric centers as they were formed. Bachmann's equilenin synthesis involved only two asymmetric carbons, but cholesterol has eight and cortisone has six. Thus stereochemical requirements for total synthesis need both a knowledge of the stereochemical course of many reactions as well as careful choice of reactions and conditions to yield the desired configuration. The growth of conformational analysis, due to Barton's impetus and Winstein's important work on the stereochemistry of neighboring group reactions, contributed to the background of information necessary to devise stereospecific syntheses. These are only two of many possible examples of the interaction of physical organic and synthetic chemistry. New reagents, such as the metal hydride reducing agents, were discovered and applied to special cases. The synthesis of naturally occurring polypeptides like oxytocin demanded extensive knowledge of the reactions and stereochemistry of amino acid derivatives. More research on carbohydrate chemistry and its stereochemistry underlay the synthetic work on various antibiotics and the nucleotides.

In some cases, more numerous after 1955, biologically important materials could be isolated from natural sources in amounts too small to allow suitable experimental work on physiology and pharmacology. This was true of aldosterone, the active compound of the adrenal cortex, of the prostaglandins, and of many polypeptide hormones. For these materials total synthesis was a necessity and not an exercise in chemical virtuosity. The synthesis of potential antagonists to natural intermediates in biochemical processes was explored in great detail, especially in connection with cancer and malaria. Although some of this work was not, strictly speaking, "natural product" work, because the attempt was to obtain biologically useful "unnatural products", the character of the syntheses showed many of the features mentioned above. Furthermore, clinically useful compounds emerged from this work.

The total synthesis of steroids, spurred by Bachmann's success with equilenin in 1939 and particularly by the discovery of the antiinflammatory properties of cortisone, was pursued actively in this country and abroad. It is beyond the scope of our work to give details of these syntheses, but many outstanding chemists from many countries made striking contributions. To biochemistry, pharmacology, and clinical medicine this work was of great importance. On organic chemistry as a science, however, its impact was even greater; the art of organic synthesis and the understanding of physical organic chemistry were powerfully advanced.

Sir Robert Robinson (1886-1975) at Oxford had been developing methods for synthesizing steroids as early as the 1930s. Some of his reactions, which eventually led him to a total steroid synthesis, became a standard part of the steroid repertory to be successfully used with modifications by many other workers.

Cortisone was first obtained for clinical use by an elaborate transformation of bile acids, which were available in large supply from ox bile. These conversions, involving thirty to forty steps, were worked out at Merck by Max Tishler and Lewis H. Sarrett with many collaborators and put into industrial production.[1]

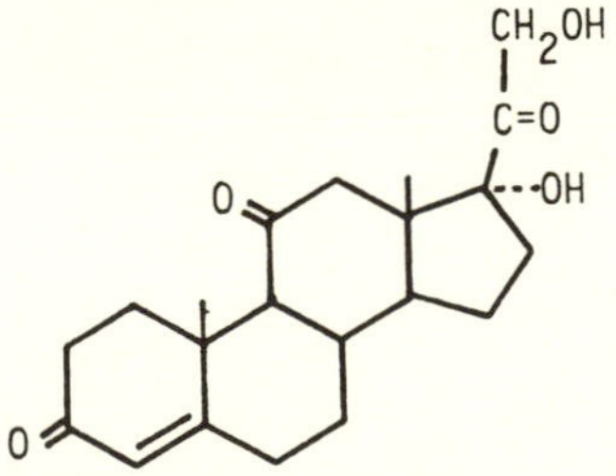

Cortisone

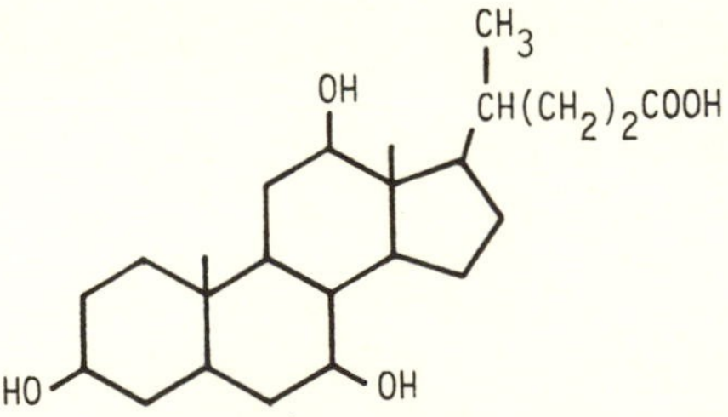

Cholic acid (a bile acid)

Cortisone was later produced much more readily from progesterone, which Syntex Corporation supplied in quantity from plant steroids prepared in their laboratories in Mexico. The key step was an oxidation of progesterone to the 11-hydroxy compound by a soil microorganism, achieved by a group under D. H. Peterson at the Upjohn laboratories in Kalamazoo.[2] The large number of trial experiments which led to this method was possible because of the development at Syntex of a paper chromatographic technique for separating mixtures of steroids. Conversion of 11-hydroxyprogesterone to cortisone was relatively simple, based on the work of the Merck group and others.

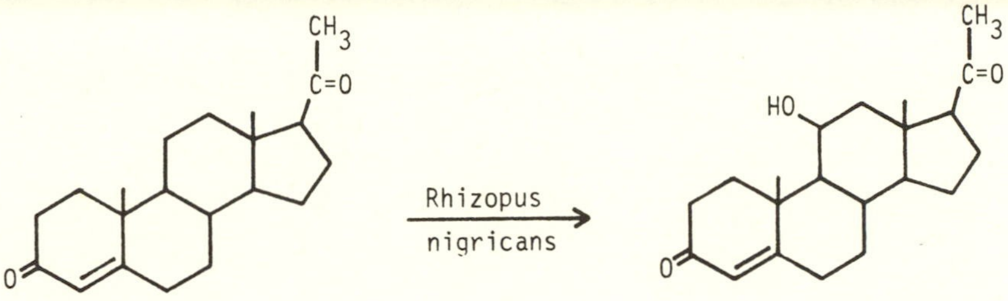

Progesterone

11-Hydroxyprogesterone
(80-90%)

A total synthesis of cortisone, and of other steroids as well, was reported[3] by R. B. Woodward in 1951. Woodward (1917-1978, B.S. 1936, Ph.D. 1937, both at MIT) was a member of the Harvard chemistry department from 1937 until his death. He was generally regarded as the outstanding organic chemist of his generation. His total syntheses of quinine[4] in 1944 with W. E. Doering established his reputation. Later syntheses of strychnine, chlorophyll, and other complex natural products, and his work in deducing the structure of the tetracyclines enhanced his prestige. He was the sole recipient of the Nobel Prize in 1965. The brilliance of his research was matched by the polish and uncompromising length of his lectures.

Another total synthesis of cortisone, one of the most elegant in all steroid chemistry, was due to Sarett[5] in 1952. L. H. Sarett (1917- , B.S. 1939, Northwestern; Ph.D. 1942, Princeton with E. S. Wallis), son of a well-known poet, started at Merck in 1942. His outstanding steroid work made him Tishler's successor as the president of Merck, Sharp and Dohme Research Laboratories in 1969.

Estrone, next to equilenin the simplest of the steroids stereochemically with only four asymmetric carbons, was synthesized totally by G. Anna and K. Miescher in Switzerland[6] in 1948. W. S. Johnson at Wisconsin developed two new estrone syntheses, the first[7] in 1950, and he later prepared all of the remaining racemates of the estrone structure.[8]

Johnson (1913- , B.A. 1936, Amherst; Ph.D. 1940, Harvard with L. F. Fieser) taught at Wisconsin from 1940 to 1960 and then became chairman at Stanford, where he built up one of the country's outstanding chemistry departments. His later work developed successful cyclizations of open-chain polyunsaturated compounds to form the steroid ring system.[9]

The realization that the use of cortisone in treating rheumatoid arthritis was attended by very serious side effects led to the synthesis of many variations on the cortisone structure in the hope of avoiding such effects. Some fluorinated derivatives appeared to be more satisfactory medicinal agents.

Of the host of other steroid syntheses, Barton's synthesis of the active principle of the adrenal cortex gland, aldosterone, is notable. Using an intermediate from Sarett's cortisone route, Barton introduced the aldehyde group by an original photochemical free-radical reaction and prepared 70g of dl-aldosterone.[10] European chemists had isolated the natural material and determined its structure with a few milligrams obtained from a ton of beef adrenal glands.

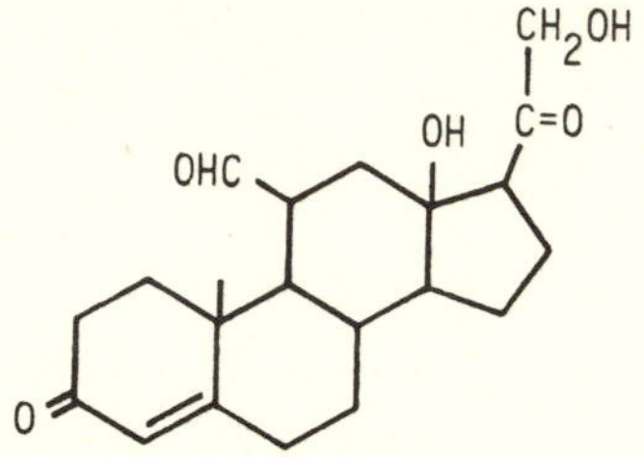

Aldosterone

The discovery of the antibacterial activity of penicillin was the most important event in the chemotherapy of infections diseases since the emergence of the sulfa drugs. In 1929 Alexander Fleming, a British pathologist, first observed penicillin (or its effects).[11] Serious work on its isolation and dramatic proof of its clinical value against bacterial infections were furnished by H. W. Florey, E. P. Abraham, and E. B. Chain at Oxford[12] in 1940-1941. Efforts to produce it in useful amounts by fermentation of the mold *Penicillin notatum* and to establish the drug's structure to make it more readily available by synthesis were pushed energetically in Britain and this country during World War II. This was an informal cooperative arrangement carried out in strict secrecy. The penicillin molecule was sensitive to acid and rearranged with great facility to a variety of products without antibacterial activity.

After very extensive studies by a large number of chemists from pharmaceutical companies and university laboratories, the structure was agreed on as shown, mainly on the basis of X-ray crystallography and IR measurements.[13] The structural evidence from chemical reactions and even from the first laboratory syntheses was not conclusive.

Penicillins, R = ArCH$_2$ or

an open-chain group

The industrial production of penicillin progressed rapidly from the use of milk bottles to the "deep culture" fermentation, using fermentation kettles of several thousand gallons capacity. This technological achievement resulted from the discovery in United States laboratories that "corn steep liquor", the solution in which cornstarch had been hydrolyzed to lower molecular weight sugars, provided an excellent and cheap medium for penicillin fermentation. Particularly valuable in this development was the experience of a few American firms, especially Chas. Pfizer and Co., which had been engaged for many years in large-scale production of compounds like citric acid by fermentation. Most other American pharmaceutical firms were inexperienced in large-scale fermentation processes.

As the penicillins became cheaper and more available, their use increased enormously, sometimes for trivial reasons, and the phenomenon of penicillin-resistant strains of bacteria became serious. It was found that free carboxylic acids added to the fermentation medium gave rise to new penicillins which contained the added acid in the RCO groups. Some of these "semi-synthetic" penicillins were useful against penicillin-resistant strains. Research verified that penicillin-resistant strains produce an enzyme, penicillinase, which catalyzes the hydrolysis of the four-membered β-lactam ring and thus deactivates the penicillin, because the β-lactam ring is necessary for antibacterial activity. J. Strominger delineated the mode of action of the penicillins, showing in elegant studies that they inhibit one stage in the biosynthesis of the bacterial cell wall.[14] The total chemical synthesis of a penicillin was achieved[15] by J. C. Sheehan at MIT, but in this case, as with many other antibiotics, the chemical synthesis could not compete with the fermentation process for clinically useful antibiotics.

John C. Sheehan (1915-   , B.S. 1937, Battle Creek College; Ph.D. 1941, Michigan with W. E. Bachmann), during wartime research with Bachmann, helped develop a practical synthesis for the explosive cyclonite RDX. After a period in the Merck research laboratories, he joined the MIT chemistry department with A. C. Cope, where he remained. In addition to work on penicillin and other antibiotics, he introduced the very useful carbodiimide reagent into peptide synthesis; this reagent had been used earlier by H. G. Khorana for the synthesis of phosphate ester linkage in nucleotides.

The spectacular clinical success of the penicillins set off a veritable gold-rush search for new antibiotics in pharmaceutical and some university laboratories. Many thousands of samples of earth and other materials which might contain molds were screened for antibacterial action, and many new mold metabolites were discovered, some of which became useful clinical agents. (Rediscovery of known antibiotics was a common occurrence in these screening programs.) Our account of these compounds will be limited to a few representatives of the leading chemical types.

Selman A. Waksman (1888-1973) had studied soil microorganisms for many years in the agriculture department of Rutgers. After examination of 10,000 samples of soil microbes for antibacterial agents, he isolated the antibiotics actinomycin and streptothricin (1940-1942). These had antibacterial activity, but were too toxic for human use. In 1943 Waksman and Schatz isolated streptomycin[16] from an actinomycetes which they named *Streptomyces griseus*.

Streptomycin is an active antibacterial agent against gram-negative bacteria, in contrast to penicillin, which is active only against gram-positive organisms. These terms refer to the staining characteristics of bacteria; the gram-negative class has a fatty, chemically resistant sheath, and the causal agent of human tuberculosis, *M. tuberculosis*, is gram-negative. Streptomycin and its dihydro derivative were found to be very effective clinically in human tuberculosis and have virtually revolutionized the treatment of this once dreaded disease.

Streptomycin is a trisaccharide, and its structure was established by the groups directed by K. Folkers at Merck, O. Wintersteiner at Squibb, H. E. Carter at Illinois, and M. L. Wolfrom at Ohio State. The elucidation of the structure, which was confirmed by synthesis of the three sugar units, was complicated and difficult.[17]

The antibiotic chloromycetin, discovered in 1947 in the Parke Davis research laboratories[18], had a nitro group and a dichloroacetyl group, novel chemical features at that time for a natural product. Because of its relatively simple chemical structure with only two asymmetric carbon atoms, it was soon synthesized.[19] It is a *wide spectrum* antibiotic, active against gram-positive and gram-negative organisms, some viruses, and several *Rickettsiae* (the class of organisms causing typhus, Rocky Mountain spotted fever, tularaemia, and some other diseases). Clinical experience has shown that use in humans may lead to blood disorders, and the antibiotic is used only in selected cases under careful supervision.

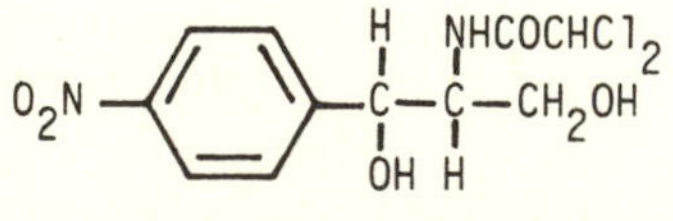

Chloromycetin

The neomycin group, also discovered by Waksman[20] from a strain of *S. fradiae*, was a mixture of two tetrasaccharides whose structures were completely established by 1962, mainly by impressive research of groups at Illinois led by K. L. Rinehart, Jr., and at Rutgers.[21] The complex polysaccharides, which contain nineteen asymmetric carbons, are representative of a sizable group of antibiotics. The neomycins, although antibacterial, are not clinically useful.

Many polypeptide antibiotics have been isolated and their structures established; few of them have shown much clinical usefulness. Many of them contain amino acids of the D-configurational series, in contrast to mammalian proteins and polypeptides which are of the L-configurational series. Why molds and microorganisms produce and utilize some D-amino acids is a mystery.

The tetracyclines, highly substituted, partially reduced products of the parent naphthacene, a tetracyclic aromatic hydrocarbon, were isolated by several pharmaceutical laboratories and found to be very valuable broad spectrum antibiotics. The extremely complicated structures were worked out primarily by the group at Chas. Pfizer and Co. with the collaboration of Woodward.[22] The relationships among the various tetracyclines became clear when the basic structure was known. H. Muxfeldt achieved a total synthesis of a tetracycline years later.[23]

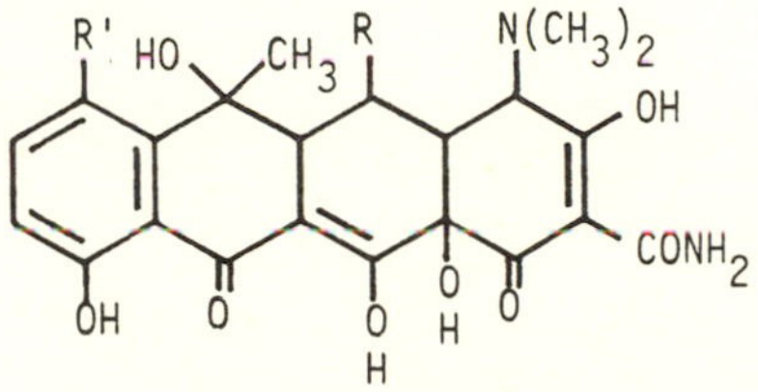

Tetracycline, R = R' = H
Chlorotetracycline, R = H, R' = Cl
Oxytetracycline, R = OH, R' = H

Numerous antibiotics containing a large lactone ring with several sugar residues attached, the *macrolides* (the name is due to Woodward), were isolated in many pharmaceutical laboratories. The broad spectrum compound erythromycin, very useful clinically, was produced by a strain of *S. erythreus* present in a sample of Philippine soil.[24] P. F. Wiley at Lilly established the structure,[25] and the configurations were shown by X-ray diffraction. Hydrolysis of the erythromycins yields two sugars, desosamine (A) and cladinose (B), and the aglycone erythronolide, containing the fourteen-membered lactrone ring. The preferred conformation of erythronolide in solution has been deduced[25] from NMR, both ^{1}H and ^{13}C. Other macrolides, with lactone rings from twelve to thirty-eight members, have been isolated.[26] Some of these contain a conjugated polyene system in the ring.

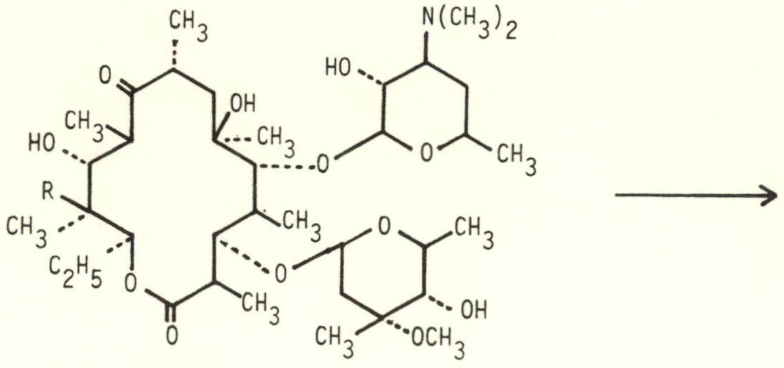

Erythromycin A, R = OH
Erythromycin B, R = H

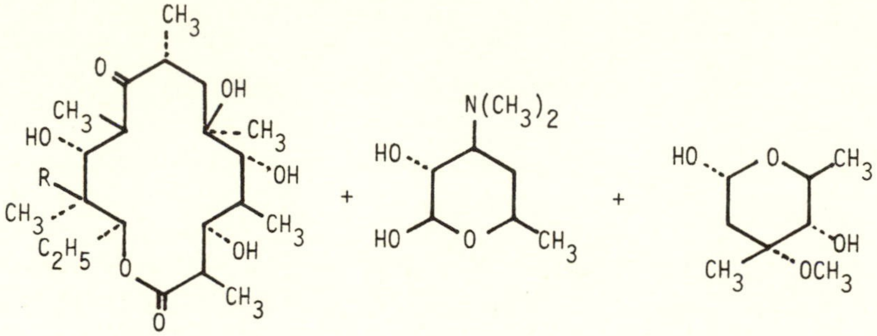

Erythronolide   Desosamine   Cladinose

Of antibiotics derived from sesquiterpenes (composed of three five-carbon isoprene units), fumagillin isolated from *Aspergillus fumigatus* at the Upjohn Laboratories may serve as example; it has amoebacidal and anticancer activity, but so far has not proved clinically useful.[27] D. S. Tarbell at Rochester established the structure by degradation; X-ray crystallography and, much later, a total synthesis by E. J. Corey at Harvard confirmed it. The antibiotic ovalicin, closely related to fumagillin, shows some antifungal properties.[28]

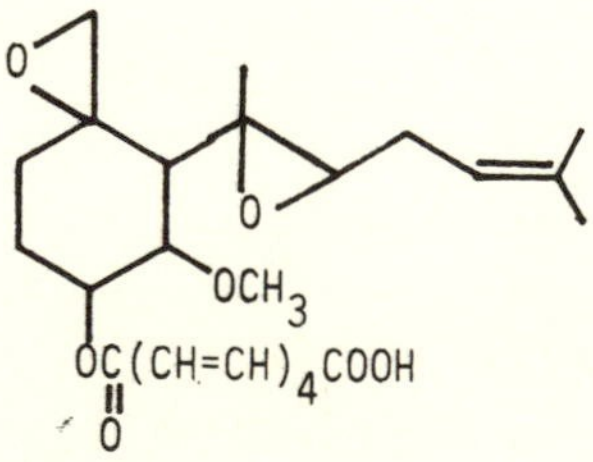

Fumagillin 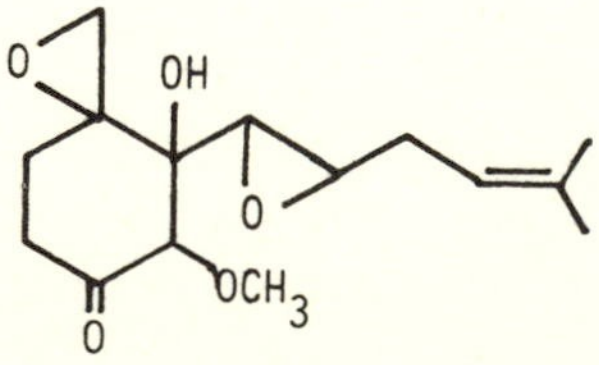   Ovalicin

During the 1940-1955 period, the chemical and instrumental foundations were laid for a dramatic advance in the knowledge of polypeptides and proteins. Determination of protein structure and synthesis of polypeptides were salient factors in the explosion of understanding of processes in living systems. (*Polypeptide* refers to compounds made up of amino acid groups connected by *peptide* bonds, -COHN-; *proteins* have the same structural feature, but the word implies a large number of amino acid units, hundreds or thousands.)

Information about protein-hydrolyzing (*proteolytic*) enzymes and the specific peptide linkages which they hydrolyzed increased greatly, starting with the work of Max Bergmann (particularly with J. S. Fruton) at the Rockefeller Institute in the 1930s. This work was possible because pure or purified enzymes were available, and pure simple polypeptides could be synthesized by methods due to Bergmann.[29]

Max Bergmann (1886-1944) took his Ph.D with Emil Fischer at Berlin in 1911, was director of the Kaiser Wilhelm Institut for Leather Research at Dresden from 1921 to 1934, and came to Rockefeller when Hitler seized power in Germany. His work at Rockefeller from 1934 to 1944 must be regarded as the starting point for modern polypeptide and protein chemistry.

Joseph F. Fruton (1912-   , B.A. 1931, Ph.D. 1934, with H. T. Clarke, both at Columbia) was a research worker with Bergmann at Rockefeller from 1934 to 1945, then member of the Yale department of biochemistry from 1945 until his retirement. His research has been on proteolytic enzymes. With his wife, Sofia Simmonds, he is author of an outstanding text in biochemistry. He has also published a book on the development of biochemistry.[30]

The determination of structure of a polypeptide or protein involves determination of the number and identity of the amino acids and then their order of linking (the *sequence*). A simple tripeptide is methionylalanylglycine (Met-Ala-Gly), whose structure can be rapidly determined by hydrolysis and quantitative analysis of the mixture of amino acids by the Moore-Stein method[31]; the sequence is determined by removing one amino acid residue at a time from the amino end by the Edman procedure.[32] Methods used in this simple case are applicable to very much larger polypeptides, particularly since both the Moore-Stein and Edman procedures can be carried out by automated instruments.

$$CH_3SCH_2CH_2\underset{\underset{NH_2}{|}}{C}H\overset{\overset{O}{\|}}{C}-NH\underset{\underset{CH_3}{|}}{C}H\overset{\overset{O}{\|}}{C}-NHCH_2COOH$$

methionine   alanine   glycine

The synthesis of naturally occuring polypeptides requires methods of forming the desired peptide linkages without forming undesired ones. This means that some amino groups must be protected from reaction by groups that can be removed later by methods that do not affect peptide linkages. Furthermore, the new peptide linkages must be formed under conditions that give no racemization of the asymmetric carbon present in all amino acids except glycine. This requirement is a stringent one because all naturally occurring polypeptides are optically active, and this problem was emphasized many years ago by H. D. Dakin. A specific example given illustrates the problem. Here COOR is a protecting group.

$$CH_3\overset{*}{C}HCOOR' \;+\; H_2N\overset{*}{C}HCOOR \quad\xrightarrow[\text{no racemization}]{-R'OH}$$

$$\underset{\substack{|\\NHCOOR}}{} \qquad \underset{\substack{|\\CH_2C_6H_5}}{}$$

$$CH_3\overset{*}{\overset{O}{C}}\!H\!C\!-\!NH\overset{*}{C}HCOOR' \quad\xrightarrow[\text{no racemization}]{\text{mild conditions}}\quad CH_3\overset{*}{\overset{O}{C}}\!H\!C\!-\!NH\overset{*}{C}HCOOR'$$

$$\underset{\substack{|\\NHCOOR}}{}\ \underset{\substack{|\\CH_2C_6H_5}}{} \qquad\qquad \underset{\substack{|\\NH_2}}{}\ \underset{\substack{|\\CH_2C_6H_5}}{}$$

Max Bergmann's great contribution was introduction of the *carbobenzoxy* protecting group, -$COOCH_2C_6H_5$, which can be removed by catalytic hydrogenation or by mild acid treatment.[33] Many other protecting groups have been used, one of the most useful being -$COOC(CH_3)_3$, *t*-BOC for *tert*-butoxycarbonyl.[34] The condensation to form the new -CONH- group may use a reactive COOR' group or an external condensing agent, such as a carbodiimide,[35] RN=C=NR.

The ingenuity of chemists and biochemists has been devoted for decades to the purification, elucidation of structure, and synthesis of polypeptides, and dramatic advances have been made in all these areas. We shall not go into further detail, except to say that the solid-phase synthetic procedure of R. B. Merrifield[36] at Rockefeller was a landmark, making possible automated syntheses of polypeptides of moderate length (twenty-five to fifty amino acid units) and of an enzyme, ribonuclease.

A decisive step in polypeptide chemistry was the purification, structure determination, and synthesis in 1953 by Vincent du Vigneaud and his group at Cornell Medical College, New York, of the hormone oxytocin and later vasopressin, present in the pituitary gland.[37] Oxytocin causes contractions of the uterus, and vasopressin raises the blood pressure. Early work in 1928 on purification of the hormones by Oliver Kamm[38] at Parke Davis has been mentioned previously. Du Vigneaud purified the hormones by L. C. Craig's countercurrent distribution technique[39] and determined the amino acid composition by the Moore-Stein procedure in its early form.[31] The amino acid sequence determination involved a number of degradative methods, including the Edman procedure. The synthesis was the first demonstration that a natural polypeptide hormone could be synthesized with no racemization and with all the qualitative and quantitative biological properties of the naturally occurring hormone, in this case a cyclic polypeptide containing nine amino acid residues. Du Vigneaud received the Nobel Prize in chemistry in 1955.

Following his work, syntheses of polypeptides, including enzymes, developed rapidly, and with these syntheses a new era in understanding biochemistry. The structure determination and synthesis of insulin followed in a few years. It became clear that polypeptides played key roles as hormones in functioning of the brain, and that further work would be richly rewarding in fundamental understanding of life processes and probably in useful clinical applications. The use of X-ray crystallography and mass spectroscopy in determining polypeptide structures has been mentioned above.

The alkaloids are complex nitrogen-containing compounds isolated from plants, and some of them have marked physiological activity, such as the uterus-contracting effect of some of the ergot alkaloids, the analgesic action of morphine and its congeners, the convulsant effect of strychnine, and the muscle-paralyzing effect of the curare alkaloids. Some of the alkaloids have been known as pure crystalline materials for many years and have had wide use as medicinals. Most of them contain complex nitrogen-carrying ring systems, whose intricacies made them more suitable as research problems in European laboratories than in American ones, because of the greater availability of experienced research workers in the large European centers. It is significant that the leading American work on alkaloids before 1940 was that of Craig and Jacobs on the ergot alkaloids at Rockefeller and a few years later by N. J. Leonard and Roger Adams at Illinois on the Senecio series. Both laboratories had experienced workers and a tradition in natural product chemistry; they were thus somewhat like the European laboratories which had done most of the work on alkaloid chemistry prior to 1940.

After 1945, American work on alkaloids expanded greatly, partly due to increased research support for postdoctorates and equipment, and partly due to the upsurge of activity on complex natural products which was becoming evident in the 1930s. In general, the period 1940-1955 was notable in this country for total synthesis of alkaloids long known, such as morphine and strychnine; for the establishing of structures for alkaloids; and for experimental work on the pathways of biogenesis of alkaloids in plants. All of these phases were actively investigated in overseas laboratories and in Canada, and the American work attained a comparable level of excellence.

Ideas of the mode of formation (*biogenesis*) had been current for many years, early fruitful speculation by Robinson in England being notable. The availability of isotopic labels, particularly ^{14}C and other isotopes, made this a very active experimental field all over the world, especially after 1945.

Chromatographic separation techniques and other instrumental methods were indispensable in this work. It was possible to show by feeding labeled possible simple precursors, or by enzyme experiments in a few cases, that plants *could* utilize some simple units in producing the complex alkaloid molecules. It was not always easy to show that a *possible* path was the *actual* path employed by the plant under natural conditions, and indeed the importance of the distinction was not always clearly drawn.

By 1960 it had been demonstrated that plants will incorporate acetate, amino acids, mevalonolactone (formed from three acetate units), as well as larger molecules into alkaloids.[40] An admirable account of the whole field of alkaloid biosynthesis as of 1969 was given by T. A. Geissman and D. H. G. Crout[41]. It should be emphasized that this work on alkaloid biogenesis, in which Edward Leete of Minnesota, Henry Rapoport of Berkeley and Woodward of Harvard were leaders in this country, was of importance in increasing the knowledge of enzymatic reactions, thus contributing to biochemistry in general. Henry Rapoport (1918-   , S.B. 1940, Ph.D. 1943, both at MIT) has taught at Berkeley from 1946. He has worked on the structure and synthesis of natural products, including colchicine, antibiotics, natural toxins, and on biogenesis of natural products.

Edward Leete (1928-   , B.Sc. 1948, Ph.D. 1950, with W. Bradley, both at Leeds) has taught at UCLA and the University of Minnesota. His research has been on the biogenesis of alkaloids and isolation of enzymes from plants.

T. A. Geissmann (1908-1974, B.S. 1930, Wisconsin; Ph.D. 1937, Minnesota with C. F. Koelsch) was a postdoctorate at Illinois and taught at UCLA from 1939 on. His research on natural products included biogenesis and the chemical genetics of flower color variation.

It is not our purpose to go into the intricacies of structure determination or of synthesis of the alkaloids; as in other cases, X-ray crystallography became the easiest way to establish structures, and the extraordinarily involved degradations and deductions of structure from fragments became a thing of the past.[42] The elegant work of Craig, Wintersteiner, and others on the veratrum alkaloids[43], Leonard and Adams on the Senecio series,[42] and V. Boekelheide on the erythroidine series[44] is representative of American alkaloid chemistry.

Nelson J. Leonard (1916-   , B. S. 1937, Lehigh; B. Sc. 1940, Oxford, as Rhodes Scholar; Ph.D. 1942, Columbia with R. C. Elderfield) was a postdoctorate and faculty member at Illinois from 1942 on. His research has included alkaloids, nitrogen heterocycles, and synthetic work on analogs of nucleic acid bases.

Virgil Boekelheide (1919-   , A.B. 1939, Ph.D. 1943, both at Minnesota with C. F. Koelsch) has taught at Illinois, Rochester, and Oregon. He made important contributions to the chemistry of the curare alkaloids in addition to the erythroidine series. He synthesized some novel nitrogen heterocycles (the cyclazines) and some derivatives of aromatic hydrocarbons (the dialkyldihydropyrenes) of great interest in connection with theories of aromaticity.

The best-known alkaloid, morphine, isolated from the mixture of akaloids from the opium poppy by Sertürner in 1805, is still one of the best analgesics, in spite of its addicting properties which are even worse in its diacetyl derivative, heroin. Its correct structure was proposed by Sir Robert Robinson and J. Gulland[45] in 1925, but its total synthesis was not achieved until 1952 by M. Gates and G. Tschudi at Rochester.[46] Marshall Gates (1915- , B.S. 1936, Rice; Ph.D. 1941, Harvard with L. F. Fieser) taught at Bryn Mawr and Rochester, where he was editor of JACS for many years. His morphine synthesis was started at Bryn Mawr, completed at Rochester, and largely done with his own hands.

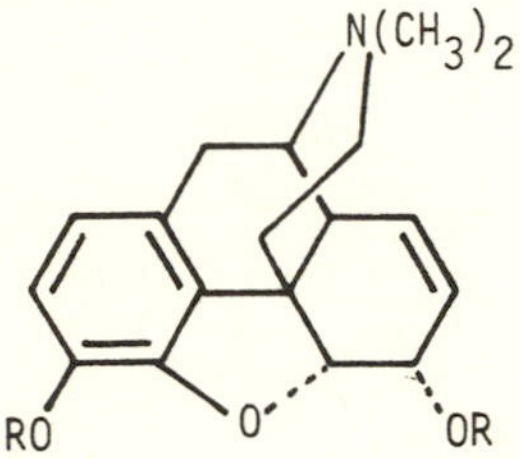

Morphine, R = H
Heroin, R = $COCH_3$

Strychnine, obtained from *Strychnos nux vomica* seeds, was discovered by Pelletier and Caventou in 1917; its structure was not settled until 1947 by Robinson and by Woodard, independently. Woodward and his group synthesized it in 1954.[47] Strychnine has been used in tonics as well as for killing rodents; it is highly toxic because it causes convulsions by acting on the central nervous system. It is a representative of the indole alkaloids, a large and complex group having tryptamine, a substituted indole derived from the amino acid tryptophane, as one of the building blocks.[41]

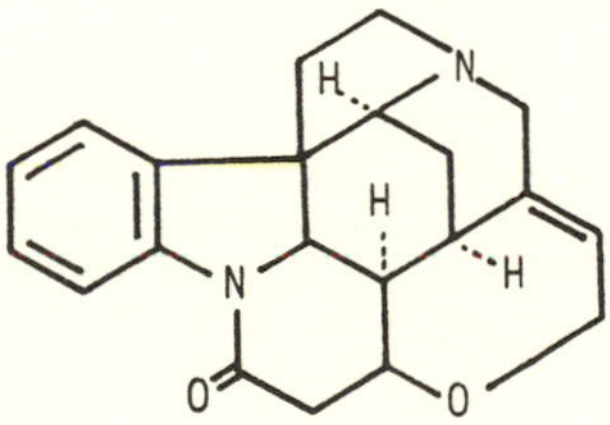

Strychnine

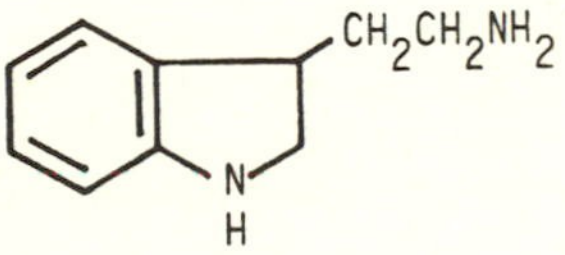

Tryptamine

With the total synthesis of morphine and strychnine and of many less well-known alkaloids, it was clear that with enough ingenuity, patience, and hard work any ring system could be synthesized. Although many difficult syntheses were to be carried out where the compounds occurred naturally in minute amounts and were needed for special studies, synthetic organic chemistry had reached a level of accomplishment such that further total syntheses merely emphasized the point. The Instrumental Revolution greatly aided synthetic work; intermediates were purified and their structures confirmed by the techniques we have already discussed. The increased knowledge of reaction mechanisms and of stereochemistry was also indispensable for outstanding synthetic work. American structural chemistry had advanced to a point of world leadership by 1955.

Able studies in natural products continued, and three chemists whose work came in large part after 1955 and compared favorably with Woodward's in originality and elegance merit mention: Gilbert Stork, George Buchi, and E. J. Corey.

Gilbert Stork (1921- , B.S. 1942, Florida; Ph.D. 1945, Wisconsin with S. M. McElvain) taught at Harvard and since 1953 at Columbia, where he is Higgins Professor. He has studied terpenes, steroids, alkaloids, antibiotics, and the prostaglandins.

George Buchi (1921- , D. Sc. 1947, Federal Technical Institute, Zurich, with O. Jeger and L. Ruzicka) after three years at Chicago, has taught at MIT since 1951, where he is Dreyfus Professor. His researches include photochemical studies, terpenes and the aflatoxins, a highly toxic series of compounds obtained from moldy peanuts.

E. J. Corey (1928- , B. S. 1948, Ph.D. 1951, at MIT with J. C. Sheehan) taught at Illinois from 1951 to 1959 and thereafter at Harvard. His investigations cover terpene chemistry, extended syntheses on the biochemically important prostaglandin series, macrolides, and computer-aided syntheses.

All three probed reaction mechanisms and skillfully applied their ideas from this field to generate new approaches to structural chemistry and create new synthetic methods, some based on the use of copper, nickel, sulfur, and silicon compounds. Their work carried on the high quality of American research in explaining biochemical processes.

LITERATURE CITED

1. L. H. Sarett JACS, **70**, 1454 (1948); **71**, 2443 (1949); particulars of these and other steroid syntheses are given by L. F. and M. Fieser, *Steroids*, New York, 1959, the standard monograph. Tishler's share in the commercial production of cortisone from bile acids is described, loc. cit., pp. 643 ff.; cortisone from bile acids was made available for clinical use in 1949.
2. D. H. Peterson et al., JACS, **74**, 5933 (1952); **75**, 408 (1953).
3. R. B. Woodward et al., ibid., **73**, 4057 (1951); **74**, 4223 (1952).
4. R. B. Woodward and W. E. Doering, ibid., **66**, 849 (1944); **67**, 860 (1945). The best accounts of Woodward's career and work are D.S.M. Wheeler, Chemie in Unserer Zeit, **18**, 109 (1984) and A. R. Todd and J. W. Cornforth, *Biog. Mem. Royal Soc.*, **27**, 629 (1981).
5. L. H. Sarett et al., JACS, **74**, 4974 (1952) and later papers.
6. G. Anna and K. Miescher, Helv. Chim. Acta., **31**, 2173 (1948) and later papers.
7. W. S. Johnson et al., JACS, **74**, 2832 (1952).
8. Summarizing paper, W. S. Johnson et al., ibid., **80**, 661 (1958).
9. W. S. Johnson, Accounts Chem. Res., **1**, 1 (1968).
10. D. H. R. Barton et al., JACS, **83**, 4077, 4083 (1961).
11. A. Fleming, Brit. J. Exp. Path., **10**, 226 (1929).

12. *The Chemistry of Penicillin*, H. T. Clarke, J. R. Johnson and Sir Robert Robinson, Eds., Princeton, 1949.
13. Dorothy Crowfoot et al., ibid., pp. 310 ff; H. W. Thompson et al., IR measurements, ibid., pp. 382 ff.
14. J. L. Strominger et. al., Proc. Nat. Acad. Scis., **68**, 2814 (1971) and many earlier papers in JBC.
15. J. C. Sheehan and K. R. Henery-Logan, JACS, **79**, 1262 (1957). Sheehan's book, *The Enchanted Ring*, Boston, 1982, is an informal and entertaining account of his experiences with the penicillin problem over many years.
16. A. Schatz, E. Bugie and S. A. Waksman, Proc. Soc. Exp. Biol. Med., **55**, 66 (1944).
17. Several reviews are available; e.g., N. G. Brink and K. Folkers, in *Streptomycin*, S. A. Waksman, Ed., Baltimore, 1949, pp. 55 ff.
18. Q. R. Bartz et al., Science, **106**, 417 (1947).
19. J. Controulis, Marian C. Rebstock and H. M. Crooks, Jr., JACS, **71**, 2463 (1949).
20. S. A. Waksman and H. A. Lechavalier, Science, **109**, 305 (1949).
21. Summarizing paper: K. L. Rinehart et al., JACS, **84**, 3218 (1962); K. L. Rinehart, *The Neomycins and Related Antibiotics*, New York, 1964.
22. The monumental nature of this paper deserves listing of all collaborators: F. A. Hochstein, C. R. Stephens, L. H. Conover, P. P. Regna, R. Pasternack, P. N. Gordon, F. J. Pilgrim, K. J. Brunings and R. B. Woodward, JACS, **75**, 5455 (1953).
23. H. Muxfeldt et al., ibid., **90**, 6534 (1968).
24. Isolation: J. M. McGuire et al., Antibiotics and Chemotherapy, **2**, 281 (1952); patents were issued to Eli Lilly and Abbott Laboratories.
25. P. F. Wiley et al., JACS, **79**, 6062 (1957) and nine earlier papers; absolute configuration by X-ray, D. R. Harris et al., Tetrahedron Lett., 679 (1965); conformation of erythromycin derivatives by ^{1}H and ^{13}C NMR, R. S. Egan, T. J. Perun et al., Tetrahedron, **29**, 2525 (1973) and earlier papers; JACS, **97**, 4578 (1975); J. G. Nourse and J. D. Roberts, ibid., p. 4585.
26. Review of macrolides: M. Berry, Quart. Rev., **17**, 343 (1963); for amphotericin B, with a 38-membered lactone ring, see A. C. Cope et al., JACS, **88**, 4228 (1966).
27. Structure: D. S. Tarbell et al., ibid., **82**, 1005, 1009 (1959); **83**, 3096 (1961); absolute configuration, J. R. Turner and D. S. Tarbell, Proc. Nat. Acad. Scis., **48**, 733 (1962); X-ray crystallography, N. J. McCorkindale and J. G. Sime, Proc. Chem. Soc., 331 (1961); total synthesis, E. J. Corey and B. B. Snider, JACS, **94**, 2549 (1972).
28. H. P. Sigg and H. P. Weber, Helv. Chim. Acta. **51**, 1395 (1968).
29. Review, J. S. Fruton, *Advances in Protein Chemistry*, 5, 1 (1949).
30. J. S. Fruton, *Molecules and Life*, New York, 1972.
31. S. Moore and W. H. Stein, JBC, **178**, 79 (1949); D. H. Spackman, W. H. Stein and S. Moore, Anal. Chem., **30**, 1190 (1958).
32. P. Edman, Acta Chem. Scand., **4**, 283 (1950).
33. M. Bergmann and L. Zervas, Ber., **65**, 1192 (1932), and many later papers in JBC.
34. F. C. McKay and N. F. Albertson, JACS, **79**, 4686 (1957); G. W. Anderson and A. C. McGregor, ibid., p. 6180.
35. J. C. Sheehan and G. P. Hess, ibid., **77**, 1067 (1955); other methods of forming peptide bonds are reviewed by N. F. Albertson, *Organic Reactions*, **12**, 157 (1962).
36. R. B. Merrifield, JACS, **85**, 2149 (1963) and later papers; Merrifield received the Nobel Prize in 1984 for this work.
37. V. du Vigneaud et al., JACS, **75**, 4879 (1953); **76**, 3115 (1954).
38. O. Kamm et al., ibid., **50**, 573 (1928).
39. L. C. Craig, JBC, **155**, 519 (1944).
40. A. R. Battersby, Quart. Rev., **15**, 259 (1961).

41. T. A. Geissman and D. H. G. Crout, *Organic Chemistry of Secondary Plant Metabolism*, San Francisco, 1969, pp. 428-572.

42. The classical book on alkaloids is T. A. Henry, *The Plant Alkaloids*, Philadelphia, 4th ed., 1949. A series of monographs, *The Alkaloids*, R. H. F. Manske and H. L. Holmes, Eds., New York, is standard; vol.1 (1950) contains an excellent account by N. J. Leonard of the Senecio alkaloids (pp. 108-164).

43. K. J. Morgan and J. A. Barltrop, Quart. Rev., **12**, 34 (1958).

44. V. Boekelheide et al, JACS, **75**, 2550 (1953) and later papers.

45. J. M. Gulland and R. Robinson, JCS, **123**, 980 (1925).

46. M. Gates and G. Tschudi, JACS, **74**, 1109 (1952); **78**, 1380 (1956).

47. Structure: R. B. Woodward et al., ibid., 69, 2250 (1947); **70**, 2107 (1948); R. Robinson, Experientia, **2**, 28 (1946); synthesis, Woodward et al., JACS, **76**, 4749 (1954); Tetrahedron, **19**, 247 (1963).

24

Polyisoprenes, Fats, Carbohydrates
1940-1955

One of the great scientific achievements since 1940 has been the elucidation of the chemical pathways by which acetic acid, CH_3COOH, is converted to a wide variety of compounds with essential biological functions. These products include the vitamin A series (20 carbons), the triterpenes (30 carbons), the carotenes (40 carbons), the steroid derivatives, including lanosterol (30 carbons, from wool fat), cholesterol, the bile acids, the sex hormones, the adrenal cortical hormones, the vitamin D series, and parts of other significant biochemical materials, such as vitamin K and vitamin E.

Long ago it had been postulated that many compounds of the terpene series, which exist in plants and are usually multiples of 5 carbon atoms, are formed by combination of isoprene units, $CH_2=C(CH_3)CH=CH_2$. Common terpenes are camphor, citral, and pinene (all 10-carbon compounds). Indeed, structures were sometimes assigned to such *polyisoprenoids* on the assumption that they were built up by linking isoprene units through some unspecified mechanism without rearrangement.

During the period from 1940 to 1955, the biochemical mechanism by which acetate becomes isoprene derivatives, and these in turn become larger compounds of 10, 20, 30, or 40 carbons, has become known. The use of isotopic tracers and instrumental techniques has made this possible. The results, which are constantly being enlarged, illuminate a whole area of plant and animal biochemistry. They have extended far beyond the explanation of the formation of simple terpenes.

Konrad Bloch in this country and J. W. Cornforth in Britain experimented with the isotopic labeling of acetic acids.[1,2] They showed the origin of each of the 27 carbon atoms of cholesterol from the CH_3 or the $COOH$ of acetic acid. Feeding trials and enzymatic synthesis of necessary compounds were followed by laborious degradation of the cholesterol molecule.

Konrad Bloch (1912-   ,Chem Eng., 1934, Technische Hochschule, Munich; Ph.D. 1938, Columbia with David Rittenberg) taught at Chicago and Harvard. He received the Nobel Prize in 1964 for his work on biogenesis.

J. W. Cornforth (1917-   , B. Sc. 1937, Sydney; D. Phil. 1941, Oxford with Sir Robert Robinson) did research at the Medical Research Council and Shell Institute of Enzymology in Britain. He has worked on penicillin, biogenesis, and mechanism of enzyme reactions. He received the Nobel Prize in 1975.

I. M. Heilbron at Manchester, England, suggested in 1926 that cholesterol was formed in animals from squalene, $C_{30}H_{50}$, a polyisoprenoid hydrocarbon isolated from fish liver oil[3], and this idea was supported by feeding squalene to rats. R. B. Woodward and Bloch, and W. G. Dauben proposed [4] a pathway of cyclization and rearrangements leading from the open chain squalene to the 30-carbon steroid lanosterol; it was shown that squalene was formed biochemically from acetate and that lanosterol was converted *in vivo* to cholesterol. Very ingenious labeling experiments by Bloch and by Cornforth supported the Woodward-Bloch-Dauben hypothesis, and all further evidence agrees with it.

W. G. Dauben (1919-   , A. B. 1941, Ohio State; Ph. D. 1944, Harvard with L. F. Fieser) has taught at Berkeley since 1945. His research has included structure of natural products, organic photochemistry, and synthesis.

As an illustration of how the accepted history of science may be in disagreement with the actual facts, it is instructive to investigate further the origin of the hypothesis about the cyclization of squalene to the steroid nucleus. Reviews of this topic always quote Sir Robert Robinson's letter of 1934[5] in which he suggested a mode of cyclization different from the later Woodward-Bloch-Dauben mechanism. Examination, however, of Robinson's publication shows that he was actually proposing some changes in a mechanism suggested a few months earlier in the same journal by M. Vanghelovici.[6] Robinson was deservedly world famous, and Vanghelovici was unknown, so that Robinson has always received sole credit for the idea. Vanghelovici actually suggested the following course for the biogenesis of cholesterol: carbohydrates → saturated fatty acids → unsaturated fatty acids → poli-isoprenic carotinoid compounds → cholesterol. It is clear from his paper that he believed that squalene or similar structures were the poli-isoprenic-carotinoid compounds (sic). Both papers were titled "Structure of Cholesterol", because in 1934 the old, erroneous structures for the sterols and bile acids proposed by Wieland and by Windaus were being revised to the modern correct structures.

Karl Folkers at Merck isolated the key intermediate in the conversion of acetate to isoprene. This is mevalonic acid, a 6-carbon compound. The Folkers group proved the structures of mevalonic acid and its lactone during a study of growth factors for lactobacilli.[7] The enzymes in rat liver homogenate convert mevalonic acid or its lactone almost completely into cholesterol,[8] the mevalonate being more efficiently utilized than any other cholesterol precursor.

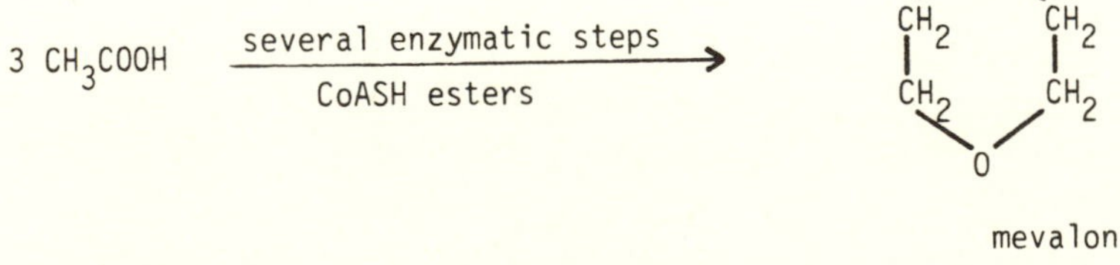

$$3 \ CH_3COOH \xrightarrow[\text{CoASH esters}]{\text{several enzymatic steps}}$$

mevalonolactone

Mevalonic acid results from condensation of three moles of acetic acid by enzymatic processes involving esters of the SH group of Coenzyme A. Coenzyme A was isolated in research entirely unconnected with the polyisoprene problem by the biochemists F. Lynen in Munich and F. Lipmann in this country. English workers established its structure, and its synthesis was an elegant feat by J. G. Moffatt and H. G. Khorana, then at British Columbia.[9] The general scheme of biosynthesis from acetate to cholesterol is indicated very briefly in the scheme shown.[10]

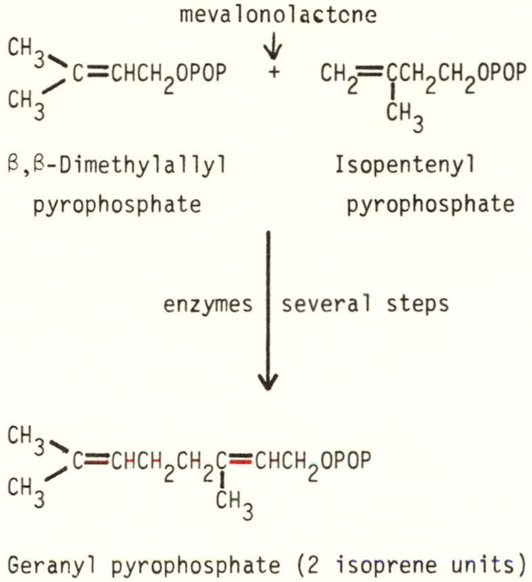

mevalonolactone

$$CH_3\!\!\diagdown_{\!\!CH_3}\!\!\diagup C\!\!=\!\!CHCH_2OPOP \quad + \quad CH_2\!\!=\!\!CCH_2CH_2OPOP$$
$$CH_3$$

β,β-Dimethylallyl pyrophosphate   Isopentenyl pyrophosphate

enzymes | several steps

$$CH_3\!\!\diagdown_{\!\!CH_3}\!\!\diagup C\!\!=\!\!CHCH_2CH_2C\!\!=\!\!CHCH_2OPOP$$
$$CH_3$$

Geranyl pyrophosphate (2 isoprene units)

Addition of another isoprene unit to the geranyl phosphate yields a 15-carbon compound; this is then dimerized to squalene, $C_{30}H_{50}$. Squalene through its epoxide gives lanosterol by enzymatic cyclization and rearrangement.[10]

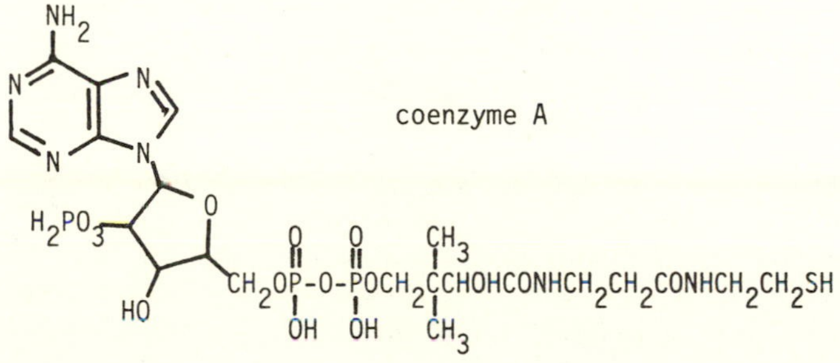

coenzyme A

$$CH_2OP\text{-}O\text{-}POCH_2CCHOHCONHCH_2CH_2CONHCH_2CH_2SH$$

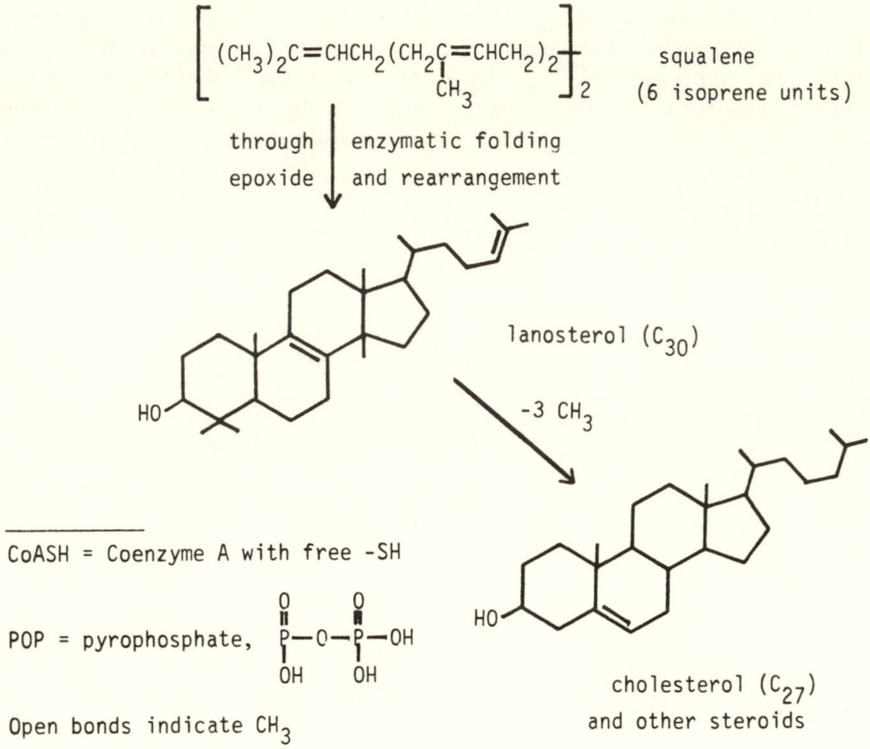

$$\left[(CH_3)_2C{=}CHCH_2(CH_2\underset{\underset{CH_3}{|}}{C}{=}CHCH_2)_2 \right]_2 \qquad \text{squalene}$$

squalene
(6 isoprene units)

through enzymatic folding
epoxide and rearrangement

lanosterol (C_{30})

−3 CH_3

CoASH = Coenzyme A with free −SH

POP = pyrophosphate, $\underset{\underset{OH}{|}}{\overset{\overset{O}{\|}}{P}}{-}O{-}\underset{\underset{OH}{|}}{\overset{\overset{O}{\|}}{P}}{-}OH$

Open bonds indicate CH_3

cholesterol (C_{27})
and other steroids

The steroid biogenesis scheme illustrates the intersection of several lines of research undertaken with different purposes: the labeling studies on acetate incorporation into steroids; the isolation and proof of function of coenzyme A, which has many biochemical roles other than the ones shown here; the isolation and function of mevalonate in forming polyisoprenoids; the cyclization of squalene though its epoxide to lanosterol; and the conversion of lanosterol to cholesterol and other steroids. Mevalonate is a stage in the biosynthesis of many other types of polyisoprenoid natural products in other than steroids.[11] Fundamental to all the biogenetic work on steroids and other natural products is the knowledge of the structures and chemical reactions of these compounds, accumulated by organic chemists over many years.

Most animal fats and many vegetable oils (such as cottonseed or linseed oil) have long been recognized as esters of long-chain acids (13-15 carbons) and glycerol. One of the earliest important organic reactions was the alkaline hydrolysis of fats (*saponification*) to form the salts of the acids (*soaps*) and glycerol.

$$\begin{array}{c} \text{RCOOCH}_2 \\ | \\ \text{RCOOCH} \\ | \\ \text{RCOOCH}_2 \end{array} \quad \xrightarrow[\substack{\text{KOH-K}_2\text{CO}_3 \\ \text{(from wood} \\ \text{ashes)}}]{3\text{H}_2\text{O}} \quad 3\text{RCOOK} \quad + \quad \begin{array}{c} \text{HOCH}_2 \\ | \\ \text{HOCH} \\ | \\ \text{HOCH}_2 \end{array}$$

Fats are physiologically important triglycerides and of great commercial value. They serve as indispensable sources of soaps, edible vegetable oils and shortenings, and the more unsaturated glycerides contained in linseed and tung oil as paint base.

Research from 1940 to 1955 worked out the pathways of metabolism of long-chain acids in living systems. Acetyl coenzyme A, $CH_3COSCoA$, is the key intermediate in a series of enzymatic reactions converting acetic acid to long-chain acids. A reversal of these steps breaks down long-chain acids to $CH_3COSCoA$, which can interact in several biosynthetic series including cholesterol formation.[12] Oxidation via the citric acid (Krebs) cycle can also burn the acetic acid moiety to carbon dioxide and water, furnishing energy to the organism.

Experimental work on triglycerides, their derivatives, and long-chain compounds in general was fraught with difficulties. Separations of closely related compounds were imperfect, the traditional criteria of purity, such as the melting point, were unsatisfactory, and identification of compounds was troublesome. In spite of these problems, the technical and biochemical importance of these classes of compounds was so great that by 1940 workers in many countries had amassed a large amount of information about fats and many of their derivatives, such as the phosphatides (or phospholipids).[13] P. A. Levene at Rockefeller was an early worker in this field.

Rudolph J. Anderson (1879-1961), editor of JBC for many years[14], worked at Yale from 1926 to 1948 on fat isolated from the human tubercle bacillus. He obtained numerous fractions, some of which produced lesions similar to those of human tuberculosis. Anderson's identification of these acids as methyl-substituted long-chain acids seems to have been correct, but later work at Oxford has revised some of his structures.[15] James Cason, a student of Anderson, developed synthetic methods for methyl-branched long-chain acids[16] in the 1940s. One of Anderson's fractions, assigned the formula $C_{38}H_{176}O_4$, was shown by E. Lederer in France to be a mixture of compounds of the type shown.[17] Further progress in this difficult field clearly required modern instrumental techniques, particularly mass spectrometry and NMR.

$$\begin{array}{l} \text{CH}_3(\text{CH}_2)_{33}\text{CHCOOH} \\ \qquad\qquad\quad | \\ \qquad\qquad \text{RCHOH} \end{array} \qquad R = C_{60}H_{121}, \ C_{60}H_{120}OCH_3, \ \text{or} \ C_{60}H_{120}OH$$

Klaus Hofmann at Pittsburgh isolated a long-chain acid containing a cyclopropane ring[18], and J. R. Nunn reported[19] a similiar acid with the unusual cyclopropene ring. Klaus Hofmann (1911-   , Ph. D. 1936, Federal Technical Institute, Zurich) was a post doctorate with du Vigneaud and since 1944 on the faculty at Pittsburgh. His research contributed to the chemistry of biotin, to long-chain acids, and to the synthesis and mode of action of polypeptide hormones.

The study of unsaturated fats and the related unsaturated acids was aided by the development of low temperature crystallization by J. B. Brown at Ohio State[20] and by chromatography. F. M. Strong at Wisconsin, W. J. Gensler at Boston University, and others[21] carried out synthesis of unsaturated acids via acetylenic intermediates leading to the naturally occurring *cis* forms of the unsaturated acids.

The phosphatides (or phospholipids) were gradually separated and their structures established.[22] R and R' may be long-chain saturated or unsaturated hydrocarbon groups of varying size, and the phosphate may be esterified with choline, $[HOCH_2CH_2N^+(CH_3)_3]Cl^-$, serine, $(HOCH_2CH(NH_2)COOH)$, or other hydroxyl-containing nitrogen compounds.

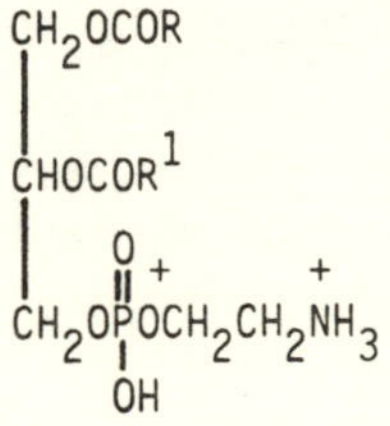

$$CH_2OCOR$$
$$CHOCOR^1$$
$$CH_2O\overset{\overset{O}{\|}}{P}OCH_2CH_2\overset{+}{N}H_3$$
$$OH$$

Constituents from the central nervous system have been isolated and their structures elucidated; Carter's extensive studies on sphingosine at Illinois are an example.[23,24]

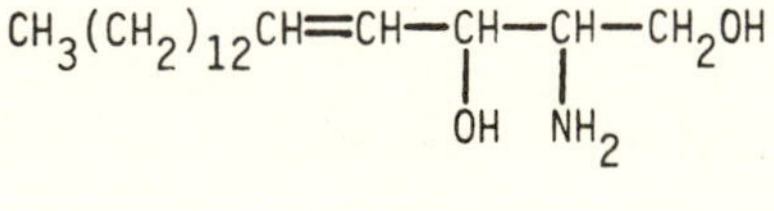

$$CH_3(CH_2)_{12}CH{=}CH{-}CH{-}CH{-}CH_2OH$$
$$\underset{OH}{|}\quad\underset{NH_2}{|}$$

sphingosine

The specialized field of carbohydrate chemistry maintained its continuity with classical methods and ideas in research during the years 1940 to 1955. Periodate oxidation and the roughly chemically equivalent lead tetraacetate were widely used for structure determination of monosaccharides and polysaccharides. Methylation of free hydroxyl groups followed by hydrolysis and isolation of the products continued as a standard procedure for polysaccharide elucidation. Nevertheless, carbohydrate chemistry moved into the mainstream of organic chemistry, utilizing new chemical and physical techniques and new ideas of reaction mechanism. Research in all countries clearly showed this change. Furthermore, sugars of novel structure were continually isolated as constituents of antibiotics or as bacterial metabolities. M. L. Wolfrom's brilliant synthesis at Ohio State of streptidine, a component of streptomycin, typifies this aspect of carbohydrate chemistry.[25]

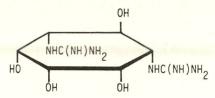

streptidine

Structural problems posed by the nucleic acids were solved during this period, particularly by A. R. Todd at Cambridge, England. Because D-ribose is a structural unit in RNA (ribonucleic acid) and 2-deoxy-D-ribose similarly in DNA (deoxyribonucleic acid), many laboratories intensively studied the chemistry of these and related sugars. Amino sugars, such as glucosamine (2-deoxy-2-amino-D-glucose), were found in physiologically important polysaccharides; these discoveries led to significant research on amino sugars in general. Phosphate esters of pentoses (usually D-ribose or deoxyribose) as parts of coenzymes and nucleic acids stimulated detailed study of sugar phosphates.

As in other fields, the Instrumental Revolution profoundly affected carbohydrate chemistry. Paper chromatography, ion exchange columns, particularly for sugar phosphates, NMR, IR, optical rotatory dispersion (ORD), mass spectroscopy, and labeling, particularly in Melvin Calvin's classic work on pathways of photosynthesis, became increasingly useful. For the naturally occurring polysaccharides and related compounds, such as nucleic acids, the ultracentrifuge, light scattering, and other physical methods became indispensable, just as they were to synthetic polymers.[25]

Melvin Calvin (1911-   , B.S. 1931, Michigan College of Mining and Technology; Ph. D. 1935, Minnesota with George Glockler) has been at Berkeley since 1937. With G. E. K . Branch he wrote the influential monograph, *The Theory of Organic Chemistry*, New York, 1940; with A. E. Martell, *Chemistry of Metal Chelate Compounds*, New York, 1952. His book with J. A. Bassham, *The Pathway of Carbon in Photosynthesis*, Englewood Cliffs, N. J., 1957, summarized his research on this problem, for which he received the Nobel Prize in 1961.

Carbohydrates contain a large number of functional groups, OH, sometimes C=O, NH_2, or derivatives of these, and hence Saul Winstein's work on the influence of neighboring groups on reactions of functional groups was applicable in controlling the course of syntheses of sugars. In turn, the carbohydrates offered good experimental cases for testing neighboring group effects and other mechanistic ideas.

Bernard R. Baker was a leader in utilizing modern ideas of reaction mechanisms in carbohydrate chemistry. Baker (1915-1974, A. B. 1937, UCLA; Ph. D. 1940, Illinois with Roger Adams) worked at Lederle Laboratories for many years on the structure and synthesis of natural products, including antibiotics and sugars. At the University of California at Santa Barbara, he did outstanding work on determining receptor sites for enzymes by studying enzyme inhibitors. He was one of the leading figures in developing medicinal chemistry as a recognized discipline.

As discussed in an earlier section, the Sachse-Mohr boat and chair conformations of 6-membered rings were early applied to pyranose sugars (6-membered rings containing one oxygen) by W. N. Haworth in England in the 1920s. The chair conformation of pyranose rings was considered to have lower energy than the boat form,[26] and the application of conformational analysis to pyranose sugars in the 1930s and 1940s before Barton's classical papers was noteworthy. Richard E. Reeves contributed greatly to studies of sugar conformations by correlating the stabilities of complexes of sugar OH groups and cupric ammonium ion with the arrangement of OH groups in the favored conformation.[27] Reeves considered boat forms to be required in some cases in which later work has shown the chair form to be correct; this was not a major error.[28] What is significant is that the ideas of conformational analysis became a valuable tool in predicting and understanding the properties of carbohydrates.

Reeves (1912-   , A. B. 1933, Doane; Ph.D. 1936, Yale with R. J. Anderson) worked on conformation of sugars at the Southern Regional Laboratory of the Department of Agriculture at New Orleans. Later as professor of biochemistry in Louisiana State University Medical School, he specialized in tropical medicine.

With the improved techniques and concepts, it became possible after 1940 to unravel the structures of complex and biologically important polysaccharides, including the difficult mucopolysaccharides, such as heparin and hyaluronic acid. Heparin, isolated from beef liver and lung, is a clinically useful blood anti-coagulant; its structure was almost completely worked out by M. L. Wolfrom at Ohio State over a period of many years.[29] It is a polymer of glucosamine and glucuronic acid units with sulfate groups attached to oxygen and nitrogen. It is significant that successful degradation of heparin involved reduction of the carboxy group by the then new boron hydride reagents.

A somewhat similar mucopolysaccharide, hyaluronic acid, occurs in synovial fluid (in joints) and many other tissues; its structure has been elucidated by Karl Meyer at Columbia.[30] Meyer (1899-   , M. D. and Ph.D., Germany) came to this country in 1930. He taught biochemistry at Columbia from 1932 to 1967, then at Yeshiva University. He studied mainly mucopolysaccharides, related enzymes, such as hyaluronidase, and glycoproteins.

Structures of many other high molecular weight polysaccharides occurring in plants and animals, such as xylans from oat hulls and the fruit pectins, were successfully unraveled.

The solution of a classical problem in carbohydrate chemistry, the chemical synthesis of sucrose (cane sugar), involved a process based on interaction of groups within a molecule to control the configuration of the product. Although sucrose is isolated and sold in far larger quantities than any other pure organic compound, its chemical synthesis was not achieved until 1953, the accomplishment of the brilliant Canadian chemist Raymond U. Lemieux.[31] The key step is the reaction of tetraacetylfructose with a 1,2-epoxide derivative of glucose. Good evidence exists that the latter reacts with alcohols to give ring opening of the epoxide without inversion of configuration at the carbon-oxygen bond being broken. This can only occur if cleavage of the bond involves two inversions, and that must involve the cyclic oxonium ion intermediate of the Winstein type.[32]

Raymond U. Lemieux (1920-   , B. Sc. 1943, Alberta; Ph. D. 1946, McGill with C. B. Purves) was first a research associate with Wolfrom at Ohio State, then research officer at the Prairie Regional Laboratory in Saskatoon, Saskatchewan, where he synthesized sucrose and maltose. After several years at Ottawa University, he returned to the University of Alberta and worked on carbohydrates, polypeptides, NMR, and conformational effects.

The recognition of the critical importance of nucleosides containing D-ribose or the 2-deoxy derivative in nucleic acids led to an enormous amount of work in synthesis of nucleosides derived from other sugars, some of them not naturally occurring sugars. It was hoped that these "unnatural nucleosides" would interfere with the biosynthesis of normal nucleic acids and thus act as anticancer agents, antimalarial drugs, or serve other clinically useful purposes. The nucleoside "Ara-A", which differs from adenosine, a component of natural RNA, only in the configuration at the starred carbon in the sugar, shows anticancer and antiviral activity. Synthesized by B. R. Baker's group,[33] it is representative of many analogous compounds with altered sugar structures or base structures (in this case adenine) or both.

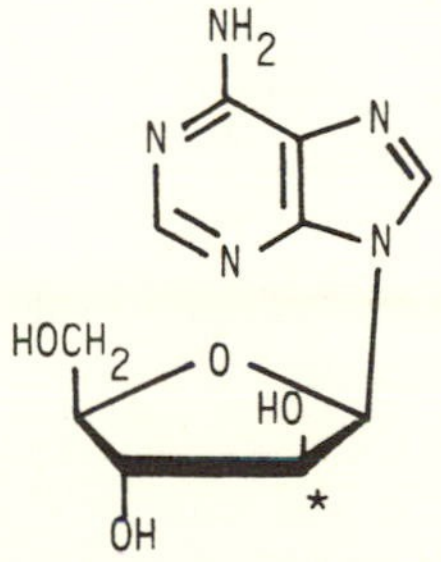

"Ara-A", or 9-(β-D-arabinofuranosyl)adenine

Since this period, striking discoveries in physiological chemistry and molecular biology have soared. American organic chemists are among the leaders in the whirlwind of new ideas and fundamental research.

LITERATURE CITED

1. Typical papers: K. Bloch et al., JBC, **195**, 439 (1952); J. W. Cornforth, G. D. Hunter and G. Popjak, Biochem. J., **54**, 590, 597 (1953).
2. Reviews: R. B. Clayton, Quart. Rev., **19**, 168, 201 (1965); G. Popjak and J. W. Cornforth, Advances in Enzymology, **22**, 281 (1952).
3. I. M. Heilbron et al., JCS, 1630 (1926).
4. R. B. Woodward and K. Bloch, JACS, **75**, 2023 (1953); W. G. Dauben, A. H. Soloway et al., ibid., p. 3038.
5. R. Robinson, Chem. and Ind., **53**, 1062 (1934).
6. M. Vanghelovici, ibid., p. 998.
7. K. Folkers et al., JACS, **78**, 4499 (1956).
8. P. A. Tavormina, Margaret H. Gibbs and J. W. Huff, ibid., p. 4498.
9. Synthesis, J. G. Moffatt and H. G. Khorana, ibid., **81**, 1265 (1959); **83**, 663 (1961); isolation, F. Lipmann et al., JBC, **186**, 235 (1950); structure, J. Baddiley et al., Nature, **171**, 76 (1953).
10. Clayton, ref. 2; Ciba Foundation, *Biosynthesis of Terpenes and Steroids*, G. E. W. Walstenholme and M. O'Connor, Eds., London, 1959; this reference gives a picture of the field when activity was at its height but not all the stages had been worked out, even to the extent we describe. See J. H. Richards and J. B. Hendrickson, *The Biosynthesis of Steroids, Terpenes, and Acetogenins*, New York, 1964, for general review.
11. Biogenesis of polyisoprenoids in plants: T. A. Geissman and D. H. G. Grout, *Organic Chemistry of Secondary Plant Metabolism*, San Francisco, 1969, passim.
12. For review, see Richards and Hendrickson, op. cit., pp. 121 ff.
13. A symposium in Chem. Rev., **29**, 199-438 (1941), with 11 papers by leading American and Canadian workers, gives the then current state of knowledge. H. O. L. Fischer (the son of Emil Fischer) then in Toronto, was doing elegant work on the synthesis of optically active glycerides. Fischer later moved to the Virus Laboratory at Berkeley, and work was continued with success by Baer in Toronto and Fischer in Berkeley.
14. For Anderson's varied career and a description of his work, see H. B. Vickery, *Biog. Mem. Nat. Acad. Scis.*, **36**, 18 (1962).
15. N. Polgar, JCS, 1008, 1011 (1954).
16. J. Cason et al., JACS, **66**, 1764 (1944) and later papers.
17. J. Asselineau and E. Lederer, Compt. rendus., **228**, 1892 (1949) and later papers.
18. K. Hofmann and R. A. Lucas, JACS, **72**, 4328 (1950) and later papers.
19. J. R. Nunn, JCS, 313 (1952).
20. J. B. Brown, Chem. Rev., **29**, 333 (1941).
21. W. R. Taylor and F. M. Strong, JACS, **72**, 4263 (1950) and other papers. Review on long-chain acids, F. D. Gunstone, Quart. Rev., **7**, 175 (1953); W. J. Gensler and H. N. Schlein, JACS, **77**, 4846 (1955); **78**, 169 (1956) and other papers by Gensler, particularly **83**, 3080 (1961) for synthesis of the important unsaturated acid arachidonic acid.
22. E.g., J. Folch, JBC, **174**, 439 (1948); **177**, 497, 505 (1949).
23. H. E. Carter et al., ibid., **142**, 449 (1947); **191**, 727 (1951); **199**, 283 (1952). The last paper describes structural proof of a more complex derivative related to sphingosine.
24. For other types of phosphatides, the plasmalogens, see G. V. Marinetti et al., JACS, **80**, 1624 (1958); general review, G. B. Ansell and J. N. Hawthorne, *Phospholipids*, New York 1964.
25. The broadening activity in carbohydrate chemistry is reflected in the annual volumes of *Advances in Carbohydrate Chemistry*, Vol. 1, 1945 and following; streptidine synthesis, M. L. Wolfrom et al., JACS, **70**, 1672 (1948); **72**, 1724 (1950).

26. E. Gorin, W. Kauzmann and J. Walter, J. Chem. Phys., **7**, 327 (1939); A. Scattergood and E. Pacsu, JACS, **62**, 903 (1940); H. S. Isbell, J. Res. Natl. Bur. Std., **18**, 505 (1937).
27. R. E. Reeves, JACS, **71**, 215, 1737, 2116 (1949), and later papers.
28. Conformations of carbohydrates are reviewed in detail by E. L. Eliel, N. L. Allinger, S. J. Angyal and G. A. Morrison, *Conformational Analysis*, New York, 1965, p. 351-432; also R. J. Ferrier and W. G. Overend, Quart. Rev., **13**, 265 (1959).
29. Heparin; review by A. B. Foster and A. J. Huggard, *Adv. Carb. Chem.*, **10**, 335 (1955); M. L. Wolfrom et al., JOC, **31**, 1173 (1966); JACS, **72**, 5796 (1950) and later papers; chemists in other countries contributed to the heparin problem.
30. B. Weissmann and Karl Meyer, JACS, **76**, 1753 (1954) and later papers by Meyer.
31. R. U. Lemieux and G. Huber, ibid., **75**, 4118 (1953); **78**, 4117 (1956); Lemieux, Can. J. Chem., **31**, 949 (1953).
32. S. Winstein and R. M. Roberts, JACS, **75**, 2297 (1953). Another application of group interaction is the synthesis of the important constitutent of DNA, 2'-deoxyadenosine (B. R. Baker et al., ibid., **81**, 3967 (1959).
33. B. R. Baker et al., ibid., **82**, 2648 (1960).

ROBERT B. WOODWARD

25

Aromatic Character, 1940-1955

In an earlier essay on aromatic compounds we considered the description of aromatic character by the resonance (valence bond, v.b.) and the molecular orbital (m.o.) approach. New experimental results during the 1940-1955 period broadened the understanding of aromatic character.[1]

E. Hückel's development of the m.o. treatment of aromatic compounds implied that aromatic character required the presence of six π electrons in a single ring. Benzene, the prototype of aromatic compounds, meets this requirement, as do thiophene, furan, and pyridine. Hückel's (4n + 2) rule, where n is an integer, for the number of π electrons necessary for aromatic character has not always been understood by organic chemists to apply *only* to one-ring systems, and there has been a good deal of extrapolation to poly-cyclic systems where the (4n + 2) rule does not apply. Hückel's rule[2] gave a theoretical basis to generalizations by E. Bamberger, Sir Robert Robinson, and others that regarded aromatic character as resulting from six centrally directed valences or π electrons.[1]

This concept was a commonplace by 1940, but its application to new structures after that time increased its significance and generalized the idea of aromatic character. A striking case was the recognition of the tropylium ion[3] as an aromatic system by W. E. Doering and L. H. Knox. Extending old work by Merling, they showed that loss of HBr from dibromotropylidene gives tropylium bromide, an ionic bromide soluble in water and converted by catalytic reduction to cycloheptane.

$$C_7H_8Br_2 \xrightarrow{-HBr} \text{(+)} \quad Br^- \xrightarrow[Pt]{4H_2}$$

Dibromotropylidene   Tropylium bromide   Cycloheptane

Spectroscopic data supported the symmetrical structure for the tropy-lium ion, and it is clear that the structure has a high degree of aromatic character with its six π electrons. It is thus a member of an aromatic triad, all with six π electrons: the cyclopentadienide anion, benzene, and tropylium ion.

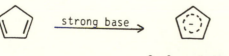

Cyclopentadienide ion   Benzene

William E. Doering (1917- , B.S. 1938, Ph.D. 1943 with R. P. Linstead, both at Harvard) collaborated with R. B. Woodward on the synthesis of quinine. Professor at Columbia, Yale, and Harvard, his work has been highly original on a number of fundamental problems, including fluxional or degenerate structures like bullvalene, which are continuously isomerizing. Much of his later work is unpublished. He first introduced the use of the inscribed circle to indicate aromatic character.

The aromatic character of the tropylium ion is supported by the properties of tropone, synthesized[4] by Doering and by H. J. Dauben at the University of Washington simultaneously; the importance of the dipolar ion resonance form is shown by the infrared spectrum, which again illustrates the aromatic character of the tropylium nucleus with its six π electrons.

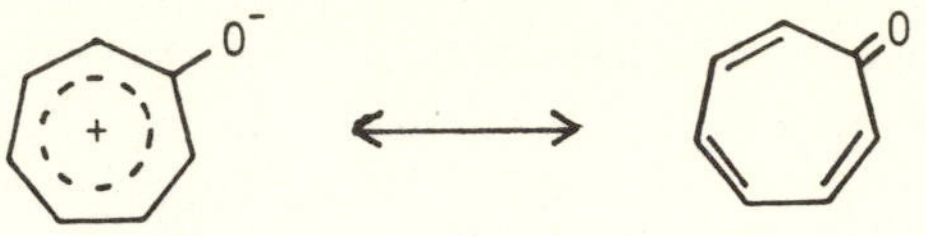

Tropone

Michael J. S. Dewar's brilliant suggestion that some mold metabolites and the alkaloid colchicine contained the "tropolone" ring with various substituents[5] brought the seven-membered ring compounds of the tropone-tropylium type into prominence. Tarbell produced experimental support for the tropolone ring system in colchicine. Several groups showed its presence in compounds where it had been postulated by Dewar, and the tropolone ring system was synthesized.[6] The rapid development of tropolone chemistry was important both in the natural product field and for enlarging ideas of aromatic character. Tropolones are much stronger acids than phenols, and their IR spectra are indicative of the aromatic character in the six π electron seven-membered ring.[7]

Tropolone

The novel aromatic structure ferrocene, derived from cyclopentadienide ion and ferrous ion[8], was assigned the "sandwich" structure by Wilkinson and Woodward.[9]

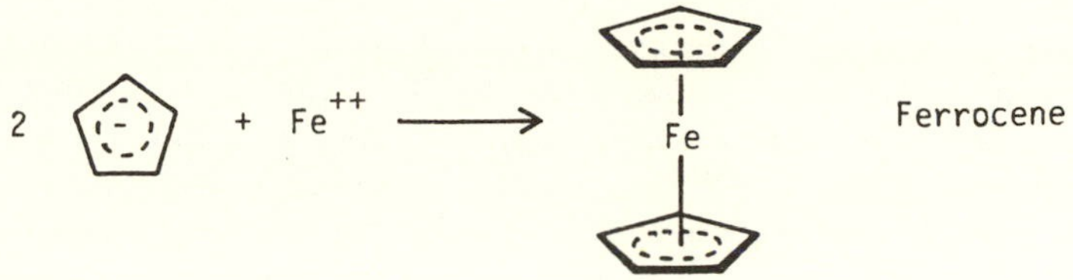

$$2 \quad \bigcirc \quad + \quad Fe^{++} \quad \longrightarrow \quad Fe \qquad \text{Ferrocene}$$

The aromatic nature of ferrocene is shown by its stability, the occurrence of the Friedel-Crafts reaction, and X-ray crystallography which shows the two rings to lie in parallel planes and all carbons to be equidistant from the iron atom.[10] The symmetry of the ferrocene molecule indicates that the six π electrons of each five-membered ring are bonded through the iron, so that the aromatic character is due to the total electronic structure. Ferrocene is the prototype for a series of analogous aromatic compounds with other metallic atoms in place of iron.[10] The whole class of sandwich compounds represents an imaginative addition to the class of aromatic structures.

If, as some chemists assumed, Hückel's six π electron condition for aromaticity could be generalized to a system with $(4n + 2)$ π electrons, one might expect that systems with ten, fourteen, and eighteen π electrons $(n = 2, 3, 4)$ would show aromatic character. However, the restrictions on this condition must be pointed out: the structure must have one ring, and the atoms must be coplanar, so that resonance stabilization can occur. The most elegant demonstration of aromatic character, with fourteen π electrons, is that of V. Boekelheide's dihydropyrenes, in which the π electron system in the periphery is coplanar and monocyclic, the saturated interior carbons not being part of the π system. The uncharged hydrocarbon has fourteen π electrons $(4 \times 3 + 2)$ and should be aromatic. Treatment with potassium adds two electrons, thus the dianion has sixteen π electrons (4×4) and should not be aromatic. Experimental observations match the predictions in striking fashion: the neutral hydrocarbon is aromatic and the dianion is non-aromatic.[11] Other $(4n + 2)$ π electron systems with ten π electrons show some aromatic character.[12]

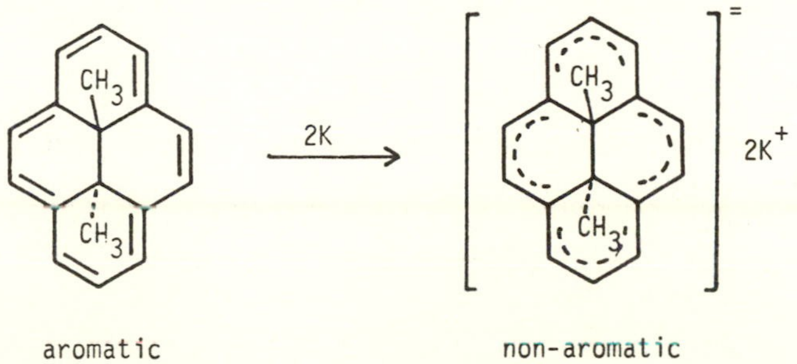

aromatic   non-aromatic

Willstätter's synthesis of cyclooctatetraene (COT), its lack of aromatic character, and the confirmation[13] of this work by A. C. Cope and C. G. Overberger are described elsewhere. After World War II, it was found that Walter Reppe in Germany had developed the commercial synthesis of COT from metal acetylides and had used it as starting material for many classes of organic compounds.[14] COT has eight π electrons, not (4n + 2), and furthermore it is not coplanar, so it has no aromatic character. The chemistry of COT and related compounds has been explored in elegant researches by Cope and his group at MIT.[15]

Another cyclic structure on which an enormous amount of work has been expended is cyclobutadiene, which would have 4n (n=1) π electrons. Rowland Pettit at Texas finally proved that cyclobutadiene could be isolated as a stable iron tricarbonyl compound; this complex had aromatic properties, somewhat similar to ferrocene, but free cyclobutadiene itself showed no aromatic properties.[16] It behaves like a very reactive diene. This conclusive work on COT and cyclobutadiene cleared up two of the outstanding problems in aromatic character.

Fe(CO)$_3$

aromatic

cyclobutadiene

non-aromatic

Rowland Pettit (1927-1981, B.Sc. 1949; Ph.D 1953, both at Adelaide with G. M. Badger) taught at the University of Texas from 1957 until his death. He worked on a variety of metal carbonyl complexes and aromatic or supposed aromatic compounds.

The fused-ring compounds, such as the azulene, pentalene, and heptalene, which have been intensively studied, showed slight aromatic character in the first, and none in the others.[17]

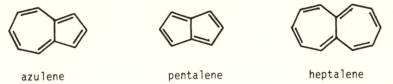

azulene   pentalene   heptalene

The whole field of aromatic character was a major problem for many years, and to the work of Doering, Boekelheide, and Pettit should be added that of Ronald Breslow of Columbia, who prepared the cyclopropenium system, which follows Hückel's rule for n = 0, with 2 π electrons.[18] J. D. Roberts had made calculations of the resonance stabilization of such systems.[19] Spectroscopic and X-ray measurements of Breslow's compounds indicate the aromatic nature of the cyclopropenium ring.[18]

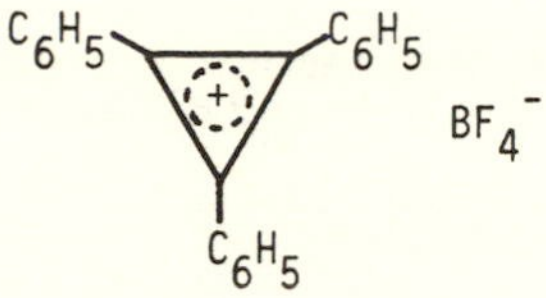

Ronald Breslow (1931- , A.B. 1952; Ph.D. 1955, both at Harvard with Gilbert Stork) has taught at Columbia since 1956. He has worked on carbonium ions, mechanism of enzymatic reactions, aromaticity, and model systems for enzymes.

Two types of aromatic substitution have been developed in detail during this period. One is nucleophilic replacement of a leaving group, such as halogen in an aromatic ring which has an electron-attracting group in the *ortho* or *para* position, such as nitro. Almost since organic chemistry became a coherent science, this has been a known reaction, but the subject was vitalized in a detailed review[20] by J. F. Bunnett and R. E. Zahler, published from Reed College.

Joseph F. Bunnett (1921- , B.A. 1942, Reed; Ph.D. 1945, Rochester with D. S. Tarbell) has taught at Reed, Chapel Hill, Brown, and the University of California at Santa Cruz. He has done research on nucleophilic aromatic substitution, action of strong base on aromatic halides, and ion-radical reactions and has edited *Accounts of Chemical Research* since its beginning.

The course of nucleophilic aromatic substitution has been best explained by an addition-elimination mechanism[21], the negative charge being stabilized by resonance with the electron-withdrawing group.

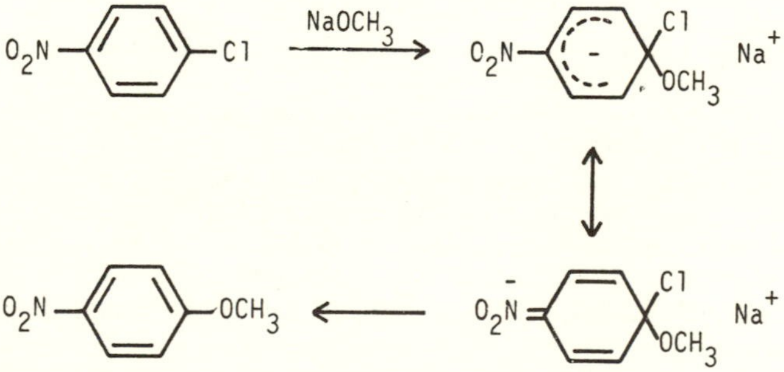

The second type occurs in the treatment of aromatic halides by very strong bases, such as sodamide, $Na^+NH_2^-$. It was frequently observed that the entering NH_2 group was in a different position from the leaving halogen, in cases where substitution allowed this to be observed. Bunnett and Zahler collected numerous examples and called this cine-substitution.[22] Gilman observed many examples such as the following.[23]

A brilliant study of this reaction by John D. Roberts at MIT and Caltech, using kinetics, isotope effects, and tracer techniques with deuterium
and ^{14}C, showed that a symmetrical neutral intermediate is formed, called
a "benzyne", written with a triple bond, which then adds NH_3 to form
aniline.[24] This idea has been a very fruitful one; Roberts showed that 1-
chlorocyclohexene treated with strong base formed cyclohexyne.[25]

John D. Roberts (1918-   , A.B. 1941; Ph.D. 1944, both at UCLA with
W. G. Young) was instructor at UCLA, National Research Fellow at Harvard,
a faculty member at MIT from 1946 to 1952, and since that time at Caltech.
Roberts was one of the first to apply ^{14}C to mechanistic studies, such as rearrangements of alkyl chlorides by aluminum chloride. He was one of the first
organic chemists to recognize the importance of NMR for establishing structures and measuring reaction rates of hydrogen shifts and rotations around
single bonds. He pioneered ^{1}H, ^{13}C, and ^{15}N NMR as a technique in organic
chemistry, and he carried out synthetic and mechanistic studies on small ring
compounds.

LITERATURE CITED

1. General reviews: *Non-Benzenoid Aromatic Compounds*, David Ginsburg, Ed., New York,
   1959; N. C. Rose in *The Kekulé Centennial*, O. T. Benfey, Ed., Washington, 1966, pp.
   91 ff; G. W. Wheland, *Resonance in Organic Chemistry*, New York, 1955, pp. 135 ff;
   C. K. Ingold, *Structure and Mechanism in Organic Chemistry*, 2nd ed., Ithaca, 1969, pp.
   173-207. Ingold's account contains a history of the subject.
2. E. Hückel, Z. für Physik, **70**, 204 (1931); his (4n + 2) rule for the number of π electrons
   required for aromatic character is not given explicitly.
3. W. E. Doering and L. H. Knox, JACS, **76**, 3203 (1954).
4. H. J. Dauben, Jr., and H. J. Ringold, ibid., **73**, 876 (1951); Doering and F. L. Detert,
   ibid., same page.
5. M. J. S. Dewar, Nature, **155**, 50, 141, 479 (1945); on the tropolone ring in colchicine,
   H. R. V. Arnstein, D. S. Tarbell et al., JACS, **70**, 1669 (1948); **71**, 2448 (1949).
6. Reviews on tropolones: J. W. Cook and J. D. Loudon, Quart. Rev., 5, 99 (1951);
   T. Nozoe in *Non-Benzenoid Aromatic Compounds*, pp. 339-464. Nozoe was a very active
   worker in the field. Syntheses of tropolones: J. W. Cook and A. R. Somerville, Nature,
   163, 410 (1949); D. S. Tarbell, G. P. Scott and A. D. Kemp, JACS, **72**, 379 (1950);
   Doering and Knox, ibid., 2305; Cooke et al., Chem. and Ind., 427 (1950); R. D.
   Haworth and J. D. Hobson, p. 441.
7. G. P. Scott and D. S. Tarbell, JACS, **72**, 240 (1950).

8. T. J. Kealy and P. L. Pauson, Nature, **168**, 1039 (1951).

9. G. Wilkinson, M. Rosenblum, M. C. Whiting and R. B. Woodward, JACS, **74**, 2125, 3458 (1952).

10. P. L. Pauson, Quart. Rev., **9**, 391 (1955); *Non-Benzenoid Aromatic Compounds*, pp. 107-140.

11. R. H. Mitchell, C. E. Klopfenstein and V. Boekelheide, JACS, **91**, 4931 (1969). The criterion of aromaticity used here is the effect of an aromatic π electron system on the NMR spectrum of protons in the plane of the system.

12. T. J. Katz, ibid., **82**, 3784 (1960); **85**, 2852 (1963); E. A. LaLancette and R. E. Benson, p. 2853.

13. A. C. Cope and C. G. Overberger, ibid., **69**, 976 (1947); **70**, 1433 (1948).

14. See R. A. Raphael in *Non-Benzenoid Aromatic Compounds*, pp. 465 ff.

15. Review, L. E. Craig, Chem. Rev., **49**, 103 (1951); evidence for non-planar structure of COT is reviewed by W. B. Person, G. C. Pimentel and K. S. Pitzer, JACS, **74**, 3437 (1952).

16. R. Pettit et al., ibid., **87**, 3253, 3254 (1965).

17. Azulenes: E. Heilbronner in *Non-Benzenoid Aromatic Compounds*, pp. 171-276; pentalene and heptalene, E. D. Bergmann, ibid., pp. 141-169; synthesis of heptalene, H. J. Dauben, Jr., and D. S. Bertelli, JACS, **83**, 4659 (1961).

18. R. Breslow, ibid., **79**, 5318 (1957) and many later papers for other substituents than aryl on the three-membered rings; cf. Rose, *Kekulé Centennial*, pp. 97-100.

19. J. D. Roberts, A. Streitwieser and C. M. Regan, JACS, **74**, 4579 (1952).

20. J. F. Bunnett and R. E. Zahler, Chem. Rev., **49**, 273-412 (1951).

21. Ibid., id., pp. 295 ff. Bunnett's discussion is accepted by C. K. Ingold, *Structure and Mechanism*, 2nd. ed., 1969, p. 395.

22. Bunnett and Zahler, p. 384; they give early refs.

23. H. Gilman and S. Avakian, JACS, **67**, 349 (1945); Gilman and R. H. Kyle, ibid., **70**, 3945 (1948); **74**, 3027 (1952).

24. J. D. Roberts, Dorothy A. Semenow et al., ibid., **75**, 3290 (1953); **78**, 601 (1956) and other papers.

25. F. Scardiglia and J. D. Roberts, Tetrahedron, **1**, 343 (1957).

SAUL WINSTEIN

26

Aspects of Physical
Organic Chemistry, 1940-1955

In our earlier essay on stereochemistry we discussed the energy barrier to rotation in ethane and other open chain compounds. We traced the Sachse-Mohr structures of saturated six-membered rings and made brief mention of the fundamental advance in understanding conformation and reactivity due to D. H. R. Barton[1] in the early 1950s. Like other major contributions, Barton's synthesis had its roots in various past observations and ideas, but it pointed the way to new fields of significant work for the future.

Stereochemistry and reaction mechanisms are intimately connected; the course which reactions take is affected, and in many cases determined, by the spatial arrangement of atoms or groups. In this essay we discuss some significant work where stereochemistry is a controlling factor or where stereochemistry is elucidated by new methods. Comprehensive reviews by E. L. Eliel and others cover the relation of stereochemistry to reaction mechanisms.[2]

Barton's classic papers dealt mainly, although not exclusively, with fused ring systems in which the axial or equatorial conformation of functional groups was frozen by the ring system. He correlated relative reactivity of groups with their conformations, reactions at the equatorial position in general being faster than those at the axial position. This was because the axial bonds are parallel to each other in the 1,3,5-positions on each side of the ring, and axial groups are shielded compared to equatorial ones. The equatorial bonds radiate out from the ring, at angles of roughly 60° for the 1,2-position, 120° for the 1,3,-and 180° for the 1,4-positions.

S. Winstein and N. J. Holness at UCLA studied effects of conformation on reactions of groups on a single cyclohexane ring[3] in 1955. The single ring could, of course, undergo ring flips to change an axial group to equatorial, and vice versa. However, by putting a *tert*-butyl group on the ring, the conformations were effectively frozen, because the bulky *tert*-butyl group was accommodated better in the equatorial position. A flip to make the *tert*-butyl group axial would introduce strong steric repulsions between it and the axial hydrogens and carbons in the 3- and 5-positions. The equatorial conformation for the *tert*-butyl group is favored by an estimated 5.5 kcal/mole, which is also the estimated figure for the stability of the chair form of cyclohexane compared to the boat form.

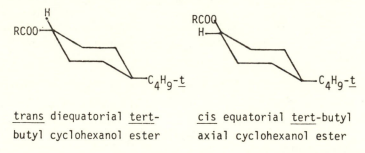

equatorial <u>tert</u>-butyl   axial <u>tert</u>-butyl

Winstein found that the *trans-tert*-butylcyclohexanol ester, in which both *tert*-butyl and OCOR are equatorial, is hydrolyzed ten times as fast as the *cis-tert*-butyl compound, where the *tert*-butyl must stay equatorial, but the OCOR is axial. The axial ester, as in Barton's examples, is hydrolyzed more slowly than the equatorial. Similar relationships are observed in the *cis* and *trans* 3-*tert*-butylcyclohexanol esters, where in each case the ester is of known conformation because the *tert*-butyl group stays equatorial. A number of other relative rates of reactions were determined and in each case are in the order predicted, on the basis of the *tert*-butyl group being locked into the equatorial position.

<u>trans</u> diequatorial <u>tert</u>-   <u>cis</u> equatorial <u>tert</u>-butyl
butyl cyclohexanol ester   axial cyclohexanol ester

In this paper also, an important relationship was derived, the Winstein-Holness principle, which says that the *total rate* of reaction of a mixture of conformational isomers is proportional to the number of molecules in each conformation. With the *tert*-butyl group present, all molecules are in essentially the same conformation.[3,4] This principle is to be distinguished from the Curtin-Hammett principle,[5] which says that the *proportion of products* from a reaction of a mixture of conformational isomers is independent of population in each conformation if energy of interconversion of the conformations is small.

The Winstein-Holness paper combines a good idea, thorough conceptual analysis, and a wide range of experimental measurements, so that it stands as a second classic after Barton's in modern stereochemistry. Winstein's use of the *tert*-butyl group as a means of freezing conformation has been employed by many subsequent workers, and quantitative data are now available about conformation and reactivity for a variety of groups.[2]

Saul Winstein (1912-1969, AB, 1935, UCLA; Ph.D. 1938, Caltech with

Lucas) was a postdoctorate with P. D. Bartlett at Harvard and was on the staff at UCLA from 1941 until his untimely death in 1969. He was one of the leaders of modern physical organic chemistry. His early work on neighboring groups branched out in many directions, to solvolysis, ion pairs, free-radical reactions, explanations of new reactions of steroids, conformational analysis, mechanism of organomercurial reactions, homoaromaticity, and many more. His personal influence, both directly and through his many students and postdoctorates, was perhaps as great as that of any individual.[6]

Ernest L. Eliel (1921- , Dr. Phys. Chem. Scis. 1946, University of Havana; Ph.D. 1948, Illinois with H.R. Snyder) taught at Notre Dame from 1948 to 1972, and since then at Chapel Hill, North Carolina. His research has included organosulfur compounds, carbanions, and heterocyclic chemistry.

Two comprehensive studies on the stereochemistry of elements other than carbon deserve mention. The first is by Leo Sommer on the preparation and mechanism of reaction of optically active compounds with asymmetric silicon atoms. In general, the reactions show a similarity to those of optically active carbon compounds, but there are striking differences due to the larger size of silicon and its failure to form stable chains or stable multiple linkages with itself. The fact that silicon is in the same group in the periodic table as carbon gives it a normal valence of four, but its chemistry is different and usually more complicated than that of carbon in many ways.[7]

Leo H. Sommer (1917- , B.S. 1942, Ph.D. 1945, Pennsylvania State University with Frank C. Whitmore) was a faculty member at Penn State from 1947 to 1965, where he did most of the silicon work. From 1965 he has been at the University of California at Davis.

The stereochemistry of phosphorus compounds, on which little conclusive work had been done, was investigated systematically by W. E. McEwen starting in the 1950s.[8] He worked out reliable methods of obtaining optically active phosphorus compounds of various types and studied the mechanisms of reactions involving the optically active phosphorus center. Both series of papers naturally benefited from the detailed knowledge of the stereochemical course of reactions involving asymmetric carbon; they also utilized new instrumental methods available. NMR spectroscopy proved to be particularly useful in studying some phosphorus compounds.

William E. McEwen (1922- , A.B. 1943, Ph.D. 1947, at Columbia with W. E. Doering) taught at the University of Kansas from 1947 to 1962, where he started research on the stereochemistry of phosphorus. He has been at the University of Massachusetts, Amherst, since 1962. Sommer's and McEwen's contributions to stereochemistry deserve more attention than they have received.

The use of optical rotatory dispersion for obtaining absolute configurations has been mentioned. This technique, developed by Carl Djerassi at Stanford, makes it possible to determine the configuration of a halogen next to a carbonyl group[9]; many other applications of ORD to determine configuration are known.

The importance of NMR for determining conformations, configuration, and rate of change of conformation cannot be overemphasized. It allows measurements of reaction rates not accessible by any other means and permits assignment of the conformation of molecules in solution.[10] X-ray crystallography, important as it is, gives information only about the solid state. H.S. Gutowsky at Illinois pioneered measurements of barriers to rotation around single bonds by NMR.[11]

The Ingold classification of replacement reactions at a saturated carbon atom was widely used in this country after 1940; the Robinson-Ingold ideas, however, were by no means accepted uncritically. Although Ingold's work was known to the initiated in the 1930s, A.E. Remick's book of 1943 gave a thoughtful and critical review of physical organic chemistry, including Ingold's system. Less physical and magisterial in approach than Hammett's classic, Remick's book was a useful disseminator of current ideas of reaction mechanisms to organic chemists in general.[12] Elementary texts in organic chemistry after 1945 incorporated much material on reaction mechanisms.

The rate of the S_N2 reaction is first order in the nucleophile or replacing agent, and first order in the substance undergoing replacement. The classical case is the displacement with inversion of 2-octyl iodide with radioactive iodide ion, quoted previously. Inversion is so universal an accompaniment of direct displacement that when an apparent retention of configuration does occur, it is normal to look for two consecutive reactions, each going with inversion. (Inversion cannot be recognized if replacement involves a primary carbon, such as RCH_2Cl.)

Extended studies on relative reactivity of displacing groups ("bases" or "nucleophiles") are tabulated by Streitwieser and others.[13] In general, the more polarizable the nucleophile, the more rapid its reactions; sulfur nucleophiles react much faster than oxygen analogs, while iodide ion is usually a better nucleophile than other halide ions. Greater substitution around the leaving group slows down the reaction by hindering the approach of the nucleophile from the rear.[14] Tertiary halides usually do not undergo displacement reactions but instead follow the S_N1 or solvolysis path.[15]

Many of the fundamental synthetic reactions are of the direct displacement type, and one of the important discoveries from the synthetic side has been that direct displacement reactions go very much faster in *dipolar aprotic* solvents than in many of the traditional solvents, such as water and alcohols.[16] These *protic* solvents furnish a proton which can hydrogen-bond to nucleophiles and greatly slow the reaction. J.C. Sheehan at MIT apparently first pointed out the great value of DMF (dimethylformamide) in accelerating displacement reactions, and the use of DMF, DMSO (dimethyl sulfoxide), and HMP (hexamethylphosphoramide) became very general.[16] (The last solvent must be used with great care in a hood, because it is a potent carcinogen to animals.) The increase in basic or proton removing strength by $KOC(CH_3)_3$ in DMSO, as shown by Cram[17], is due to preferential solvation of K^+ by DMSO, breaking up ion pairs and making the $^-OC(CH_3)_3$ freer to extract a proton. Where the nucleophile is a salt of a carbanion, the dipolar aprotic solvent breaks up the ion pair, or larger aggregate, of the carbanion and its cation, thus making the former more available for reaction. Zaugg's studies[16] support the idea that the dipolar aprotic solvent solvates the cation preferentially. He found that carbanions from a malonic ester and the metallic ion, $RC(COOR)_2^- Na^+$, existed in benzene as complexes of about forty units. Addition of DMF broke up the complexes, and the displacement reaction of the carbanion with $n\text{-}C_4H_9Br$ was greatly accelerated.

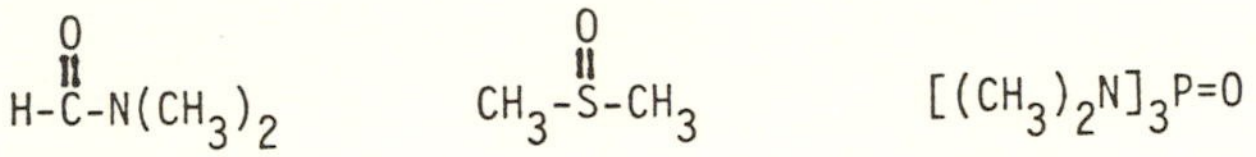

$$\underset{\text{DMF}}{\overset{\displaystyle O \atop \displaystyle \|}{H-C-N(CH_3)_2}} \qquad \underset{\text{DMSO}}{\overset{\displaystyle O \atop \displaystyle \|}{CH_3-S-CH_3}} \qquad \underset{\text{HMP}}{[(CH_3)_2N]_3P{=}O}$$

Donald J. Cram (1919-   , B.S. 1941, Rollins; M.S. 1942, Nebraska; Ph.D. 1947, Harvard with L.F. Fieser) has taught at UCLA since 1948. His research has included stereochemistry of carbanions[17], macro rings (cyclophanes), and model enzyme systems.

Harold E. Zaugg (1916-   , A.B. 1937, Oberlin; Ph.D. 1941, Minnesota with R. T. Arnold) did research at Abbott Laboratories from 1941 to 1981, being a senior research fellow from 1972. His work has included synthesis of drugs affecting the central nervous system, solvent effects, and reactions of delocalized anions.

A more recent development, and one of very great practical usefulness, is the effect of a small amount of substituted quarternary ammonium salts, such as $C_6H_5N^+(CH_3)_2CH_2C_6H_5Cl^-$, in increasing the rate of displacement and other reactions between organic compounds virtually insoluble in water and a nucleophile in aqueous solution.[18] This "phase transfer" process apparently works by increasing the mutual solubility of the two phases, and probably also by some selective solvation of the components of the reaction.

A significant theoretical contribution to the direct displacement reaction was Saul Winstein's demonstration of the generality of participation by neighboring groups, the reaction which he and Lucas first identified. Thus displacement on a *trans*-acetoxybromide with silver acetate gave a cyclic acetoxonium ion, which could react further to yield a product of inverted or retained configuration, depending on the conditions.[19]

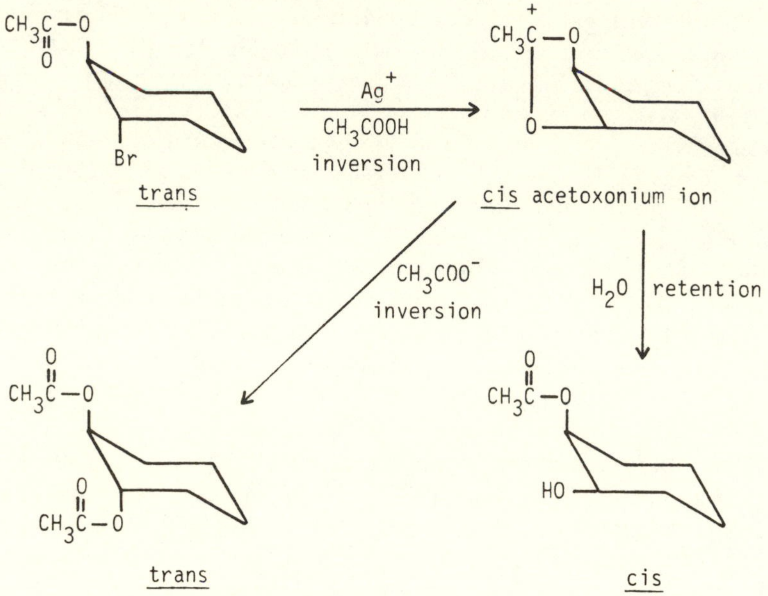

Winstein later showed neighboring group participation for OH, O⁻, OR, CONH₂, COO⁻, and C=C. These ideas found synthetic application in carbohydrate chemistry by B. R. Baker, R. U. Lemieux, and others. In some cases, the neighboring groups are not on adjacent carbons but are farther removed. A striking example is the isolation and characterization of the quinone-like intermediate shown below, which is a relatively stable compound, *not* a transition state, and is formed by participation of the π electrons of the phenyl group.[20]

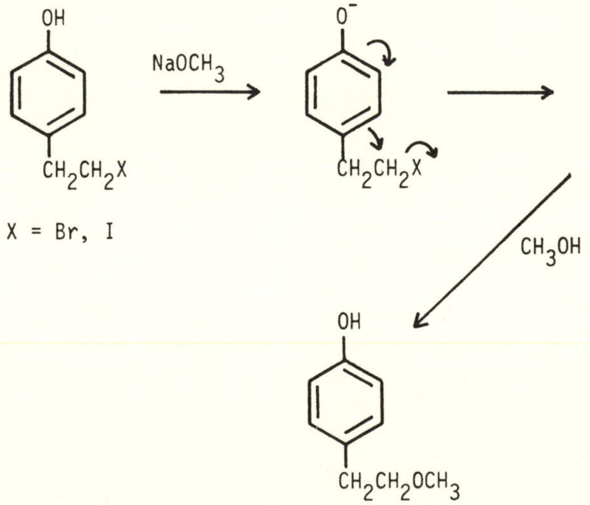

A displacement reaction assisted by a neighboring group is an important aspect of many enzymatic processes; the energy necessary to break bonds in the substrate is decreased by interaction with nearby groups or solvent molecules on the enzyme surface.[21] Such assisted direct displacement is very closely related to solvent-assisted $S_N 1$ or solvolytic replacement. In the solvolysis process, or $S_N 1$ reaction, the rate of reaction is proportional only to the concentration of the *tert*-butyl chloride, for example, and is unaffected by nucleophiles, such as hydroxyl ions.

$$(CH_3)_3CCl \xrightarrow[\text{slow}]{H_2O} (CH_3)_3C^+ \quad Cl^- \xrightarrow[\text{fast}]{H_2O} (CH_3)_3COH$$

The Ingold school attributed the rate-determining ionization of *tert*-butyl chloride to the electron-donating effect of the alkyl groups, which pushed off the chloride ion leaving a reactive carbonium ion. This then reacted rapidly with solvent or some dissolved nucleophile. In other systems showing similar behavior, such as some benzyl or allyl compounds, the resonance stabilization of the resulting carbonium ion caused the rapid ionization.

American workers have made significant changes to these rather simplistic notions of carbonium ion intermediates. Hammett in his first work emphasized the role of the solvent in promoting ionization and stabilizing the carbonium ion. The importance of steric strains in promoting carbonium formation was emphasized by P. D. Bartlett and by H. C. Brown. Bartlett's work on bridgehead halogens showed that carbonium ions formed only if they could assume a planar structure. Rearrangements of the Whitmore type frequently followed formation of highly branched carbonium ions.

John D. Roberts, in some of the first tracer studies using ^{14}C at MIT in 1949 showed that skeletal rearrangement of carbon occurred with *tert*-amyl and *tert*-butyl chloride and aluminum chloride or bromide.[22] The results could be classified under the Whitmore rearrangement scheme. Roberts also found that cyclopropylmethyl chloride solvolyzed twenty-seven times as rapidly as cyclobutyl chloride and forty times as fast as β-methylallyl chloride. The cyclopropyl derivative rearranged on solvolysis to cyclobutyl and allylmethyl derivatives.[23] In this way the cyclopropyl group shows its resemblance to the carbon-carbon double bond, a resemblance also seen in its spectroscopic behavior by E. P. Carr.

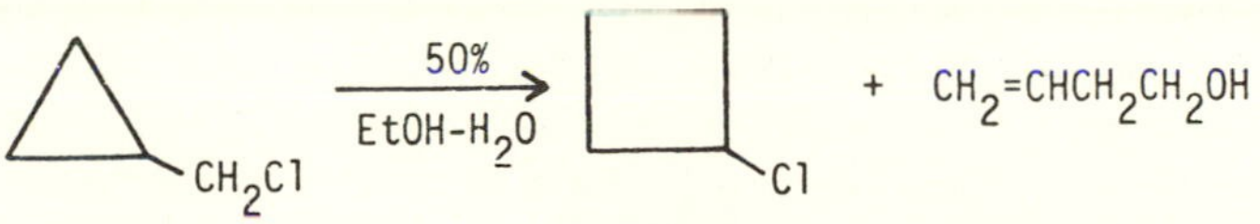

$$\triangle\text{-}CH_2Cl \xrightarrow[\text{EtOH-H}_2O]{50\%} \square\text{-}Cl \quad + \quad CH_2=CHCH_2CH_2OH$$

| cyclopropylmethyl chloride | cyclobutyl chloride | "homoallylic" alcohol |

Winstein and R. Adams[24] unraveled an analogous case of homoallylic systems, which explained some novel reactions in steroids. Here the double bond is in the 3,4-position (homoallylic) with respect to the leaving group, and the cyclopropyl derivative (the "i-steroid") readily reforms the unsaturated steroid. This requires a driving force to form a stabilized ion from both directions and is in agreement with the high rate of solvolysis found by

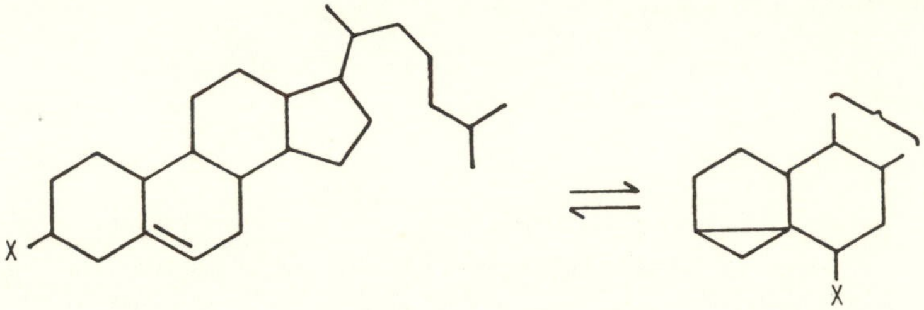

"i-steroid"

Roberts for cyclopropylmethyl derivatives. This study of the relation between cyclopropylmethyl and homoallylic systems was extended by Winstein to further cases.

The most striking example of the promoting effect of a double bond on homoallylic solvolysis is *anti*-7-norbornenyl tosylate, which solvolyzes in acetic acid 10^{11} times faster than the saturated 7-norbornyl compound.[25] In the unsaturated compound the π electrons assist ionization, presumably by stabilizing the carbonium ion. Phenonium ion stabilization of solvolysis was reported by Cram and later by Winstein.[26]

In the solvolysis reaction, if part of the driving force comes from solvation of the carbonium ion, it is pertinent to examine the stereochemistry of the product when an optically active tertiary alcohol derivative is solvolyzed. W. E. Doering and H. H. Zeiss found that methanolysis of an optically active tertiary phthalate gave sixty percent inversion.[27]

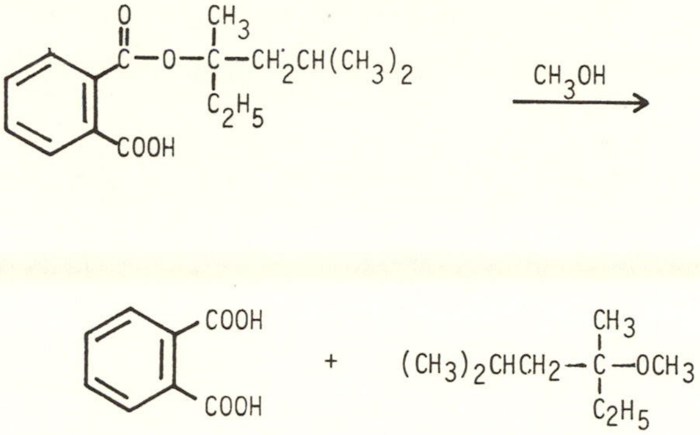

Obviously here the solvent attaches to the carbonium ion in a manner which preserves asymmetry, and in general the stereochemistry of the overall solvolysis must depend on the relative rate constants for several processes involving solvent and the carbonium ion. E. Grunwald and Winstein[28] showed that the ionizing power of a solvent in solvolytic reactions can be expressed quantitatively in an empirical equation. This equation is reasonably successful in predicting solvolysis rates by a number of solvents and mixtures. A more elaborate equation[29] contains a term representing the nucleophilicity of the solvent in addition to its ionizing power; this reduces to the first equation for solvents of similar nucleophilicity, such as formic and acetic acids.

Further detailed experimental analyses of solvolyses of optically active brosylates (*p*-bromobenzenesulfonates) led Winstein to the concept of *intimate ion pairs*, in which the carbonium ion and its counter ion, the brosylate, are enclosed by the same sphere of solvent molecules. This intimate ion pair can collapse to starting material or the racemic form of it. The intimate ion pair can change to a *solvent separated* ion pair, in which the original ions are now in two different spheres of solvent. The two types of ion pairs can be recognized by experimental criteria[30], and these criteria are sharpened by the use of isotopic labels in the sulfonate esters.[31] Effects of added salts can be used for distinguishing the two types of ion pairs.

It is worth recalling that S. F. Acree at Johns Hopkins in 1910 had made a kinetic distinction between ionized and unionized reagents[32], which corresponds in principle to the Winstein scheme of ion pairs, although Acree did not have the experimental or conceptual tools available fifty years later.

We have given examples of the facilitating effect or anchimeric effect (the term is due to Winstein[33], although the idea was suggested by Ingold in 1939) of neighboring groups including π electrons on direct displacement or solvolysis. The classical case of the so-called "non-classical carbonium ion" is the solvolysis of exo-norbornyl brosylates.[34] The enhanced rate, stereochemistry, and labeling pattern of this reaction are attributed to the σ-bond-stablilized carbonium ion. The ensuing controversy over the necessity for invoking the σ-bonded carbonium ion can be followed by the curious in the relevant reviews.[35]

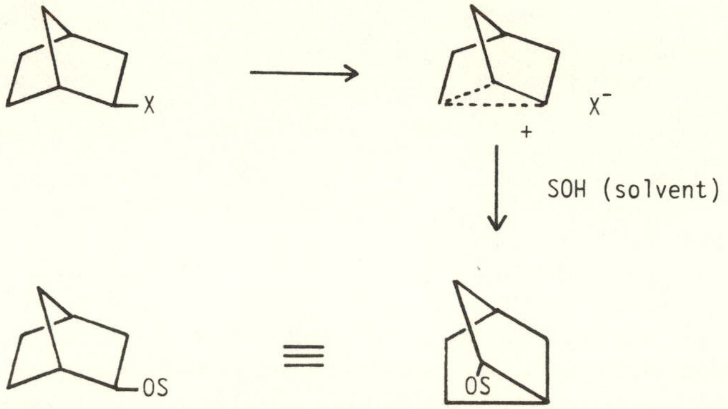

The overall significance of the solvolysis reaction is that it gives information, fragmentary to be sure, about the interactions of solvent molecules with themselves and their interaction with solutes undergoing reactions. Theoretical knowledge of the liquid state is far behind that of the solid or gaseous state, and hence any additional information about reactions in solution is valuable.

Although there are few if any recognized solvolysis reactions among enzymatic processes, the importance of the solvent and of solvation in such reactions scarcely needs emphasis. One of the intermediates in the synthesis of polyisoprenes, $(CH_3)_2C=CHCH_2OPOP$, mentioned earlier, would be expected to solvolyze rapidly, but in its role as a biochemical reactant with $CH_2=C(CH_3)CH_2CH_2OPOP$, it probably reacts at an enzyme surface and not via a carbonium ion.

Intensive study of free-radical reactions and of the free-radical chain reaction of polymerization continued to yield results of fundamental significance and practical usefulness during this period. The research of Morris S. Kharasch at Chicago, first on the free-radical addition of HBr to olefins in the 1930s (the peroxide effect) and then on a wide variety of free-radical reactions, established the field, and many workers were attracted to it.[36] With an exceptional group of collaborators, including F. R. Mayo, Cheves Walling, H. C. Brown, and others, Kharasch continued to detect and elucidate whole families of free-radical reactions which form carboxylic acids, halides, sulfonic acids, and other derivatives from saturated hydrocarbons.[37] Kharasch liked to point out that the textbook picture of saturated hydrocarbons as unreactive applied only to ionic reactions; the saturated hydrocarbons are highly reactive toward a variety of free-radical reagents.

Kharasch developed free-radical routes to formation of carbon-carbon bonds, as by radical addition of alkyl halides to carbon-carbon double bonds.[37] He also studied radical decomposition of hydroperoxides, radical addition of thiols to olefines and a modification of Grignard reactions by a free-radical chain process initiated by cobaltous ions. His last reaction, published posthumously[38], describes the preparation of allylic esters by a radical process from peresters.

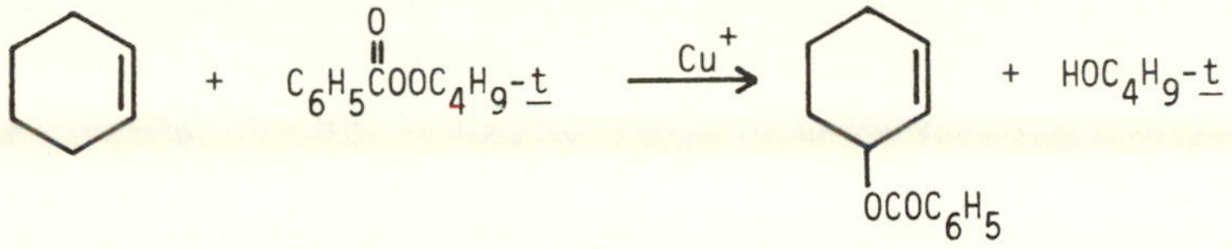

Frank R. Mayo (1908-   , B.S. 1929, Ph.D. with Kharasch, both at Chicago) was on the faculty of Chicago for several years, collaborated with Walling at U.S. Rubber Co. on fundamental studies on free-radical polymerization, and later worked at General Electric and Stanford Research Institute. He also published excellent work on the acid-catalyzed cleavage of ethers.

Cheves Walling (1916-   , A.B. 1937, Harvard; Ph.D. Chicago, with Kharasch) worked at Du Pont and then at U. S. Rubber (1943-1949), where much of the classical work on copolymerization was done in collaboration with Mayo and M. S. Matheson. He was professor at Columbia from 1952 to 1966, where he worked on reactions occurring at very high pressures, and at Utah from 1970 on. He was editor of JACS from 1975 to 1981.

The stereochemistry of free-radical addition is not as clear-cut as that of ionic addition.[37] In some cases addition to *cis* and to *trans* isomers gives the same final products, indicating loss of configuration of an intermediate. Molecular rearrangements during free-radical additions are much rarer than during ionic addition.

Free-radical chain reactions were shown to occur in addition polymerization at about the same time that Kharasch's work was becoming influential. Polymerization is a process of great practical importance, as well as of fundamental chemical interest; it attracted able researchers in industrial and academic laboratories, and several outstanding reviews were written in addition to the one by Carothers.[39,40,41] During World War II, the shortage of natural rubber made the development of synthetic rubber by free-radical polymerization of crucial importance. Marvel's group at Illinois made contributions to the solution of the rubber crisis, and academic and industrial research was stimulated and supported.

One of the problems pretty well worked out by 1955 was the question of copolymerization; if a mixture of monomers (M_1 and M_2) is polymerized, what proportion of each monomer appears in the polymer molecule? This reduces to the monomer reactivity ratio; it must be determined for a growing polymer chain ending in the M_1 radical whether it adds as next monomer unit M_1 or M_2. Similarly for growing chains ending in M_2, does it add M_1 or M_2? This problem required experimental solution[42], because the answer could not be inferred from the rates of polymerization of M_1 or M_2 by themselves. Furthermore, the properties of the polymer may be markedly affected by presence of a second monomer unit; thus the synthetic rubber made in largest amount during World War II was GR-S (Government Reserve-Synthetic), a copolymer of seventy-five percent butadiene and twenty-five percent styrene.

The copolymerization equation assumes that relative rates do not depend on the size of the growing polymer chain and are also unaffected by the nature of the monomer unit next to the free-radical end. Experiment showed that these assumptions are valid. The most reasonable explanation for the monomer reactivity ratios is that, in the transition state, the radical and the reacting monomer may be stabilized by resonance forms involving a transfer of an electron from the radical to the monomer.[43]

$$-M_1\cdot \quad + \quad CH_2=CHCl \longrightarrow -M_1^+CH_2\overset{\cdot}{C}HCl^- \longleftrightarrow -M_1^-CH_2\overset{\cdot}{C}HCl^+$$

Charles C. Price (1913-   , A.B. 1934, Swarthmore; Ph.D. 1936, Harvard with L. F. Fieser) was on the faculty at Illinois, Notre Dame, and Pennsylvania. There he was chairman and Benjamin Franklin Professor. He has worked on a variety of mechanistic problems, addition polymerization, polypropylene, polyurethans, and organic sulfur compounds. He has taken an active interest in politics and public affairs and was president of the American Chemical Society in 1965.

The actual existence of free radicals as intermediates in addition polymerization was shown directly by electron spin measurements on a polymerizing system by Walling.[44] The measurements of individual rates for free-radical polymerizations is possible by initiating the radical chain by intermittent illumination. Such experiments, done by Bartlett and by Matheson in this country, and by H. W. Melville in Britain[45] show with vinyl acetate that a polymer molecule of 10,000 monomer units is formed in about 1-4 sec., and free-radical concentrations are about 0.5×10^{-8} molar. The lifetime of an individual radical is thus of the order of 10^{-4} sec.

The growing polymer chain may be terminated by reaction of the radical with some compound to transfer an atom or group to the chain, making a saturated molecule and forming a new radical. This process of *chain transfer* allows control of the average molecular weight of the polymer, depending on the effectiveness of the chain transfer reaction, the concentration of the agent, and the effectiveness of R at producing new chains. Many workers have studied it.[46,47]

$$-M\cdot \quad + \quad XR \longrightarrow -MX \quad + \quad R\cdot$$

The poor polymerization of allyl acetate is due to chain transfer with the monomer; it forms an unreactive resonance-stabilized allylic radical. Benzyl radicals are likewise unreactive and inhibit formation of long polymer chains. Thus a growing polymer chain may react with allyl acetate in two ways, to add a monomer unit in the usual way, or react by chain transfer to remove an allylic hydrogen.[46] The rate constant for transfer is larger than for propagation.

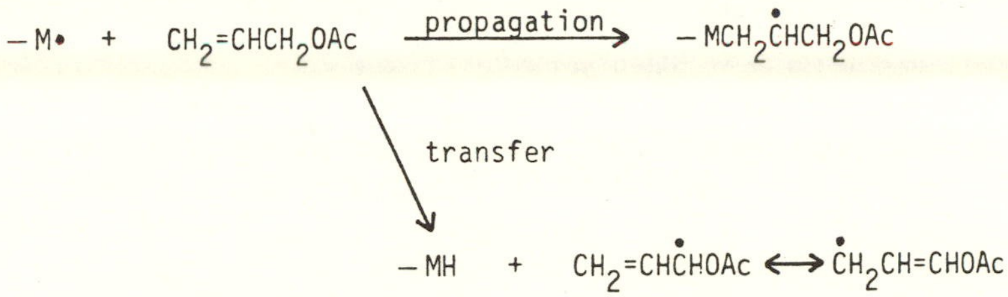

$$-M\bullet \; + \; CH_2=CHCH_2OAc \xrightarrow{\;propagation\;} -MCH_2\overset{\bullet}{C}HCH_2OAc$$

$$\searrow transfer$$

$$-MH \; + \; CH_2=CH\overset{\bullet}{C}HOAc \longleftrightarrow \overset{\bullet}{C}H_2CH=CHOAc$$

The chain initiation and termination processes have been studied in great detail. The azoisobutyronitriles, $R_2C(CN)-N=N-(CN)CR_2$ which lose nitrogen to form R_2CCN radicals, are easily prepared with wide variation in the R group and are good initiators; they were studied particularly by C. G. Overberger.[48]

We have said enough to indicate the extensive range of research in polymer chemistry. The Instrumental Revolution had its impact in this field as in the rest of chemistry. Physical chemists H. Mark and P. Debye (two distinguished workers of European origin), W. O. Baker, F. T. Wall, P. M. Doty, P. J. Flory, Turner Alfrey, W. Kauzmann, W. H. Stockmayer, and many more of this country have played leading roles. Although an organic chemist, C. S. Marvel has had a shrewd feeling for applications of physical chemistry to the polymer field and is responsible for increasing the interest of many physical chemists in this work.

We have mentioned some of the leading academic polymer chemists. However, most of the science of free-radical polymerization was developed in industrial research laboratories by scientists too numerous to mention, and these contributions are not reviewed here.

There was a mass of important qualitative and quantitative data on steric effects published in the period 1940 to 1955. These data showed that the space-filling effect, of alkyl groups particularly, could give an adequate explanation of many observations, without invoking electronic effects of such groups. The principal, but not the only protagonist of this point of view, was Herbert C. Brown.

Brown (1912- , B. S. 1936, Ph.D. 1938, both at Chicago with H. I. Schlesinger) worked with Kharasch as a postdoctorate, taught at Wayne State from 1943 to 1947 and at Purdue from 1947 to 1978, retiring as R. B. Wetherill Research Professor.[49] His voluminous researches include steric effects, reactions of complex hydrides, syntheses using boranes, non-classical carbonium ions, and free-radical reactions. He received the Nobel Prize in chemistry in 1979. His work shows indomitable industry, originality, and an instinct for detecting promising leads.

The Ingold school had been a strong proponent of the electronic (or polar) effect of alkyl groups. However, the work of Kharasch on addition of HBr to unsymmetrical C=C had shown that the Ingold theories must be amended in this case. Kharasch strongly criticized Ingold's speculations about ionic control of addition reactions.[50]

Brown notes that in the 1930s the tendency was to give electronic explanations rather than steric ones for the effects on reactivity of highly branched groups. Nevertheless, the many resolutions of substituted biphenyls by Roger Adams showed that size rather than electronic character was the decisive factor in preventing free rotation in biphenyls.[51]

In a long series of papers, Brown showed by measuring heat of dissociation that the reaction of a tertiary amine with a triborane was sensitive to changes in steric bulk of both R and R'; the *tert*-butyl group has a much larger effect than would be predicted from the ethyl and isopropyl groups. Triethylamine:trimethylborane is very unstable, but the corresponding compound from quinuclidine, where the alkyl groups are constrained in a ring, is extremely stable.[52] The three ethyl groups in triethylamine shield the nitrogen atom effectively because of the large number of positions they can assume. The carbon groups in quinuclidine cannot shield the nitrogen, and therefore its compound with trimethylborane is more stable.

$$R_3N: \quad + \quad BR'_3 \quad \longrightarrow \quad R_3N:BR'_3$$

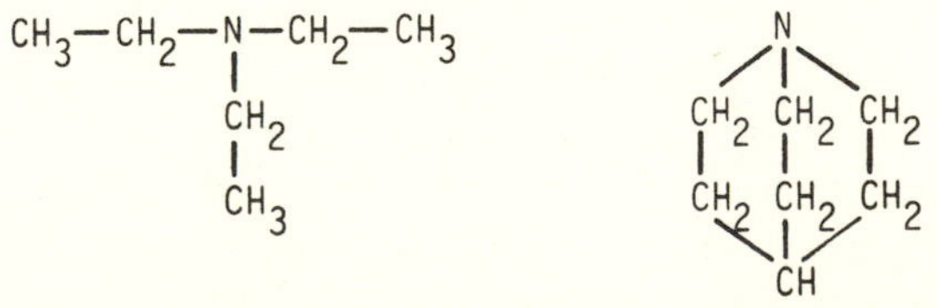

A most striking demonstration of steric effects is the observation that 2,6-di-*tert*-butylpyridine is a much weaker base to a proton than 2,6-diisopropylpyridine; further, the *tert*-butyl compound does not react with the Lewis acid boron trifluoride.[53]

Effects of steric interactions give a reasonable explanation of the greater rate of addition of cyclohexanones compared to cyclopentanones.[54] Steric effects also control the course of elimination reactions in some cases.[55]

Solvolysis of more highly branched chlorides or *p*-nitrobenzoates is much more rapid than of *tert*-butyl chloride.[56] The highly branched alcohols became available from the action of tertiary or secondary lithium compounds on branched esters, ketones, or acid chlorides.[57]

Solvolysis of these highly branched compounds is first order and is accompanied by rearrangement with formation of unsaturated hydrocarbons. The acceleration of rate in the highly branched compounds was attributed by Brown and by Bartlett[56] to relief of strain when the leaving group (chloride or p-nitrobenzoate) is expelled yielding the planar carbonium ion, in which the bulky branched groups are not as constrained as they are around a tetrahedral carbon.[56] Rate accelerations over *tert*-butyl chloride of many thousand were observed[58], and explanations other than the steric one were unsatisfactory. When NMR and other instrumentation became available, analysis of the reaction mixtures and conclusive structure determination of the products were possible.[58]

F. H. Westheimer at Chicago carried out two admirable series of researches on the mechanism of oxidation.[59] In the first, he showed that the mechanism of chromic acid oxidation of isopropyl alcohol takes the course below. The slow step was identified by showing that the reaction is slower by a factor of six, if the H on the carbon of isopropyl alcohol is replaced by deuterium. The slower rate of breaking a C-D bond compared to a C-H bond is in accord with theory and has proved to be widely useful in studying mechanism.

$$HCrO_4^- \quad + \quad (CH_3)_2CHOH \quad + \quad H^+ \quad \underset{\longleftarrow}{\overset{fast}{\longrightarrow}}$$

$$(CH_3)_2CHOCrO_3H \quad \xrightarrow[base]{slow} \quad (CH_3)_2CO \quad + \quad HCrO_3^- \quad + \quad base \quad + \quad H^+$$

In the second series[60], Westheimer and Birgit Vennesland demonstrated in detail the role of the coenzyme diphosphopyridine nucleotide in oxidation of ethanol to acetaldehyde in the presence of the enzyme yeast alcohol dehydrogenase. By suitable sequences they prepared the two mirror images of CH_3CHDOH, where the asymmetric carbon is due to the difference between H and D. This work remains unsurpassed in elegance and usefulness, both for the information it gives on the enzymatic process and for its synthetic applications.

Frank H. Westheimer (1912-   , A.B. 1932, Dartmouth; Ph.D. 1935, Harvard with J. B. Conant) was a research fellow with Hammett at Columbia for one year and taught at Chicago from 1936 to 1953. From then until 1982 he was professor at Harvard. His researches include electrostatic effects in chemistry, chemical and enzymatic oxidation, steric effects, mechanism of chemical and enzymatic decarboxylation, and phosphate esters. He has been a pioneer in studying the mechanism of enzymatic processes.

The topics discussed above give an idea of some representative problems in reaction mechanisms investigated in the period 1940 to 1955. American work in this field set a world-wide standard of performance.

LITERATURE CITED

1. D. H. R. Barton, Experientia, 6, 316(1950); JCS, 1027 (1953).
2. E. L. Eliel, *The Stereochemistry of Carbon Compounds*, New York, 1962; this edition roughly corresponds with our upper time limits. A brief summary of stereochemistry based on use of symmetry properties is Kurt Mislow, *Introduction to Stereochemistry*, New York, 1965; also E. L. Eliel et al., *Conformational Analysis*, New York, 1965; W. G. Dauben and K. S. Pitzer in *Steric Effects in Organic Chemistry*, M. S. Newman, Ed., New York, 1956, pp. 1-60.
3. S. Winstein and N. J. Holness, JACS, 77, 5562 (1955).
4. This relationship was derived independently by E. L. Eliel and C. A. Lukach, ibid., 79, 5986 (1957).
5. Eliel, *Stereochemistry*, pp. 151-153; an analytical solution to the Winstein-Holness/Curtin-Hammett kinetic system is given by J. I. Seeman and W. A. Farone, JOC, 43, 1854(1978).
6. See biography by W. G. Young and D. J. Cram, *Biog. Mem. Nat. Acad. Scis.*, 43, 320 (1973); on Winstein's work, P. D. Bartlett, JACS, 94, 2161 (1972).
7. L. H. Sommer, *Stereochemistry, Mechanism and Silicon*, New York, 1965; this reviews Sommer's work, much of which was unpublished in journal articles in 1965.
8. Review: W. E. McEwen, in *Topics in Phosphorus Chemistry*, M. Griffith, Ed., New York, 1965, Vol. 2, pp. 1-42. Key early papers: W. E. McEwen et al., JACS, 78, 3061, 3063 (1956); K. Kumli, W. E. McEwen and C. A. Vander Werf, ibid., 81, 248 (1959).
9. Carl Djerassi, *Optical Rotatory Dispersion*, New York, 1960; the axial haloketone rule, ibid., pp. 120-124.
10. R. J. Gillespie and R. White in *Progress in Stereochemistry*, Vol. 3, P. B. D. de la Mare and W. Klyne, Eds., New York, 1962, Chap. 2.
11. H. S. Gutowsky and C. H. Holm, J. Chem. Phys., 25, 1228 (1956) and other papers.
12. A. E. Remick, *Electronic Interpretations of Organic Chemistry*, New York, 1943, 1949; reviewed by P. D. Bartlett, JACS, 67, 158 (1945).
13. The best review of the displacement and solvolysis reactions is A. Streitwieser, *Solvolytic Displacement Reactions*, New York, 1962, based on a Chem. Rev. paper, 56, 571, (1956). Streitwieser calls the S_N2 reaction the "direct displacement reaction".
14. Streitwieser, loc. cit., pp. 9-15. "Basicity" is not really a satisfactory term for displacing activity at carbon, because it implies basicity toward a proton, which is not necessarily parallel to nucleophilicity toward carbon.
15. Useful reviews are in E. S. Gould, *Mechanism and Structure in Organic Chemistry*, New York, 1959, particularly for topics in reaction mechanisms for the 1940-1955 period.
16. J. C. Sheehan and W. A. Bolhofer, JACS, 72, 2786 (1950); H. E. Zaugg et al., ibid., 82, 2895 (1960); JOC, 26, 644 (1961); later physicochemical studies revealed the actual nature of sodio malonates in less polar solvents, H. E. Zaugg et al., ibid., 37, 2246, 2249, 2253 (1972). Examples of DMSO: R. A. Smiley and C. Arnold, ibid., 25, 257 (1960); L. Friedman and H. Shechter, ibid., 25, 877 (1960). HMP: L. A. Paquette and J. C. Philips, Tetr. Letters, 4645 (1967).
17. D. J. Cram, *Fundamentals of Carbanion Chemistry*, New York, 1965, pp. 32-38.
18. Reviews: J. Dockx, Synthesis, 441(1973); E. V. Dehmlow, Angew. Chem. Int. Ed., 13, 170 (1974), and refs. therein.
19. S. Winstein and R. E. Buckles, JACS, 64, 2780, 2787, 2796 (1942).
20. R. Baird and S. Winstein, ibid., 79, 756, 4238 (1957).
21. W. P. Jencks, *Catalysis in Chemistry and Enzymology*, New York, 1969, gives an survey of this rapidly developing field.
22. J. D. Roberts et al., JACS, 71, 1896 (1949) and later papers.
23. J. D. Roberts and R. H. Mazur, ibid., 73, 2509 (1951).
24. S. Winstein and R. Adams, ibid., 70, 838 (1948) and later papers.

25. S. Winstein, R. B. Woodward et al., ibid., **77**, 4183 (1955); Winstein and M. Shatavasky, ibid., **78**, 592 (1956).
26. D. J. Cram, ibid., **71**, 3863 (1949) and later papers; Winstein et al., ibid., **74**, 1113 (1952) and accompanying papers.
27. W. E. Doering and H. H. Zeiss, ibid., **75**, 4733 (1953); A. Streitwieser, op. cit., pp. 69-72.
28. E. Grunwald and S. Winstein, JACS, **70**, 846 (1948).
29. Winstein, Grunwald and H. W. Jones, ibid., **73**, 2700 (1951).
30. D. J. Cram, ibid., **74**, 2129 (1952); Winstein and K. Schreiber, **74**, 2165 (1952) and later papers by Winstein.
31. D. B. Denney and B. Goldstein, ibid., **79**, 4948 (1957); H. L. Goering and J. F. Levy, ibid., **84**, 3853 (1962).
32. S. F. Acree, ACJ, **48**, 352 (1912); Acree and G. H. Shadinger, ibid., **39**, 226 (1908).
33. Winstein et al., JACS, **75**, 147 (1953).
34. Winstein and D. Trifan, ibid., **71**, 2953 (1949); Streitwieser, op. cit., pp. 130 ff.
35. P. D. Bartlett, *Nonclassical Ions*, New York, 1965; H. C. Brown, *Boranes in Organic Chemistry*, Ithaca, 1972, pp. 131-205; H. C. Brown, *The Nonclassical Ion Problem*, with comments by P. v. R. Schleyer, New York, 1977; H. C. Brown, Accts. Chem. Res., **16**, 432 (1983); G. Olah, et al., ibid., 440; C. Walling, ibid., 448.
36. Kharasch's world standing in free-radical organic chemistry at his death in 1957 is shown by the memorial volume, *Vistas in Free-Radical Chemistry*, W. A. Waters, Ed., New York, 1959. Waters had played an important role in the development of free-radical chemistry. H. C. Brown, *Boranes in Organic Chemistry*, pp. 24 ff., gives some autobiographical pictures of his work with Kharasch; the biography of Kharasch by F. H. Westheimer, *Biog. Mem. Nat. Acad. Scis.*, **34**, 123 (1960) is excellent. H. C. Brown and M. S. Kharasch, JCE, **18**, 589 (1941) on reactivity of saturated hydrocarbons toward free radicals.
37. Review by Cheves Walling and E. S. Huyser, Organic Reactions, **3**, 91-149 (1963). A review of addition of radicals other than carbon ones to olefins is given by F. W. Stacey and J. F. Harris, Jr., ibid., p. 150-377.
38. M. S. Kharasch et al., JACS, **81**, 5819 (1959).
39. C. Walling, *Free Radicals in Solution*, New York, 1957.
40. P. J. Flory, *Principles of Polymer Chemistry*, Ithaca, 1953.
41. F. R. Mayo and C. Walling, Chem. Rev., **46**, 191 (1950).
42. Ref. 39, pp. 99 ff.
43. Ref. 39, p. 134; C. C. Price, J. Polymer Sci., **1**, 83 (1946) and later papers.
44. G. K. Fraenkel, J. M. Hirshon and C. Walling, JACS, **76**, 3606 (1954).
45. P. D. Bartlett and C. G. Swain, ibid., **67**, 2273 (1945); **68**, 2381 (1946); P. D. Bartlett, et al., ibid., **72**, 1060 (1950); G. M. Burnett and H. W. Melville, Nature, **156**, 661 (1945); M. S. Matheson et al., JACS, **71**, 2610 (1949).
46. R. B. Carlin and Nancy Shakespeare, ibid., **68**, 876 (1946); Walling, *Free Radicals*, and Flory, *Polymer Chemistry*, review chain transfer in detail.
47. P. D. Bartlett and R. Altschul, JACS, **67**, 812, 816 (1945).
48. C. G. Overberger et al., ibid., **71**, 2661 (1949) and later papers.
49. Brown, *Boranes*, ref 36. Brown's collaborators give informal accounts of his career in *Remembering HCB*, Shelton Bank, Ed., West Lafayette, 1978.
50. M. S. Kharasch and W. M. Potts, JACS, **58**, 57 (1936).
51. Brown (ref. 49) does not refer to the Treffers-Hammett paper, JACS, **59**, 1708 (1937), on the surprising behavior of diortho-substituted benzoic esters.
52. Brown and S. Sujishi, ibid., **70**, 2878 (1948).
53. Brown and B. Kanner, ibid., **88**, 986 (1966) and refs. therein.
54. Brown, J. H. Brewster and H. Shechter, ibid., **76**, 467 (1954).

55. W. H. Saunders, Jr., and A. F. Cockerill, *Mechanisms of Elimination Reactions*, New York, 1973, pp. 170-171.

56. P. D. Bartlett and L. H. Knox, JACS, **61**, 3184 (1939); Brown and Roslyn S. Fletcher, ibid., **71**, 1845 (1949); Brown and H. L. Berneis, ibid., **75**, 10 (1953); Bartlett and M. Stiles, ibid., **77**, 2806 (1955); Bartlett and Marguerite S. Swain, ibid., p. 2801.

57. Triisopropylcarbinol, W. G. Young and J. D. Roberts, ibid., **66**, 1444 (1944); tri-*tert*-butylcarbinol: Bartlett et al., ibid., **67**, 141 (1945); **77**, 2804 (1955); Brown et al., ref. 56.

58. V. J. Shiner, Jr., and G. F. Meier, JOC, **31**, 137 (1966); Bartlett and T. T. Tidwell, JACS, **90**, 4421 (1968).

59. F. H. Westheimer et al., ibid., **73**, 65 (1951); Westheimer, Chem. Rev., **45**, 419 (1949).

60. F. A. Loewus, F. H. Westheimer and Birgit Vennesland, JACS, **75**, 5018 (1953) and refs. therein.

Epilogue

As we have seen, the steady evolution of scientific thought and capability in organic chemistry in America, starting from so little knowledge and practice in 1875, had produced striking results in the discipline by 1955. Strong currents, expected to reveal even more in future years, can be identified.

Natural product synthesis continued to triumph over still more complicated structures and has expertly dealt with control over complex stereochemical problems. New reagents and new procedures, many of them now involving elements scattered through the Periodic Table—silicon, boron, thallium, selenium, sulfur, and metallic elements like palladium and copper—have made their appearance. Metallic coordination compounds are increasingly useful synthetic tools, and complex metal hydrides indispensable.

Automation has opened the way to practical synthesis of polynucleotides, and the techniques of recombinant DNA allow the relatively smooth preparation of polypeptides like insulin.

Increasingly sophisticated instruments determine chemical structures, X-ray crystallography and NMR being the most powerful tools. Since 1955, NMR spectrometers, scaled up by superconducting magnets yielding many times the resolving power of earlier instruments, can determine structure and conformation of complicated compounds. Their advantage over X-ray crystallography, which requires a suitable crystal, is the ability of NMR to analyze a compound in solution. Other instruments and analytical techniques are constantly improving in sensitivity and effectiveness.

Biochemical processes involving very minute concentrations of materials can now be fruitfully investigated, particularly by greater use of isotopic labels and NMR, and the picture of pathways in enzymatic reactions is clearer.

Physical organic chemistry has profited from the advance in instrumentation. With more powerful computers it is possible to make more accurate quantum mechanical calculations of resonance energy, to calculate the energy of transition states, and hence to decide in some cases the details of reaction mechanisms. Inorganic chemistry plays a greater role in organic research, and increased knowledge clarifies free-radical chemistry and heterogeneous catalysis.

Finally, the history of chemistry is engaging the interest of many more scholars than in 1955. Now a recognized discipline, it encompasses broad studies of academic research and the history of industrial chemical research and production.

Indeed, "the little band of workers at Johns Hopkins" and the teaching and research chemists who followed could justifiably be proud of America's world position in fundamental research and brilliant advances in organic chemistry.

INDEX OF NAMES

Pages numbered in bold face give biographical notes. In general, only the first author of a publication is listed.

INDEX OF SUBJECTS

V

W

X

About the Authors

D. Stanley Tarbell and Ann Tracy Tarbell have been actively engaged for over a decade in the study and writing about the development of organic chemistry in the United States. They have published numerous research articles on detailed problems in this field, as well as the biography *Roger Adams: Scientist and Statesman* (1981). The basis of the present volume is a detailed study of American research publications and extensive related background material.

D. Stanley Tarbell took his Ph.D. at Harvard in 1937, at the age of 23, with Paul D. Bartlett. He was a postdoctoral fellow with Roger Adams at Illinois in 1937-1938. He was faculty member at the University of Rochester from 1938 to 1967, where he directed an active research program in organic chemistry with the collaboration of many Ph.D. candidates and postdoctoral students. Since 1967 he has been Distinguished Professor of Chemistry at Vanderbilt, becoming Emeritus in 1981.

Ann Tarbell took her Ph.D. at Columbia with Robert C. Elderfield and was a postdoctoral fellow at Radcliffe in 1940-42 and research fellow at Rochester on the Manhattan Project from 1943 to 1945.

The Tarbells have three children, a psychiatric social worker, a mathematician, and an astrophysicist. Ann is an amateur ornithologist, and they are both interested in travel, history, music, and the outdoors.